MATHEMATICS IN ACTION

ALGEBRAIC, GRAPHICAL, AND TRIGONOMETRIC PROBLEM SOLVING

SECOND EDITION

The Consortium for Foundation Mathematics

Ralph Bertelle	*Columbia-Greene Community College*
Judith Bloch	*University of Rochester*
Roy Cameron	*SUNY Cobleskill*
Carolyn Curley	*Erie Community College—South Campus*
Ernie Danforth	*Corning Community College*
Brian Gray	*Howard Community College*
Arlene Kleinstein	*SUNY Farmingdale*
Kathleen Milligan	*Monroe Community College*
Patricia Pacitti	*SUNY Oswego*
Rick Patrick	*Adirondack Community College*
Renan Sezer	*LaGuardia Community College*
Patricia Shuart	*Polk Community College—Winter Haven, Florida*
Sylvia Svitak	*Queensborough Community College*
Assad J. Thompson	*LaGuardia Community College*

PEARSON

Addison
Wesley

Boston San Francisco New York
London Toronto Sydney Tokyo Singapore Madrid
Mexico City Munich Paris Cape Town Hong Kong Montreal

Editor in Chief: Maureen O'Connor
Acquisitions Editor: Jennifer Crum
Associate Editor: Greg Erb
Managing Editor: Ron Hampton
Text Designer: Susan C. Raymond
Cover Designer: Leslie Haimes
Composition: Lynn L'Heureux
Media Producer: Sara Anderson
Software Development: TestGen: Kathleen Bowler; InterAct: Alicia Anderson
Marketing Manager: Dona Kenly
Marketing Coordinator: Lindsay Skay
Prepress Services Buyer: Caroline Fell
First Print Buyer: Hugh Crawford

Library of Congress Cataloging-in-Publication Data

Consortium for Foundation Mathematics
     Mathematics in action: algebraic, graphical, and trigonometric problem solving / the Consortium for Foundation Mathematics 2nd ed.
     p. cm.
Includes index.

     ISBN 0-321-14920-3 (pbk.)
     1. Mathematics.   I. Title.

QA152.3 .M38 2003
510—dc21   2002074647

ISBN: 0-321-14920-3

   5 6 7 8 9 10 CRK 07 06

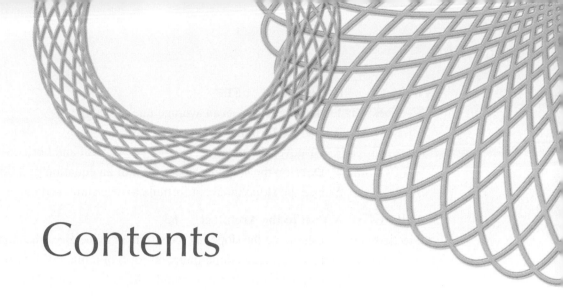

Contents

CHAPTER 4 QUADRATIC AND HIGHER-ORDER POLYNOMIAL FUNCTIONS 411

Cluster 1 Introduction to Quadratic Functions 411

APPENDIXES

Preface

Our Vision

Mathematics in Action: Algebraic, Graphical, and Trigonometric Problem Solving, Second Edition, is intended to help college mathematics students gain mathematical literacy in the real world and simultaneously help them build a solid foundation for future study in mathematics and other disciplines.

Our team of fourteen faculty, primarily from the State University of New York and the City University of New York systems, used the AMATYC *Crossroads* standards to develop this three-book series to serve a very large population of college students at the pre-precalculus level. Many of our students have had previous exposure to mathematics at this level. It became apparent to us that teaching the same content in the same way to students who have not previously comprehended it is not effective, and this realization motivated us to develop a new approach.

Mathematics in Action is based on the principle that students learn mathematics best by doing mathematics within a meaningful context. In keeping with this premise, students solve problems in a series of realistic situations from which the crucial need for mathematics arises. *Mathematics in Action* guides students toward developing a sense of independence and taking responsibility for their own learning. Students are encouraged to construct, reflect on, apply, and describe their own mathematical models, which they use to solve meaningful problems. We see this as the key to bridging the gap between abstraction and application and as the basis for transfer learning. Appropriate technology is integrated throughout the books, allowing students to interpret real-life data verbally, numerically, symbolically, and graphically.

We expect that by using the *Mathematics in Action* series, all students will be able to achieve the following goals:

- Develop mathematical intuition and a relevant base of mathematical knowledge.
- Gain experiences that connect classroom learning with real-world applications.
- Prepare effectively for further college work in mathematics and related disciplines.
- Learn to work in groups as well as independently.
- Increase knowledge of mathematics through explorations with appropriate technology.
- Develop a positive attitude about learning and using mathematics.
- Build techniques of reasoning for effective problem solving.
- Learn to apply and display knowledge through alternative means of assessment, such as mathematical portfolios and journal writing.

Our vision for you is to join the growing number of students using our approaches who discover that mathematics is an essential and learnable survival skill for the 21st century.

Pedagogical Features

The pedagogical core of *Mathematics in Action* is a series of guided-discovery activities in which students work in groups to discover mathematical principles embedded in realistic situations. The key principles of each activity are highlighted and summarized at the activity's conclusion. Each activity is followed by exercises that reinforce the concepts and skills revealed in the activity.

The activities are clustered within each chapter. Each cluster contains regular activities along with project and lab activities that relate to particular topics. The lab activities require more than just paper, pencil, and calculator; they also require measurements and data collection and are ideal for in-class group work. The project activities are designed to allow students to explore specific topics in greater depth, either individually or in groups. These activities are usually self-contained and have no accompanying exercises. For specific suggestions on how to use the three types of activities, we strongly encourage instructors to refer to the *Instructor's Resource Guide* that accompanies this text.

Each cluster concludes with two sections: What Have I Learned? and How Can I Practice? The What Have I Learned? exercises are designed to help students pull together the key concepts of the cluster. The How Can I Practice? exercises are designed primarily to provide additional work with the numeric and algebraic skills of the cluster. Taken as a whole, these exercises give students the tools they need to bridge the gaps between abstraction, skills, and application.

Additionally, each chapter ends with a Summary that contains a brief description of the concepts and skills discussed in the chapter, plus examples illustrating these concepts and skills. The concepts and skills are also cross-referenced to the activity in which they appear, making the format easier to follow for those students who are unfamiliar with our approach. Each chapter also ends with a Gateway Review, providing students with an opportunity to check their understanding of the chapter's concepts and skills.

Changes from the First Edition

Instructors who have used the first edition of *Mathematics in Action* will notice that many activities throughout the text have been rewritten and new activities added to ensure that all topics are covered in depth.

New features have also been added to make definitions, procedures, and properties more visible to the student. Specific changes include the following:

- A list of objectives opens each activity. The objectives also appear in the table of contents.
- Activities now contain examples that are worked out in detail.
- Chapter Summaries have been added. Each summary lists the concepts and skills covered in the chapter with a corresponding description and example. Each concept or skill includes an activity number for reference.
- The exposition within the development of several topics has been expanded.
- All activity questions are numbered for easy reference.
- Definitions, properties, and procedures are set off from the surrounding text.
- All exercises have been carefully reviewed to ensure a thorough reinforcement of topics and an appropriate level of difficulty for an elementary or intermediate algebra course.

Supplements

Instructor Supplements

Annotated Instructor's Edition

ISBN 0-321-14921-1

This special version of the student text provides answers to all exercises directly beneath each problem.

Instructor's Resource Guide/Printed Test Bank

ISBN 0-321-15511-4

This valuable teaching resource includes the following materials:

- Sample syllabi suggesting ways to structure the course around core and supplemental activities and within different credit-hour options.
- Sample course outlines containing timelines for covering topics.
- Teaching notes for each chapter. These notes are ideal for those using the Mathematics in Action approach for the first time.
- Extra practice skills worksheets for topics with which students typically have difficulty.
- Sample chapter tests and final exams for in-class and take-home use by individual students and groups.
- Sample journal topics for the students to write comments and observations about the course are included for each chapter.
- A section discussing learning in groups provides questions and answers for instructors trying collaborative learning for the first time.
- Information about incorporating technology in the classroom, including sample graphing calculator assignments.

TestGen with QuizMaster

ISBN 0-321-15510-6

TestGen enables instructors to build, edit, print, and administer tests using a computerized bank of questions developed to cover all the objectives of the text. TestGen is algorithmically based so that multiple, yet equal, versions of the same question or test can be generated at the click of a button. Instructors can also modify test bank questions or add new questions by using the built-in question editor, which allows users to create graphs, import graphics, insert math notation, and insert variable numbers or text. Tests can be printed or administered on-line via the Web or other network. Many questions in TestGen can be expressed in a short-answer or multiple-choice form, giving instructors greater flexibility in their test preparation. TestGen comes packaged with QuizMaster, which allows students to take tests on a local area network. The software is available on a dual-platform Windows/Macintosh CD-ROM.

Instructor Training Video

ISBN 0-201-70959-7

This innovative video discusses effective ways to implement the teaching pedagogy of the *Mathematics in Action* series, focusing on how to make collaborative learning, discovery learning, and alternative means of assessment work in the classroom.

Supplements for Instructors and Students

MathXL

ISBN 0-321-22388-8 (Instructors); 0-201-72611-4 (Students)

MathXL is an on-line testing, homework, and tutorial system that uses algorithmically generated exercises correlated to your textbook.

Instructors can assign tests and homework provided by Addison-Wesley or create and customize their own tests and homework assignments. Instructors can also track their students' results and tutorial work in an on-line gradebook.

Students can take chapter tests, and receive personalized study plans that will diagnose weaknesses and link students to areas they need to study and retest. Students can also work unlimited practice problems and receive tutorial instruction for areas in which they need improvement.

MathXL can be packaged with new copies of *Mathematics in Action: Algebraic, Graphical, and Trigonometric Problem Solving.* Please contact your Addison-Wesley representative for details.

MyMathLab

ISBN 0-321-19992-8 (Instructors); 0-321-17883-1 (Students)

MyMathLab is a complete, on-line course for Addison-Wesley mathematics textbooks that provides interactive multimedia instruction correlated to the textbook content. MyMathLab is easily customizable to suit the needs of students and instructors and provides a comprehensive and efficient on-line course-management system that allows for diagnosis, assessment, and tracking of students' progress.

MyMathLab features include:

- Chapter and section folders in the on-line course mirror the textbooks' table of contents and contain a wide range of multimedia instruction.
- The actual pages of the textbook are loaded into MyMathLab, and as students work through a section of the on-line text, they can link to multimedia resources that are correlated directly to the examples and exercises in the text.
- Hyperlinks take you directly to on-line testing, diagnosis, tutorials, and tracking in MathXL—Addison-Wesley's tutorial and testing system for mathematics and statistics.
- Instructors can create, copy, edit, assign, and track all tests and homework for their course as well as track students' results and practice work.
- With push-button ease, instructors can remove, hide, or annotate Addison-Wesley preloaded content, add their own course documents, or change the order in which material is presented.
- Using the communication tools found in MyMathLab, instructors can hold on-line office hours, host a discussion board, create communication groups within their class, send e-mail, and maintain a course calendar.

For more information, visit our Web site at www.mymathlab.com or contact your Addison-Wesley sales representative for a live demonstration.

InterAct® Math Tutorial Software

ISBN 0-321-15508-4

This interactive tutorial software provides algorithmically generated practice exercises that are correlated to the chapter content of the texts. Every exercise in the program is accompanied by an example and a guided solution designed to involve students in the solution process. The software tracks students' activity and scores and can generate printed summaries of students' progress. The software also recognizes common student errors and provides appropriate feedback. Instructors can use the InterAct Math Plus course-management software to create, administer, and track on-line tests and monitor student performance during their practice sessions in InterAct Math.

Student Supplements

Addison-Wesley Math Tutor Center

ISBN 0-321-21115-4

The Addison-Wesley Math Tutor Center is staffed by qualified college mathematics instructors who tutor students on examples and exercises from the textbook. Tutoring is provided via toll-free telephone, fax, e-mail, and the Internet. Interactive Web-based technology allows students and tutors to view and listen to live instruction in real-time over the Internet. The Math Tutor Center is accessed through a registration number that can be packaged with a new textbook or purchased separately. (Note: MyMathLab students obtain access to the Math Tutor Center by using their MyMathLab access code.)

Acknowledgments

The Consortium would like to acknowledge and thank the following people for their invaluable assistance in reviewing and testing material for this text:

Mary Kay Abbey, *Montgomery College*

Jennifer Dollar, *Grand Rapids Community College*

Irene Duranczyk, *Eastern Michigan University*

Ernest East, *Northwestern Michigan College*

Maryann B. Faller, *Adirondack Community College*

Linda Green, *Santa Fe Community College*

Lois Higbie, *Brookdale Community College*

Teresa Hodge, *University of the Virgin Islands*

Maria Ilia, *Clarke College*

Ashok Kumar, *Valdosta State University*

J. Robert Malena, *Community College of Allegheny County—South Campus*

Raquel Mesa, *Xavier University of Louisiana*

Beverly K. Michael, *University of Pittsburgh*

Paula J. Mikowicz, *Howard Community College*

Adam Parr, *University of the Virgin Islands*

Debra Pharo, *Northwestern Michigan College*

Kathy Potter, *St. Ambrose University*

Dennis Risher, *Loras College*

Sandra Spears, *Jefferson Community College*

Christopher Teixeira, *Rhode Island College*

Kurt Verderber, *State University of New York Cobleskill*

Lynn Wolfmeyer, *Western Illinois University*

We would also like to thank our accuracy checkers, Vincent Koehler and Ann Ostberg.

Finally, a special thank-you to our families for their unwavering support and sacrifice, which enabled us to make this text a reality.

The Consortium for Foundation Mathematics

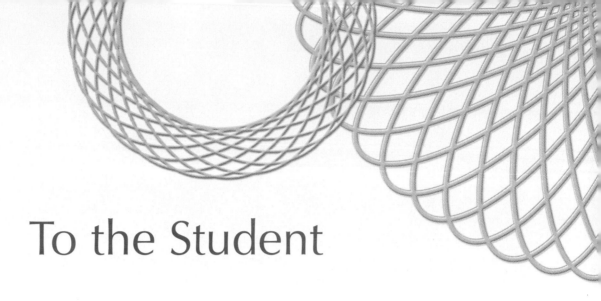

To the Student

The book in your hands is most likely very different from any mathematics textbook you have seen before. In this book, you will take an active role in developing the important ideas of arithmetic and beginning algebra. You will be expected to add your own words to the text. This will be part of your daily work, both in and out of class. It is the belief of the authors that students learn mathematics best when they are actively involved in solving problems that are meaningful to them.

The text is primarily a collection of situations drawn from real life. Each situation leads to one or more problems. By answering a series of questions and solving each part of the problem, you will be using and learning one or more ideas of introductory college mathematics. Sometimes, these will be basic skills that build on your knowledge of arithmetic. Other times, they will be new concepts that are more general and far-reaching. The important point is that you won't be asked to master a skill until you see a real need for that skill as part of solving a realistic application.

Another important aspect of this text and the course you are taking is the benefit gained by collaborating with your classmates. Much of your work in class will result from being a member of a team. Working in small groups, you will help each other work through a problem situation. While you may feel uncomfortable working this way at first, there are several reasons we believe it is appropriate in this course. First, it is part of the learning-by-doing philosophy. You will be talking about mathematics, needing to express your thoughts in words. This is a key to learning. Secondly, you will be developing skills that will be very valuable when you leave the classroom. Currently, many jobs and careers require the ability to collaborate within a team environment. Your instructor will provide you with more specific information about this collaboration.

One more fundamental part of this course is that you will have access to appropriate technology at all times. You will have access to calculators and some form of graphics tool—either a calculator or computer. Technology is a part of our modern world, and learning to use technology goes hand in hand with learning mathematics. Your work in this course will help prepare you for whatever you pursue in your working life.

This course will help you develop both the mathematical and general skills necessary in today's workplace, such as organization, problem solving, communication, and collaborative skills. By keeping up with your work and following the suggested organization of the text, you will gain a valuable resource that will serve you well in the future. With hard work and dedication, you will be ready for the next step.

The Consortium for Foundation Mathematics

CHAPTER 1

Function Sense

Modeling with Functions

ACTIVITY 1.1
Parking
Problems

OBJECTIVES

1. Distinguish between input and output.

2. Define a function.

3. Represent a function numerically and graphically.

4. Write a function using function notation.

Did you have trouble finding a parking space this morning? Was the time that you arrived on campus a factor? As part of a reconstruction project at a small community college, the number of cars in the parking lot was counted each hour from 7:00 A.M. to 10:00 P.M. on a particular day. The results are shown in the following table.

Park It

TIME OF DAY	NUMBER OF CARS
7 A.M.	24
8 A.M.	212
9 A.M.	384
10 A.M.	426
11 A.M.	538
12 P.M.	497
1 P.M.	384
2 P.M.	337
3 P.M.	285
4 P.M.	278
5 P.M.	302
6 P.M.	427
7 P.M.	384
8 P.M.	315
9 P.M.	187
10 P.M.	56

This situation involves two variables, the time of day and the number of cars in the parking lot. A **variable**, usually represented by a letter, is a quantity that may change in value from one particular instance to another. Typically, one variable is designated as the **input** and the other is called the **output**. The input is the value given first, and the output is the value that corresponds to, or is determined by, the given input value.

1. In the parking lot situation, identify the input variable and the output variable.

2. a. For an input of 10:00 A.M., how many cars are in the parking lot (output)?

   b. For an input of 5:00 P.M., how many cars are in the parking lot (output)?

   c. For each value of input (time of day), how many different outputs (number of cars) are there?

The set of data in the table is an example of a mathematical function.

> **DEFINITION**
> A **function** is a correspondence between an input variable and an output variable that assigns a single, unique output value to each input value. Therefore, for a function, any given input value has exactly one corresponding output value.

3. Explain how the data in the table on page 1 fit the description of a function.

A functional relationship is stated as follows: "The output variable is a function of the input variable." Because the input for the parking lot function is the time of day and the output is the number of cars in the lot at that time, you write that the number of cars in the parking lot is a function of the time of day.

Example 1 *Consider the following table listing the official high temperature (in °F) in the village of Lake Placid, New York, during the first week of January. Note that the date has been designated the input and the high temperature on that date the output. Is the high temperature a function of the date?*

Almost Freezing

DATE (INPUT)	1	2	3	4	5	6	7
TEMPERATURE (OUTPUT)	25	30	32	24	23	27	30

SOLUTION

From this table, you observe that the high temperature is a function of the date. For each date there is exactly one high temperature.

The relationships in Example 1 can be visualized as follows:

DATE (INPUT)	TEMPERATURE (OUTPUT)
1 ⟶	25
2 ⟶	30
3 ⟶	32
4 ⟶	24
5 ⟶	23
6 ⟶	27
7 ⟶	30

If the input and output are switched, the daily high temperature becomes the input and the date becomes the output. The date is not a function of the high temperature. The input value 30 has two output values, 2 and 7.

TEMPERATURE (INPUT)	DATE (OUTPUT)
25 ⟶	1
30 ⟶	2
32 ⟶	3
24 ⟶	4
23 ⟶	5
27 ⟶	6
	7

4. Interchange the input and the output in the parking lot situation. Let the number of cars in the lot be the input and the time of day be the output. Is the time of day a function of the number of cars in the lot? Write a sentence explaining why this switch does or does not fit the description of a function.

Example 2 *Determine whether or not the following situation describes a function. Give a reason for your answer.*

The amount of postage for a letter is a function of the weight of the letter.

SOLUTION

Yes, this statement does describe a function. The weight of the letter is the input, and the amount of postage is the output. Each letter has one weight. This weight determines the postage necessary for the letter. There is only one amount of postage for each letter. Therefore, for each value of input (weight of the letter) there is one output (postage).

5. Determine whether or not each situation describes a function. Give a reason for your answer.

   a. The amount of property tax you have to pay is a function of the assessed value of the house.

   b. The weight of a letter in ounces is a function of the postage paid for mailing the letter.

Defining Functions Numerically

The input/output pairing in the parking lot function on page 1 is presented as a **table of matched pairs**. In such a situation, the function is defined **numerically**. Another way to define a function numerically is as a set of ordered pairs.

DEFINITION

An **ordered pair** of numbers consists of two numbers written in the form

(input value, output value).

The order in which they are listed is significant.

Example 3 *The ordered pair (3, 4) is distinct from the ordered pair (4, 3). In the ordered pair (3, 4), 3 is the input and 4 is the output. In the ordered pair (4, 3), 4 is the input and 3 is the output.*

DEFINITION

A function may be defined **numerically** as a set of ordered pairs in which the first number of each pair represents the input value and the second number represents the corresponding output value. No two ordered pairs have the same input value and different output values.

Example 4   *(9:00 A.M., 384) or (0900, 384) (using a 24-hour clock) is an ordered pair that is part of the parking lot function.*

6. Using a 24-hour clock, write three other ordered pairs for the parking lot function.

Defining Functions Graphically

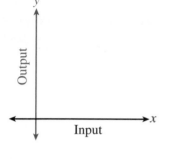

You may have seen an ordered pair before as the coordinates of a point in a **rectangular coordinate system**, typically using ordered pairs of the form (x, y), where x is the input and y is the output. The first value, the horizontal coordinate, indicates the directed distance (right or left) from the vertical axis. The second value, the vertical coordinate, indicates the directed distance (up or down) from the horizontal axis.

The variables may not always be represented by x and y, but the horizontal axis will always be the input axis, and the vertical axis will always be the output axis.

7. Using a 24-hour clock, convert to ordered pairs all the values in the Park It table on page 1. Plot each ordered pair on the following grid. Set your axes and scales by noting the smallest and largest values for both input and output. Label each axis by both the variable name and its designation as input or output. *Remember, the scale for the horizontal axis does not have to be the same as the scale for the vertical axis. However, each scale (vertical and horizontal) must be divided into equal intervals.*

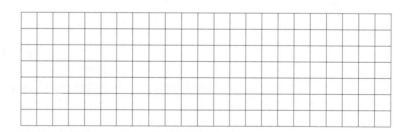

The preceding graph, which consists of a set of labeled axes and 16 points, presents the same information that is in the Park It table, but in a different way. It shows the information as a graph and therefore defines the function **graphically**.

8. In Example 1, the high temperature in Lake Placid is a function of the date.

    a. Convert to ordered pairs all the values in the Almost Freezing table on page 2.

b. Plot each ordered pair as a point on an appropriately scaled and labeled set of coordinate axes.

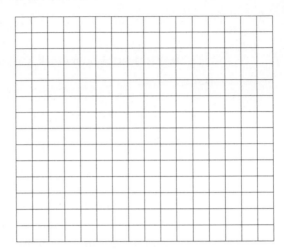

Function Notation

There is a special notation for functions in which the function itself is represented by a name or letter. For example, the function that relates the time of day to the number of cars can be represented by the letter f. Let t represent time, the input variable, and let c represent the number of cars, the output variable. The following simplification (really an abbreviation) is now possible.

The number of cars in the parking lot is a function of the time of day.

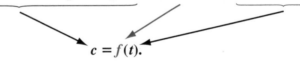

$$c = f(t).$$

The final function notation is read "c equals f of t."

In general, function notation is written as follows:

Output variable = name of function (*input variable*).

The input variable or input value is also called the **argument** of the function.

Example 5 *Values from the table or ordered pairs for the parking function can be written as follows using function notation:*

$$212 = f(800), \quad 302 = f(1700), \quad f(2100) = 187$$

9. a. Rewrite the three examples given in Example 5 as three ordered pairs. Pay attention to which is the input value and which is the output value.

b. Write a sentence explaining the meaning of $f(1600) = 278$ in the parking lot situation.

10. a. Referring to the table in Example 1 (Almost Freezing), determine $g(3)$, where g is the name of the temperature function.

b. Write a sentence explaining the meaning of $g(5) = 23$.

Gross Pay Function

11. If you work for an hourly wage, your gross pay is a function of the number of hours that you work.

a. Identify the input and output.

b. If you earn $8 per hour, complete the following table.

NUMBER OF HOURS	0	3	5	7	10	12
GROSS PAY						

c. Plot the ordered pairs determined in part b on an appropriately scaled and labeled set of axes.

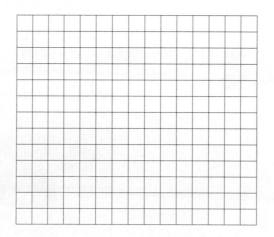

d. Let n represent the number of hours worked and $f(n)$ represent the gross pay. Use the table or the graph to determine $f(5)$.

e. Write a sentence explaining the meaning of $f(10) = 80$.

SUMMARY
Activity 1.1

1. A **variable**, usually represented by a letter, is a quantity that may change in value from one particular instance to another.

2. In a situation involving two variables, one variable is designated the **input** and the other the **output**. The input is the value given first, and the output is the value that corresponds to or is determined by the given input value.

3. A **function** is a rule relating an input variable (sometimes called the argument) and an output variable so that a single, unique output value is assigned to each input value. In such a case, you state that the output variable is a function of the input variable.

4. An **ordered pair** of numbers consists of two numbers written in the form

   (input value, output value).

   The order in which they are listed is significant.

5. Functions may be defined **numerically** using ordered pairs of numbers. The ordered pairs are always given in the form (input value, output value). These can be displayed as a table of values or points on a graph. For each input value, there is one and only one corresponding output value.

6. When a function is defined **graphically**, the input variable will be represented on the horizontal axis and the output variable on the vertical axis.

7. The function relationship is often defined using function notation:

   output variable = name of function(input variable).

   If y represents the output variable, f is the name of the function, and x represents the input variable, then

   $y = f(x)$ is read "y equals f of x."

EXERCISES
Activity 1.1

1. The weights and heights of six mathematics students are given in the following table.

Weighty Issues

WEIGHT (IN POUNDS)	HEIGHT (IN CENTIMETERS)
165	172
123	157
212	183
175	178
165	163
147	167

Exercise numbers appearing in color are answered in the Selected Answers appendix.

a. In the statement "Height is a function of weight," which variable is the input and which is the output?

b. Is height a function of weight for the six students? Explain using the definition of function.

c. In the statement "Weight is a function of height," which variable is the input and which is the output?

d. Is weight a function of height for the six students? Explain using the definition of function.

e. For all students, is weight a function of height? Explain.

For Exercises 2–7, determine whether or not each of the situations describes a function. Give a reason for your answer.

2. a. On a given night, your blood-alcohol level is a function of the number of beers you drink in a two-hour time period.

b. The number of beers you drink in a two-hour period is a function of your blood-alcohol level.

3. a. The letter grade in this course is a function of your numerical grade.

b. The numerical grade in this course is a function of the letter grade.

4. a. The input is any number and the output is the square of the number.

b. The square of a number is the input and the output is the number.

5. a. In the following table, elevation is the input and amount of snowfall is the output.

ELEVATION (IN FEET)	SNOWFALL (IN INCHES)
2000	4
3000	6
4000	9
5000	12

b. In the preceding table, snowfall is the input and elevation is the output.

6. Number of hours using the Internet is the input, and the monthly cost for the Internet service is $19.95.

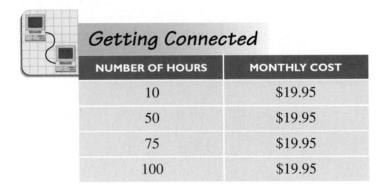

Getting Connected

NUMBER OF HOURS	MONTHLY COST
10	$19.95
50	$19.95
75	$19.95
100	$19.95

7. a. $(2, 5), (-3, 5), (10, 5), (\pi, 5)$ **b.** $(5, 2), (5, -3), (5, 10), (5, \pi)$

8. In Exercise 5, the amount of snowfall is a function of the elevation.

a. Plot the ordered pairs on an appropriately scaled and labeled set of coordinate axes.

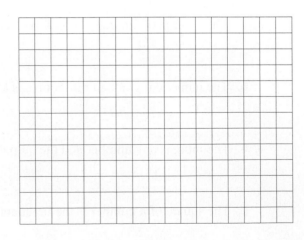

b. Let x represent the elevation in feet and $a(x)$ represent the amount of snowfall in inches. Determine $a(4000)$.

c. Write $a(5000) = 12$ as an ordered pair.

d. Write a sentence explaining the meaning of $a(5000) = 12$.

9. Identify the input, the output, and the name of the function. For each of the functions, write in words the equation as you would say it.

   a. $y = g(x)$

   b. $h(a) = b$

   c. $f(6) = 3.527$

   d. $520 = g(t)$

   e. sales tax $= T(\text{price})$

10. Your college community service organization has volunteered to help with Spring Cleanup Day at a youth summer camp. You have been assigned the job of supplying paint for the exterior of the bunk houses. You discover that 1 gallon of paint will cover 400 square feet of flat surface.

    a. If n represents the number of gallons of paint you supply and s represents the number of square feet you can cover with the paint, complete the following table.

Painting by Numbers

n, NUMBER OF GALLONS OF PAINT	1	2	4	6
s, SQUARE FEET COVERED BY THE PAINT	400	800		

b. Let n be the input variable and s be the output. Plot the ordered pairs determined in part a on an appropriately scaled and labeled set of axes.

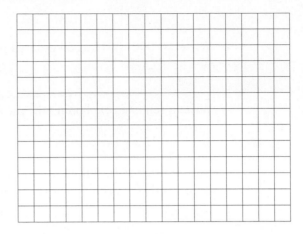

c. Let s be represented by $f(n)$, where f is the name of the function. Determine $f(6)$.

d. Write a sentence explaining the meaning of $f(4) = 1600$.

11. a. Give an example of a function that you may encounter in your daily life or that describes something about the world around you.

   i. Identify the input and the output variables.

   ii. Write the function in the form "output is a function of the input."

   iii. Explain how the example fits the definition of a function.

b. Switch the input and the output of the function you determined in part a.

   i. Identify the input and the output.

   ii. Explain how the example fits, or does not fit, the definition of a function.

c. Write the function you listed in part a in function notation. Represent the input variable, the output variable, and the function itself by letters.

ACTIVITY 1.2

Fill 'er Up

OBJECTIVES

1. Determine the equation (symbolic representation) that defines a function.

2. Write the equation to define a function.

3. Determine the domain and range of a function.

4. Identify the independent and the dependent variables of a function.

You probably need to fill your car with gas more often than you would like, so you drive around looking for the best price per gallon.

1. There are two input variables that determine the cost (output) of a fill-up. What are they? Be specific.

2. Assume that the price of gas is 1.36\frac{9}{10}$. Now one of the input variables in Problem 1 will become a constant. The value of a constant will not vary throughout the problem. The cost of a fill-up is now dependent on only one variable, the number of gallons of gas pumped.

   a. Complete the following table.

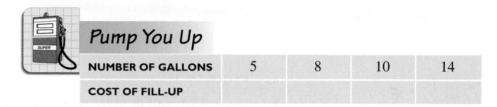

Pump You Up				
NUMBER OF GALLONS	5	8	10	14
COST OF FILL-UP				

   b. Is the cost of a fill-up a function of the number of gallons pumped? Explain.

3. a. Write a verbal statement that describes how the cost of a fill-up is determined.

   b. Let g represent the number of gallons of gasoline pumped (output) and c represent the cost of the fill-up. Translate the verbal statement in part a into a symbolic statement (an equation) that expresses c in terms of g.

Defining Functions by a Symbolic Rule

The symbolic rule (equation) $c = 1.369g$ represents the third method of defining a function. Recall that the other two methods are numerically (tables and ordered pairs) and graphically.

4. a. Use the given equation to determine the cost of a fill-up that required 8 gallons of gas.

   b. Explain the steps that you used to determine the cost in part a.

Recall from Activity 1.1 that function notation is an efficient and convenient way of representing the output variable. The equation $c = 1.369g$ may be written using the function notation by replacing c with $f(g)$ as follows:

$$f(g) = 1.369g$$

Now, the cost of 5 gallons of gas can be represented by $f(5)$. To evaluate $f(5)$, substitute 5 for g in $f(g) = 1.369g$ as follows:

$$f(5) = 1.369(5) = 6.845$$

The results can be written as $f(5) = 6.845$ or as the ordered pair $(5, 6.845)$. Therefore, 5 gallons of gas will cost \$6.85 (rounded to the nearest cent).

5. a. Write the cost of 8 gallons of gas using function notation and evaluate. Write the result as an ordered pair.

b. Use the equation for the cost-of-fill-up function to evaluate $f(12)$, and write a sentence describing its meaning. Write the result as an ordered pair.

Real Numbers

The numbers that you will be using as input and output values in this text will be real numbers. A real number is any rational or irrational number.

A rational number is any number that can be expressed as the quotient of two integers (negative and positive counting numbers as well as zero).

> **Example 1** *Rational numbers include the following:*
>
> $\frac{3}{4}$ $-\frac{7}{8}$ $2\frac{1}{3} = \frac{7}{3}$ $5 = \frac{5}{1}$ $0 = \frac{0}{1}$ $-3\frac{1}{4} = -3.25$ $\frac{2}{3} = 0.666\ldots = 0.\overline{6}$

An irrational number is a real number that cannot be expressed as a quotient of two integers.

> **Example 2** *Irrational numbers include* $\sqrt{2}, -\sqrt{7}, \sqrt[3]{5}, \pi$.

All of the numbers in Examples 1 and 2 are real numbers. A real number can be represented as a point on the number line.

Domain and Range

6. Can any real number be substituted for the input variable g in the cost-of-fill-up function? Describe the values of g that make sense, and explain why they do.

> **DEFINITION**
> The collection of all possible values of the input variable is called the **domain** of the function. The **practical domain** is the collection of replacement values of the input variable that makes practical sense in the context of the problem.

Example 3 *The practical domain in Problem 6 is the real numbers 0 through 20, assuming that 20 gallons is the maximum capacity of your gas tank. The value of g would be 0 if you do not need any gas and 20 if your gas tank is totally empty.*

Following is the graph of the cost-of-fill-up function defined by $c = 1.369g$ over its practical domain.

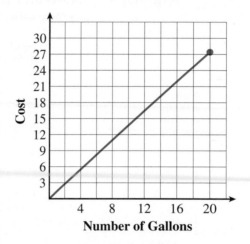

The domain for the general function defined by $c = 1.369g$, with no connection to the context of the problem, is the set of all real numbers, since any real number can be substituted for g in $1.369g$. Following is a graph of $c = 1.369g$ for any real number g.

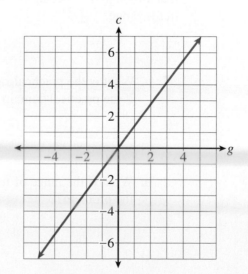

> **DEFINITION**
>
> In all functions the output variable is called the **dependent** variable, and the input variable is called the **independent** variable.

When a value from the domain is substituted for the independent variable and the corresponding output is evaluated, the result is a value of the dependent variable.

> **DEFINITION**
>
> The collection of all possible values of the output or dependent variable is the **range** of the function. The practical range corresponds to the practical domain.

Example 4   *Consider the following table that gives the percentage of mothers in the workforce with children under the age of 6 from 1995 to 2000.*

YEAR	PERCENTAGE
1995	62.3
1996	62.3
1997	65.0
1998	64.9
1999	64.8
2000	64.6

Although the table contains only six pairs of numbers, the table represents a function. The input or independent variable is the year, and the output or dependent variable is the percentage. The domain of the function is {1995, 1996, 1997, 1998, 1999, 2000} because these are all the input values. The range of the function is {62.3, 64.6, 64.8, 64.9, 65.0} because this is the set of all of the output values. Note that although 62.3 occurs twice in the table as an output value, it is listed only once in the range.

7. a. What is the practical range for the cost function defined by $f(g) = 1.369g$ if the practical domain is 0 to 20?

b. What is the range of this function if it has no connection to the context of the problem?

Constructing Tables of Input/Output Values

8. Use the symbolic form of the gas cost-of-fill-up function, $f(g) = 1.369g$, to evaluate $f(0)$, $f(5)$, $f(10)$, $f(15)$, and $f(20)$, and complete the following table. Note that the input variable g increases by 5 units. In such a case, you say the input increases by an **increment** of 5 units.

NUMBER OF GALLONS OF GAS NEEDED, g	COST OF FILL-UP, f(g)
0	
5	
10	
15	
20	

Appendix

A numerical form of the cost-of-fill-up function is a table or a collection of ordered pairs. When a function is defined in symbolic form, you can use technology to generate the table. The TI-83 Plus calculator is a function grapher. The y variables Y_1, Y_2, and so on represent function output (dependent) variables. The input, or independent variable, is x. The steps to build tables with the TI-83 Plus can be found in Appendix C.

9. Use your grapher to generate a table of values for the function represented by $f(g) = 1.369g$ to check your values in the table in Problem 8. The screens on your grapher should appear as follows:

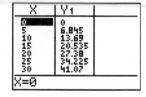

Roasting a Turkey

The directions for roasting a turkey indicate that it should be cooked for 20 minutes per pound at 350°F.

10. **a.** Complete the following table.

Turkey Time

WEIGHT OF TURKEY (IN POUNDS) w	COOKING TIME (IN MINUTES) t
10	
12	
14	
16	
18	

b. Write an equation to determine the cooking time t (in minutes) of the turkey weighing w pounds.

c. Is the cooking time a function of the weight of the turkey? Explain using the definition of a function.

d. What is the independent variable? What is the dependent variable?

e. Verify the results in part a by generating a table using your grapher.

f. Using h for the name of the function, the output variable t can be written as $t = h(w)$. Rewrite the equation in part b using the function notation $h(w)$ for cooking time.

g. What are the practical domain and the practical range of the function? Explain.

h. Evaluate $h(14)$ and write a sentence describing its meaning.

SUMMARY
Activity 1.2

1. **Independent variable** is another name for the input variable of a function.

2. **Dependent variable** is another name for the output variable of a function.

3. The collection of all possible replacement values for the independent or input variable is called the **domain of the function.** The **practical domain** is the collection of replacement values of the input variable that makes practical sense in the context of the problem.

4. The collection of all output values of a function is called the **range of the function.** When a function describes a real situation or phenomenon, its range is often called the **practical range** of the function.

5. When a function is represented by an equation, the function may also be written in function notation. For example, given $y = 2x + 3$, you can replace y with $f(x)$ and rewrite the equation as $f(x) = 2x + 3$.

EXERCISES
Activity 1.2

In Exercises 1 and 2,

a. *Identify the input and output.*

b. *Let x represent the input variable. Use function notation to represent the output variable.*

c. *Translate the verbal statement into a symbolic statement.*

1. Sales tax is a function of the price of an item. The amount of sales tax is 0.08 times the price of the item. Use h to represent the function.

2. The Fahrenheit measure of temperature is a function of the Celsius measure. The Fahrenheit measure is 32 more than 9/5 times the Celsius measure. Use g to represent the function.

For each function in Exercises 3–5, evaluate $f(2)$, $f(-3.2)$, $f(\pi)$, and $f(a)$.

3. $f(d) = 2d - 5$

4. $f(t) = -16t^2 + 7.8t + 12$

5. $f(x) = 4$

In Exercises 6–8, construct a table of values of four ordered pairs for the given function. Check your results using the table feature of your grapher.

6. $g(x) = x^2$. Start the inputs at 3 and use an increment of 2.

x	g(x)

7. $h(x) = \dfrac{1}{x}$. Start the inputs at 10 and use an increment of 10.

x	h(x)

8. $f(x) = \sqrt{100 - x}$. Start the inputs at 0 and use an increment of 5.

x	f(x)

9. **a.** The distance you travel while hiking is a function of how fast you hike and how long you hike at this rate. You usually maintain a speed of three miles per hour while hiking. Write a verbal statement that describes how the distance that you travel is determined.

   b. Identify the input and output variables of this function.

   c. Write the verbal statement in part a using function notation for the input variable. Let t represent the input variable. Let h represent the function, and $h(t)$ the output variable.

   d. Which variable is the dependent variable? Explain.

   e. Use the equation from part c to determine the distance traveled in four hours.

   f. Evaluate $h(7)$ and write a sentence describing its meaning. Write the result as an ordered pair.

   g. Determine the domain and range of the general function.

    h. Determine the practical domain and the practical range of the function.

    i. Use your calculator to generate a table of values beginning at zero with an increment of 0.5.

x	g(x)

10. Determine the domain and range of each function.

a.

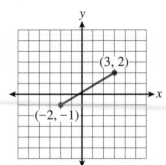

b.

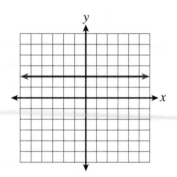

c.

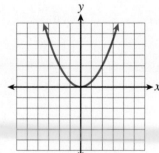

d.

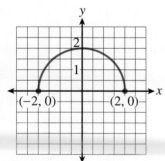

11. Determine the domain and range of each function.

    a. $\{(-2, 4), (0, 3), (5, 8), (8, 11)\}$

    b. $\{(-6, 5), (-2, 5), (0, 5), (3, 5)\}$

12. According to the Federal Highway Administration, it costs approximately $0.36 per mile to own and operate a car. The total cost is a function of the number of miles driven and can be represented by the function $C = 0.36m$. When you finally take your car to the junkyard, the odometer reads 157,200.

    a. What is the practical domain of this function?

    b. What is the practical range for this situation?

13. To change a Celsius temperature to Fahrenheit, use the formula $F = 1.8C + 32$. You are concerned only with temperatures between freezing and boiling.

    a. What is the practical domain of the function?

    b. What is the practical range of the function?

ACTIVITY 1.3
Stopping Short

OBJECTIVES

1. Use a function as a mathematical model.

2. Determine when a function is increasing, decreasing, or constant.

3. Use the vertical line test to determine if a graph represents a function.

After an automobile accident, the investigating police officers often estimate the speed of the vehicle by measuring the length of the skid mark. The following table gives the average skid distances for an automobile with good tires on dry pavement.

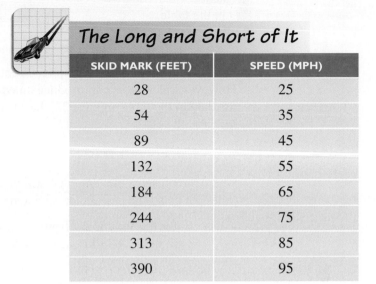

The Long and Short of It

SKID MARK (FEET)	SPEED (MPH)
28	25
54	35
89	45
132	55
184	65
244	75
313	85
390	95

1. Does the table data define speed as a function of skid mark? Explain using the definition of function.

2. Identify the independent variable and the dependent variable.

3. Plot the (input, output) ordered pairs on an appropriately scaled and labeled set of coordinate axes. Remember, the input values appear along the horizontal axis, and the output values appear along the vertical axis.

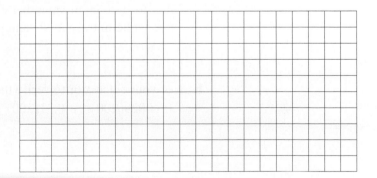

After a particular accident, a skid mark was measured to be 200 feet. From the table, the investigating officer knows that the speed of the vehicle was between 65 and 75 miles per hour. She would, however, like to be able to be more precise in reporting the speed. One way of getting values that are not listed in the table is to use a graph or equation of a function that best fits the actual data.

> **DEFINITION**
>
> A **mathematical model** is an equation or a graph that fits or approximates the actual data. The model can be used to predict output values for input values not in the table.

Note that the points on the graph in Problem 3 do not lie exactly on a specific curve. However, calculators can produce an equation that best models actual data. From the data in the table on page 23, the TI-83 Plus can be used to generate the following model,

$$f(x) = -0.00029x^2 + 0.31x + 18.6 \text{ (coefficients are rounded)}$$

where x represents the length of skid marks in feet and $f(x)$ represents the speed in miles per hour. The process for generating these equations is covered in later activities.

Appendix

4. a. Enter the function f from the model above into your calculator. For help with the TI-83 Plus, see Appendix C.

b. Use the values in the table on page 23 to set appropriate window values in your grapher to view the graph.

c. Display your graph using the window settings from part b. Your graph should resemble the following:

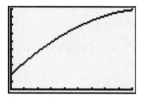

5. a. By pressing the trace button and the left and right arrow keys, you can display the input/output values of points on the graph on the bottom of the display. Use the trace feature of your calculator to approximate the speed of the car when the length of the skid marks is 200 feet. You can obtain the exact value for any input value between Xmin and Xmax by entering the input value while in trace mode and pressing ENTER.

b. Use your calculator to verify the result in part a by evaluating $-0.00029(200)^2 + 0.31(200) + 18.6$.

Increasing and Decreasing Functions

There are many other advantages to having the function in graphical form. For example, you are often interested in determining how the output values change as the input values increase.

> **DEFINITION**
>
> A function is **increasing** if its graph goes up to the right, **decreasing** if its graph goes down to the right, and **constant** if its graph is horizontal. In each case, you are viewing the graph as a point moves along the curve from left to right, that is, as the input values increase.

Example 1

a. *The graph of the function $y = 3x + 1$ is always increasing. The graph of this function goes up from left to right.*

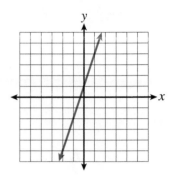

b. *The graph of the function $f(x) = -x + 2$ is decreasing because its graph goes down from left to right.*

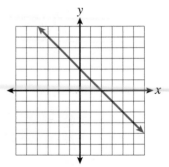

c. *The graph of $y = 3$ is constant, because the graph goes neither up nor down. The output value is always 3 no matter what the input value is.*

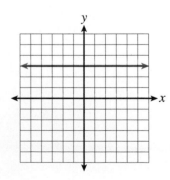

6. Is the function $f(x) = -0.00029x^2 + 0.31x + 18.6$ in Problem 4 increasing, decreasing, or constant over the domain displayed in the window?

Vertical Line Test

7. To cover your weekly expenses while going to school, you earn $100 each week as a part-time aide in your college's health center. The following table represents a typical month.

x, WEEKLY SALARY ($)	100	100	100	100
y, WEEKLY EXPENSES ($)	50	70	90	60

a. Plot the data points on the following grid.

b. Are weekly expenses a function of the weekly salary? Explain using the definition of function.

c. Do all four points lie on the same straight line?

d. Is the line horizontal, vertical, or slanted?

e. How can you determine from the graph that it does not represent a function?

8. A unit circle is a circle having a center at the origin and a radius of 1. Such circles are used in the study of trigonometry. The graph of a unit circle is

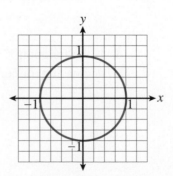

Does the graph of a unit circle represent a function? Explain how you can determine from the graph whether or not it represents a function.

Problems 7 and 8 illustrate what is known as the vertical line test. You know that a function can have only one output value for each input value. Since any two points with the same first coordinate lie on a vertical line, a graph defines a function if any vertical line drawn through the graph intersects the graph no more than once. This is called the **vertical line test.**

9. Use the vertical line test to verify that the graphs displayed in Problems 4 and 7 define a function.

10. Use the vertical line test to determine which of the following graphs represent functions.

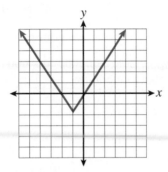   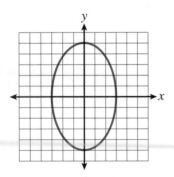

11. If a graph is increasing everywhere, does the graph represent a function? Explain.

SUMMARY
Activity 1.3

1. A **mathematical model** is an equation or a graph that fits or approximates the actual data. The model can be used to predict output values for input values not in the table.

2. A function is **increasing** if its graph goes up to the right, **decreasing** if its graph goes down to the right, and **constant** if its graph is horizontal.

3. In the **vertical line test**, a graph defines a function if any vertical line drawn through the graph intersects the graph no more than once.

1. The following table defines snowfall as a function of elevation for a recent snow storm in upstate New York.

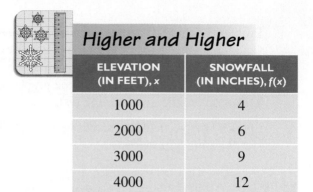

Higher and Higher

ELEVATION (IN FEET), x	SNOWFALL (IN INCHES), $f(x)$
1000	4
2000	6
3000	9
4000	12

A function that closely models the data in the table is

$$f(x) = 0.0027x + 1.$$

Enter this function into a y variable on your calculator.

a. Complete the following table using the function f. Verify your answers using the table feature of your graphing calculator.

ELEVATION	1000	2000	3000	4000
$f(x)$				

b. Determine $f(2500)$ and explain its practical meaning in this situation.

c. Determine $f(-2000)$. Does this have any meaning in this context?

d. Use the table values to set appropriate window values to view the graph. Graph the function f on your calculator. Identify the window you used.

e. Does the graph pass the vertical line test for a function? Explain.

f. Does the graph indicate that the function is increasing, decreasing, or constant?

g. Use the trace function of your calculator to determine $f(2500)$. Compare your answer to your result in part b.

2. The number of new hotels opening in the United States between 1997 and 2001 can be modeled by the function

$$h(x) = 9.75x^3 - 95.86x^2 + 120.18x + 1477.79,$$

where x is the number of years after 1997 and $h(x)$ is the number of new hotels.

a. Using your calculator, complete the following table. Round your results to the nearest whole number.

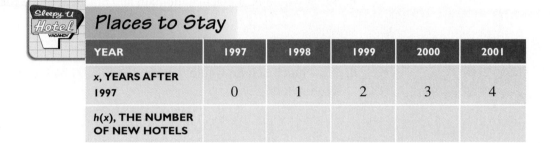

Places to Stay

YEAR	1997	1998	1999	2000	2001
x, YEARS AFTER 1997	0	1	2	3	4
h(x), THE NUMBER OF NEW HOTELS					

b. Evaluate $h(2)$, and explain its practical meaning in this situation.

c. Use the table values to set appropriate window values to view the graph of the function $h(x)$. Graph the function on your calculator. Identify the window you used.

d. Does the graph pass the vertical line test for a function? Explain.

e. Use the trace function of your calculator to determine $h(3)$. Compare your answer to your result in part a.

3. You're having a party, and you want to fill balloons with helium gas. The volume of helium needed for one balloon is a function of the size of the balloon. You can write an equation, $V = f(r)$ to represent this function. If you assume the balloon is approximately the shape of a sphere, the volume, V, of a spherical balloon is given by the formula $V = \dfrac{4}{3}\pi r^3$, where r is the radius of the balloon. The balloon will pop when the radius is approximately 25 centimeters.

a. Complete the following table.

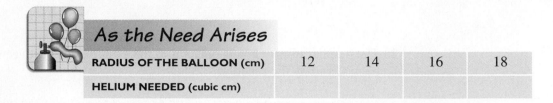

As the Need Arises

RADIUS OF THE BALLOON (cm)	12	14	16	18
HELIUM NEEDED (cubic cm)				

b. Identify the dependent (output) and independent (input) variables.

c. What would be a practical domain and practical range for this function?

d. How much helium would you need to fill 100 balloons if each has a radius of 15 centimeters?

e. Rewrite the volume formula using function notation. Use f to name the function and $f(r)$ to represent the output variable.

f. Determine $f(20)$, and write the result as an ordered pair.

g. Enter the function into your grapher. Use the practical domain and range determined in part c to determine an appropriate window for your graph.

h. Is the graph increasing, decreasing, or constant? Explain.

i. How can you determine, just by looking at the graph, that it is a function?

j. Use the trace feature on your calculator, and determine $f(20)$. Does the result check with your answer in part f?

4. a. Sketch the graph of a function that is everywhere decreasing.

b. What does the graph of a constant function look like?

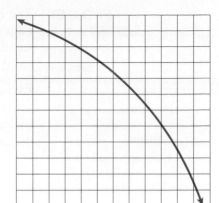

5. Use the vertical line test to determine whether either graph represents a function.

a.

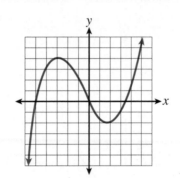

b.

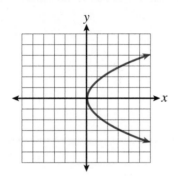

In Exercises 6 and 7, enter the given function into a y variable on your calculator. Match the graph given for the function by determining the appropriate window. (Hint: Use a table to help determine a window.) Give the Xmin, Xmax, Ymin, and Ymax window values that you use.

6. $f(x) = 5400x + 3600$

7. $g(x) = 2x^4 - 5x - 2$

**PROJECT
ACTIVITY 1.4**
Graphs Tell
Stories

OBJECTIVES

1. Describe in words what a graph tells you about a given situation.

2. Sketch a graph that best represents the situation described in words.

The expression "A picture is worth a thousand words" is a cliché, but it is true. Functions are often easier to understand when presented in visual form. To understand such pictures, you need to practice going back and forth between graphs and word descriptions.

Every graph shows how the inputs and outputs change in relation to one another. As you read a graph from left to right, the input variable is increasing in value. The graph indicates the change in the output values (increasing, decreasing, or constant) as the input values increase.

Example 1   *The following graph describes your walk from the parking lot to the library.*

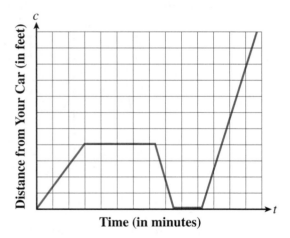

As you read the graph in Example 1 from left to right, it shows how your distance from your car changes as time passes. One possible scenario this graph describes is:

You leave your car and walk at a steady pace toward the library. You meet some friends and stop to chat for a while. You realize that you forgot something and quickly return to your car. After rummaging around for a while, you hurry off to the library.

How did anyone come up with this from the graph? Look at the graph in sections.

a. The first increasing line segment indicates you are moving away from your car because the time and the distance are increasing.

b. The first horizontal section indicates that your distance from the car is constant, so you are standing still.

c. The decreasing line segment indicates your distance from the car is decreasing. When it reaches the horizontal axis, it tells you that you are back at your car.

d. The second horizontal segment indicates you stay at your car for a time.

e. The final increasing segment is steeper and longer than the first, so you are moving away from the car faster and farther than in the first segment.

For Problems 1–3, describe in words what the graph tells you about the situation. Indicate what happens to the output as the input increases and if the function reaches either a minimum or a maximum point.

1. Elevation of a hot-air balloon is a function of time.

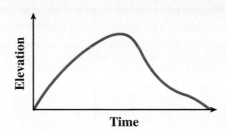

2. Profits at a juice bar stand are a function of daily temperature.

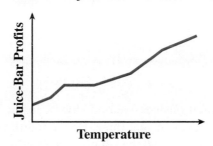

3. Average annual income is a function of years of education.

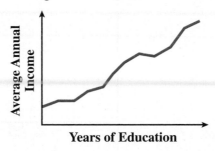

In Problems 4–11, sketch a graph that best represents the given situation. Many times you won't know the actual values, so estimate what seems reasonable to you. Your graphs of these problems will be more qualitative than quantitative. Be sure to label your axes, with the input variable always on the horizontal axis. Provide numerical scales when appropriate.

4. You leave home on Friday afternoon for your weekend getaway. Heavy traffic slows you down for the first half of your trip, but you make good time by the end. Express your distance from home as a function of time.

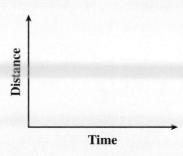

5. Five years ago, you started a job that paid $8.00 per hour. After six months, you got a 50-cent raise and three years ago you got another raise of $1.00. For the past year, you have been making your current wage of $12.30 per hour. Express your hourly wage as the dependent variable and time as the independent variable.

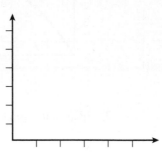

6. Your small business started slowly, losing money in its first two years, then breaking even in year three. By the fourth year, you made as much as you lost in the first year and then doubled your profits each of the next two years. Graph your profit as the output and time as the input.

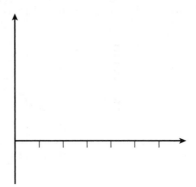

7. Hair grows at a steady rate. Suppose you get your hair cut every month. Measuring the longest hair on your head, graph your hair length over the course of six months.

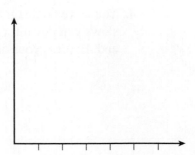

8. Hours of daylight depend upon the day of the year.

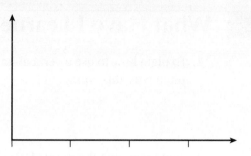

9. The distance traveled is a function of speed in a fixed time interval.

10. The sale price of a computer is a function of the percent of discount.

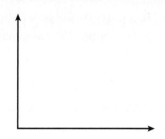

11. The area of a square is a function of the length of one side of the square.

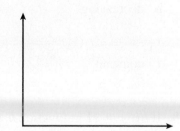

What Have I Learned?

1. Explain how to use a vertical line to determine if a graph represents a function, and explain why this works.

2. You know that the point (1, 2) is on the graph of a function, f. Give the coordinates of a point that you know is not on the graph of f. Explain.

3. Is it possible for an input/output relationship to represent a function if two different inputs produce the same output? Explain.

4. Functions are defined four different ways in this cluster. List them.

5. Is there ever a difference between the domain of a function and practical domain of a function if their symbolic representations are identical? Explain.

6. What are two other words or phrases used in this cluster to represent the input variable and output variable?

7. If you are given the graph of a function how can you tell if the function is
   a. increasing?

   b. decreasing?

   c. constant?

8. If $h(5) = 11$ in function h, what is the relationship between the numbers 5 and 11 and the graph of h?

9. If $f(x) = -3x + 2$, explain how to determine $f(4)$.

CLUSTER 1   How Can I Practice?

1. The following table shows the total number of points accumulated by each student and the numerical grade in the course.

STUDENT	TOTAL POINTS	NUMERICAL GRADE
Tom	432	86.4
Jen	394	78.8
Kathy	495	99
Michael	330	66
Brady	213	42.6

a. Is the numerical grade a function of the total number of points? Explain.

b. Is the total number of points for these five students a function of the numerical grade? Explain.

c. Using the total points as the input and the numerical grade as the output, write the ordered pairs that represent each student. Call this function f.

d. Plot the ordered pairs determined in part c on an appropriately scaled and labeled set of axes.

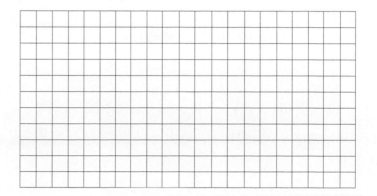

e. What is the value of $f(394)$?

    f. What is the practical meaning of $f(394)$?

    g. What is the value of $f(213)$?

    h. What is the practical meaning of $f(213)$?

    i. Determine the numerical value n, given that $f(n) = 66$.

In Exercises 2–10, determine which of the given relationships represent functions.

2. The money you earn is a function of the number of hours you work.

3. Your heart rate is a function of your level of activity.

4. The cost of daycare depends on the number of hours a child stays at the facility.

5. The number of children in a family is a function of the parents' last name.

6. $\{(2, 3), (4, 3), (5, -5)\}$

7. $\{(-3, 4), (-3, 6), (2, 6)\}$

8.

x	-3	5	7
$f(x)$	0	-5	9

9.

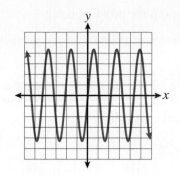

10.

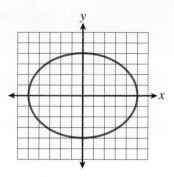

11. a. For a part-time student, the cost of college tuition is a function of the number of hours for which the student is registered. Write an equation to represent the tuition cost, c, if the cost per hour is $107 and h represents the number of hours taken for the current semester.

b. Use $f(h)$ to represent the cost, and rewrite the equation in part a using function notation.

c. Complete the table.

h	2	4	7	8	11
$f(h)$					

d. Evaluate $f(3)$, and write a sentence describing its meaning. Write the results as an ordered pair.

e. Given $f(h) = \$535$, determine the value of h.

f. Which variable is the output? Explain.

g. Which is the independent variable? Explain.

h. Explain (using the table in part c) how you know that the data represents a function.

i. What is the practical domain for this function?

j. Plot the ordered pairs on an appropriately scaled and labeled coordinate system. Which axis represents the input values?

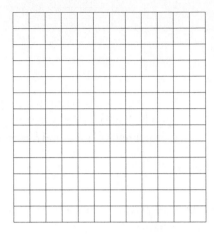

k. Explain from the graph how you know that f is a function.

l. Use your grapher to verify your answers to part c and j.

m. Use the trace and table features to determine the cost of nine credit hours.

12. Given $p(x) = 2x + 7$, determine each of the following.

    a. $p(3)$ **b.** $p(-4)$

    c. $p\left(\dfrac{1}{2}\right)$ **d.** $p(0)$

13. Given $t(z) = 2z^2 - 3z - 5$, determine each of the following.

    a. $t(2)$ **b.** $t(-3)$

14. According to the U.S. Department of Health, the average life expectancy for males (from birth) in the U.S. may be modeled by the function $f(x) = 0.27x + 48.3$, where x is the number of years since 1900.

a. Use your calculator to complete the following table. Round your results to the nearest whole number.

| | YEAR | | | | |
	1900	1940	1970	1990	2000
x, YEARS SINCE 1900	0	40	70	90	100
f(x), LIFE EXPECTANCY					

b. Evaluate $f(85)$, and explain its practical meaning in this situation.

c. Use the table values to set appropriate window values to view the graph of f. Graph the function on your calculator. Identify the window you used.

d. Is the graph increasing, decreasing, or constant? Explain.

e. Use the trace feature of your calculator to determine $f(85)$. Compare your answer to your result in part b.

15. Determine the domain and the range of each of the following functions.

a. $\{(3, 5), (4, 5), (5, 8), (6, 10)\}$

b.

CELLULAR PHONE CHARGES (MONTHLY)					
NUMBER OF MINUTES INPUT	0	50	100	150	200
MONTHLY COST (DOLLARS) OUTPUT	19.95	23.95	26.95	30.45	33.95

c.

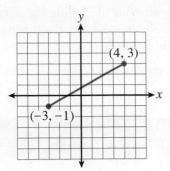

d.

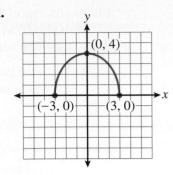

e. $y = 3x$

16. Each of the following graphs show how the inputs and outputs change in relation to each other. Describe in words what the graph is telling you about the situation. Provide a reasonable explanation for the behavior you describe.

a.

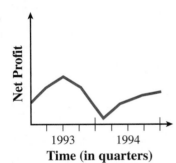

b.

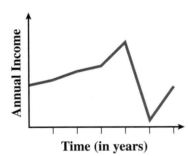

CLUSTER 2   Linear Functions

ACTIVITY 1.5
Walking for
Fitness

OBJECTIVE

1. Determine the average rate of change.

Suppose you are a member of a health-and-fitness club. A special diet and exercise program has been developed for you by your personal trainer. At the beginning of the program, and once a week thereafter, you are tested on the treadmill. The test consists of how many minutes it takes you to walk, jog, or run three miles on the treadmill. The following data gives your time, t, over an eight-week period.

END OF WEEK, w	0	1	2	3	4	5	6	7	8
TIME, t, (in minutes)	45	42	40	39	38	38	37	39	36

Note that $w = 0$ corresponds to the first time on the treadmill, $w = 1$ is the end of the first week, $w = 2$ is the end of the second week, and so on.

1. **a.** Is time, t, a function of weeks, w? If so, what are the input and output variables?

   b. Plot the data points using ordered pairs of the form (w, t).

DEFINITION

A set of points in the plane whose coordinate pairs represent input/output pairs of a data set is called a **scatterplot**.

Appendix

Example 1   *The points plotted in Problem 1b are a scatterplot of the treadmill data. Your graphing calculator can generate a scatterplot of data points. Refer to Appendix C for instructions. The final screen would appear as follows:*

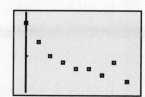

2. a. What was your treadmill time at the beginning of the program?

   b. What was your treadmill time at the end of the first week?

An important question that can be asked about this situation is, how did your time change from one week to the next?

3. a. During which week(s) did your time increase?

   b. During which week(s) did your time decrease?

   c. During which week(s) did your time remain unchanged?

4. Your time decreased during each week of the first four weeks of the program.

   a. Determine the total change in time, t, during the first four weeks of the program (i.e., from $t = 45$ to $t = 38$). Why should your answer contain a negative sign? Explain.

   b. Determine the change in weeks, w, during this period (that is, from $w = 0$ to $w = 4$).

PROCEDURE: **Determining Total Change**

The change in time, t, is represented by the symbol Δt. The symbol Δ (delta) is used to represent "change in." You generally calculate the change in time, t, from a first (initial) value to a second (final) value of t. The first time is represented by t_1 (read "t sub 1"), and the second time is represented by t_2 (read "t sub 2"). The change in t is then calculated by subtracting the first (initial) value from the second (final) value. This is symbolically represented by

$$\Delta t = t_2 - t_1 \quad \text{or} \quad \Delta t = \text{final time} - \text{initial time}.$$

Because t is the output variable, Δt is the change in output.

Similarly, Δw represents the change in weeks, w; w_1 represents the first (initial) value of w; and w_2 represents the second (final) value of w. Symbolically,

$$\Delta w = w_2 - w_1, \quad \text{or} \quad \Delta w = \text{final week} - \text{initial week}.$$

Because w is the input variable, Δw is the change in input.

5. Use the Δ notation to express your results in Problems 4a and 4b.

Average Rate of Change

Neither the change in treadmill time nor the change in the number of weeks completely describes your progress during the first four weeks. The ratio of the change in t, Δt, to the change in w, Δw, written $\frac{\Delta t}{\Delta w}$, provides more relevant information about the effect of the exercise program over time. This ratio, $\frac{\Delta t}{\Delta w}$, shows how the time changed on average over the four-week period.

6. Use your results from Problem 5 to determine the ratio $\frac{\Delta t}{\Delta w}$, during the first four-week period. Interpret your answer.

7. a. What are the units of measurement of the ratio determined in Problem 6?

b. On your graph from Problem 1, connect the points $(0, 45)$ and $(4, 38)$ with a line segment. Does the output increase, decrease, or remain unchanged over the interval?

DEFINITION

The ratio $\frac{\Delta t}{\Delta w}$ is called the **average rate of change** of time, t, with respect to weeks, w. In general, the average rate of change is

$$\frac{\text{change in output}}{\text{change in input}}.$$

The rate of change of a function gives the change in the output per unit increase of the input.

8. a. Determine the average rate of change of t with respect to w during the sixth and seventh weeks (from the point where $w = 5$ to the point where $w = 7$).

b. What is the significance of the positive sign of the average rate of change in this situation?

c. Connect the data points (5, 38) and (7, 39) on your graph from Problem 1, using a line segment. Is the output increasing, decreasing, or constant on the interval?

9. a. At what average rate did your time change during the fifth week (from $w = 4$ to $w = 5$)?

b. Interpret your answer in this situation.

c. Connect the data points (4, 38) and (5, 38) on the graph from Problem 1 using a line segment. Is the output increasing, decreasing, or constant over this interval?

10. a. At what average rate is your time changing as w increases from $w = 3$ to $w = 7$?

b. Does your answer mean that your time did not change in this four-week period? Interpret your answer in this situation.

11. As part of your special diet and exercise program, you record your weight at the beginning of the program and each week thereafter. The following data gives your weight, w, over a five-week period.

WEEKS, w	0	1	2	3	4	5
WEIGHT, y, (pounds)	196	183	180	177	174	171

a. Determine the average rate of change of your weight during the first three weeks.

b. Determine the average rate of change during the five-week period.

c. Determine the change in weight during each week of your exercise program.

d. What are the units of measure of the average rate of change?

e. What is the practical meaning of the average rate of change in this situation?

f. What can you say about the average rate of change of weight during any time interval in this situation?

SUMMARY
Activity 1.5

1. The **average rate of change** of a function over a specified input interval is the ratio

$$\frac{\text{change in output}}{\text{change in input}}.$$

2. The rate of change indicates how much, and in which direction, the output changes when the input increases by a single unit. It measures how the output changes on average.

EXERCISES
Activity 1.5

The following table of data from the United States Bureau of the Census gives the median age of an American man at the time of his first marriage.

YEAR	1910	1920	1930	1940	1950	1960	1970	1980	1990	2000
MEDIAN AGE	25.1	24.6	24.3	24.3	22.8	22.8	23.2	24.7	26.1	26.8

Use this data to answer Exercises 1–6.

1. a. Determine the average rate of change in median age per year from 1950 to 1990.

b. Describe what the average rate of change in part a represents in this situation.

Exercise numbers appearing in color are answered in the Selected Answers appendix.

2. Determine the average rate of change in median age per year from 1930 to 1960.

3. What is the average rate of change over the ninety-year period described in the table?

4. During what ten-year period did the average age increase the most?

5. a. What does it mean in this situation if the average rate of change is negative?

   b. Determine at least one ten-year period when the average rate of change is negative.

   c. What trend would you observe in the graph if the average rate of change were negative? That is, would the graph go up, go down, or remain constant?

6. a. Is the average rate of change zero over any ten-year period? If so, when?

   b. What does a rate of change of zero mean in this situation?

   c. What trend would you observe in the graph during this period? That is, would the graph go up, go down, or be horizontal?

The following table gives information about new hotel construction in the U.S. from 1997 to 2001.

YEAR, t	1997	1998	1999	2000	2001
HOTELS CONSTRUCTED, h	1476	1519	1402	1246	1047

Use the table of data to answer Exercises 7 and 8.

7. Plot the data points using ordered pairs of the form (t, h).

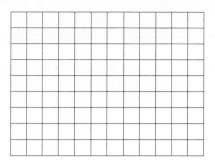

8. a. Determine the average rate of change of new hotel construction from 1997 to 1998.

b. Determine the average rate of change from 1998 to 1999.

c. Compare the average rate of change from 1997 to 1998 with the rate of change from 1998 to 1999.

d. When the average rate of change is negative, what trend will you observe in the graph? What does that mean in this situation?

9. Between 1960 and 1999, the size and shape of automobiles in the United States has changed almost annually. The fuel consumed by these vehicles has also changed. The following table describes the average fuel consumed per year per passenger car in gallons of gasoline.

YEAR, t	1960	1970	1980	1990	1995	1997	1998	1999
GALLONS CONSUMED PER PASSENGER CAR (average)	668	760	576	520	530	538	544	552

a. Determine the average rate of change, in gallons of fuel used per passenger car, from 1960 to 1970.

b. Determine the average rate of change, in gallons of gas per year, from 1960 to 1990.

c. Determine the average rate of change, in gallons of gas per year, from 1995 to 1997.

d. Determine the average rate of change, in gallons of gas per year, between 1960 and 1999.

e. What does the result in part d mean in this situation?

ACTIVITY 1.6

Depreciation

OBJECTIVES

1. Interpret slope as an average rate of change.

2. Use the formula to determine slope.

3. Discover the practical meaning of vertical and horizontal intercepts.

4. Develop the slope/intercept form of an equation of a line.

5. Use the slope/intercept formula to determine vertical and horizontal intercepts.

You have decided to buy a new Honda Accord LX, but you are concerned about the value of the car depreciating over time. You search the Internet and obtain the following information at www.intellichoice.com:

2004 Accord LX

- Suggested retail price $18,790
- Depreciation per year $1,385 (assume constant)

1. a. Complete the following table in which V represents the value of the car after n years of ownership.

Accordingly

n, YEARS	V, VALUE IN DOLLARS
0	
1	
2	
3	
5	
8	

b. Is the value of the car a function of the number of years of ownership? Explain.

c. What is the input? What is the output?

2. a. Select two ordered pairs of the form (n, V) from the table in Problem 1 and determine the average rate of change.

b. What are the units of measure of the average rate of change?

c. What is the practical meaning of the sign of the average rate of change?

d. Select two different ordered pairs and compute the average rate of change.

e. Select two ordered pairs not used in parts a or d, and compute the average rate of change.

f. Using the results in parts a, d, and e, what can you infer about the average rate of change over any interval of time?

If the computation of the average rate of change using any two ordered pairs yields the same result, the average rate of change is said to be constant.

> **DEFINITION**
>
> Any function in which the average rate of change, $\dfrac{\text{change in output}}{\text{change in input}}$, is constant is called a **linear function**.

3. a. Is the value, V, of the car a linear function of the number of years, n, of ownership? Explain using the definition of linear function.

b. Is this function increasing, decreasing, or constant?

Graph of a Linear Function

4. Consider the ordered pairs of the form (n, V), and plot each ordered pair in Problem 1 on an appropriately scaled and labeled set of axes. Connect the points to see if there is a pattern.

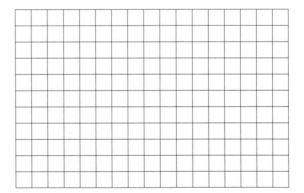

The graph of a linear function is a nonvertical line. The constant average rate of change is called the **slope** of the line and is denoted by the letter m.

> **DEFINITION**
>
> If x represents the input variable and y represents the output variable, then the **slope m** is given by
>
> $$m = \frac{\Delta y}{\Delta x} = \frac{y_2 - y_1}{x_2 - x_1}, \text{ where } x_1 \neq x_2.$$

5. a. What is the slope of the line graphed in Problem 4?

b. What is the relationship between the slope of the line and the average rate of change?

c. What is the practical meaning of slope in this situation?

Vertical Intercept

> **DEFINITION**
>
> The **vertical intercept** is the point where the graph crosses, or intercepts, the vertical axis. The input value of a vertical intercept is always zero. If the output variable is represented by y, the vertical intercept is referred to as the y-intercept.

6. a. Using the table of data in Problem 1 or the graph in Problem 4, determine the vertical intercept (V-intercept).

b. What is the practical meaning of the vertical intercept in this situation? Include units.

Slope-Intercept Form of a Linear Equation

7 a. Review how you determined the value, V, of the car in Problem 1 for a given number of years, n, of ownership. Write an equation for V in terms of n.

b. Use your graphing calculator to sketch a graph of this equation. Use the window Xmin $= -2$, Xmax $= 15$, Ymin $= -500$, and Ymax $= 20{,}000$.

c. How does this graph compare to your graph in Problem 4?

8. Recall that the slope of your line is $m = -1385$ and the vertical intercept is (0, 18,790). How is this information contained in the equation of the line you determined in Problem 7a?

DEFINITION

The coordinates of all points (x, y) on the line with slope m and vertical intercept $(0, b)$ satisfy the equation

$$y = mx + b \quad \text{or} \quad y = b + mx.$$

This is called the **slope-intercept form** of the equation of a line.

Note that the coefficient of x, which is m, is the *slope* of the line. The constant term, b, is the *y-coordinate of the vertical intercept*. If $f(x)$ replaces y, the equation $y = mx + b$ can be written as

$$f(x) = mx + b.$$

Example 1 *The slope-intercept form of the equation of the line with slope 3 and vertical intercept $(0, -6)$ is $y = 3x - 6$. Using function notation and replacing y with $f(x)$, the equation becomes $f(x) = 3x - 6$.*

9. Identify the slope and vertical intercept of the line whose equation is given. Write the vertical intercept as an ordered pair.

a. $y = -2x + 5$
                                        b. $s = \frac{3}{4}t + 2$

c. $q = 2 - r$
                                        d. $y = \frac{5}{6} + \frac{x}{3}$

Horizontal Intercepts

DEFINITION

A **horizontal intercept** of a graph is a point where the graph meets or crosses the horizontal axis. The output value of a horizontal intercept is always zero. If the input variable is represented by x, the horizontal intercept is referred to as the x-intercept.

Example 2   *Consider the equation* $y = 2x - 10$.

a. The vertical intercept (y-intercept) occurs where the line crosses the vertical axis, i.e., where $x = 0$. Letting $x = 0$, $y = 2(0) - 10$ or $y = -10$. The vertical intercept is $(0, -10)$.

b. The horizontal intercept (x-intercept) occurs where the line crosses the horizontal axis, that is, where $y = 0$.

Letting $y = 0$,

$$0 = 2x - 10$$
$$\underline{+\ 10 \qquad\quad +\ 10} \qquad \text{Add 10 to each side.}$$
$$10 = 2x$$
$$\frac{10}{2} = \frac{2x}{2} \qquad\qquad \text{Divide each side by 2.}$$
$$5 = x$$

Therefore, the horizontal intercept is $(5, 0)$.

c. You can now sketch a graph of $y = 2x - 10$ by plotting the horizontal and vertical intercepts and connecting the points.

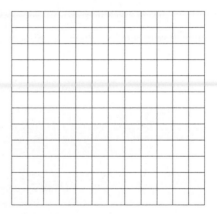

10. a. Determine the horizontal intercept (n-intercept) of the graph of the car value equation $V = -1385n + 18{,}790$.

b. What is the practical significance of the horizontal intercept? Include units.

Converting Celsius to Fahrenheit

11. The temperature on a warm summer day may be 27°. Your refrigerator is set at 5°. You set your oven to 180° to cook a turkey. If these temperatures seem strange, it is because they are measured on the Celsius scale. If the TV weather forecaster says it is going to be 65°F tomorrow, you know what that temperature will feel like and you can dress accordingly. How should you dress if the reporter says it will be 10°C? Most of us need to convert Celsius to Fahrenheit so we can better understand the temperature. The equation that relates Celsius measure to Fahrenheit measure is:

$$F = 1.8C + 32,$$

where C is the Celsius measure and F is the Fahrenheit measure.

a. Use the equation to complete the following table.

C	−10	0	5	10	27	100
F						

b. Sketch a graph of the equation $F = 1.8C + 32$ by plotting the ordered pairs in part a.

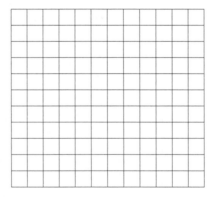

c. Determine the average rate of change of F as C increases from 0 to 100.

d. What is the slope of the line? How does the slope compare to the average rate of change in part c?

e. Is the function increasing, decreasing, or constant?

f. Determine the vertical and horizontal intercepts of the graph using the equation of the line. Verify your results using the graph of the line.

g. What is the practical significance of each intercept in this situation?

SUMMARY
Activity 1.6

1. A function for which the average rate of change between any pair of points remains constant is called a **linear function**.

2. The graph of a linear function is a nonvertical line. The constant average rate of change is called the **slope** and is denoted by the letter m (from the French verb *monter*, "to climb" or "rise").

3. The **slope** of a line segment joining two points (x_1, y_1) and (x_2, y_2) is denoted by m and is defined by $m = \dfrac{y_2 - y_1}{x_2 - x_1}$, where $x_1 \neq x_2$.

4. The **vertical intercept** $(0, b)$ of a graph is the point where the graph crosses the vertical axis. The **horizontal intercept** $(a, 0)$ of a graph is the point where the graph crosses the horizontal axis.

5. The slope-intercept form of the equation of a line is $y = mx + b$.

6. To determine the b-value of the vertical intercept $(0, b)$ from $y = mx + b$, set $x = 0$ and solve for y.

7. To determine the a-value of the horizontal intercept $(a, 0)$ from $y = mx + b$, set $y = 0$ and solve for x.

8. If the **slope** of a linear function is **positive**, the graph of the function rises to the right.

9. If the **slope** of a linear function is **negative**, the graph of the function falls to the right.

EXERCISES
Activity 1.6

1. Determine whether the following functions are linear. If they are, determine the constant average rate of change, called the slope of the line.

    a.

x	y
0	−1
1	9
2	19

    b.

x	2	4	6
y	11	8	−1

    c. $\{(-2, 18), (2, 9), (6, 0)\}$

2. Determine whether the following tables contain data that represent a linear function. Assume that the first row of the table is the input and the second row is the output. Explain your reasoning.

a. You owe your grandmother $1000. She paid for your first semester at community college. The conditions of the loan are that you must pay her back the whole amount in one payment using a simple interest rate of 6% per year. She doesn't care in which year you pay her. The table contains input and output values to represent how much money you will owe your grandmother one, two, three, or four years later.

YEAR	1	2	3	4
AMOUNT OWED (in $)	1060	1120	1180	1240

b. You decide to invest $1000 of your 401 (k) funds into an account that pays 5.5% interest compounded continuously. The table contains input and output values that represent the amount an initial investment of $1000 is worth at the end of each year.

YEAR	1	2	3	4	5
TOTAL INVESTMENT (in $)	1057	1116	1179	1246	1317

c. For a fee of $20 per month you may have breakfast (all you can eat) in the college snack bar each day. The table contains input and output values that represent the total number of breakfasts consumed each month and the amount you pay each month.

NUMBER OF BREAKFASTS	10	22	16	13
COST (in $)	20	20	20	20

3. You belong to a health-and-fitness center. You and your friends are enrolled in the center's weight-loss program. The charts contain input and output values (assume that week is input and weight is output) that represent the weight over a four-week period for you and your two friends. Determine which charts contain data that is linear and explain why.

a.

WEEK	1	2	3	4
WEIGHT (in lb)	150	147	144	141

b.

WEEK	1	2	3	4
WEIGHT (in lb)	183	178	174	171

c.

WEEK	1	2	3	4
WEIGHT (in lb)	160	160	160	160

4. Consider the equation $y = -2x + 5$.

a. Construct a table of five ordered pairs that satisfy the equation.

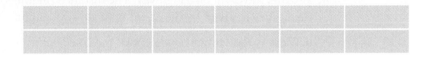

b. What is the slope of the line represented by the equation?

c. What is the vertical intercept?

d. What is the horizontal intercept?

e. Sketch a graph of the line using each of the following methods:
Method 1: Plot the five ordered pairs.
Method 2: Plot one point and use the slope to obtain additional points on the line.
Method 3: Plot the intercepts.

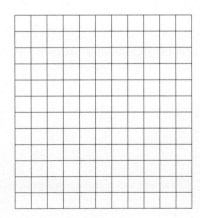 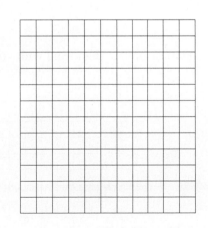 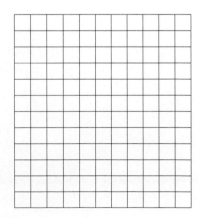

5. a. Determine the slope of the line that goes through the points $(2, -7)$ and $(0, 5)$.

b. Determine the vertical intercept of this line.

  c. What is the equation of the line through these points? Write the equation in function notation.

  d. What is the horizontal intercept?

6. A car is traveling on a highway. The distance (in miles) from its destination and the time (in hours) is given by the equation $d = 420 - 65t$.

  a. What is the vertical intercept of the line?

  b. What is the practical meaning of the vertical intercept?

  c. What is the slope of the line represented by the equation?

  d. What is the practical meaning of the slope determined in part c?

  e. What is the horizontal intercept?

  f. What is the practical domain of this function?

  g. Graph the equation, both by hand and with your graphing calculator, to verify your answers.

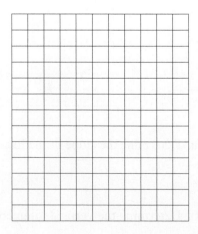

7. The following table gives a jet's height above the ground (in feet) as a function of time (in seconds) as the jet makes its landing approach to the runway.

t (in seconds)	h (in feet)
0	3500
5	3000
10	2500
15	2000
20	1500
25	1000

a. Is this function linear? Explain.

b. Calculate the slope using the formula $m = \dfrac{\Delta h}{\Delta t}$.

c. What is the significance of the sign of the slope in part b?

d. Determine where the graph crosses the vertical axis.

e. Write the equation in slope-intercept form.

f. Determine the horizontal intercept. What is its significance in this situation?

8. Determine the horizontal intercept of the line whose equation is given.

a. $y = 4x + 2$　　　　　　　　　　b. $y = \dfrac{x}{2} - 3$

9. a. Use your graphing calculator to graph the linear functions defined by the following equations: $y = 2x - 3$, $y = 2x$, $y = 2x + 2$, $y = 2x + 5$. Discuss the similarities and the differences of the graphs.

b. Use your graphing calculator to graph the linear functions defined by the following equations: $y = x - 2$, $y = 2x - 2$, $y = -x - 2$, $y = -2$. Discuss the similarities and the differences in the graphs.

c. Use your graphing calculator to graph the linear functions defined by the following equations: $y = 3x$, $y = -2x$, $y = \frac{1}{2}x$, $y = -5x$. Discuss the similarities and differences in the graphs.

ACTIVITY 1.7
A Visit to the
Architect

OBJECTIVES

1. Determine the slope-intercept form of a line given the slope and vertical intercept.

2. Determine the slope-intercept form of a line given two points on the line.

3. Compare the slopes of parallel lines.

You are traveling by car to visit your architect to make final decisions on the plans for your new home. You stop at a convenience store for a cup of coffee and then leave from there. The accompanying graph defines the distance, d, in miles from home as a function of time, t, measured in hours from the time you leave the convenience store.

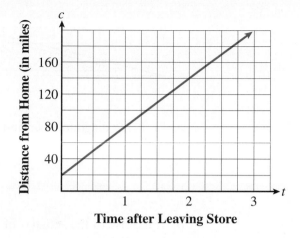

Time after Leaving Store

1. a. What is the vertical intercept? What is the practical significance of the vertical intercept?

b. Use the graph to determine the distance traveled during the first hour and during the first two hours.

c. Calculate the slope of the line using the values from part b and the formula

$$m = \frac{\Delta d}{\Delta t}.$$

d. What is the practical meaning of slope in this situation? Include units of measure in your explanation.

e. Use the slope found in part c and the vertical intercept determined in part a to write the equation of the line of the graph.

f. It takes you 2.5 hours to reach the architect's office from the convenience store. How far from home do you travel?

2. The architect shows you the plans for your roof. You must make a decision about the roof's pitch. The pitch is the slope of the roof expressed as a ratio of the vertical (rise) to the horizontal (run). The diagram shows a rafter of this roof with each tick mark representing one unit.

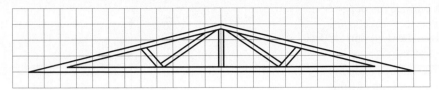

a. From the diagram, determine the slope (pitch) of the roof. Express this as a ratio.

b. If the rise of the actual roof must be 5 feet, then how long must the actual run be?

c. You decide that you would prefer a pitch of 5 to 16. On the grid below, draw a diagram of half of a rafter that shows this pitch.

3. Your architect will charge you a flat fee of $2000 for the plans for your home. The cost of your home is estimated by the square footage. The following table gives the total estimated cost of your home, c, including the architect's fees, as a function of the square footage, h. Assume that the total cost is a linear function of square footage.

TOTAL SQUARE FEET, h	TOTAL COST, c
0	2000
2500	152,000

a. What is the vertical intercept of the line containing these points? Explain how you determined this intercept.

b. Using the data in the table and the formula $m = \dfrac{\Delta c}{\Delta h}$, calculate the slope. What is the practical meaning of the slope in this situation?

c. Use the results from parts a and b to write the equation of the line in slope-intercept form that can be used to determine the cost for any given square footage.

d. You decide that you cannot afford a house with 2500 square feet. Using the equation from part c, determine the cost of your home if you decrease its size to 2000 square feet.

4. You leave the architect's office and travel farther from home to your favorite vacation spot. The following table gives the distance, d, in miles from your home as a function of time, t, the hours since you left the architect's office.

t (in hours)	d (in miles)
2	126
4	234

Assume that the distance, d, is a linear function of time, t. You want to determine the equation of the line from the data in this table. Because d is the output and t is the input variable, the slope-intercept form of the line is $d = mt + b$.

a. Calculate the slope $m = \dfrac{\Delta d}{\Delta t}$ from the given data. What is the practical meaning of the slope in this situation?

b. Are you given the vertical intercept in the table? Explain.

c. If you were to graph the two points, would you be able to determine the vertical intercept from the graph? Explain.

Given any two points on a line, you can determine the value of the vertical intercept using an algebraic approach. Example 1 illustrates this procedure.

Example 1 *Determine the vertical intercept of the line containing the points (2, −3) and (4, 3).*

SOLUTION

Step 1. Determine the slope of the line.

$$m = \frac{\Delta y}{\Delta x} = \frac{-3 - 3}{2 - 4} = \frac{-6}{-2} = 3$$

Step 2. Substitute the slope m into the slope-intercept form equation, $y = mx + b$.

$$y = 3x + b$$

The coordinates of any point on the given line will satisfy the equation $y = 3x + b$. This leads to the next step.

Step 3. Choose a point on the line and substitute its coordinates for x and y in the equation $y = 3x + b$: Selecting (2, −3), substitute 2 for x and −3, for y, as follows:

$$y = 3x + b$$
$$-3 = 3(2) + b$$

Step 4. Solve for b.

$$-3 = 3(2) + b$$
$$-3 = 6 + b$$
$$-9 = b$$

Therefore, the vertical intercept is (0, −9). Since the slope of the line is 3, the equation of the line containing (2, −3) and (4, 3) is $y = 3x − 9$.

5. a. Referring back to Problem 4, determine the vertical intercept of the line containing the points (2, 126) and (4, 234). Recall that the slope was determined in Problem 4a.

b. Write the equation of the line in slope-intercept form.

c. Check your equation by graphing on your grapher. Are the two data points on your line? If not, try to determine the source of your error.

Parallel Lines

You leave your favorite vacation spot and travel back home. The following table gives the distance, d, from the vacation spot as a linear function of time, t, the number of hours since you left your vacation spot.

t	d
0	0
3	162

6. a. Determine the equation of the line containing the points in the table.

b. Graph this line on the same coordinate axis as the line having equation $d = 54t + 18$ determined in Problem 5. What do you observe?

c. Will the lines ever intersect? Explain why or why not.

> **DEFINITION**
>
> Two lines are parallel if the lines have equal slopes but different vertical intercepts.

Example 2 *The lines having equation $Y_1 = -3x + 4$ and $Y_2 = -3x - 5$ are parallel because each line has the same slope, $m = -3$, but different vertical intercepts, namely $(0, 4)$ and $(0, -5)$.*

7. Write the equation of the line having vertical intercept $(0, -2.3)$ that is parallel to the line having the equation $y = 0.25x + 7.25$.

SUMMARY
Activity 1.7

1. To determine the equation of a line when two points on the line are known:

Step 1. Determine the slope of the line, $m = \dfrac{\Delta y}{\Delta x} = \dfrac{change\ in\ output}{change\ in\ input}$.

Step 2. Choose either of the given points (x, y).

Step 3. Use the slope intercept form, $y = mx + b$, and replace m, x, and y with the values found from steps 1 and 2.

Step 4. Solve the resulting equation for the only unknown, b.

Step 5. Substitute the values for m and b into the slope-intercept form, $y = mx + b$.

2. Parallel lines have equal slopes.

EXERCISES
Activity 1.7

1. The following table gives the cost, c, of a car rental as a function of miles driven, x.

x (in miles)	c (in dollars)
0	35
100	40

   Assume that the function is linear and that you want to determine the equation of the line from the table.

   a. What is the vertical intercept? How do you know this?

   b. Calculate the slope $m = \dfrac{\Delta c}{\Delta x}$ from the data in the table. What is the practical meaning of this slope?

   c. Use your results from parts a and b to write the equation of the line in slope-intercept form.

2. The following table gives the distance, d, in miles from the marina as a linear function of time, t, in hours.

t (in hours)	d (in miles)
2	75
4	145

   a. Determine the slope of the line. What is the practical meaning of slope in this situation?

   b. Write the equation of the line in slope-intercept form.

Exercise numbers appearing in color are answered in the Selected Answers appendix.

3. For each graph below,

    i. Determine the slope of the graph.

    ii. Determine the *y*-intercept of each graph.

    iii. Write the equation of the line whose graph is given.

Each tick mark denotes one unit.

a.

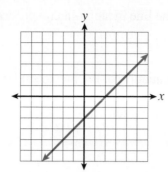

b.

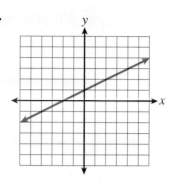

c.

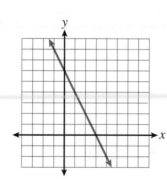

d.

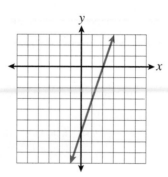

4. The following graph represents the distance from home of a car as a function of time (in hours).

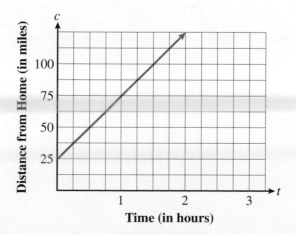

    a. How fast is the car traveling?

    b. Determine the vertical intercept.

    c. Write the equation of the line in slope-intercept form.

5. Write the slope-intercept equation of the line to satisfy the given conditions.

    a. The slope $= \dfrac{1}{2}$, the vertical intercept $= -1$.

    b. The line contains the points $(-3, 5)$ and $(0, 1)$.

    c. The line contains the points $(-4, -3)$ and $(2, 6)$.

    d. The line runs parallel to the line $y = 3x - 7$ and passes through the point $(2, -5)$. Recall that parallel lines have the same slope.

6. a. Graph a line with a slope of 4 that goes through the point $(1, 5)$. Write the equation of the line.

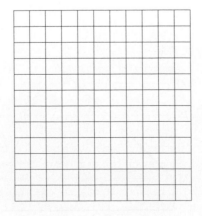

b. Graph a line with a slope of $-\frac{1}{2}$ that goes through the point $(-2, 3)$. Write the equation of the line.

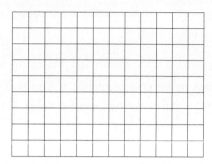

7. The data in the following table shows the circumference of a circle as a function of its radius.

r (radius in feet)	0	5	10	15
C (circumference in feet)	0	31.42	62.83	94.25

a. Assuming that this function is linear, determine the slope to the nearest hundredth place.

b. From the table, determine the vertical intercept.

c. Write the equation of the linear function.

d. What is the formula of the circumference of a circle?

e. Does your equation approximate this formula? Explain.

8. You own a kayak company and open only during the summer months. You discover that if you sell a certain type of kayak for $400, your sales per day average $5200. If you raise the price of the kayak to $450, the sales fall to approximately $3600.

a. Assume that the sales per day is a function of the price of the kayak. Write two ordered pairs that describe this situation.

b. Assume that the sales per day is a linear function of the price of the kayak. Write an equation describing this relationship.

c. You cannot make enough profit if you sell the kayak for less than $375. What would be the average sales per day if you change the price to $375?

9. According to the Hudson Institute, from 1980 to 2000, the number of U.S. lawyers increased in thousands from 506 to 1026.

a. The number of U.S. lawyers is a function of the number of years since 1980 (use 0 to represent the year 1980, 1 to represent 1981, and so forth). Write the two ordered pairs describing this situation.

b. Assume that this function is linear, and write the slope-intercept equation of the line that represents this data.

c. What is the practical meaning of the slope in this equation? (Remember, your slope is in thousands.)

d. How many lawyers were there in the U.S. in 1995?

10. Straight-line depreciation helps spread the cost of new equipment over a number of years. The value of your company's copy machine after one year will be $14,700 and after four years will be $4800.

a. Write a linear function that will determine the value of the copy machine for any specified year.

b. The salvage value is the value of the equipment when it gets replaced. What will be the salvage value of the copier if you plan to replace it after five years?

ACTIVITY 1.8

Skateboard
Heaven

OBJECTIVES

1. Write an equation of a line in general form $Ax + By = C$.

2. Write the slope-intercept form of a linear equation given the general form.

Your town has just authorized funding to build a new ramp and pathways for skateboarding. For security, the ramp and pathways must have a rectangular fence surrounding them. The money allocated in the budget for fencing will be enough to purchase 350 feet of fence. The only stipulation is that the width must be between 35 and 60 feet to properly enclose the new ramp. The length will depend on the width you choose. Your task is to determine the length and width of the rectangular region so that you use all of the fencing.

1. a. What does the value of 350 represent with regard to the rectangular region?

b. Using x to represent the width and y to represent the length, write an equation for the perimeter of this rectangular region.

c. The linear equation in part b should be in the form of $Ax + By = C$. Identify the values of the constants A, B, and C in the equation.

> **DEFINITION**
>
> When a linear equation is written in the form $Ax + By = C$, it is said to be in general form.

Example 1 *Sometimes, it is advantageous to rewrite a linear equation given in general form as its equivalent slope-intercept form. For example, consider the equation $3x + 7y = 5$. To write this equation in the slope-intercept form, you need to solve for y as follows.*

$$3x + 7y = 5$$
$$\underline{-3x \qquad\quad -3x} \qquad \text{Add } -3x \text{ to each side of the equation.}$$
$$7y = -3x + 5$$
$$\frac{7y}{7} = \frac{-3x}{7} + \frac{5}{7} \qquad \text{Divide each side of the equation by 7, the coefficient of } y.$$
$$y = -\frac{3}{7}x + \frac{5}{7}$$

2. a. Rewrite the equation from Problem 1b in slope-intercept form by solving the equation for y in terms of x.

b. What is the slope of the line represented by the equation you wrote in part a? What is the practical meaning of the slope in this situation?

c. What is the vertical intercept of the line represented by the equation you determined in part a? What is the practical meaning of the vertical intercept in this situation?

d. What is the practical domain and range of this function?

3. a. Determine the corresponding y-value for $x = 35$ using $2x + 2y = 350$ (general form of the equation of a line).

b. Determine the corresponding y-value for $x = 35$ using $y = -x + 175$ (slope-intercept form of the equation of a line).

c. Was it more convenient to use the general form or the slope-intercept form to determine a y-value given $x = 35$? Explain.

d. Complete the table to determine some possible lengths and widths for the rectangular region.

x	35	40	50	60
y				

4. Use function notation to write the length as a function of the width, letting f represent the function.

Equations of a Horizontal Line

If either A, the coefficient of x, or B, the coefficient of y, equals zero, a special situation arises. You will explore this in the following problems.

5. a. When the ramp for skateboarding is complete, you and your friends will be able to pay a monthly fee of $12.50 to use the ramp for as many hours as you wish. Complete the table of values below where x is the number of hours per month that each person who pays the fee uses the ramp and y is the total cost per month for each person.

x, TIME (in hours)	5	10	15	20
y, COST (in $)				

b. Use the data points from the table to sketch the graph.

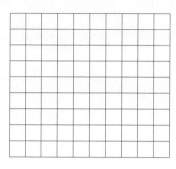

c. Describe the graph in words.

d. Choose two ordered pairs from part a and determine the slope, *m*, of the line. What is the practical meaning of the slope in this situation?

e. What are the intercepts (vertical and horizontal) of the line (if they exist)?

f. Write the equation of this line in slope-intercept form, $y = mx + b$.

g. Change the equation from part f to the general form $Ax + By = C$ and identify *A*, *B*, and *C*. Note that when there is no input variable appearing in an equation, its coefficient is understood to be zero.

DEFINITION

A graph in which the output *y* is a constant or, equivalently, $f(x)$ is a constant, is a horizontal line. The equation of a horizontal line is $y = c$ (or $f(x) = c$), where *c* is any real number. The slope of any horizontal line is zero.

6. Determine three ordered pairs that satisfy each of the following equations, and then sketch each graph on the same coordinate axes.

a. $y = -2$ **b.** $f(x) = 1$

c. $g(x) = \frac{5}{2}$

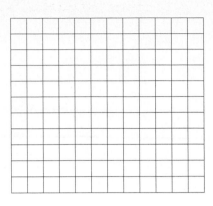

Equation of a Vertical Line

7. You apply for a part-time job at the skateboarding rink to help cover your weekly expenses while going to college. The following table gives your weekly salary, x, and corresponding weekly expenses, y, for a typical month.

x, WEEKLY SALARY IN DOLLARS	70	70	70	70
y, WEEKLY EXPENSES IN DOLLARS	45	35	50	60

a. Sketch a graph of the given data points.

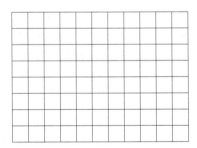

b. Describe the graph in words.

c. Choose two ordered pairs from the table and determine the slope, m.

d. What are the intercepts (vertical and horizontal) of the line (if they exist)?

e. Is *y* a function of *x*? Explain using the definition of a function.

f. Do the ordered pairs in the table satisfy the equation $x = 70$? Explain why or why not.

g. Write the equation $x = 70$ in $Ax + By = C$ form, and identify *A*, *B*, and *C*.

h. Can you graph the equation $x = 70$ using your grapher? Explain.

DEFINITION

A graph in which *x* is a constant is a **vertical line**. The equation of a vertical line is $x = a$, where *a* is any real number. The slope of a vertical line is undefined.

---

8. Determine three ordered pairs that satisfy each of the following equations, and then sketch each graph.

a. $x = -2$ **b.** $x = 4$ **c.** $x = \dfrac{5}{2}$

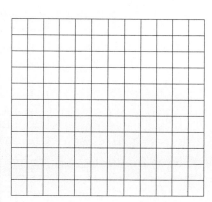

SUMMARY
Activity 1.8

1. The general form of a linear equation is $Ax + By = C$, where A, B, and C are constants.

2. The graph of $y = c$ or $f(x) = c$ is a **horizontal line**. In this case, f is called a constant function. Every point on this line has a y-coordinate equal to c. A horizontal line has slope of zero.

3. The graph of $x = a$ is a **vertical line**. Every point on this line has an x-coordinate equal to a. The slope of a vertical line is undefined.

EXERCISES
Activity 1.8

1. Write the following linear equations in slope-intercept form. Determine the slope and vertical intercept of each line.

    a. $2x - y = 3$
                      b. $x + y = -2$

    c. $2x - 3y = 7$
                    d. $-x + 2y = 4$

    e. $0x + 3y = 12$

2. **a.** Sketch the graph of the horizontal line through the point $(-2, 3)$.

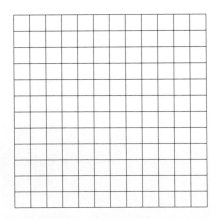

    b. Write the equation of a horizontal line through the point $(-2, 3)$.

   c. What is the slope of the line?

   d. What are the vertical and horizontal intercepts of the line?

   e. Does the graph represent a function? Explain.

3. a. Sketch the graph of the vertical line through $(-2, 3)$.

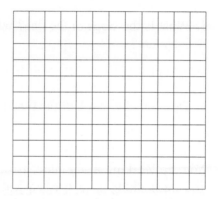

   b. Does the graph represent a function? Explain.

   c. Write the equation of a vertical line through the point $(-2, 3)$.

   d. What is the slope of the line?

   e. What are the vertical and horizontal intercepts of the line?

4. Explain the difference between a line with a zero slope and a line with an undefined slope.

5. You are retained as a consultant for a major computer company. You receive $2000 per month as a fee no matter how many hours you work.

   a. Using x to represent the number of hours you work each month, write a function, f, in symbolic form to represent the total amount received from the company each month.

b. Complete the following table of values.

HOURS WORKED PER MONTH	15	25	35
FEE PER MONTH (in $)			

c. Use your graphing calculator to sketch the graph of this function.

d. What is the slope of the line? What is the practical meaning of the slope in this situation?

e. Describe the graph of the function.

6. You are working in the purchasing department of an appliance retailer. This month you are stocking up on washers and dryers. Your supervisor informs you that your budget this month is $10,000. You know that the average wholesale cost of the washer over the past year has been $250, while the average wholesale cost of a dryer has been $200.

a. If w represents the number of washers you can purchase, write an expression that represents the amount you can spend on washers.

b. If d represents the number of dryers you can purchase, write an expression that represents the amount you can spend on dryers.

c. Write a linear equation in general form that relates the number of washers and dryers you can expect to purchase with your budget.

d. Solve your equation in part c for d. In other words, express the number of dryers you can expect to purchase as a function of the number of washers you can expect to purchase.

e. What is the horizontal intercept for this function? What is its practical meaning in this situation?

f. What is the vertical intercept for this function? What is its practical meaning in this situation?

g. What is the slope of this function? What is its significance in this situation?

h. Use your graphing calculator to graph the function in part d. What part of this graph is relevant to this situation?

i. What are the practical domain and range of this situation?

ACTIVITY 1.9
College Tuition

The following graphic contains the average tuition and required fees for full-time matriculated students at private four-year colleges over the past two decades.

OBJECTIVES

1. Determine a line of best fit with a straightedge.

2. Determine the equation of a regression line using a graphing calculator.

3. Use the regression equation to interpolate and extrapolate.

College Costs

YEAR	1989	1990	1992	1994	1996	2000
COST	$10,350	$11,380	$10,290	$11,480	$12,990	$15,030

Source: National Center for Educational Statistics, U.S. Department of Education

1. Let t, the number of years since 1989, represent your input variable and c, the average cost, your output variable. Determine an appropriate scale, and plot the data points from the accompanying table. Therefore, $t = 0$ corresponds to the year 1989, $t = 1$ to 1990, $t = 2$ to 1991, and so on.

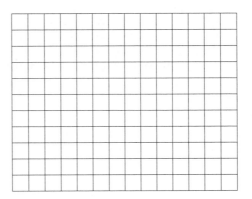

2. Using a straightedge (a taut string or an uncooked piece of spaghetti is helpful in positioning the line), draw a line that has as many data points as close to the line as possible. The line you are drawing is called the **line of best fit**.

The accepted method for determining the line of best fit to a set of data is called the **method of least squares**. This method is used to calculate a slope and y-intercept. The line with this slope and y-intercept is called the **regression line** for your data. Your graphing calculator is programmed with an algorithm that determines the regression line.

3. **a.** Enter the tuition data into your calculator by pressing STAT and choosing EDIT. See Appendix C for help operating the TI-83 Plus graphing calculator.

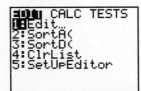

b. Enter the year data in L1 and the tuition data in L2.

c. Determine the regression line by pressing $\boxed{\text{STAT}}$, choosing CALC and option 4:LinReg($ax + b$).

d. Write the result, rounding the coefficients to three decimal places.

4. Now use the regression equation you determined in Problem 3 to model the tuition data. Use the table feature on your graphing calculator to determine the model's predicted output in the following table.

INPUT, t	ACTUAL OUTPUT, c	MODEL'S PREDICTED OUTPUT
0	10,350	
1	11,380	
3	10,290	
5	11,480	
7	12,990	
11	15,030	

DEFINITION

Using a regression model to predict an output within the boundaries of the input values of the given data is called **interpolation**. Using a regression model to predict an output outside boundaries of the input values of the given data is called **extrapolation**. In general, interpolation is more reliable than extrapolation.

5. **a.** Use the regression equation to predict the average tuition and fees at private four-year colleges in 1993 ($t = 4$). Is this an example of interpolation or extrapolation?

b. Use the regression equation to predict the average tuition and fees at private four-year colleges in 2005 and in 2010. Is this an example of interpolation or extrapolation?

c. Which prediction do you believe would be more accurate? Explain.

d. Use the regression line to estimate the year in which average tuition and fees will be at least $20,000.

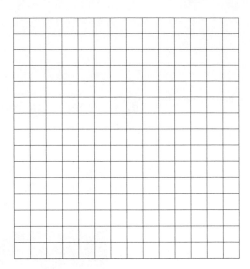

6. Over the past quarter century, the number of bachelor's degrees conferred by degree-granting institutions has steadily increased. The following table contains data on the number of bachelor's degrees (in thousands) earned by women in a given year. The input t represents the number of years since 1977.

	YEAR					
	1977	1981	1985	1990	1995	2000
t, NUMBER OF YEARS SINCE 1977	0	4	8	13	18	23
$f(t)$, NUMBER OF DEGREES IN THOUSANDS	423	465	492	558	634	708

a. Sketch a scatterplot of the given data.

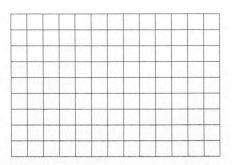

b. Enter the data into your graphing calculator, and determine a linear regression model to represent the data. Write the result here.

c. What is the slope of the line? What is the practical meaning of the slope in this situation?

d. Use the regression model to predict the number of bachelor's degrees that will be granted in the year 2010.

SUMMARY
Activity 1.9

1. The linear regression equation is the linear equation that "best fits" a set of data.

2. The regression line is a mathematical model for the data.

3. **Interpolation** is the process of using a regression equation to predict a value of output for an input value that lies within the boundaries of the given data.

4. **Extrapolation** is the process of using a regression equation to predict a value of output for an input value that lies outside the boundaries of the given data.

EXERCISES
Activity 1.9

1. a. Plot the following data.

x	0	3	6	9	12	15	18
y	−0.8	6.3	13.1	19.6	27.0	33.5	40.8

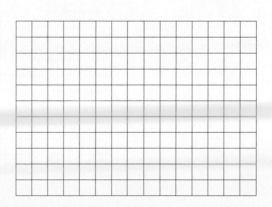

b. With a straightedge, draw a line that you think looks like the line of best fit. Does this data appear to be linear?

c. Use your graphing calculator to determine the equation of the regression line. Write the result below.

d. Use your equation from part c to predict the value of y when $x = 10$.

e. Use your equation from part c to predict the value of y when $x = 25$.

f. Which prediction, $f(10)$ or $f(25)$, would be more accurate? Explain.

2. Public debt increased at a relatively constant rate from 1986 to 1996. The following table gives the average debt per capita (in thousands of dollars) for selected years, where $t = 0$ corresponds to the year 1986.

t (in years since 1986)	0	2	4	6	8	10
d (in thousands of dollars)	8.77	10.53	13.00	15.85	18.02	19.81

Source: Bureau of Public Debt, U.S. Department of the Treasury

a. Plot the data.

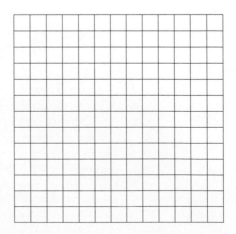

b. Use your graphing calculator to determine the equation of the regression line. Write the result here.

c. What is the slope of the line? What is the practical meaning of slope in this situation?

d. Use your regression line to determine the average debt of an individual in 1990 ($t = 4$). Compare your result with the actual value of $13,000 in 1990.

e. Use the regression equation to predict the average debt of an individual in 2002.

f. What is the process called that you used to make the prediction in part e?

What Have I Learned?

1. The coordinates of all points (x, y) on the line with slope m and vertical intercept $(0, b)$ satisfy the equation

$$y = mx + b.$$

a. The point $(2, 3)$ is on the line whose equation is $y = 4x - 5$. Show how the coordinates of the point $(2, 3)$ satisfy the equation $y = 4x - 5$.

b. Determine another point on the line whose coordinates satisfy the equation $y = 4x - 5$.

2. Given an input/output table that represents a function, how can you determine if the function is linear?

3. Write a procedure for determining the equation of a line, given two points on the line.

4. Write a procedure for determining the horizontal intercept of a graph, given its equation.

5. Consider two linear functions $f(x) = m_1x + b_1$ and $g(x) = m_2x + b_2$, whose graphs are distinct and parallel. What can you say about m_1 and m_2 and about b_1 and b_2? Explain.

6. If the graph of $Ax + By = C$ is a vertical line, what can you conclude about the values of A and B? Explain.

7. a. Describe the graph of a line with a positive slope.

b. Describe the graph of a line with a negative slope.

c. Describe the graph of a line with a slope of zero.

d. Describe the graph of a line with undefined slope.

8. Explain the difference between interpolation and extrapolation.

CLUSTER 2 **How Can I Practice?**

1. Match each of the following functions or equations with its corresponding graph below. Use your graphing calculator for checking purposes only.

$$f(x) = -2x + 3 \qquad g(x) = 2x - 3 \qquad y = 2$$

$$h(x) = -2x - 3 \qquad y = -2x \qquad x = 2$$

a.

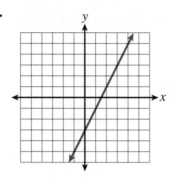

b.

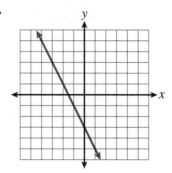

c.

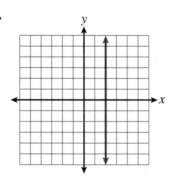

d.

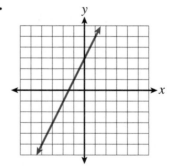

e.

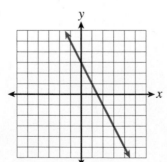

f.

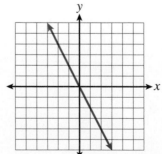

g.

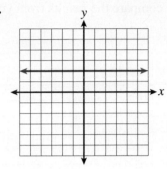

h.

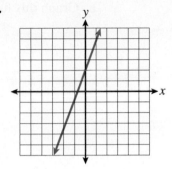

i.

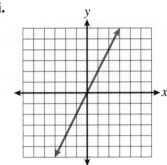

2. The cost of renting a grapher from the bookstore is $20, plus $4 per month for as long as you wish to rent.

   a. Complete the table.

MONTHS, m	2	5	8	10	12
COST IN $, c					

   b. Is the cost, c, a linear function of months, m?

   c. What is the slope?

   d. Write the equation for the function.

e. Graph this function and compare the results from your graphing calculator.

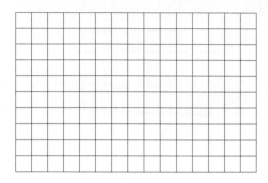

f. What is the practical meaning of the slope in this situation?

g. What is the vertical intercept of the graph of the function? What is the practical meaning of the vertical intercept in this situation?

h. What is the horizontal intercept of the graph of the function? What practical meaning does this have in this situation?

i. Approximately how many months will you be able to keep your graphing calculator if you have $65 budgeted for this expense?

3. Given the table of values:

t	10	20	40
$s(t)$	0	15	45

a. Determine the average rate of change from $t = 10$ to $t = 20$.

b. Determine the average rate of change from $t = 20$ to $t = 40$.

c. Is $s(t)$ a linear function? Explain.

4. Determine the slope of the line through the points $(3, 8)$ and $(-5, 12)$.

5. Determine the slope of the line represented by the equation $y = -4x + 2$.

6. Determine the slope of the line having equation $2x - 5y = 9$.

7. Write an equation of a line with a slope of -7 and a vertical intercept of 4.

8. Write an equation of a line that has a slope of 2 and goes through the point $(0, 10)$.

9. Write an equation of a line that has a slope of 0 and goes through the point $(-4, 5)$.

10. Write an equation of a vertical line that goes through the point $(-3, -5)$.

11. Write an equation of a line that goes through the points $(2, -3)$ and $(-4, 0)$.

12. Write an equation of a line parallel to $y = \dfrac{1}{3}x - 7$ and through the point $(6, -1)$.

13. Sketch a graph of the line through $(2, -3)$ with slope of $\dfrac{1}{2}$.

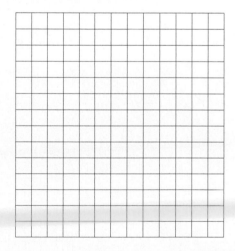

In Exercises 14–16, graph each of the functions by hand. Then compare your results using your graphing calculator.

14. $y = \frac{4}{5}x - 3$

15. $4x - 5y = 20$

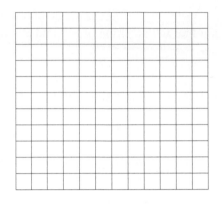
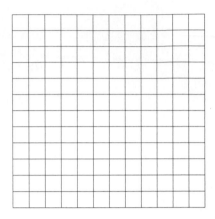

16. $y = -4$

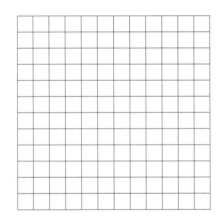

17. **a.** Plot the data on the following grid using an appropriate scale.

x	0	2	4	6	8	10	12
y	23.76	24.78	25.93	26.24	26.93	27.04	27.93

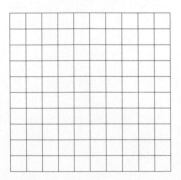

b. Use your graphing calculator to determine the equation of the regression line. Write the result.

c. Use the result from part b to predict the value of y when $x = 9$.

d. Use the result from part b to predict the value of y when $x = 20$.

18. The following table shows the percentage of elected women state legislators from the years 1969 to 1997, as tabulated by the Center for American Women and Politics.

YEAR	1969	1973	1977	1981	1985	1989	1993	1997
PERCENTAGE OF WOMEN LEGISLATORS	6.6	7.6	9.9	10.5	13.3	14.3	22.2	25.4

Let x represent the number of years since 1969 and y represent the percentage of women state legislators.

a. Plot the data from the table on the following grid.

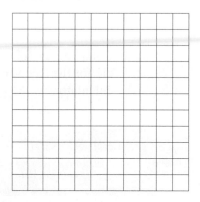

b. Use your graphing calculator to produce the linear regression equation for the data in the graph, and record it here.

c. Use the equation from part b to estimate the percentage of women state legislators in 1986 and 2010.

d. In which of the values found in part c do you have more confidence? Explain.

CLUSTER 3 **Systems of Linear Equations, Inequalities, and Absolute Value Functions**

ACTIVITY 1.10

Ride for Less

OBJECTIVES

1. Solve a system of 2 × 2 linear equations numerically and graphically.

2. Solve a system of 2 × 2 linear equations using the substitution method.

3. Solve an equation of the form $ax + b = cx + d$ for x.

Finals are over, and you are moving back home for the summer. You need to rent a truck to move your possessions from the college residence hall back to your home. You contact two local rental companies and obtain the following information for the one-day cost of renting a truck:

Company 1: $40.95 per day, plus $0.19 per mile.

Company 2: $19.95 per day, plus $0.49 per mile.

The total cost of renting a truck for one day is a function of the number of miles driven.

1. Identify the input and output in this situation.

2. Let n represent the total number of miles driven in one day.

   a. Write an equation to determine the total cost, C, of renting a truck for one day from company 1.

   b. Write an equation to determine the total cost, C, of renting a truck for one day from company 2.

3. a. Complete the following table to compare the total cost of renting the vehicle for the day. Verify your results using the table feature of your graphing calculator.

Driven to Rent

n, NUMBER OF MILES DRIVEN	TOTAL COST, C COMPANY 1	TOTAL COST, C COMPANY 2
0		
10		
20		
30		
40		
50		
60		
70		
80		

   b. For what mileage is the one-day rental cost the same?

c. Which company should you use?

4. a. Graph the two cost functions, for *n* between 0 and 120 miles, on the same coordinate axes below.

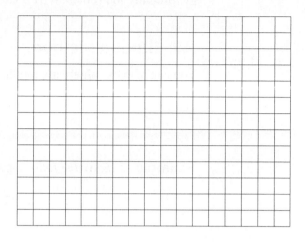

b. Determine the point where the lines in part a intersect. What is the significance of the point in this situation?

Appendix

c. Verify your results from part b using your graphing calculator. Use the intersect feature of your grapher. See Appendix C for the procedure for the TI-83 Plus. Your final screens should appear as follows:

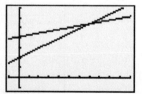

DEFINITIONS

The cost of renting situation is an example of a 2 × 2 **system of linear equations**. Such a linear system of two equations in two variables can be written

$$y = ax + b$$
$$y = cx + d.$$

A **solution of the system**, if it exists, is the ordered pair of numbers (x, y) that makes both equations true. If the system has a solution, the system is called **consistent**.

The system in this activity (see Problem 2) can be written

$$C = 0.19n + 40.95$$
$$C = 0.49n + 19.95.$$

You first solved the system **numerically** in Problem 3 by completing the table and noting the value of the input that produced the same output. You then solved this system **graphically** in Problem 4 by locating the point of intersection of the two lines.

Substitution Method

You can also use **algebraic** methods to solve systems. An algebraic method to solve a system of equations is by substitution. The following example illustrates this method.

Example 1   *Solve the following system algebraically using the substitution method.*

$$y = 3x - 10 \quad \text{(equation 1)}$$
$$y = 5x + 14 \quad \text{(equation 2)}$$

SOLUTION

After substituting $3x - 10$ for y into equation 2, you have

$$3x - 10 = 5x + 14.$$

To solve the equation for x, you need to isolate the variable. Rewrite the equation so that all terms involving x are on one side and all other terms are on the other side. This is generally accomplished by adding and/or subtracting the appropriate terms to each side of the equation.

$$
\begin{array}{rcl}
3x - 10 &=& 5x + 14 \\
-5x & & -\ 5x \\
\hline
-2x - 10 &=& 14 \\
+ 10 & & +\ 10 \\
\hline
-2x &=& 24 \\
\overline{-2} & & \overline{-2} \\
x &=& -12
\end{array}
$$

Subtract $5x$ from both sides and collect like terms.

Add 10 to each side and collect like terms.

Divide each side by -2, the coefficient of the variable x.

To determine the corresponding y-value, substitute -12 for x into either of the original equations and solve for y. After substituting into equation 1, you have

$$y = 3(-12) - 10 = -36 - 10 = -46.$$

Remember that the ordered pair $(-12, -46)$ represents the point of intersection of the two lines as well as the solution to the system. This is verified by the following screen.

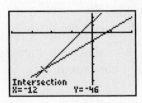

In this activity, you want C (total cost) to be the same for both functions. Because each expression is solved for C, set the expressions equal to each other. You are substituting an expression for C from one equation into the other equation.

5. a. Solve the introductory problem involving the one-day cost of renting a truck by using the substitution method to solve the following system for n.

$$C = 0.19n + 40.95$$
$$C = 0.49n + 19.95$$

b. What does the value of n in part a represent in the rent-a-truck situation?

c. Determine the value of C by substituting the value of n from part b into one of the original equations and solving for C.

d. Check to see if your solution satisfies both equations.

6. You are going to graduate and are interested in purchasing a new car. You have narrowed the choice to a Honda Accord LX and a Passat GLS. You are concerned about the value of the car depreciating over time. You search the Internet and obtain the following information. The depreciation per year is the amount by which the value of the car will decrease each year.

MODEL	SUGGESTED RETAIL PRICE ($)	DEPRECIATION PER YEAR (ASSUME CONSTANT)
Accord LX	18,790	1385
Passat GLX	21,450	1790

a. Complete the following table.

Under Appreciated

YEARS THE CAR IS OWNED	VALUE OF ACCORD LX ($)	VALUE OF PASSAT GLS ($)
0		
1		
2		
3		
4		

b. Will the value of the Passat GLS ever be lower than the value of the Accord LX? Explain.

c. Let V represent the value of the car after x years of ownership. Write an equation to determine V in terms of x for the Accord LX. Write another equation to determine V in terms of x for the Passat GLS.

d. The two equations in part c form a 2×2 system of linear equations. Solve this system using the substitution method.

e. Use your graphing calculator to solve this system graphically. How does your solution compare to your solution in part d?

f. The solution to this system is an ordered pair of the form (x, V). What do the values x and V represent in this solution?

g. Which car has the better resale value? Explain.

Types of Linear Systems

7. You and your friend are traveling in the same direction, but in different cars, on the New York State Thruway.

   a. Let d represent the distance you are from the common starting point. Write an equation for d in terms of time, t, if you are traveling at 65 miles per hour.

   b. Write another equation to determine the distance, d, in terms of time, t, that your friend is from the starting point. She is traveling at 60 miles per hour and has a 30-mile head start.

8. The linear equations in Problem 7 form a 2×2 system of linear equations:

$$d = 65t$$
$$d = 30 + 60t$$

   a. Solve the given system using each of the following methods.

Numerically

t (in hour)	YOUR DISTANCE	YOUR FRIEND'S DISTANCE
0		
1		
2		
3		
4		
5		
6		

Graphically   **Algebraically**

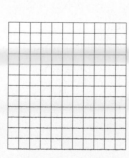

    b. What is the solution of this system? What is the meaning of this solution in the context of this problem?

The linear system in Problem 8 has exactly one solution. Such a system is said to be **consistent**.

Do all systems have exactly one solution? Consider the following.

  9. a. You and your friend are traveling again. She still has a 30-mile head start, but this time both of you are traveling at 60 miles per hour. When will you catch up with your friend? Explain.

    b. The situation just described can be represented by the following system:

$$d = 60t + 30$$
$$d = 60t$$

Try to solve this system graphically. Do the lines intersect? What is the solution to the system?

    c. Try to solve the system algebraically using the substitution method. What type of equation do you obtain?

The linear system in Problem 9 is said to be **inconsistent**. There is no solution because the lines never intersect. Graphically, the slopes of the lines are equal, but the vertical intercepts are different. Therefore, the graphs are parallel lines. Solving such an equation algebraically results in a false equation such as $30 = 0$.

  10. a. You and your friend are taking one last trip. This time she does not have a head start. You both leave from your house, both travel in the same direction, and both travel at 60 miles per hour. When will you both be at the same point?

    b. The system for this situation is

$$d = 60t$$
$$d = 60t.$$

Try to solve this system graphically. Do the lines intersect? What is the solution to the system?

c. Try to solve this system algebraically. What will be the result from a dependent system if you attempt to solve it algebraically?

The system in Problem 10 is an example of a **dependent** system. Graphically, in such a system, both equations represent the same line. The system has an infinite number of solutions. Solving a dependent system algebraically results in a true equation such as $0 = 0$.

SUMMARY
Activity 1.10

1. A 2×2 system of linear equations consists of two equations with two variables. The graph of each equation is a line.

2. A **solution** to a 2×2 system of equations is an ordered pair that satisfies both equations of the system.

3. Solutions can be found in three different ways:
   - **numerically**, by examining tables of values for both functions
   - **graphically**, by graphing each equation and finding the point of intersection
   - **algebraically**, by combining the two equations to form a single equation in one variable, which can then be solved. This is called the substitution method.

4. A linear system is **consistent** if there is at least one solution, the points of intersection of the graphs.

5. A linear system is **inconsistent** if there is no solution; the lines are parallel.

6. A linear system is **dependent** if there are infinitely many solutions. The equations represent the same line.

EXERCISES
Activity 1.10

Solve the following systems of linear equations numerically, graphically, and algebraically (substitution method).

1. a. $y = 2x + 3$

   $y = -x + 6$

Numerically   **Graphically**   **Algebraically (substitution method)**

x	y₁	y₂
−2		
−1		
0		
1		
2		
3		

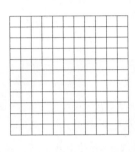

   b. $y = x + 9$

   $y = - 17$

Numerically   **Graphically**   **Algebraically (substitution method)**

x	y₁	y₂
−2		
−1		
0		
1		
2		
3		

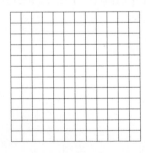

c. $y = 5x - 3$

   $y = 5x + 7$

| Numerically | Graphically | Algebraically (substitution method) |

x	y_1	y_2
0		
1		
2		
3		
4		

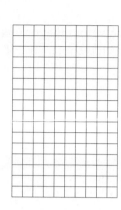

2. Two companies sell software products. In 2002, company A had total sales of $17.2 million. Its marketing department projects that sales will increase by $1.5 million per year for the next several years. Company B had total sales of $9.6 million of software products in 2002 and projects that its sales will increase by $2.3 million each year.

   Let n represent the number of years since 2002.

   a. Write an equation that represents the total sales, s, of company A since 2002.

   b. Write an equation that represents the total sales, s, of company B since 2002.

   c. The two equations in parts a and b form a system. Solve this system to determine the year in which the total sales of both companies will be the same.

3. You are considering installing a security system in your new house. You gather the following information from two local home security dealers for similar security systems:

   Dealer 1: $3560 to install and $15 per month monitoring fee

   Dealer 2: $2850 to install and $28 per month monitoring fee

   Although the initial fee of dealer 1 is much higher than dealer 2, dealer 1's monitoring fee is lower.

   Let n represent the number of months you have the security system.

    a. Write an equation that represents the total cost, c, of the system with dealer 1.

    b. Write an equation that represents the total cost, c, of the system with dealer 2.

    c. Solve the system of equations that results from parts a and b to determine the number of months for which the total cost of the systems will be equal.

    d. If you plan to live in the house and use the system for ten years, which system would be less expensive?

4. You can run a 400-meter race at an average rate of 6 meters per second. Your friend can run the race at a rate of 5 meters per second. You give your friend a 40-meter head start. She then runs 360 meters.

    a. Write an equation for your distance, in meters, from the starting point as a function of time in seconds.

    b. Write an equation for your friend's distance, in meters, from the starting point as a function of time in seconds.

    c. How long does it take you to catch up with your friend?

    d. How far from the finish line do you meet?

5. For many years, the life expectancy for women has been longer than the life expectancy for men. In the past few years, the life expectancy for men has been increasing at a faster rate than that for women. The life expectancy, E, for men and women in the United States can be modeled by

    Women: $E = 0.126t + 76.74$

    Men: $E = 0.169t + 69.11,$

where t represents the number of years since 1975.

    a. Solve the system numerically by completing the following table. Approximate the value of t to the nearest year. Use your calculator.

t, NUMBER OF YEARS SINCE 1975	LIFE EXPECTANCY FOR WOMEN	LIFE EXPECTANCY FOR MEN
0		
25		

b. What does the solution to this system represent in the context of this problem?

c. Solve this problem graphically using your calculator. Use the window Xmin = 0, Xmax = 225, Ymin = 0, and Ymax = 125.

d. Solve the system algebraically using the substitution method.

6. When the cost for running a business equals the revenue taken in by that business, a company breaks even. You sell ceramic planters at craft shows across the country. If x equals the number of planters created and sold, the monthly costs are represented by $C = 3x + 1287$. The monthly revenue is $R = 12x$.

a. In the cost equation, what is the practical meaning of the slope?

b. In the cost equation, what is the practical meaning of the vertical intercept?

c. In the revenue equation, what is the practical meaning of the slope?

d. Write the equation that must be solved to determine how many planters you must sell to break even.

e. How many planters must you sell to break even?

f. If your expenses increase to $2286, how many planters will you have to sell in a month to break even?

7. Solve the following systems of equations algebraically using the substitution method. Check your solution by using your grapher.

a. $y = -x + 6$

   $y = x + 2$

b. $x = 9 - 3y$

   $x = -7 - 2y$

c. $2x = 5y - 1$

   $2x = y + 1$

d. $y = 3x + 1$

   $y = 3x - 3$

ACTIVITY 1.11
Healthy Lifestyle

OBJECTIVES

1. Solve a 2 × 2 linear system algebraically using the substitution method and the addition method.

2. Solve equations containing parentheses.

You are trying to maintain a healthy lifestyle. You eat a well-balanced diet and follow a regular schedule of exercise. One of your favorite exercise activities is a combination of walking and jogging in your nearby park.

On one particular day, it takes you 1.3 hours to walk and jog a total of 5.5 miles in the park. You are curious about the amount of time you spent walking and the amount of time you spent jogging during the workout.

Let x represent the time you walked and y represent the time you jogged.

1. Write an equation, using x and y, for the total time of your walk/jog workout in the park.

2. **a.** If you walk at 3 mph, write an expression that represents the distance you walked.

   b. If you jog at 5 mph, write an expression that represents the distance you jogged.

   c. Write an equation for the total distance you walked/jogged in the park.

The situation just described can be represented by the following system:

$$x + y = 1.3$$
$$3x + 5y = 5.5$$

Note that each equation in this system is in general form. One approach to solving this system is to solve each equation for one variable in terms of the other and then use the substitution method.

3. **a.** Solve each of the equations in the system above for y.

   b. Solve the system in part a using the substitution method.

   c. Check your answer graphically using your graphing calculator. You may want to use the window Xmin = −2.5, Xmax = 2.5, Ymin = −2.5, and Ymax = 2.5.

Addition Method

Sometimes it is more convenient to leave each equation in the linear system in general form $(Ax + By = C)$ rather than solving for one variable in terms of the other. Look again at the original system:

$$x + y = 1.3 \quad (1)$$
$$3x + 5y = 5.5 \quad (2)$$

If you apply the addition principle of algebra by adding the two equations (left side to left side and right side to right side), you may be able to obtain a single equation containing only one variable.

$$x + y = 1.3 \quad (1)$$
$$\underline{3x + 5y = 5.5 \quad (2)}$$
$$4x + 6y = 6.8$$

In this case, adding the equations does not eliminate a variable. But if you multiply both sides of the equation 1 by -5, the coefficients of y will be opposite, and the variable y can be eliminated.

$$-5(x + y) = -5(1.3) \longrightarrow \quad -5x - 5y = -6.5$$
$$3x + 5y = 5.5 \qquad\qquad\qquad \underline{3x + 5y = 5.5}$$
$$-2x + 0 = -1$$

Add the corresponding sides of the equation.

Solving the resulting equation, $-2x = -1$, for x, you have

$$-2x = -1$$
$$x = \tfrac{1}{2} = 0.5$$

Substituting for x in $x + y = 1.3$, you have

$$0.5 + y = 1.3$$
$$\text{or } y = 0.8$$

This method of solving systems algebraically is called the **addition method**.

4. Solve the following system again using the addition method. Multiply the appropriate equation by the appropriate factor to eliminate x and solve for y first.

$$x + y = 1.3$$
$$3x + 5y = 5.5$$

Not all systems will have convenient coefficients, and you may need to multiply one or both equations by a factor that will produce coefficients that are additive inverses, or opposites of the same variable.

5. Consider solving the following system with the addition method.

$$-2x + 5y = -16$$
$$3x + 2y = 5$$

a. Identify which variable you wish to eliminate. Multiply the appropriate equation by the appropriate factor, so that the coefficients of your chosen variable are opposite. Show the two equations after you multiply by the factor. (Remember to multiply both sides of the equation by the factor.)

b. Add the two equations to eliminate the chosen variable.

c. Solve the resulting linear equation.

d. Determine the complete solution. Remember to check by substituting into both of the original equations.

Substitution Method Revisited

The substitution method of solving a 2×2 system of linear equations is generally used when each equation in the system is solved for one variable in terms of the other. However, it can be convenient to use the substitution method when only one of the equations is solved for a variable.

Example 1   *In the walk/jog system,*

$$x + y = 1.3 \quad (1)$$
$$3x + 5y = 5.5 \quad (2)$$

you could solve equation 1 for y and then substitute for y in equation 2 as follows

Step 1. Solve $x + y = 1.3$ for y.

$$y = 1.3 - x$$

Step 2. Substitute $1.3 - x$ for y in equation 2.

$$3x + 5(1.3 - x) = 5.5$$

Step 3. Solve the resulting equation for x.

$3x + 5(1.3 - x) = 5.5$	Remove the parentheses by applying the distributive property.
$3x + 6.5 - 5x = 5.5$	Collect like terms on the same side.
$-2x + 6.5 = 5.5$	
$\underline{ - 6.5 - 6.5}$	Add the opposite of 6.5 to both sides.
$-2x = -1$	Divide each side by -2.
$\frac{-2x}{-2} = \frac{-1}{-2}$	
$x = 0.5$	

6. a. Solve the following linear system using the substitution method in which you solve only one of the equations for a variable.

$$x - y = 5$$
$$4x + 5y = -7$$

b. Check your answer in part a by solving the system using the addition method.

SUMMARY
Activity 1.11

- There are two methods for solving a 2×2 system of linear equations algebraically:
  1. the substitution method and
  2. the addition method.

- To solve linear systems by substitution,
  1. solve one or both equations for a variable.
  2. substitute the expression that represents the variable in one equation for that variable in the other equation.

3. solve the resulting equation for the remaining variable.

4. substitute the value from part 3 into one of the original equations, and solve for the other variable.

- To solve linear systems by addition with equations in general form,

1. multiply one equation by the number that will make the coefficients of one of the variables opposites.

2. add the two equations to eliminate one variable and solve the resulting equation.

3. substitute the value from part 2 into one of the original equations, and solve for the other variable.

EXERCISES
Activity 1.11

1. Solve each of the following equations.

   a. $5(x + 3) + 4 = 6x - 1 - 5x$

   b. $-3(1 - 2x) - 3(x - 4) = -5 - 4x$

   c. $2(x + 3) - 4x = 5x + 2$

2. Solve the following systems algebraically using the substitution method.

   a. $y = 3x + 1$
   $y = 6x - 0.5$

   b. $y = 3x + 7$
   $2x - 5y = 4$

c. $2x + 3y = 5$

$-2x + y = -9$

d. $4x + y = 10$

$2x + 3y = -5$

3. Solve the following systems algebraically using the addition method.

a. $y = 2x + 3$

$y = -x + 6$

b. $2x + y = 1$

$-x + y = -5$

4. Solve the system both graphically and algebraically.

$3x + y = -18$

$5x - 2y = -8$

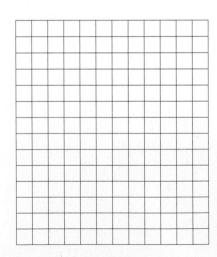

5. A catering service placed an order for eight centerpieces and five glasses, and the bill was $106. For the wedding reception, they were short one centerpiece and six glasses and had to reorder. This order came to $24. Let x represent the cost of one centerpiece, and let y represent the cost of one glass.

   a. Write a system of equations that represents both orders.

   b. Solve the system using the substitution method. Interpret your solution.

   c. Check your result in part b using the addition method.

   d. Use your graphing calculator to solve the system.

ACTIVITY 1.12

Sam's Cafe

OBJECTIVE

1. Solve a 3 × 3 linear system of equations.

In your job as buyer for Sam's Cafe, a nationwide coffee bar, you need to buy three grades of coffee bean that will be blended with various flavors to make Sam's well-known coffee drinks. This week the three grades of beans are selling for $0.80, $1.20, and $1.80 per pound. You want to know how much of each grade bean to buy, based on the following assumptions.

- The total weight of beans needed is 15,000 pounds.
- You wish to spend $20,000.
- You need 2000 more pounds of the most expensive grade than the least expensive grade.

Use x, y, and z to represent the weights of each grade you are to buy, with x the least expensive and z the most expensive grade.

1. Write an equation for the total weight of coffee beans.

2. Write an equation for the total cost.

3. Write an equation for the last of the three assumptions.

Taken together, the equations in Problems 1–3 form a **3 × 3 system of linear equations**.

The solution to this system is the ordered triple of numbers (x, y, z) that satisfies all three equations. The strategy for solving such a system is typically to reduce the system to a 2 × 2 linear system, and then proceed to solve this smaller system.

4. One of the equations already has only two variables. Combine the other two equations (by either the substitution or addition method of the previous activity) to form a new equation with the same two variables.

5. Now solve this new 2 × 2 system.

6. Substitute the solutions from Problem 5 into one of the original three equations. Now solve for the third variable.

7. How much of each grade of coffee should you order? Check to make sure your solution agrees with each of the three original assumptions.

8. Explain why it is not possible to solve this 3 × 3 system by using your graphing calculator.

> All of the equations in your 3 × 3 system are called *linear equations*, even though they cannot all be graphed as single lines. In this case, linearity refers to each variable being linear, that is, raised to the first power.

In general, each equation of a 3 × 3 system of equations will have three variables. The strategy remains the same, namely, to reduce the system to a 2 × 2 linear system of equations. This smaller system can then be solved by any of the methods learned in Activities 1.10 and 1.11.

9. Solve the following 3 × 3 linear system.

Equation I: $x - 2y + z = -5$
Equation II: $2x + y - z = 6$
Equation III: $3x + 3y - z = 11$

If you are not sure where to start, follow these steps.

Step 1. Is it possible to add two of the equations (right side to right side and left side to left side) so that one of the variables is eliminated? (Add equation I to equation II.)

Step 2. Is it possible to add a different pair of equations to eliminate the same variable? (Add equation I to equation III.)

Step 3. Notice that your equations from parts a and b form a 2 × 2 linear system. Solve this 2 × 2 system any way you can.

Step 4. Substitute your solution from step 3 into any one of the three original equations, and solve the resulting equation for the remaining variable.

Step 5. The final step is to substitute your potential solution into each of the three original equations. This is the only way you can be confident that your solution is correct.

Check: Equation I:

Equation II:

Equation III:

Most 3 × 3 systems will not have coefficients that are quite so convenient as the ones you just encountered. The following application provides a case in point.

10. Your sister works for Sam's competitor, Java Works, and has a similar job. She needs to purchase the same three grades of coffee, at the same prices you paid. But she must make the purchase using the assumptions given in parts a, b, and c below. Determine the equation that corresponds to each assumption.

   a. The total weight of beans needed is 11,400 pounds.

   b. She can only spend $13,000.

   c. She needs 500 more pounds of the least expensive grade than the other two grades combined.

11. Solve the system you determined in Problem 10. State your solution in terms of the application. Verify that all three assumptions are satisfied.

Further examples and practice in solving 3 × 3 linear systems of equations can be found in Appendix A.

SUMMARY
Activity 1.12

- A **3 × 3 system of linear equations** consists of three equations involving three variables.

- A **linear equation in the three variables** x, y, and z is of the form $Ax + By + Cz = D$, where A, B, C, and D are any constants. The equation is linear because the variables are all raised to the first power.

- To solve a 3 × 3 system algebraically:
    1. reduce the system in size to a 2 × 2 system,
    2. solve the 2 × 2 system,
    3. substitute the 2 × 2 solution into any original equation to solve for the third unknown,
    4. check the solution by substituting into all three of the original equations.

- Note that 3 × 3 systems may also be inconsistent (have no solution) or dependent (have infinitely many solutions).

EXERCISES
Activity 1.12

In Exercises 1–4, solve the 3 × 3 linear systems. Be sure to check your solution in all three of the original equations.

1. $x + y - z = -8$
   $-x + y + z = 2$
   $2x - y + z = 8$

2. $2x - 3y + z = 7$
   $x + 2y - 2z = -5$
   $-2x + y + z = -1$

3. $x + 2y - z = 0$
   $3x + 2y + z = -8$
   $2x + 3y + z = 0$

4. $x + 2y - 3z = 5$
   $-x + y + 2z = 0$
   $2x - y + z = -1$

5. Recall that some 2×2 linear systems do not have unique solutions. Try to solve these 3×3 linear systems. Identify each system as either dependent or inconsistent.

a.
$$x + 2y + z = 4$$
$$2x - y + 3z = 2$$
$$3x + y + 4z = 6$$

b.
$$2x - y + 3z = 3$$
$$-x + 2y - z = 1$$
$$x + y + 2z = 2$$

6. You are responsible for buying parts from a wholesale distributor. There are three types of comparable switches that are needed. The cost per switch is $1.20, $1.90, and $2.30. You need all three types and will place your order as dictated by the following facts.

i. You need a total of 12,000 switches in this order.

ii. Your budget will allow an expenditure of $23,400.

iii. You need three times as many of the most expensive switches as the least expensive switches.

Let x, y, and z represent the number of the first type, second type, and third type of switch, respectively.

a. Write a 3×3 linear system of equations to model this problem.

b. Solve the system. Be sure to check your solution.

ACTIVITY 1.13
How Long Can
You Live?

OBJECTIVES

1. Solve linear inequalities numerically and graphically.

2. Use properties of inequalities to solve linear inequalities algebraically.

3. Solve compound inequalities algebraically and graphically.

Life expectancy in the United States is steadily increasing, and the number of Americans aged 100 or older will exceed 850,000 by the middle of this century. Medical advancements have been a primary reason for Americans living longer. Another factor has been the increased awareness of maintaining a healthy lifestyle.

The life expectancies at birth for men and women in the United States can be modeled by the following functions,

$$W(x) = 0.126x + 76.74$$
$$M(x) = 0.169x + 69.11,$$

where $W(x)$ represents the life expectancy for women, $M(x)$ represents the life expectancy for men, and x represents the number of years since 1975. That is, $x = 0$ corresponds to the year 1975, $x = 5$ corresponds to 1980, and so forth.

1. a. Complete the following table:

Counting the Years

	YEAR					
	1975	1980	1985	1990	1995	2000
x, **YEARS SINCE 1975**	0	5	10	15	20	25
$W(x)$						
$M(x)$						

b. For people born between 1975 and 2000, do men or women have the greater life expectancy?

c. Is the life expectancy of men or women increasing more rapidly? Explain using slope.

You would like to determine in what years the life expectancy of men is greater than that of women. The phrase "greater than" indicates a mathematical relationship called an inequality. Symbolically, the relationship can be represented by

$$\underbrace{M(x)}_{\substack{\text{life expectancy} \\ \text{for men}}} \quad \underbrace{>}_{\substack{\text{is greater} \\ \text{than}}} \quad \underbrace{W(x)}_{\substack{\text{life expectancy} \\ \text{for women}}}$$

Other commonly used phrases that indicate inequalities are given in the following example.

Example 1

STATEMENT, WHERE x REPRESENTS A REAL NUMBER	TRANSLATION TO AN INEQUALITY
x is greater than 10	$x > 10$ or $10 < x$
x is less than 10	$x < 10$ or $10 > x$
x is at least 10	$x \geq 10$ (also read, x is greater than or equal to 10)
x is at most 10	$x \leq 10$ (also read, x is less than or equal to 10)

2. Substitute for $M(x)$ and $W(x)$ to obtain an inequality involving x that can be used to determine the birth years for which the life expectancy of men is greater than that of women.

Solving Inequalities Numerically and Graphically

> **DEFINITION**
>
> **Solving an inequality** is the process of determining the values of the variable that make the inequality a true statement. These values are called the **solutions** of the inequality.

3. Solve the inequality in Problem 2 numerically. That is, continue to construct a table of values (see Problem 1) until you determine the values of the years x (inputs) for which $0.169x + 69.11 > 0.126x + 76.74$. Use the table feature of your graphing calculator.

Therefore, if the trends given by the equations for $M(x)$ and $W(x)$ continue, the approximate solution to the inequality $M(x) > W(x)$ is $x > 177$. That is, according to the models, after the year 2152, men will live longer than women.

4. Now, solve the inequality $0.169x + 69.11 > 0.126x + 76.74$ graphically.

   a. Use your graphing calculator to sketch a graph of $M(x) = 0.169x + 69.11$ and $W(x) = 0.126x + 76.74$ on the same coordinate axis.

   b. Determine the point of intersection of the two graphs using the intersect feature of your grapher. What does the point represent in this situation?

To solve the inequality $M(x) > W(x)$ graphically, you need to determine the values of x for which the graph of $M(x) = 0.169x + 69.11$ is above the graph of

$$W(x) = 0.126x + 76.74.$$

   c. Use the graph to solve $M(x) > W(x)$. How does your solution compare to the solution in Problem 3?

5. a. Write an inequality to determine the birth years of women whose life expectancy is at least 85.

   b. Solve the inequality numerically, using the table feature of your graphing calculator.

   c. Use your graphing calculator to solve this inequality graphically.

Solving Inequalities Algebraically

The process of solving an inequality algebraically is very similar to solving an equation algebraically. Your goal is to isolate the variable on one side of the inequality symbol. You isolate the variable in an **equation** by performing the same operations to both sides of the equation so as not to upset the **balance**. You isolate the variable in an **inequality** by performing the same operations to both sides so as not to upset the **imbalance**.

6. a. Write the statement "15 is greater than 6" as an inequality.

b. Add 5 to each side of $15 > 6$. Is the resulting inequality a true statement? (That is, is the left side still greater than the right side?)

c. Subtract 10 from each side of $15 > 6$. Is the resulting inequality a true statement?

d. Multiply each side of $15 > 6$ by 4. Is the resulting inequality true?

e. Multiply each side of $15 > 6$ by -2. Is the left side still greater than the right side?

f. Reverse the direction of the inequality symbol in part e. Is the new inequality a true statement?

Problem 6 demonstrates two very important properties of inequalities.

Property 1 If $a < b$ represents a true inequality, then if

    i. The same quantity is added to or subtracted from both sides

or

    ii. both sides are multiplied or divided by the same *positive number*, then the resulting inequality remains a true statement and the direction of the inequality symbol remains the same.

For example, because $-4 < 10$, then

    i. $-4 + 5 < 10 + 5$ or $1 < 15$ is true.

        $-4 - 3 < 10 - 3$ or $-7 < 7$ is true.

    ii. $-4(6) < 10(6)$ or $-24 < 60$ is true.

        $\frac{-4}{2} < \frac{10}{2}$ or $-2 < 5$ is true.

Property 2 If $a < b$ represents a true inequality, then if both sides are multiplied or divided by the same *negative number*, then the inequality symbol in the resulting inequality statement must be reversed ($<$ to $>$ or $>$ to $<$) in order for the resulting statement to be true.

For example, because $-4 < 10$, then $-4(-5) > 10(-5)$ or $20 > -50$.

Because $-4 < 10$, then $\dfrac{-4}{-2} > \dfrac{10}{-2}$ or $2 > -5$.

These properties will be true if $a < b$ is replaced by $a \le b$, $a > b$, or $a \ge b$.

The following example demonstrates how properties of inequalities can be used to solve an inequality algebraically.

Example 2 *Solve* $3(x - 4) > 5(x - 2) - 8.$

SOLUTION

$$3(x - 4) > 5(x - 2) - 8 \qquad \text{Apply the distributive property.}$$
$$3x - 12 > 5x - 10 - 8 \qquad \text{Combine like terms on the right side.}$$
$$3x - 12 > 5x - 18$$
$$\underline{-5x \qquad\quad -5x} \qquad \text{Subtract } 5x \text{ from both sides; the direction of the inequality}$$
$$-2x - 12 > -18 \qquad\quad \text{symbol remains the same.}$$
$$\underline{+ 12 \quad + 12} \qquad \text{Add 12 to both sides; the direction of the inequality does not}$$
$$\dfrac{-2x}{-2} \quad \dfrac{-6}{-2} \qquad\quad \text{change.}$$
$$ \qquad \text{Divide both sides by } -2\text{; the direction is reversed!}$$
$$x < 3$$

Therefore, from Example 2, any number less than 3 is a solution to the inequality $3(x - 4) > 5(x - 2) - 8$. The solution set can be represented on a number line by shading all points to the left of 3:

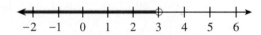

The open circle at 3 indicates that 3 is *not* a solution. A closed circle indicates that the number beneath the closed circle *is* a solution. The arrow shows that the solutions extend indefinitely to the left.

7. Solve the inequality $0.169x + 69.11 > 0.126x + 76.74$ algebraically to determine the birth years in which men will be expected to live longer than women. How does your solution compare to the solutions determined numerically and graphically in Problems 3 and 4c?

Compound Inequality

You have joined a health-and-fitness club. Your aerobics instructor recommends that to achieve the most cardiovascular benefit from your workout, you should maintain your pulse rate between a lower and upper range of values. These values depend on your age.

8. If the variable *a* represents your age, then the lower and upper values for your pulse rate are determined by the following:

$$\text{lower value: } 0.72(220 - a)$$
$$\text{upper value: } 0.87(220 - a)$$

   a. Determine your lower value.

   b. Determine your upper value.

For the most cardiovascular benefit, a 20-year-old's pulse rate should be between 144 and 174. The phrase "between 144 and 174" means the pulse rate should be greater than 144 *and* less than 174. Symbolically, this combination or **compound inequality** is written as:

$$144 < \text{pulse rate } and \text{ pulse rate} < 174$$

This statement is written more compactly as: $144 < \text{pulse rate} < 174$.

The numbers that satisfy this compound inequality can be represented on a number line as:

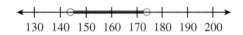

Other commonly used phrases that indicate compound inequalities involving the word "and" are given in the following example.

Example 3

STATEMENT, WHERE *x* REPRESENTS A REAL NUMBER	TRANSLATION TO A COMPOUND INEQUALITY
x is greater than or equal to 10 and less than 20	$10 \leq x < 20$
x is greater than 10 and less than or equal to 20	$10 < x \leq 20$
x is from 10 to 20 inclusive	$10 \leq x \leq 20$

9. Recall that the life expectancy for men is given by the expression $0.169x + 69.11$, where *x* represents the number of years since 1975. Use this expression to write a compound inequality that can be used to determine in what birth years men will be expected to live into their 80s.

The following example demonstrates how to solve a compound linear inequality algebraically and graphically.

Example 4   a. *Solve* $-4 < 3x + 5 \le 11$ *using an algebraic approach.*

SOLUTION

Note that the compound inequality has three parts: left: -4, middle: $3x + 5$, and right: 11. To solve this inequality, isolate the variable in the middle part.

$$\begin{array}{ccc} -4 < 3x + 5 \le 11 \\ \underline{-5 \qquad -5 \quad -5} & \text{Subtract 5 from each part.} \\ -9 < 3x \le 6 \\ -\dfrac{9}{3} < \dfrac{3x}{3} \le \dfrac{6}{3} & \text{Divide each part by 3.} \\ -3 < x \le 2 \end{array}$$

The solution can be represented on a number line as follows:

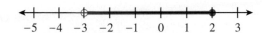

b. *Solve* $-4 < 3x + 5 \le 11$ *using a graphical approach.*

SOLUTION

First, graph each part of the inequality on the same coordinate axes.

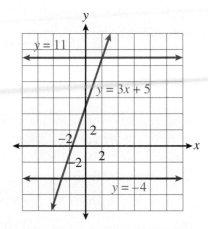

You need to determine values of x for which the graph of $y_2 = 3x + 5$ is above the graph of $y_1 = -4$ and on or below the graph of $y_3 = 11$. Using the graphs, verify that the solution is $-3 < x \le 2$.

10. a. Solve the compound inequality $80 \le 0.169x + 69.11 < 90$ from Problem 9 to determine in what birth years men will be expected to live into their 80s

b. Verify your results in part a by solving the compound inequality graphically. Use the window from Problem 4a.

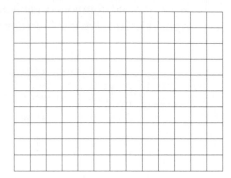

SUMMARY
Activity 1.13

1. The **solution set** of an inequality is the set of all values of the variable that satisfy the inequality.

2. The direction of an inequality is not changed when

   i. the same quantity is added to or subtracted from both sides of the inequality. Stated algebraically ,

   if $a < b$, then $a + c < b + c$ and $a - c < b - c$.

   ii. both sides of an inequality are multiplied or divided by the same positive number.

   If $a < b$, then $ac < bc$, where $c > 0$ and $\frac{a}{c} < \frac{b}{c}$, where $c > 0$.

3. The direction of an inequality is reversed if both sides of an inequality are multiplied by or divided by the same negative number. These properties can be written symbolically as:

   i. If $a < b$, then $ac > bc$, where $c < 0$.

   ii. If $a < b$, then $\frac{a}{c} > \frac{b}{c}$, where $c < 0$.

   The two properties of inequalities above will still be true if $a < b$ is replaced by $a \leq b, a > b$, or $a \geq b$.

4. Inequalities such as $f(x) < g(x)$ can be solved using three different methods:

   i. a **numerical approach**, in which a table of input-output pairs is used to determine values of x for which $f(x) < g(x)$.

   ii. a **graphical approach**, in which values of x are located so that the graph of f is below the graph of g.

   iii. an **algebraic approach**, in which the properties of inequalities are used to isolate the variable.

   Similar statements can be made for solving inequalities of the form $f(x) \leq g(x), f(x) > g(x)$, and $f(x) \geq g(x)$.

EXERCISES
Activity 1.13

In Exercises 1–6, translate the given statement into an algebraic inequality or compound inequality.

1. To avoid an additional charge, the sum of the length, l, width, w, and depth, d, of a piece of luggage to be checked on a commercial airline can be at most 61 inches.

2. A PG-13 movie rating means that your age, a, must be at least 13 years for you to view the movie.

3. The cost, $C(A)$, of renting a car from company A is less expensive than the cost $C(B)$ of renting from company B.

4. The label on a bottle of film developer states that the temperature, t, of the contents must be kept between 68° and 77° Fahrenheit.

5. You are in a certain tax bracket if your taxable income, i, is over \$24,650, but not over \$59,750.

6. The range of temperature, t, on the surface of Mars is from 28°C to 140°C.

Solve Exercises 7–14 graphically and algebraically.

7. $3x > -6$

8. $3 - 2x \leq 5$

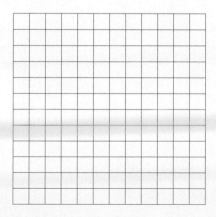

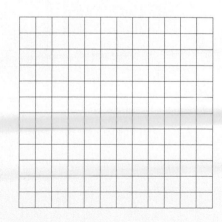

Exercise numbers appearing in color are answered in the Selected Answers appendix.

9. $x + 2 > 3x - 8$ **10.** $5x - 1 < 2x + 11$

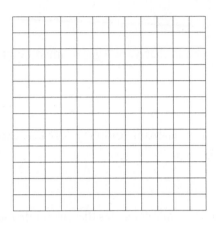 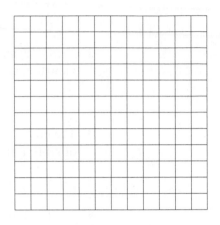

11. $8 - x \geq 5(8 - x)$ **12.** $5 - x < 2(x - 3) + 5$

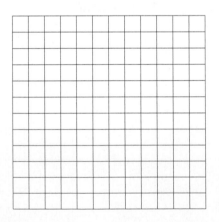 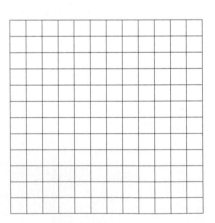

13. $\frac{x}{2} + 1 \leq 3x + 2$ **14.** $0.5x + 3 \geq 2x - 1.5$

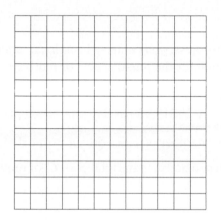   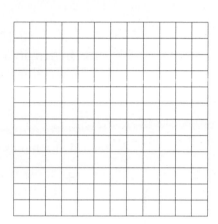

Solve Exercises 15–16 graphically and algebraically.

15. $1 < 3x - 2 < 4$ **16.** $-2 < \frac{x}{3} + 1 < 5$

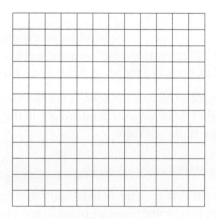   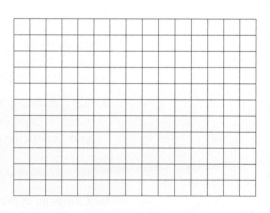

17. The consumption of cigarettes is declining. If t represents the number of years since 1985, then the consumption, C, is modeled by

$$C = -14.25t + 598.69,$$

where C represents the number of billions of cigarettes smoked per year.

a. Write an inequality that can be used to determine the first year in which cigarette consumption is less than 200 billion cigarettes per year.

b. Solve the inequality in part a using an algebraic as well as a graphical approach.

18. In Activity 1.10, Ride for Less, you contacted two local rental companies and obtained the following information for the one-day cost of renting a truck:

   Company 1: $40.95 per day plus $0.19 per mile
   Company 2: $19.95 per day plus $0.49 per mile

   Let n represent the total number of miles driven in one day.

   a. Write an expression to determine the total cost, C, of renting a truck for n days from company 1.

   b. Write an expression to determine the total cost, C, of renting a truck for n days from company 2.

   c. Use the expressions in parts a and b to write an inequality that can be used to determine for what number of miles it is less expensive to rent the truck from company 2.

   d. Solve the inequality.

19. The sign on the elevator in a seven-story building on campus states that the maximum weight it can carry is 1200 pounds. As part of your work-study program, you need to move a large shipment of books to the sixth floor. Each box weighs 60 pounds.

   a. Let n represent the number of boxes placed in the elevator. If you weigh 150 pounds, write an expression that represents the total weight in the elevator. Assume that only you and the boxes are in the elevator.

   b. Using the expression in part a, write an inequality that can be used to determine the maximum number of boxes that you can place in the elevator at one time.

   c. Solve the inequality.

20. The following equation is used in meteorology to determine the temperature humidity index:

$$T = \tfrac{2}{5}(w + 80) + 15,$$

where w represent the wet-bulb thermometer reading. For what values of w would T range from 70 to 75?

21. The temperature readings in the United States have ranged from a record low of $-79.8°F$ (Alaska, January 23, 1971) to a record high of $134°F$ (California, July 10, 1913).

a. If F represents the Fahrenheit temperature, write a compound inequality that represents the interval of temperatures (in °F) in the United States.

b. Recall that Fahrenheit and Celsius temperatures are related by the formula

$$F = 1.8C + 32.$$

Rewrite the compound inequality in part a to determine the temperature range in degrees Celsius.

c. Solve the compound inequality.

22. You are enrolled in a wellness course at your college. You achieved grades of 70, 86, 81, and 83 on the first four exams. The final exam counts the same as an exam given during the semester.

a. If x represents the grade on the final exam, write an expression that represents your course average.

b. If your average is greater than or equal to 80 and less than 90, you will earn a B in the course. Using the expression from part a for your course average, write a compound inequality that must be satisfied to earn a B.

c. Solve the inequality.

ACTIVITY 1.14

Long Distance
by Phone

OBJECTIVES

1. Graph a piecewise linear function.

2. Write a piecewise linear function to represent a given situation.

3. Graph a function defined by $y = |x - c|$.

A certain long-distance telephone carrier offers the following rates for calls outside the state in which you live: $0.15 per minute for the first 10 minutes, $0.08 per minute for each minute thereafter. Also, whereas most companies round up to the next minute for any fraction of a minute, this company charges for the exact duration of your call.

1. Complete the following table using time, t, in minutes, and cost, c, in dollars.

t (in minutes)	1	2	5	10	15	20	30
c (in dollars)							

2. a. Write an equation that gives the cost, c, of a call in terms of time, t, if the call lasts 10 minutes or less.

b. Write an equation that gives the cost, c, of a call, if the call lasts longer than 10 minutes. Remember, the input, t, is the total time of the call.

3. The cost is determined in two different ways, depending on the length of the call. In other words, the cost function is defined in pieces. At what input value does the definition change from the first to the second piece?

The resulting cost function is called a **piecewise function**.

This cost function is written in the following way:

$$c = f(t) = \begin{cases} 0.15t & \text{if } t \le 10 \\ 1.50 + 0.08(t - 10), & \text{if } t > 10 \end{cases}$$

The expression $0.15t$ is used to determine the total cost if the length of the call is less than or equal to 10 minutes (see Problem 2a). The expression $1.50 + 0.08(t - 10)$ is used to determine the total cost if the length of the call is more than 10 minutes (see Problem 2b).

4. a. What is the cost of a 6-minute call?

b. Determine $f(23)$, and interpret the result in the context of the situation.

5. Sketch the graph of the cost function over the domain 0 to 30 minutes. Be sure to use an appropriate scale for the horizontal and vertical axes.

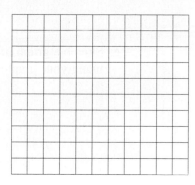

6. You can use your grapher to graph the cost function. To accomplish this, you must enclose each piece of the function in parentheses with its domain also in parentheses. The inequality symbols are in the Test (a 2nd function) menu. The pieces are connected by plus signs as follows:

$$Y_1 = (0.15x)(x \leq 10) + (1.5 + 0.08(x - 10))(x > 10)$$

Using the indicated window, your graph should appear as follows.

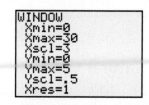 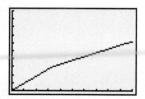

 For more information on graphing piecewise functions, see Appendix C.

7. Another company gives the following promotion for your long-distance service:

$0.99 for the first 20 minutes and $0.07 for each minute thereafter.

a. Complete the following table using total time, t, in minutes and cost, c, in dollars.

t (in minutes)	1	5	10	20	25	30
c (in dollars)						

b. Write an equation that gives the cost, c, of a call that lasts 20 minutes or less.

c. Write an equation that gives the cost, c, if the call lasts longer than 20 minutes. Remember, t is the total time of the call.

d. Write a piecewise function for the cost of the long-distance service.

c. Sketch a graph of the cost function over the domain 0 to 30 minutes. Verify using your graphing calculator.

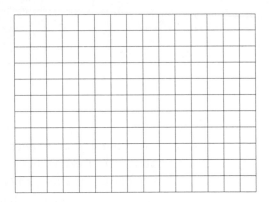

f. Determine the cost of a call that lasts 33 minutes.

Cost of Natural Gas

In a region of upstate New York, the natural gas bills for rural customers are calculated on gas meter readings that are taken every month. A therm is a unit that measures the amount of natural gas needed to produce 100,000 Btus of heat. Each month, a customer's gas bill will state how many therms were used.

For residential customers, the gas delivery charge for natural gas is billed in the following manner:

$14.55 for 0 to 3 therms,

$0.3529 for each therm over 3 therms but less than or equal to 50 therms, and

$0.0536 for each therm over 50 therms.

You can write a function that represents the total charge of natural gas delivery to your home. The output value will be calculated differently depending on whether you use 3 therms or less, more than 3 and less than or equal to 50 therms, or more than 50 therms. This piecewise function is determined in Problems 8–12.

8. a. Complete the following table, using x to represent the total number of therms used and C to represent the total monthly charge for natural gas delivery.

It's a Gas

x, NUMBER OF THERMS	0	1	2	5	10	50	100	200
C, TOTAL MONTHLY COST								

b. Write an equation to represent the delivery charge if you use 3 therms or less.

c. Write an equation to represent the delivery charge if the number of therms used is greater than 3 and less than or equal to 50. Remember, x represents the total number of therms used.

d. Write an equation to represent the delivery charge if the number of therms used is more than 50.

9. Use the results from Problem 8 to write a piecewise delivery charge function C representing the total cost of using x therms over the following three intervals: $0 \le x \le 3$, $3 < x \le 50$, and $x > 50$

10. Use your grapher to graph the piecewise function in Problem 9. To indicate $(3 < x \le 50)$, you must use two separate inequalities, $x > 3$ and $x \le 50$. Use the dot mode on the grapher. Your graph should appear as follows:

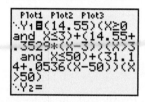 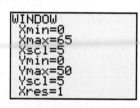 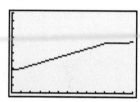

11. **a.** What is the delivery charge if you use 157 therms of gas during the month?

b. If it costs $0.6070 for each therm of gas used (gas supply charge), then what is the cost of the 157 therms used?

c. What was your total natural gas bill?

Absolute Value Function

12. Consider the piecewise function defined by

$$f(x) = \begin{cases} -x & \text{if } x < 0 \\ x & \text{if } x \ge 0 \end{cases}.$$

a. What is the domain of the function f?

b. Complete the following table.

x	−4	−3	−2	−1	0	1	2	3	4
f(x)									

c. Sketch a graph of the function f. Use your graphing calculator to verify the graph.

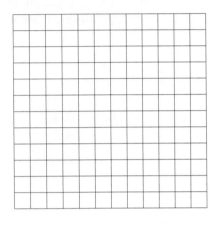

d. Describe the shape of the graph of function f.

e. What is the range of the function?

The function f defined in Problem 12 may look familiar. This function is the piecewise definition of the absolute value function defined by $f(x) = |x|$.

13. a. Use your graphing calculator to obtain the graph of $y = |x|$. The absolute value function is located in the Math menu under the Num submenu.

b. How does the graph compare to the graph in Problem 12?

14. a. Use your graphing calculator to sketch a graph of $y = |x - 5|$.

b. Describe how the graph of $y = |x - 3|$ can be obtained from the graph of $y = |x|$.

SUMMARY
Activity 1.14

1. A **piecewise** function is a function that is defined differently for certain "pieces" of its domain.

2. The absolute value function is a special piecewise function defined by

$$|x| = \begin{cases} x \text{ if } x \geq 0 \\ -x \text{ if } x < 0 \end{cases}$$

3. The absolute value of a linear function $g(x) = |x - c|$, always has a V-shaped graph with a vertex at $(c, 0)$.

EXERCISES
Activity 1.14

1. To travel outside the city limits, a certain taxicab company charges \$1.20 for the first mile or less of travel. After the first mile, the charge is an additional \$0.90 per mile.

   Complete the following table.

x, NUMBER OF MILES OUTSIDE CITY LIMITS	0	0.5	1	2	5	10
C, TOTAL COST (IN \$)						

   a. Write an equation for the total cost, c, of the first mile or less of travel?

   b. Write an equation that gives the total cost, C, if you travel more than one mile, all outside city limits.

   c. Write a piecewise function for the total cost, C.

d. Sketch a graph of the cost function over the domain 0 to 10 miles. Verify using your graphing calculator.

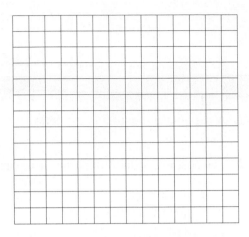

e. Determine the cost of a 12-mile taxi ride, all outside the city limits.

2. a. Sketch the graph of the piecewise function.

$$H(x) = \begin{cases} -2x + 3 & \text{if } x < -2 \\ 4 & \text{if } -2 \le x < 1 \\ x - 1 & \text{if } x \ge 1 \end{cases}$$

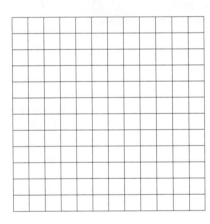

b. Use your graphing calculator and compare the result with your sketch.

3. You are an author about to publish your first novel, *Hockey in the 90s*. The book will sell for $25. You will be paid royalties of 10% on the first 15,000 copies sold; 12% on the next 6000 copies, and 16% on any additional copies.

a. Write a piecewise function, *f*, that specifies the total royalties if *x* copies are sold.

b. Graph the function on your graphing calculator.

c. Use your function from part a to determine if the royalty is $46,500 when 18,000 books are sold. Is the royalty $71,500 when 25,000 copies are sold? If you do not obtain these answers when substituting into the function, then discover where you went wrong.

d. With the royalties from your book, you would like to pay for your advanced degree in journalism. It will cost $65,000. How many books must sell to cover the cost of the degree?

4. You receive a bill each month for your credit card use. The bill indicates the minimum amount that is due by a certain date. The minimum amount due depends upon your unpaid balance. One credit card company uses the following criteria to determine your bill:

- The entire amount is due if the balance is less than $10.
- A minimum of $10 is due if the balance is $10 or more but less than $500.
- A minimum of $30 is due if the balance is $500 or more but less than $1000.
- A minimum of $50 is due if the balance is $1000 or more but less than $1500.
- A minimum of $70 is due if the balance is $1500 or more.

a. Let x represent the dollar amount of the unpaid balance. Complete the following table.

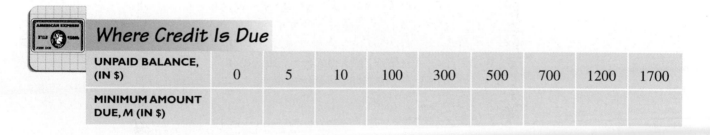

Where Credit Is Due

UNPAID BALANCE, (IN $)	0	5	10	100	300	500	700	1200	1700
MINIMUM AMOUNT DUE, M (IN $)									

b. Write a piecewise function for the minimum payment due.

c. Sketch a graph of the minimum payment function. Verify using your graphing calculator.

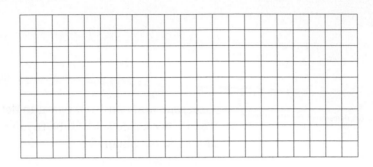

5. a. Complete the following table where $f(x) = |x - 3|$ and $g(x) = |x + 3|$.

x	−5	−4	−3	−2	−1	0	1	2	3	4	5
f(x)											
g(x)											

b. Sketch a graph of each of the functions f and g. Verify using your graphing calculator.

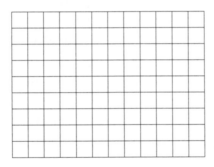

c. Describe how the graphs of f and g can be obtained from the graph of $y = |x|$.

d. Write $f(x) = |x - 3|$ as a piecewise function.

e. What is the domain of $f(x)$? What is the domain of $g(x)$?

f. What is the range of $f(x)$? What is the range of $g(x)$?

6. a. Without using a table of values or a grapher, sketch the graph of $f(x) = |x - 5|$.

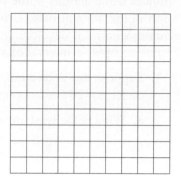

b. Describe the location and shape of the graph of $g(x) = |x + 5|$.

c. Sketch the graph of $h(x) = |x| + 5$ without a graphing calculator or a table of values.

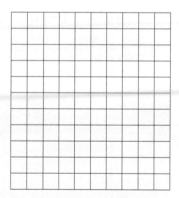

d. Describe the similarities and differences in the two graphs in parts b and c. Include shape, location, and intercepts.

ACTIVITY 1.15
How Much Can
You Tolerate?

OBJECTIVES

1. Write a compound inequality to represent a given statement.

2. Determine the error.

3. Solve an equation involving absolute value using a number line.

4. Solve an inequality involving absolute value using a number line.

5. Solve absolute value equations and inequalities using a graphing approach.

6. Interpret absolute value as distance.

7. Graph an absolute value function.

No two manufactured items are exactly the same. There is always some variation. For example, a 20-ounce bottle of diet soda may contain slightly more than 20 fluid ounces while another bottle may contain slightly less. Customers expect the variation in the amount of soda in a 20-ounce bottle to be small. One of the goals of quality control is developing a bottling process that ensures the variation is within some acceptable interval of values.

1. Suppose that a 20-ounce bottle of Diet Coke must be filled to within 0.2 fluid ounces of 20 fluid ounces to be acceptable.

   a. What is the least amount of soda the bottle can contain to be acceptable?

   b. What is the greatest amount of soda the bottle can contain to be acceptable?

   c. For the bottle of soda to be acceptable, the amount of soda must be greater than or equal to the least value and less than or equal to the greater value. Write a compound inequality that represents this situation. Let x represent the actual amount of soda in the bottle.

2. Four bottles of soda are randomly selected to be tested. The actual amount of soda in each bottle is measured and recorded in the following table. Use the results from Problem 1 to determine if the bottle is acceptable (write yes) or not acceptable (write no).

RANDOM SAMPLE FOR TESTING	x, ACTUAL NUMBER OF FLUID OUNCES	ACCEPTABLE?
Bottle 1	19.7	
Bottle 2	19.83	
Bottle 3	20.15	
Bottle 4	20.2	

For the bottle to be acceptable, the actual number of fluid ounces had to satisfy the compound inequality $19.8 \le x \le 20.2$, where x is the actual amount of soda. Represented on a number line, the acceptable number of fluid ounces must be in the shaded region:

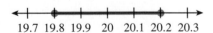

19.7 19.8 19.9 20 20.1 20.2 20.3

Absolute Value

3. a. Referring to Problems 1 and 2, complete the following table.

RANDOM SAMPLE FOR TESTING	TARGET VALUE (NUMBER OF FLUID OUNCES)	x (ACTUAL NUMBER OF FLUID OUNCES)	$x - 20$ (ACTUAL VALUE − TARGET VALUE)
Bottle 1	20	19.7	
Bottle 2	20	19.83	
Bottle 3	20	20.15	
Bottle 4	20	20.2	

b. What does the sign of the difference in the last column represent?

c. If you are determining whether the actual number of fluid ounces represented by x is acceptable or not, does it matter whether the actual value is above or below 20?

To determine whether the actual number of fluid ounces of soda is acceptable, you disregard the sign of the difference and consider just the magnitude or size of the difference. The number resulting from disregarding the sign of the difference is called the absolute error or simply error. The error is represented symbolically by

$$|x - 20|,$$

where the vertical bars represent absolute value, x is the actual amount of soda in the bottle, and 20 is the target value.

Recall from Activity 1.14 that the absolute value of a real number, n, is defined as follows:

$$|n| = \begin{cases} n \text{ if } n \geq 0 \\ -n \text{ if } n < 0 \end{cases}$$

Example 1   **a.** $|5| = 5$ **b.** $|-3| = -(-3) = 3$ **c.** $|0| = 0$

d. $|7 - 5| = |2| = 2$ For you to evaluate $|7 - 5|$, you must compute the operation inside the absolute value before you apply the absolute value.

e. $|5 - 8| = |-3| = 3$

4. a. Complete the following table.

RANDOM SAMPLE FOR TESTING	x, ACTUAL NUMBER OF x, FLUID OUNCES	|x − 20|
Bottle 1	19.7	
Bottle 2	19.83	
Bottle 3	20.15	
Bottle 4	20.2	

b. For you to determine if the actual number of fluid ounces in the test bottle is acceptable or not, what must be true about the value of $|x - 20|$?

c. Use your result from part b to determine which test bottles are acceptable.

d. Let x represent the actual amount of soda in the bottle. Write a compound inequality that represents all the acceptable values. These numbers are the solutions to the inequality $|x - 20| \leq 0.2$.

e. How do these solutions compare to your results in Problem 1c?

Absolute Value as a Distance

Absolute value can be interpreted as a distance.

Example 2 **a.** *$|5|$ represents the distance that the number 5 is from the origin on the number line. Therefore, $|5| = 5$.*

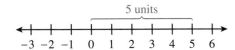

b. *$|-3|$ represents the distance that -3 is from the origin. Therefore, $|-3| = 3$.*

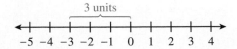

c. *$|0|$ represents the distance that 0 is from the origin. Therefore, $|0| = 0$.*

d. $|7 - 5|$ *represents the distance between 7 and 5 on the number line. Therefore,* $|7 - 5| = 2..$

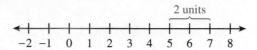

e. $|5 - 7|$ *represents the distance between 5 and 7 on the number line.*

Note that $|7 - 5| = |5 - 7| = 2.$

Example 2 demonstrates the following definitions.

> **DEFINITION**
>
> If a is any real number, then $|a|$ is the distance that a is from the origin on the number line.

> **DEFINITION**
>
> The distance between a and b on the number line is represented by $|a - b|$ or $|b - a|$.

Equations Involving Absolute Value

Interpreting absolute value as a distance can be useful when solving equations and inequalities involving absolute value.

Example 3   *Solve* $|x - 5| = 4.$

SOLUTION

Method 1: Using the number line: The equation $|x - 5| = 4$ means that the distance between x and 5 must be 4 units. That is, x must be a distance of 4 units from 5. Pictured on a number line, you have

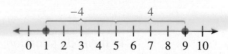

Therefore, $x = 1$ or $x = 9$.

Method 2: Algebraic approach: Because $|-4| = 4$ or $|4| = 4$, the equation $|x - 5| = 4$ is true when $x - 5 = 4$ or $x - 5 = -4$. Solving each equation separately, you have

$$x - 5 = 4 \quad \text{or} \quad x - 5 = -4$$
$$\underline{+5 \ +5} \qquad\qquad \underline{+5 \quad +5}$$
$$x = 9 \quad \text{or} \qquad\quad x = 1$$

Method 3: Graphical approach: The solution to the equation $|x - 5| = 4$ is the same as the solution to the following system of equations:

$$Y_1 = |x - 5|$$
$$Y_2 = 4$$

Graphing each equation and using the intersect feature of your graphing calculator, the final screen should appear as follows.

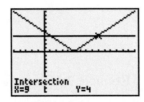

5. Recall that a 20-fluid-ounce bottle of Diet Coke must be filled within 0.2 fluid ounces of 20 to be acceptable. The lowest and the highest acceptable values are determined when the error, represented by $|x - 20|$, is equal to 0.2.

   a. Solve the equation $|x - 20| = 0.2$ using each of the three methods demonstrated in Example 3.

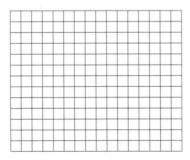

   b. Compare your results with your answers in Problem 1.

In the soda situation, 20 is called the *target value* and 0.2 is called the *tolerance* or *maximum error*.

> **DEFINITION**
>
> The largest acceptable amount above or below a target value is called the **tolerance** or **maximum error**.

Inequalities Involving Absolute Value

Example 4 *Solve the inequality $|x - 5| \leq 4$.*

SOLUTION

Method 1: Using the number line: The solutions to the inequality are all the real numbers that are a distance of four units or less from 5. Pictured on a number line, you have

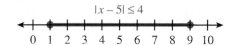

Therefore, the solutions are $1 \leq x \leq 9$.

Method 2: Algebraic approach: Stated another way, $|x - 5| \leq 4$ is true if the value of $x - 5$ is from -4 to 4. That is, $|x - 5| \leq 4$ is true if the compound inequality

$$-4 \leq x - 5 \leq 4$$

is true. Isolating x in the middle, you have $1 \leq x \leq 9$.

Method 3: Graphical approach: The solution to the equation $|x - 5| \leq 4$ are all of the x-values of the points where the graph of $Y_1 = |x - 5|$ intersects or is below the graph of $Y_2 = 4$.

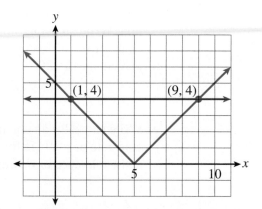

6. From Problem 4d, the acceptable numbers of fluid ounces in a 20-ounce bottle of Diet Coke are the solutions to the absolute value inequality $|x - 20| \leq 0.2$.

   a. Solve the inequality using each of the methods demonstrated in Example 4.

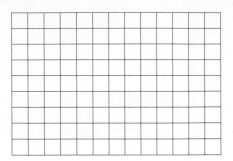

b. Compare your answer with the results from Problem 4d.

Applications

In a survey conducted prior to the 2000 presidential election, Gore was predicted to get 42% of the vote. Pollers indicate that, for their prediction to be good, the actual vote percentage for Gore needed to be within 3% of the target prediction of 42%.

7. a. If Gore actually received 40.5% of the vote, was the survey prediction a good one? Explain why or why not.

b. Let x represent the actual percent vote that Gore received in the 2000 election. Write an expression for the error. Recall that the error is the absolute value of the difference between the actual vote percentage and the target percentage of 42%.

c. Write and solve an equation that determines the lowest and highest acceptable vote percentage for the survey prediction to be a good one. Note that 3 is the tolerance or maximum error.

d. Write an inequality, and solve it algebraically and graphically to determine all the acceptable vote percentages. Represent the solution graphically.

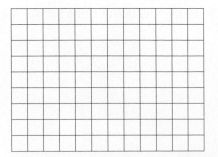

8. a. In some applications, you may be more interested in knowing when a value is farther than a tolerable amount from the target value. For example, in measuring the thickness of a precision part, the target thickness of 0.23 millimeter must be met to within 0.0015 millimeter (the tolerance). To determine all thicknesses that are unacceptable, solve $|x - 0.23| > 0.0015$ graphically.

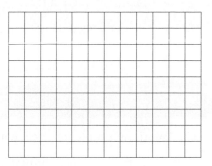

b. Because the unacceptable measurements are separated into two intervals (those too thick and those too thin), two inequalities are needed to express the solutions algebraically. What are they?

9. Normal body temperature is 98.6°F. A body temperature that is less than 1.5°F of normal is considered healthy.

a. If x represents your actual body temperature, write an inequality involving absolute value that can be used to determine healthy body temperatures.

b. Solve this inequality for x both algebraically and graphically.

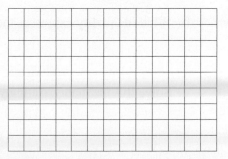

SUMMARY
Activity 1.15

1. $|a - b|$ or $|b - a|$ is the distance between a and b on the number line.

2. The basic absolute value function, $f(x) = |x| = |x - 0|$, is the distance from x to 0 on the number line.

3. The **maximum error** (or **tolerance**) determines the acceptable values for a measurement. If E is the maximum error, and a is the predicted or ideal measurement, then the acceptable values, x, are defined by $|x - a| \leq E$.

4. If $a > 0$, then the inequality $|x - a| \leq c$ is equivalent to $-c \leq x - a \leq c$.

5. If $a > 0$, then the inequality $|x - a| \geq ac$ is equivalent to $x - a \leq c$ or $x - a \geq c$.

EXERCISES
Activity 1.15

1. The length of a screw must be within 0.25 centimeter of 8 centimeters.

   a. What is the tolerance (or maximum error)?

   b. What inequality expresses the acceptable screw lengths?

   c. Solve your inequality using a number line, expressing your solution symbolically as a compound inequality.

2. You are given a budget of $24,500 for the month. Your boss will tolerate your expenditures as long as they stay within 1% of your budget.

   a. What is the tolerance for this budget problem?

   b. What inequality expresses the acceptable expenditures for the month?

   c. Solve this inequality, stating your solution within the context of the situation.

3. Express the following statements symbolically using absolute value.

   a. The distance between x and -3 on the number line is 5.

   b. The distance between x and 4 on the number line is 3.

   c. The weight must be within 5 grams of 175 grams.

   d. The difference between x and 10 must be at least 3.

   e. The difference between x and -20 is less than 5.

   f. The actual score differed by 12 points from the predicted score of 83.

   g. The predicted amount was more than $5 off from the actual $123 spent.

4. Solve each of the equations and inequalities from Exercise 3 using a number line. Express the solutions symbolically. You can check your answers using your graphing calculator.

   a.

   b.

   c.

   d.

e.

f.

g.

5. Solve each equation algebraically.

   a. $|x - 8| = 6$

   b. $|x + 9| = 15$

   c. $|x + 10| = 14$

   d. $|5 - x| = 0$

6. Solve each inequality algebraically. Represent your solution on a number line.

   a. $|x - 8| < 6$

   b. $|x + 9| \leq 15$

   c. $|x + 10| > 14$

   d. $|5 - x| \geq 10$

In Exercises 7 and 8, x represents the length in centimeters of a certain machine part. The tolerance is 0.01 centimeters for which the part is acceptable.

7. If the length of the machine part is supposed to be 9.2 centimeters, then solve the inequality $|x - 9.2| \leq 0.01$ and interpret the solution.

8. If the length of the machine part is specified to be 6.3 centimeters, then solve the inequality $|x - 6.3| \leq 0.01$ and interpret the solution.

9. A sheet of steel is to be 0.25 inch thick with a tolerance of 0.025 inch.

   a. Let x represent the thickness of the sheet of steel. Write the given specification in an inequality containing absolute value.

   b. Solve the inequality. Interpret the solution.

CLUSTER 3 — What Have I Learned?

1. You are given two linear equations in slope-intercept form. How can you tell by inspection if the system is consistent, inconsistent, or dependent? Give examples.

2. In this cluster, you solved 2×2 linear systems four ways. List them. Give an advantage of each approach.

3. Describe a procedure that will combine the following two linear equations in three variables into a single linear equation in two variables.

$$2x + 3y - 5z = 10$$
$$3x - 2y + 2z = 4$$

4. What number is its own opposite?

5. The graphs of the absolute value functions in this cluster look like a V. What are the coordinates of the point of the V of the graph of $f(x) = |x - 10|$?

6. If the tolerance (maximum acceptable error) for some measurement is 3 centimeters, what is the width of the actual interval of acceptable values?

7. Explain when the addition method would be more efficient to use than the substitution method as you solve a system of linear equations algebraically.

8. Use absolute value to write an inequality with a solution from -7 to 7, inclusive.

9. In solving an inequality, explain when you would change the direction of the inequality symbol.

10. Explain using your graphing calculator how to determine the solution of the inequality $|x - 3| < 5$.

CLUSTER 3 | How Can I Practice?

1. Solve the following systems both graphically and algebraically.

a. $x + y = -3$

  $y = x - 5$

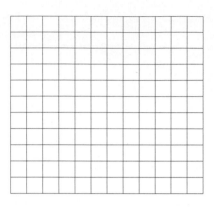

b. $x - 2y = -1$

  $4x - 3y = 6$

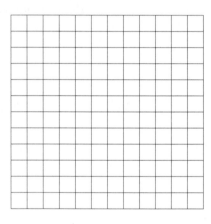

c. $2x - 3y = 7$

  $5x - 4y = 0$

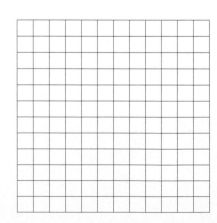

d. $x - y = 6$

  $y = x + 2$

2. Rewrite the systems in Exercise 1 in the form

$$y = ax + b$$
$$y = cx + d'$$

and check your solutions numerically using the table feature of your grapher.

a.   **b.**   **c.**   **d.**

3. Solve the 3 × 3 system algebraically.

$$2x - y + z = -5$$
$$x - 2y + 2z = -13$$
$$3x + y - 2z = 12$$

4. Graph the following piecewise function.

$$f(x) = \begin{cases} 2x - 1 & \text{if } x \le 1 \\ 3 & \text{if } 1 < x \le 4 \\ -4x & \text{if } x > 4 \end{cases}$$

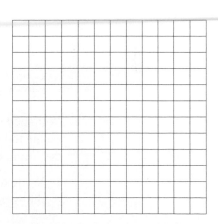

5. Solve the following graphically and algebraically.

a. $|x - 4| = 7$ **b.** $|3x - 6| \le 4$

    c. $|2x - 5| > 13$

6. Solve the following inequalities algebraically. Check your solutions graphically.

    a. $2.5x + 9.8 \geq 14.3$ **b.** $-3x + 14 < 32$ **c.** $-5 \leq 3x - 8 < 7$

7. You are going to create a garden of tulips and daffodils. You have space for approximately 80 bulbs. The florist tells you that tulips cost $0.50 per bulb and daffodils cost $0.75 per bulb. How many of each can you purchase if your budget is approximately $52?

    a. Write the system of equations.

    b. Solve the system algebraically.

    c. Check your solution graphically using your graphing calculator.

8. You need some repair work done on your truck. Towne Truck charges $80 just to examine the truck and $30 per hour for labor costs. World Transport Co. charges $50 for the initial exam and $40 per hour for the labor.

    a. Write a cost equation for each company. Use y to represent the total cost of doing the work and x to represent the number of hours of labor.

    b. Complete the table of values for the cost functions.

x (NUMBER OF HOURS)	y, TOWNE TRUCK COST	y, WORLD TRANSPORT COST
2		
4		
6		
8		

c. Graph the functions.

d. From the graph, determine after how many hours the costs will be equal. What will be the total cost?

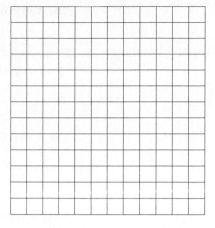

e. Check your solution in part d by solving the system algebraically.

f. You think that you have a transmission problem that will take approximately 6 hours to fix. Determine from the graph which company you will hire for this job. Explain.

9. Your friend tells you that he has 27 coins. Some coins are nickels, some are dimes, and the rest are quarters. The total value is $3.25. When your friend gives you the last clue by saying he has twice as many dimes as nickels, you can easily solve the system to tell him how many of each coin he has.

a. Write the system of equations.

b. Solve the system algebraically.

c. Check your solution.

10. Translate each of the following into an inequality statement.

  a. x is greater than -5 and at most 6.

  b. x is less than -5 or x is at least 3.

  c. x is greater than or equal to -3 and less than 4.

11. You own a hot dog cart in New York City. Your monthly profit is determined from the expression $1.50x - 50$, where x represents the number of hot dogs sold each month. The number 1.50 in the expression is the selling price for each hot dog. The cost of leasing the hot dog stand is the number 50 in the expression $1.50x - 50$.

  a. To ensure a profit of at least $2000 per month, approximately how many hot dogs do you have to sell? Write the inequality and solve.

  b. Your profit has been fluctuating between $1500 and $2200 per month. Determine approximately between what two values your hot dog sales have to be to realize this range of profit. Write the inequality and solve.

12. You are an obstetrician. Over the years, 90% of all the babies you have delivered were within 1.2 pounds of 8 pounds.

  a. Write an absolute value inequality that represents the weight of 90% of the babies you have delivered.

  b. Solve the inequality in part a.

  c. Write an absolute value inequality that represents the weight of the remaining 10% of these babies you have delivered.

  d. Solve the inequality in part c.

Summary

The bracketed numbers following each concept indicate the activity in which the concept is discussed.

CONCEPT / SKILL	DESCRIPTION	EXAMPLE
Variable [1.1]	A variable, usually represented by a letter, is a quantity or quality that may change in value from one particular instance to another.	In a survey of your class, an individual's height, weight, and gender are all variables.
Input variable [1.1]	The input variable is the value given first in a relationship.	In the relationship between the perimeter and the side of a square, $P = 4s$, s is the input variable.
Output variable [1.1]	The output is the value that corresponds to or is determined by the given input value.	In the relationship between the perimeter and the side of a square, $P = 4s$, P is the output variable.
Function [1.1]	A function is a correspondence between an input variable and an output variable that assigns a single, unique output value to each input value.	See Example 1 in Activity 1.1 (pages 2 and 3).
Ordered pair [1.1]	An ordered pair of numbers consists of two numbers written in the form (input value, output value). The order in which they are listed is significant.	(2, 3) is an ordered pair. In this pair, 2 is the input and 3 is the output.
Verbally defined function [1.1]	A function is defined verbally when it is defined using words.	The high temperature in Albany, New York, is a function of the day of the year, because for each day there is one high temperature.
Graphically defined function [1.1]	A function is defined graphically when the input variable is represented on the horizontal axis and the output variable on the vertical axis.	
Numerically defined function [1.1]	A function is defined numerically using ordered pairs.	The Park It table (page 1) in Activity 1.1 is a numerically defined function because for each hour there is only one value for the number of cars in the parking lot.
Function notation [1.1]	Output variable = name of function(input variable). $y = f(t)$ is read "y equals f of t."	See Example 5 of Activity 1.1 (page 6).

Independent variable [1.2]	Independent variable is another name for the input variable of a function.	See Example 4 of Activity 1.2 (page 16).
Dependent variable [1.2]	Dependent variable is another name for the output variable of a function.	See Example 4 of Activity 1.2 (page 16).
Domain of a function [1.2]	The domain of the function is the collection of all replacement values for the independent or input variable.	See Example 4 in Activity 1.2 (page 16).
Practical domain of a function [1.2]	The practical domain is the collection of replacement values of the input variable that makes practical sense in the context of the situatuon.	See Example 3 in Activity 1.2 (page 15).
Range of a function [1.2]	The range of a function is the collection of all output values of a function.	See Example 3 in Activity 1.2 (page 15).
Practical range of a function [1.2]	The practical range is the collection of all output values that make practical sense in the context of the situation.	See Example 3 in Activity 1.2 (page 15).
Mathematical model [1.3]	A function can be used as a mathematical model that best fits the actual data and can be used to predict output values for input values not in the table.	See Problem 4 in Activity 1.3 (page 24).
Increasing function [1.3]	A function is increasing if its graph goes up to the right.	See Example 1 in Activity 1.3 (page 25).
Decreasing function [1.3]	A function is decreasing if its graph goes down to the right.	See Example 1 in Activity 1.3 (page 25).
Constant function [1.3]	A function is constant if its graph is horizontal.	See Example 1 in Activity 1.3 (page 25).
Vertical line test [1.3]	A graph defines a function if a vertical line intersects the graph no more than once. This is called the vertical line test.	No circle can represent a function because a vertical line through the center will pass through the circle twice, indicating that there is at least one input value paired with two different output values.
Average rate of change [1.5]	The average rate of change of a function over a specified input interval is the ratio $$\frac{\text{change in output}}{\text{change in input}}.$$	See Problem 8 in Activity 1.5 (pages 45–46).

Linear function [1.6]	A function for which the rate of change between any pair of points remains constant is called a linear function.	$f(x) = 2x + 1$ defines a linear function.
Slope [1.6]	The slope of a line segment joining two points (x_1, y_1) and (x_2, y_2) is denoted by m and defined by $$m = \frac{y_2 - y_1}{x_2 - x_1}.$$	The slope of the line segment joining $(2, -1)$ and $(5, 2)$ is given by $$m = \frac{2 - (-1)}{5 - 2} = \frac{3}{3} = 1$$
Vertical intercept [1.6]	The vertical intercept $(0, b)$ of a graph is the point where the graph crosses the vertical axis.	The vertical intercept of $f(x) = 2x + 1$ is $(0, 1)$
Horizontal intercept [1.6]	The horizontal intercept $(a, 0)$ of a graph is the point where the graph crosses the horizontal axis.	The horizontal intercept of $f(x) = 2x + 1$ is $\left(-\frac{1}{2}, 0\right)$
Slope-intercept form [1.6]	The slope-intercept form of the equation of line is $f(x) = mx + b$.	$f(x) = 2x + 1$ is a linear function in slope-intercept form.
Parallel lines [1.7]	The graphs of linear functions with the same slope but different y-intercepts are parallel lines.	The graphs of $f(x) = 2x + 1$ and $g(x) = 2x - 3$ are parallel lines.
General form of a linear equation [1.8]	A linear function whose equation is in the form $Ax + By = C$, where A, B, and C are constants, is said to be written in general form.	$2x + 3y = 6$ is an equation of a linear function written in general form.
Horizontal line [1.8]	The graph of $y = c$ or $f(x) = c$ is a horizontal line.	The graph of $y = 3$ is a horizontal line 3 units above the x-axis.
Vertical line [1.8]	The graph of $x = a$ is a vertical line.	The graph of $x = 2$ is a vertical line 2 units to the right of the y-axis.
Linear regression equation [1.9]	The linear regression equation is the linear equation that best fits a set of data.	See Problem 2 in Activity 1.9 (page 82).
Interpolation [1.9]	Interpolation is the process of using a regression equation to predict a value of output for an input value that lies within the range of the original data.	See Problem 2 in Activity 1.9 (page 82).
Extrapolation [1.9]	Extrapolation is the process of using a regression equation to predict a value of output for an input value that lies outside the range of the original data.	See Problem 2 in Activity 1.9 (page 82).

2 × 2 system of linear equations [1.10]	A 2 × 2 system of linear equations consists of two linear equations with two variables.	$y = 3x - 10$ $y = 5x + 14$
Solution to a 2 × 2 linear system [1.10]	A solution to a 2 × 2 linear system is an ordered pair that solves both equations of the system.	$(-12, -46)$ is a solution to $y = 3x - 10$ $y = 5x + 14$
Consistent system [1.10]	A linear system is consistent if there is at least one solution.	$y = 3x - 10$ $y = 5x + 14$ is a consistent system.
Inconsistent system [1.10]	A linear system is inconsistent if there is no solution. The lines are parallel.	$y = 2x + 1$ $y = 2x - 3$
Dependent system [1.10]	A linear system is dependent if there are infinitely many solutions. The equations represent the same line.	$2x - 3y = 6$ $4x - 6y = 12$
Linear equation in the three variables [1.12]	A linear equation in the three variables x, y, and z is of the form $Ax + By + Cz = D$, where A, B, C, and D are any constants.	$2x + 3y - 7z = 23$ is a linear equation in three variables.
3 × 3 system of linear equations [1.12]	A 3 × 3 system of linear equations consists of three equations with a total of three variables.	$x + y - z = 8$ $-x + y + z = 2$ $2x - y + z = 8$ is a system of three linear equations.
A linear inequality in 2 variables [1.13]	A linear inequality is a statement that can be written $Ax + By < C$, where A, B, and C are constants. The $<$ symbol can be replaced with $>$, $\leq$, $\geq$, or $\neq$	$3x + 2y < 5$
A compound inequality [1.13]	A compound inequality is a statement that involves more than 1 inequality symbol $<$, $>$, $\leq$, $\geq$, or $\neq$	$-3 < x + 7 \leq 10$
Piecewise function [1.13]	A piecewise function is a function that is defined differently for certain "pieces" of its domain.	$f(x) = \begin{cases} x & \text{if } x \leq 2 \\ -x + 1 & \text{if } x > 2 \end{cases}$
Absolute value function [1.14]	The absolute value function is the function defined by $$\lvert x \rvert = \begin{cases} x & \text{if } x \geq 0 \\ -x & \text{if } x < 0 \end{cases}$$	$\lvert x \rvert = f(x) = \begin{cases} x & \text{if } x \geq 0 \\ -x & \text{if } x < 0 \end{cases}$
The absolute value of a linear function [1.14]	The absolute value of a linear function $g(x) = \lvert x - c \rvert$ has a V-shaped graph with a point at $(c, 0)$	$g(x) = \lvert x + 2 \rvert$ has a point at $(-2, 0)$

Absolute value equations and inequalities [1.15]	Absolute value equations and inequalities are mathematical statements involving at least one expression containing absolute value symbols	$\lvert 2x - 5 \rvert < 4$
Distance between a and b on the number line [1.15]	The distance between points a and b on the number line is given by $\lvert a - b \rvert$ or $\lvert b - a \rvert$.	The distance between 4 and -3 is given by $\lvert 4 - (-3) \rvert = 7$
Tolerance [1.15]	The tolerance determines the acceptable values for a measurement. If E is the tolerance and a is the predicted or ideal value, then the acceptable values x are defined by $\lvert x - a \rvert < E$	See Problem 8 Activity 1.15 (page 151).

Gateway Review

1. Determine whether each of the following is a function.

   a. The loudness of the stereo system is a function of the position of the volume dial.

   b. $\{(2, 9), (3, 10), (2, -9)\}$

   c.

   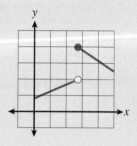

2. For an average yard, the fertilizer costs $20. You charge $8 per hour to do yard work. If x represents the number of hours worked on the yard and $f(x)$ represents the total cost, including fertilizer, complete the following table.

x	0	2	3	5	7
f(x)					

   a. Is the total cost a function of the hours worked? Explain.

   b. Which variable represents the input?

   c. Which is the dependent variable?

   d. Which value(s) of the domain would not be realistic for this situation? Explain.

   e. What is the rate of change from 0 to 3?

   f. What is the rate of change from 5 to 7?

Answers to all Gateway exercises are included in the Selected Answers appendix.

g. What can you say about the rate of change between any two of the points?

h. What kind of relationship exists between the two variables?

i. Write this relationship in the form $f(x) = mx + b$.

j. What is the practical meaning of the slope in this situation?

k. What is the vertical intercept? What is the practical meaning of this point?

l. Determine $f(4)$.

m. For what value(s) of x does $f(x) = 92$? Interpret your answer in the context of the situation.

3. Let $f(x) = x^2 - 5x$ and let $g(x) = -3x + 4$. Evaluate each of the following.
  a. $f(-2)$ and $g(-2)$

  b. $f(3) + g(3)$

  c. $f(-3) - g(-4)$

  d. $f(-4) \cdot g(2)$

4. Which of the following sets of data represent a linear function?

a.

x	0	2	4	6	8
$f(x)$	14	22	30	38	46

b.

x	5	10	15	20	25
y	4	2	0	−2	−4

c.

x	1	3	4	6	7
g(x)	10	20	30	40	50

d.

t	0	10	20	30	40
d	143	250	357	464	571

5. a. Determine the slope of the line through the points $(5, -3)$ and $(-4, 9)$.

b. From the equation $3x - 7y = 21$, determine the slope.

c. Determine the slope of the line from its graph.

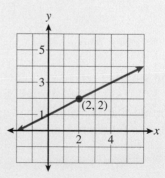

6. Write the equation of the line described in each of the following.
   a. a slope of 0 and passing through the point (2, 4)

   b. a slope of 2 and a vertical intercept of (0, 5)

c. a slope of -3 and passing through the point $(6, -14)$

d. a slope of 2 and passing through the point $(7, -2)$

e. a line with no slope passing through the point $(2, -3)$

f. a slope of -5 and a horizontal intercept of $(4, 0)$

g. a line passing through the points $(-3, -4)$ and $(2, 16)$

h. a line parallel to $y = \frac{-1}{2}x$ and passing through the point $(0, 5)$

7. Given the following graph of the linear function, determine the equation of the line.

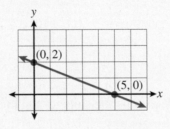

8. a. The building where your candy store is located is ten years old and has a value of \$200,000. When the building was one year old, its value was \$290,000. Assuming that the building's depreciation is linear, express the value of the building as a function, f, of its age, x, in years.

b. What is the slope of the line? What is the practical meaning of the slope in this situation?

c. What is the vertical intercept? What is the practical meaning of the vertical intercept in this situation?

d. What is the horizontal intercept? What is the practical meaning of the horizontal intercept in this situation?

9. Determine the vertical intercept of the following functions. Solve for y if necessary.

a. $y = 2x - 3$ **b.** $y = -3$ **c.** $x - y = 3$

d. What relationship do the graphs of these functions have to one another?

e. Use your grapher to graph the functions in parts a–c on the same coordinate axes. Compare your results with part d.

10. Determine the slopes and y-intercepts of each of the following functions. Solve for y if necessary.

a. $y = -2x + 1$ **b.** $2x + y = -1$

c. $-4x - 2y = 6$

d. What relationship do the graphs of these functions have to one another?

e. Use your grapher to graph the functions in parts a–c on the same coordinate axes. Compare your results with part d.

11. Determine the slopes and y-intercepts of each of the following functions. Solve for y if necessary.

a. $y = -3x + 2$ **b.** $3x + y = 2$

c. $6x + 2y = 4$

d. What relationship do the graphs of these functions have to one another?

e. For two lines to be parallel to each other, what has to be the same?

f. For two lines to lie on top of each other (coincide), what has to be the same?

g. Use your grapher to graph the functions in parts a–c on the same coordinate axes. Compare your results with part d.

12. a. Graph the function defined by $y = -2x + 150$. Indicate the vertical and horizontal intercepts. Make sure to include some negative values of x.

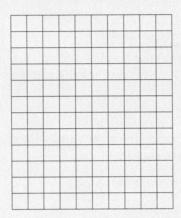

b. Using your grapher, verify the graph you have drawn in part a.

c. Using the graph, determine the domain and range of the function.

d. Assume that a 150-pound person starts a diet and loses 2 pounds per week for 4 weeks. Write the equation modeling this situation.

e. Compare the equation you found in part d with the one given in part a.

f. What is the practical meaning of the vertical and horizontal intercepts you found in part a?

g. What is the practical domain and range of this function for the situation given in part d?

13. a. You pay a flat fee of $25 per month for your trash to be picked up, and it doesn't matter how many bags of trash you have. Use x to represent the number of bags of trash, and write a function, f, in symbolic form to represent the total cost of your trash for the month.

b. Sketch the graph of this function.

c. What is the slope of the line?

14. Sketch the graph of the following piecewise function.

$$f(x) = \begin{cases} 4x & \text{if} \quad x < -2 \\ -x + 3 & \text{if} \quad -2 \leq x < 1 \\ 5 & \text{if} \quad x \geq 1 \end{cases}$$

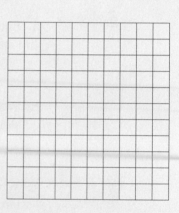

15. You work as a special events salesperson for a golf course owned by your city. Your salary is based on the following. You receive a flat salary of $1500 per month for sales of $10,000 or less; for the next $30,000 of sales, you receive your salary plus 2% of the sales over $10,000 and up to $40,000; and for any sales exceeding $40,000, you receive your salary and commission of 4% of sales over $40,000.

a. Write a piecewise function, f, that specifies the total monthly salary when x represents the amount of sales for the month.

b. Graph the function on your grapher.

c. What is your salary if your sales are $25,000?

d. You need to make $3150 to cover your expenses this month. What will your sales have to be for your salary to be that amount?

16. During the years 1994–1998, the number of finishers in the top 30 marathons worldwide increased. The following table gives the total number of finishers (to the nearest hundred) each year, where t represents the number of years after 1994.

YEARS AFTER 1994, t	0	1	2	3	4
NUMBER OF FINISHERS, n	7800	9100	10,000	10,900	12,100

a. Enter the data from your table into your calculator. Determine the linear regression equation model, and write the result.

b. What is the slope of the regression line? What is the practical meaning of the slope in this situation?

c. What is the vertical intercept? What is the practical meaning of the vertical intercept in this situation?

d. Use your graphing calculator to graph the regression line in the same screen as the scatterplot. How well do you think the line fits the data?

e. Use your regression model to determine the number of finishers worldwide in 2000.

f. Did you use interpolation or extrapolation to determine your result in part e? Explain.

g. Do you think that the prediction for the year 2024 will be as accurate as that in 2000? Explain.

17. Solve the systems of equations. Solve at least one algebraically and at least one graphically.

a. $3x - y = 10$
$5x + 2y = 13$

b. $4x + 2y = 8$
$x - 3y = -19$

c. $2x + y = 10$
$y = -2x + 13$

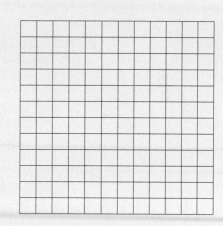

d. $2x + 6y = 4$
$x + 3y = 2$

18. The employees of a beauty salon order lunch two days in a row from the corner deli. Lunch on the first day consists of 5 small pizzas and 6 cookies for a total of $27. On the second day, 8 pizzas are ordered along with 4 cookies, totaling $39. To know how much money to pay, employees have to determine how much each pizza and each cookie cost. How much does the deli charge for each pizza and each cookie?

19. Solve these 3×3 systems of linear equations.

a. $x + y + z = 3$
$2x - y + 2z = 3$
$3x + 2y - z = 0$

b. $2x + 4y - z = -2$
$x - 2y + z = -5$
$-2x + y + 2z = 7$

c. $4x - 2y + z = 1$
$2x + 6y - z = 3$
$-3x + 4y + z = -1$

d. $3x - 4y - z = -1$
$4x + y - 5z = 10$
$-x + 2y + z = 11$

20. Your favorite photography studio advertises a family portrait special in the newspaper. There are three different print sizes available: small (3×5), medium (5×7), and large (8×10). There are three different packages that can be ordered.

PACKAGE	SMALL	MEDIUM	LARGE	TOTAL COST
A	4	2	1	$6
B	6	4	2	$11
C	10	6	4	$19

a. Normally, small prints sell for $0.65 per print, medium for $1.10, and large for $2.50. Decide whether you are getting a good deal by determining the cost per print of each size print.

b. Will you take advantage of the special? Explain.

21. a. Sketch the graph of $f(x) = |x + 2|$

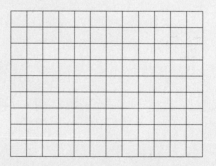

b. Determine the interval over which the function is increasing. Over which interval is the function decreasing?

c. Determine the domain of the function.

d. Determine the range of the function.

e. Graph $g(x) = -|x + 2|$. How does the graph compare with the graph in part a?

f. Let $h(x) = |x| + 2$. Explain what makes the functions f(x) and h(x) similar. How are the functions different?

22. Solve the following equations algebraically.

    a. $|x - 23| = 5$ **b.** $|x + 12| = 7$

    c. $|2x - 5| = 6$ **d.** $|3 - 5x| = 2$

23. Solve the following inequalities algebraically. Express your solutions both as inequalities and graphically on a number line.

    a. $|x - 2.5| \leq 0.2$ **b.** $|x + 5| > 2$

    c. $|2x - 9| < 3$ **d.** $|18 + 3x| \leq 2$

24. To pass inspection, a bag of strawberry twists must be within 8 grams of the target weight of 453 grams.

    a. Express this statement in symbolic form using absolute value.

    b. Determine the solution to the inequality in part a.

The Algebra of Functions

CLUSTER 1 | ## Addition, Subtraction, and Multiplication of Polynomial Functions

ACTIVITY 2.1
Spending and
Earning Money

OBJECTIVES

1. Identify a polynomial expression.
2. Identify a polynomial function.
3. Add and subtract polynomial expressions.
4. Add and subtract polynomial functions.

You are planning a trip with friends and are going to rent a van. The van you want rents for $75 per day. You are given 100 free miles each day and are charged $0.20 per mile for extra miles. The dealer claims that you can expect to average 25 miles per gallon. The Auto Club says you can expect to pay an average of $1.35 per gallon for gas on your trip. You are planning to be gone for eight days, and you know that you will be traveling at least 1000 miles on the trip. You have been assigned the job of estimating the total cost of operating the van for the trip.

1. The cost of renting the van is a function of the total number of miles driven on the trip. Note that with 100 free miles each day, you will not have to pay for 800 miles of the trip over the eight days.

   a. Complete the following table. $R(m)$ represents the cost of renting the van for a given number of miles, m, driven.

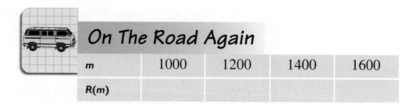

On The Road Again

m	1000	1200	1400	1600
R(m)				

   b. Write an equation for the cost, $R(m)$, of renting the van in terms of the number of miles, m, traveled.

2. The cost of the fuel is also a function of the total number of miles, m, traveled. The rental agency will start you with a full tank, but expects a full tank when you return.

   a. Complete the following table, where $F(m)$ represents the cost of the fuel for a given number of miles, m.

m	1000	1200	1400	1600
F(m)				

181

b. Write an equation for the cost, $F(m)$ of fuel in terms of m.

3. The total cost of using the van for the eight-day period is a function of the total number of miles, m, traveled.

a. Complete the following table, where $C(m)$ represents the total cost of renting. The entries for $R(m)$ and $F(m)$ were determined in Problems 1 and 2.

m	$R(m)$, RENTAL COST	$F(m)$, FUEL COST	$C(m)$, TOTAL COST
1000	640	54	
1200	680	64.80	
1400	720	75.60	
1600	760	86.40	

b. What is the relationship between the rental cost, $R(m)$, the fuel cost, $F(m)$, and the total cost, $C(m)$?

c. Add the equations in Problems 1 and 2 to determine an equation to define the total cost, $C(m)$, of using the van, as a function of miles, m, traveled.

DEFINITION

The rental van situation involves the addition of functions. The total cost function determined in Problem 3 is called the **sum function**. The notation is

$$C(m) = (R + F)(m) = R(m) + F(m).$$

Important note: The notation $(R + F)(m)$ does *not* represent multiplication by m. The notation $(R + F)(m)$ is another way of writing $R(m) + F(m)$.

Example 1 *Suppose f and g are defined by the following tables.*

x	−2	0	2	4	6	8
f(x)	0	2	6	20	42	72

x	−2	0	2	4	6	8
g(x)	7	3	7	19	39	67

Complete the following table for f + g.

SOLUTION

For any given x, $(f + g)(x) = f(x) + g(x)$.

x	−2	0	2	4	6	8
(f + g)(x)						

4. Enter the rental function, R, the fuel function, F, and the total cost function, C, into your grapher. Use the table feature to complete the following table for four input values.

m	R(m)	F(m)	C(m)
1250			
904			
1303			
1675			

Subtraction of Functions

You have returned from your trip, and now it's back to work. You are the owner of a small pet kennel. Your kennel can accommodate at most 20 dogs. Your current charge for boarding a dog is $12 per day. Utility bills are approximately $15 per day. The cost of feeding each dog, cleaning its stall, and exercising it is approximately $7.15 per day.

5. a. Suppose you let the input variable d represent the number of dogs boarding on a given day. Determine an equation that expresses the total revenue, $R(d)$, as a function of the number of dogs, d, boarding on a given day.

b. Complete the following input/output table for the revenue function R.

d	0	5	10	15	20
R(d)					

6. a. The total daily cost of operating the kennel is a function of the number of dogs boarding on a given day. If $C(d)$ represents the cost, write an equation for $C(d)$ in terms of d.

b. Complete the following table for the daily cost function, C.

d	0	5	10	15	20
C(d)					

7. The results from Problems 5 and 6 can be used to determine the profit, $P(d)$, of boarding dogs, d.

a. Use the output values from Problems 5b and 6b to complete the following table for the profit. Recall that profit = revenue − cost.

d	0	5	10	15	20
$P(d)$					

b. Using the equations for revenue, $R(d)$, and cost, $C(d)$, in Problems 5a and 6a, determine an equation for the profit, $P(d)$, as a function of d. Use the new equation to verify some of the entries in the table in part a.

DEFINITION

The kennel situation involves the subtraction of functions. The profit function determined in Problem 7b is called the **difference function**. The notation is

$$P(d) = (R - C)(d) = R(d) - C(d).$$

Example 2 *Suppose f and g are defined by the following tables.*

x	−2	0	2	4	6	8
$f(x)$	2	0	6	20	42	72

x	−2	0	2	4	6	8
$g(x)$	7	3	7	19	39	67

Complete the following table for f − g.

SOLUTION

For any given x, $(f - g)(x) = f(x) - g(x)$

x	−2	0	2	4	6	8
$(f - g)(x)$						

8. a. Sketch the graphs of the functions R, C, and P on the following grid. Label the axes with the appropriate scales.

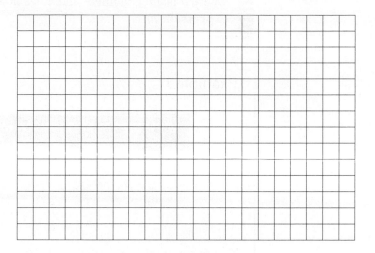

b. Use the graphs from part a to determine the break-even point for your pet-kennel business. That is, determine the number of dogs necessary for the profit to equal $0. Explain how you determined your answer.

c. Write and solve an equation to determine the break-even point.

d. For you to obtain a profit of at least $50, how many dogs must board?

Polynomial Expressions, Functions, and Terminology

The functions encountered in this activity, such as the total cost van rental function defined by $C(m) = 0.2m + 440$, are examples of a special category of functions called **polynomial functions**.

Such functions are defined by equations of the form $y = p(x)$, where x is the input variable and $p(x)$ is a polynomial expression involving the input variable. Therefore, to identify a polynomial function, you must be able to identify a polynomial expression.

DEFINITION

Any expression that is formed by adding or subtracting terms of the form ax^n, where a is a real number and n is a nonnegative integer, is called a **polynomial expression** in x.

Example 3

POLYNOMIAL EXPRESSIONS	EXPRESSIONS THAT ARE NOT POLYNOMIALS
$10, 5x, -3x^2 + 2,$ $4x^3 + 7x^2 - 3,$ and $\frac{5}{3}x^4$	$\frac{3}{x+1}, \sqrt{x+1},$ and $2x^3 + \frac{1}{x}$

The polynomial expression $4x^3 + 7x^2 - 3$ is said to be written in **descending order** because the term having the largest exponent is written first, the term having the next largest exponent is written second, etc. The same polynomial written in **ascending order** is

$$-3 + 7x^2 + 4x^3.$$

Polynomial expressions can be classified by the number of terms that are contained in the expression.

Terminology

• A monomial is a single term that consists of a constant or a constant times a variable or variables raised to nonnegative integers, such as $-3, 2x^4,$ and $\frac{1}{2}s^2$.

• A binomial is a polynomial that has two terms, such as $4x^3 + 2x$ and $3t - 4$.

• A trinomial is a polynomial that has three terms, such as $3x^4 - 5x^2 + 10$ and $5x^2 + 3x - 4$.

9. a. Write yes if the expression is a polynomial. Write no if the expression is not a polynomial.

$3x - 5$ _____ $5x^3 - 2x + 7$ _____

$\frac{5x}{3}$ _____ $\sqrt{x + 10}$ _____

$\frac{2}{x^2} - 8$ _____

b. In part a, classify any polynomial expressions as a monomial, binomial, or trinomial.

Example 4

POLYNOMIAL FUNCTIONS	NONPOLYNOMIAL FUNCTIONS
$y = 10, y = 5x$	$f(x) = \frac{3}{x + 1}$
$f(x) = -3x^2 + 2$	$g(x) = \sqrt{x + 1}$
$y = 4x^3 + 7x - 3$	$h(x) = 2x^3 + \frac{1}{x}$
$g(x) = \frac{5}{3}x^4$	

10. Write yes if the equation defines a polynomial function. Write no if it does not.

a. $y = 5x^2 + 2x - 1$

b. $f(x) = 3x + \frac{1}{x}$

c. $g(m) = 1.75m - 7$

d. $R(t) = \sqrt{t} + 7$

Addition and Subtraction of Polynomial Expressions

Operations with polynomial functions (such as addition and subtraction) involve operations with polynomial expressions. Example 5 demonstrates how to perform these operations.

Example 5 *Consider the polynomials f and g defined by*

$$f(x) = 2x^2 + 3x - 5 \text{ and } g(x) = -x^2 + 5x + 1.$$

Determine each of the following.

a. $(f + g)(x) = f(x) + g(x)$

$= (2x^2 + 3x - 5) + (-x^2 + 5x + 1)$ Remove parentheses.

$= 2x^2 + 3x - 5 - x^2 + 5x + 1$ Combine like terms.

$= x^2 + 8x - 4$

b. $(f - g)(x) = f(x) - g(x)$

$= (2x^2 + 3x - 5) - (-x^2 + 5x + 1)$ Change the sign of each term of the polynomials being subtracted.

$= 2x^2 + 3x - 5 + x^2 - 5x - 1$

$= 3x^2 - 2x - 6$

c. $-5 \cdot f(x) = -5 \cdot (2x^2 + 3x - 5)$ Apply the distributive property.

$= -10x^2 - 15x + 25$

11. Given the polynomial functions g and h, defined by

$$g(x) = 4x^2 - 3x + 10, \; h(x) = -3x^2 + 5x - 2,$$

determine each of the following:

a. $(g + h)(x)$

b. $(g - h)(x)$

c. $-2 \cdot g(x)$

SUMMARY
Activity 2.1

1. Given two functions, f and g, the **sum function**, $f + g$, is defined by

$$(f + g)(x) = f(x) + g(x)$$

and the **difference function**, $f - g$, is defined by

$$(f - g)(x) = f(x) - g(x).$$

2. Any expression that is formed by adding or subtracting terms of the form ax^n, where a is a real number and n is a nonnegative integer, is called a **polynomial expression** in x.

3. A **monomial** is a polynomial with one term. A **binomial** is a polynomial with two terms. A **trinomial** is a polynomial with three terms. Polynomials having more than three terms are not given special names.

4. A polynomial function is any function defined by an equation of the form $y = f(x)$, where $f(x)$ is a polynomial expression. For example,

$$y = \underbrace{2x^3 + 5x^2 - x + 1}_{f(x), \text{ a polynomial}}$$

EXERCISES
Activity 2.1

1. Jackie is a financial planner. In an effort to attract new customers, she sponsors a dinner at a local restaurant. The restaurant will charge $100 for the banquet room, plus $12.50 per person for each meal. Jackie will pay these expenses herself. From past experience, Jackie knows she can expect to make sales to about 15% of the people attending. She also knows that the average in sales she can expect from each new client is $750, for which she receives a 13% commission. It is clear that Jackie's personal financial success depends on how many people she can attract to this dinner.

Exercise numbers appearing in color are answered in the Selected Answers appendix.

a. Complete the following table.

Food for Thought

x, NUMBER OF PEOPLE ATTENDING	COST OF THE BANQUET HALL ($)	TOTAL MEAL COST ($)	TOTAL COST ($)
20	100		
40			
60			
80			
100			

b. Determine a formula for the total cost of restaurant expenses as a function of x, the number of attendees. Represent the total cost by $C(x)$.

c. Complete the following table.

x, NUMBER OF PEOPLE ATTENDING	NUMBER OF NEW CUSTOMERS	TOTAL SALES ($)	JACKIE'S COMMISSION ($)
20	3	2250	292.50
40			
60			
80			
100			

d. Determine a formula for the revenue that Jackie can expect to generate from this dinner as a function of x, the number of attendees. Represent the revenue by $R(x)$.

e. Combine the formulas in parts b and d to define a new function for the profit, P, that Jackie can expect from her dinner. A basic business equation is:

$$\text{profit} = \text{revenue} - \text{cost}.$$

f. What is the practical domain of this new function?

g. Use your new function to determine how many people must attend for Jackie to break even. Explain how you arrive at your decision.

h. Jackie hopes to make a profit of $500 on the dinner. How many people must attend for her to meet this goal? Explain. Write an equation that can be solved to answer this question. Then show how to solve the equation.

2. Write yes if the expression is a polynomial. Write no if the expression is not a polynomial. If the expression is a polynomial, classify it as a monomial, binomial, or trinomial.

   a. $5x^{-3} + 4$ **b.** $-3x^{10} - 2x^2 - 1$ **c.** $x^{1/2}$

   d. x **e.** $\frac{5}{4x} - 8$

3. Suppose that f and g are defined by the following tables.

x	0	2	4	6	8	10
$f(x)$	3	−5	0	7	−1	4

x	0	2	4	6	8	10
$g(x)$	1	−1	1	−1	3	4

Complete the following table for $f + g$ and $f - g$.

x	0	2	4	6	8	10
$(f + g)(x)$						
$(f - g)(x)$						

4. **a.** Suppose that $f(x) = 4x + 1$ and $g(x) = -2x + 4$. Determine an algebraic expression for $(f - g)(x)$ by subtracting $g(x)$ from $f(x)$ and combining like terms.

b. Complete the following table, using f and g from part a.

x	f(x)	g(x)	(f − g)(x)
0			
2			
4			
6			

c. Use your graphing calculator to plot all three functions, f, g, and $f - g$. Use the trace or table feature to complete the following table for four input values not used in part b.

x	f(x)	g(x)	(f − g)(x)

d. Do the results in part c agree with your understanding of $f - g$? Explain.

5. The algebraic skills necessary for determining the algebraic form of the new functions are those of simplifying expressions and combining like terms. Simplify the following.

a. $(2x + 3) + (3x - 5)$

b. $(2x^2 - 3x + 1) - (x^2 - 6x + 9)$

c. $2(x + 9) - 3(x - 4)$

d. $14x - 9 - 3(x^2 + 2x - 2)$

e. $4(3x - 2) - (7 - 3x)$

f. $6x + 5 + 3(2 - 2x)$

g. $2x^2 + 5x - 3(3 - x^2)$

h. $(5x - 2) - 2(3x^2 - 5x + 1)$

i. $2x + 5 - [3x - 4(5 - x)]$

j. $7x + 2[3x - 2(4 - 5x)] + 6$

6. Given $f(x) = 3x - 5$ and $g(x) = -x^2 + 2x - 3$, determine a formula, in simplest form, for each of the following.

   a. $f(x) + g(x)$

   b. $f(x) - g(x)$

   c. $2f(x) + 3g(x)$

   d. $f(x) - 2g(x)$

7. Given $h(x) = 6$, $p(x) = 3 - 4x$, and $r(x) = 4x^2 - x - 6$, determine an expression, in simplest form, for each of the following.

   a. $r(x) + h(x)$

   b. $p(x) + r(x) - h(x)$

   c. $h(x) - p(x)$

   d. $r(x) + p(x) + h(x)$

8. Given $f(x) = x^2 - 2$ and $g(x) = x + 4$, determine a value for each of the following.

   a. $f(2) + g(2)$

   b. $g(3) - f(3)$

   c. $f(-5) + g(-5)$

   d. $f(-2) - g(-2)$

9. Suppose f and g are defined by the following tables.

x	−6	−4	−2	0	2	4
f(x)	50	16	−2	−4	10	40

x	−6	−4	−2	0	2	4
g(x)	49	25	9	1	1	9

   a. Complete the following table for $f - g$.

x	−6	−4	−2	0	2	4
(f − g)(x)						

   b. If $f(x) = 2x^2 + 3x - 4$ and $g(x) = x^2 - 2x + 1$, determine an algebraic expression for $(f - g)(x)$.

   c. Check your answers in the table in part a by using the function you found in part b.

PROJECT ACTIVITY 2.2

Viewing the Algebra of Functions

OBJECTIVE

1. Explore adding and subtracting functions graphically.

In Activity 2.1, Spending and Earning Money, you added and subtracted functions both algebraically and numerically. In this activity, you will explore the addition and subtraction of functions graphically.

Consider two functions defined by $f(x) = x - 3$ and $g(x) = 2$. If you enter these functions into your graphing calculator with a **standard window:** $\text{Xmin} = -10$, $\text{Xmax} = 10$, $\text{Xscl} = 1$, $\text{Ymin} = -10$, $\text{Ymax} = 10$, and $\text{Yscl} = 1$, the graphs should resemble the following.

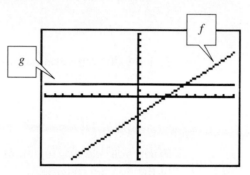

1. **a.** Determine an equation for the sum function $f + g$.

   b. Use your graphing calculator to graph the sum function $f + g$ on the same coordinate axes as the graphs of f and g. Your screen should resemble the following:

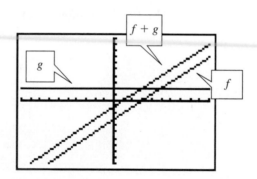

Let us investigate how the graph of the sum function $f + g$ can be determined from the graphs of f and g.

2. **a.** Use the definitions of f and g to determine each of the following.

   i. $f(2) =$ **ii.** $g(2) =$ **iii.** $(f + g)(2) =$

   b. What does the statement $f(2) = -1$ represent in the graph of f? Plot the point on the graph of f in Problem 1.

   c. Plot $g(2) = 2$ and $(f + g)(2) = 1$ as points on the appropriate graph in Problem 1.

   d. Explain how $(f + g)(2)$ can be determined using the graphs of f and g.

3. Use the trace feature of your graphing calculator and jump from f to g to $f + g$ for several values of x. What do you observe?

4. Use the table feature of your graphing calculator to investigate f, g, and $f + g$ for several different values of x. What relationship do you observe between the output values of each of the three functions?

5. How do your observations in Problems 3 and 4 compare with your conjecture in Problem 2d?

When you add two functions, you add the range (output) values for each value of the domain (input).

6. a. Now suppose that $f(x) = 1.5x + 1$ and $g(x) = 1.5$. Determine an equation for the difference function $f - g$.

b. Graph all three functions on the following grid. Label the axes with the appropriate scales.

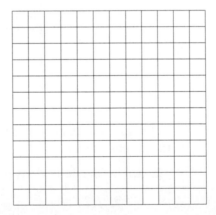

c. Is the graph of $f - g$ what you expected? Explain.

7. The graphs of *f* and *g* are given as follows:

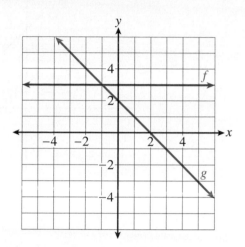

Which of the following graphs—a, b, or c—best represents the graph of $f + g$? Explain why you chose the graph you did.

a.

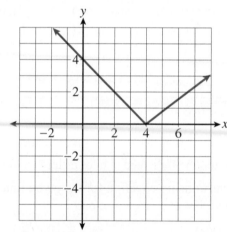

b.

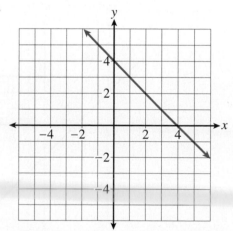

c.

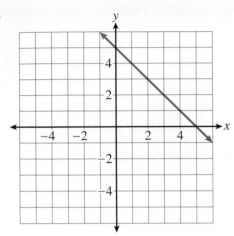

SUMMARY
Activity 2.2

When adding or subtracting two functions, f and g, graphically, for each input value x, add or subtract the output values (vertical displacements) represented by $f(x)$ and $g(x)$ to obtain the value of $(f + g)(x)$ or $(f - g)(x)$.

EXERCISES
Activity 2.2

1. Consider the functions $f(x) = -x - 2$ and $g(x) = 3$. For each of the following,

   i. sketch the indicated sum or difference function using the given graphs of f and g.

   ii. determine an equation for the given function, and use your graphing calculator to verify your sketch in part i.

   a. $(f + g)(x)$

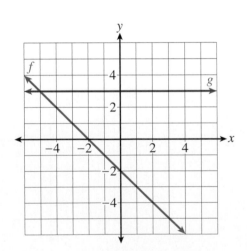

b. $(f - g)(x)$ **c.** $(g - f)(x)$

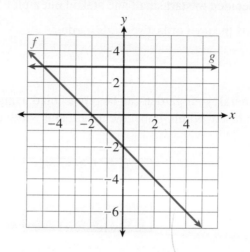 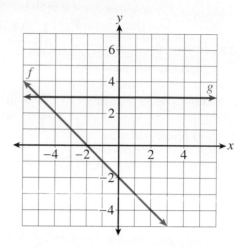

2. Consider the functions $f(x) = 2x - 3$ and $g(x) = -2$. Sketch the indicated sum or difference function using given graphs of f and g. Then determine an equation for the given function. Use your graphing calculator to verify the sketch.

   a. $(g + f)(x)$ **b.** $(g - f)(x)$

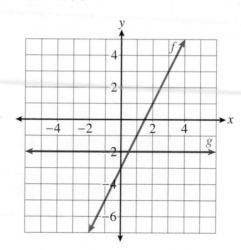 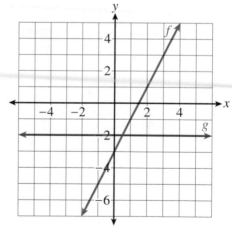

3. For the following graph of f and g, sketch a possible graph of $f + g$.

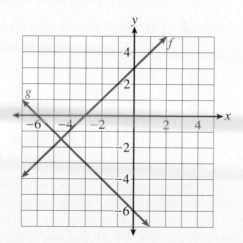

ACTIVITY 2.3

How Does Your
Garden Grow?

OBJECTIVES

1. Multiply two binomials using the FOIL method.

2. Multiply two polynomial functions.

3. Apply property of exponents to multiply powers having the same base.

Last spring you put in your first garden. Since you were taking a course load of 19 hours, you decided to start small and staked out a plot 5 feet long by 3 feet wide.

1. What was the area of last spring's garden?

2. a. In general, your garden can be represented using l to represent length and w to represent width. Label the length and width on the following diagram.

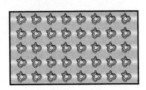

b. Write the expression for area (A) using l and w.

This spring you are only taking 15 credits of course work. You now want to expand your garden. You decide to add 4 feet to the length and 2 feet to the width.

Starting with a geometric representation of last spring's plot (5 feet by 3 feet), extend the length by 4 feet and the width by 2 feet to obtain a geometric model of the new plot.

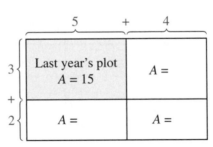

3. Determine the area of each section of this plot diagram, and then record the area in the appropriate place in the diagram above.

4. Determine the total area in two different ways.

a. Sum the areas of the sections.

b. Multiply the total length by the total width.

Next year you will have just about finished the requirements for your degree. Your garden will definitely get larger, but you are not sure by how much because you are concerned about expenses such as fertilizer and lime. Nevertheless, you decide to expand the length and width of your garden by the same number of feet, x.

5. a. Starting with a geometric representation of the current plot (9 feet by 5 feet), extend the length and the width by x feet to obtain a geometric model of the new plot.

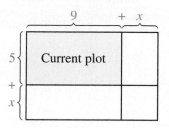

b. Complete the following table

x	LENGTH OF THE NEW PLOT	WIDTH OF THE NEW PLOT	AREA OF THE NEW PLOT
2			
5			
10			

6. a. The length of the rectangular plot in Problem 5 is a function of x. Write an equation for the length, $l(x)$, of the garden as a function of x.

b. The width of the rectangular plot is also a function of x. Write an equation for the width, $w(x)$, of the garden in terms of x.

c. The area of the garden is a function of x. If $A(x)$ represents the area of the plot, use the results in parts a and b to write an equation that defines $A(x)$ as a function of x. Do not simplify.

The area function, $A(x)$, is called a **product function** because $A(x)$ is determined by the multiplication, or product, of two functions. The notation for a product function is

$$A(x) = (lw)(x) = l(x) \cdot w(x).$$

Multiplication of Binomials

The area function is defined by $A(x) = l(x) \cdot w(x) = (9 + x)(5 + x)$. The product of the binomials $9 + x$ and $5 + x$ can be determined using a geometric model.

7. a. Return to the geometric model of the garden plot from Problem 5.

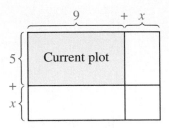

Determine the area of each section, and fill in each area in the geometric model.

b. Add the areas of all the sections in the geometric model, and simplify the expression. What does this algebraic expression represent?

As a result of Problem 7, you now know that

$$A(x) = (9 + x)(5 + x) = 45 + 14x + x^2.$$

The geometric model can be used to develop an algorithm (process or procedure) for determining the product of two binomials such as $9 + x$ and $5 + x$. The sum of the areas in Problem 7 is

$$9 \cdot 5 + 9 \cdot x + x \cdot 5 + x \cdot x.$$

These sums can be obtained from the terms of the binomial factors, as follows:

$$(9 + x)(5 + x) = 9 \cdot 5 + \underline{9 \cdot x} + \underline{x \cdot 5} + \overline{x \cdot x}$$

Combining like terms, you have $45 + 14x + x^2$. This procedure is called the FOIL method, in which, essentially, you multiply each term of the first binomial by each term of the second binomial.

Example 1 *If $f(x) = 3x + 2$ and $g(x) = 2x - 5$, determine $f(x) \cdot g(x)$.*

SOLUTION

$$f(x) \cdot g(x) = (3x + 2)(2x - 5)$$

First, Last, Inner, Outer

$= 3x \cdot 2x + 3x(-5) + 2 \cdot 2x + 2(-5)$

$= 6x^2 - 15x + 4x - 10$

$= 6x^2 - 11x - 10$

Note that the product $(3x + 2)(2x - 5)$ can be represented by the following diagram.

	$2x$	-5
$3x$	$6x^2$	$-15x$
$+2$	$4x$	-10

Therefore, $(3x + 2)(2x - 5) = 6x^2 - 15x + 4x - 10 = 6x^2 - 11x - 10$.

8. a. Given $f(x) = x + 7$ and $g(x) = x + 5$, determine a single expression for $f(x) \cdot g(x)$ by multiplying $(x + 7)(x + 5)$. Write your answer as a sum of terms.

b. Using f and g as defined in part a, complete the following table.

x	f(x)	g(x)	f(x) · g(x)
0			
1			
2			
3			
4			

c. Use the table feature of your grapher to complete the following table for four input values not used in part b.

x	f(x)	g(x)	f(x) · g(x)

d. Are the results found in part c consistent with your algebraic solutions in part b? Explain.

e. When you determine the product function, what do you multiply: domain values, range values, or both? Explain.

Multiply Powers Having the Same Base

The volume, v (in cubic feet), of a partially cylindrical storage tank of liquid fertilizer is represented by the formula

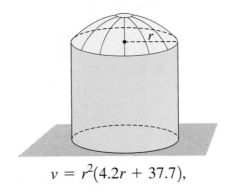

$$v = r^2(4.2r + 37.7),$$

where r is the radius (in feet) of the cylindrical part of the tank.

9. Determine the volume of the tank if its radius is 3 feet.

Suppose you were asked to write the expression $r^2(4.2r + 37.7)$ as an equivalent expression without parentheses. Using the distributive property, you would multiply each term within the parentheses by r^2. The first product is $r^2(4.2r)$. What is r^2 times r?

Recall that in the expression r^2, the exponent 2 tells you that the base r is used as factor two times. In the expression r, the exponent, 1, tells you that the base r is used as a factor once.

$$r^2 \cdot r^2 = \underbrace{r \cdot r \cdot r}_{\substack{\text{base } r \text{ is used as} \\ \text{a factor 3 times}}} = r^3$$

10. a. Complete the following table:

INPUT r	OUTPUT FOR $r^2 \cdot r$	OUTPUT FOR r^3
2		
4		
5		

b. How does the table demonstrate that $r^2 \cdot r$ is equivalent to r^3?

c. Consider the following products.

    i. $x \cdot x^4 = x^5$ **ii.** $w^2 \cdot w^5 = w^7$

  iii. $a^2 \cdot a^3 \cdot a^4 = a^9$

What pattern do you observe?

DEFINITION

Multiplication Property of Exponents

To multiply powers of the same base, keep the base and add the exponents

$$a^m \cdot a^n = a^{m+n}$$

---

11. Expressions for $p(x)$ and $q(x)$ are given in the following table. Fill in the last column of the table with a single power of x.

$p(x)$	$q(x)$	$p(x) \cdot q(x)$
x^2	x^4	
$2x^3$	x	
$-3x^5$	$4x^2$	
$5x^4$	$3x^4$	

12. Multiply $(-2a^5)(8b^3)(3a^2b)$. Explain the steps you used to determined this product.

13. If $f(x) = 2x^2 + 3$ and $g(x) = 5x^3 - 2$, determine $f(x) \cdot g(x)$.

14. Given $f(x) = 4x + 2$ and $g(x) = x^2 - 4x + 3$, determine $f(x) \cdot g(x)$. Note that $g(x)$ has three terms. Therefore, the FOIL method cannot be applied. However, the geometric principle behind the FOIL method can be used. Multiply each term of the first polynomial by each term of the second, and then collect like terms.

SUMMARY
Activity 2.3

1. A common method to multiply two binomials is the FOIL method.

   Step 1. Multiply the FIRST terms in each binomial.

   Step 2. Multiply the OUTER terms.

   Step 3. Multiply the INNER terms.

   Step 4. Multiply the LAST terms.

   Step 5. Sum the products in steps 1–4.

2. To multiply any two polynomials, multiply each term of the first by each term of the second.

3. Given two functions, f and g, the product function is defined by
   $$y = (f \cdot g)(x) = f(x) \cdot g(x).$$

4. To multiply powers of the same base, keep the base and add the exponents. Symbolically, this property of exponents is written as $a^m \cdot a^n = a^{m+n}$

EXERCISES
Activity 2.3

1. **a.** You are drawing up plans to enlarge your square patio. You want to triple the length of one side and double the length of the other side. If x represents a side of your square patio, write an expression for the new area in terms of x.

   b. You discover from the plan that after doubling one side of the patio, you must cut off 3 feet from that side to clear a bush. Write an expression in terms of x to represent the length of this side.

   c. Use the result from part b to write an expression without parentheses to represent the new area of the patio. Remember that the length of the other side of the original square patio was tripled.

2. A rectangular bin has the following dimensions:

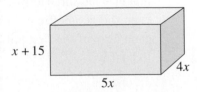

$x + 15$
$5x$
$4x$

a. Write an expression that represents the area of the base of the bin.

b. Using the result from part a, write an expression that represents the volume of the bin.

3. You are the manager of a retail clothing store. It is time to stock up on dress shirts. Your wholesaler has the shirts you want at $7.50 each for up to 50 shirts. However, if you order more than 50, each shirt will cost $0.05 less, for every shirt over 50 that you order, up to 100 shirts. So, 60 shirts will cost $7.00 each, 70 shirts will cost $6.50 each, and so on.

a. Let x be the number of *additional* shirts ordered over 50. Write an expression for the total number of shirts purchased. Assume you are going to purchase at least 50 shirts.

b. Write an expression for the cost per shirt.

c. Determine an equation for the total cost of the shirts, $C(x)$, as a function of the number of shirts, x, you purchase in excess of 50.

d. What is the domain of the cost function if you know that you will not purchase more than 100 shirts? (Be careful.)

e. Using an appropriate window, graph the cost function on your grapher.

f. Rewrite the cost function by multiplying the factors and then combining like terms.

g. Graph the new function from part f, in the same window as part e. What do you see? Compare this graph to the graph in part c.

4. Use the property of exponents $a^m \cdot a^n = a^{m+n}$ to determine the following products.

   a. $3^5 \cdot 3^7$ **b.** $t^4 \cdot t$ **c.** $x^2 y^5$

   d. $(2z^4)(3z^8)$ **e.** $(-2x)(3x^2)(-5x^3)$

   f. $(a^2 b^2)(a^3 b^4)$ **g.** $x^{2n} \cdot x^n$

5. Multiply $(x + 3)(x^2 + 3x - 5)$. Determine the appropriate products to complete the chart. Combine like terms, and write the final answer for this multiplication in descending order of the exponents.

	x^2	$3x$	-5
x	x^3	$3x^2$	
3			-15

6. Multiply $(x^2 + 2x - 3)(2x^2 + 3x - 4)$. Determine the appropriate products to complete the chart. Combine like terms, and write the final answer for this multiplication in descending order of the exponents.

	$2x^2$	$3x$	-4
x^2	$2x^4$	$3x^3$	
$2x$		$6x^2$	$-8x$
-3			12

7. Determine each product, and simplify the result.

   a. $(3x + 2)(2x + 5)$ **b.** $(3x - 2)(2x - 5)$

   c. $(x + 2)(4x - 3)$ **d.** $(x - 2)(4x + 3)$

8. Determine the following products, and simplify the results.

   a. $(2x + 5)(x - 3)$ **b.** $(4x + 3)(3x - 2)$

   c. $(x + 2)(x^2 + 4x - 3)$ **d.** $(4 - 3x + x^2)(2x^2 + x)$

e. $(x - 3)(2x^2 - 5x + 1)$ **f.** $(x - 4)(4 - x^2)$

g. $(x^2 - 3x + 1)(3x^2 - 5x + 2)$ **h.** $(2x^2 + 5x)(6 - 2x)$

9. a. Multiply $(3x - 2)^2$. Note that $(3x - 2)^2 = (3x - 2)(3x - 2)$.

b. Multiply $(5x + 2)(5x - 2)$.

c. Multiply $(x^2 + 5)(x^2 - 5)$.

d. After simplifying in parts b and c, the product contains only two terms. Explain why. HINT: Compare the first terms to each other and the second terms to each other.

10. a. Given $f(x) = x + 1$ and $g(x) = 2x - 3$, determine $f(x) \cdot g(x)$ by multiplying and combining like terms.

b. Use f and g as defined in part a to complete the following table.

x	f(x)	g(x)	f(x) · g(x)
0			
1			
2			
3			
4			

c. Use your grapher to plot all three functions, f, g, and the product of f and g. Use the trace or table feature to complete the following table for four input values not used in part b.

x	f(x)	g(x)	f(x) · g(x)
6	7	9	63
7	8	11	88
8	9	13	117
9	10	15	150

ACTIVITY 2.4

Stargazing

OBJECTIVES

1. Convert scientific notation to decimal notation.

2. Convert decimal notation to scientific notation.

3. Apply the property of exponents to divide powers having the same base.

4. Apply the property of exponents $a^0 = 1$, where $a \neq 0$.

5. Apply the property of exponents $a^{-n} = \frac{1}{a^n}$, where $a \neq 0$, and n is any real number.

On any clear evening, the sky is filled with millions of stars. Some of these are closer to Earth than others. Some are large. Some are small. All of them send light to us. The speed of light through the universe is constant. Light travels at a speed of 300,000 km per second.

1. How many kilometers does light travel in one minute?

2. How many kilometers does light travel in one hour?

3. How many kilometers does light travel in one day?

The result of Problem 3 displayed on the TI-83 Plus is 2.592E10. This is the way your calculator displays a very large number in **scientific notation**.

DEFINITION

Scientific notation is a convenient way to write a very large (or small) number. A positive number is written in scientific notation as a number (the base) between 1 and 10 times a power of 10. A negative number is written as a number (the base) between -10 and -1 times a power of 10.

Example 1   *Convert* 3.2×10^3 *and* -9.8×10^7 *from scientific notation to decimal notation.*

SOLUTION

$3.2 \times 10^3 = 3.2 \times 1000 = 3200$

$-9.8 \times 10^7 = -9.8 \times 10,000,000 = -98,000,000$

Notice that to convert a number whose absolute value is greater than 1 from scientific notation to decimal notation, you move the decimal point of the base to the right, the number of decimal places indicated by the exponent of the power of 10.

4. Convert the following numbers from scientific notation to decimal notation.

   a. $2.23 \times 10^4 =$

   b. $-4.78 \times 10^6 =$

   c. $8.37 \times 10^{12} =$

You can use the TI-83 Plus to check your answers. To input 2.23×10^4, type in the base, 2.23, and then press (2nd) (,) to access the EE command, then (4) (ENTER). Your display should look like this:

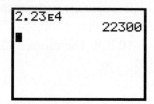

The EE button, which displays as E on the screen, is followed by the exponent of the power of 10.

The calculator display for converting -4.78×10^6 is:

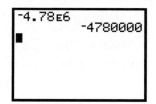

Because the calculator displays at most 10 digits, it will not convert 8.37×10^{12} to a decimal number. Try it.

Decimal Notation to Scientific Notation

In English units, light travels approximately 5,880,000,000,000 miles in one year. This distance is called the light-year. If you enter 5,880,000,000,000 into the calculator, it converts the number to scientific notation and returns 5.88E12.

5. a. Write the number 5,880,000,000,000 in scientific notation.

    b. Describe the process the calculator used to convert 5,880,000,000,000 into 5.88E12.

In general, to convert a number whose absolute value is greater than 1 from decimal notation into scientific notation, move the decimal point to the immediate right of the first nonzero digit. Count the decimal places moved. This number is the exponent on the power of 10.

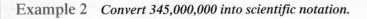

Example 2 *Convert 345,000,000 into scientific notation.*

SOLUTION

Because the first nonzero digit is 3, the decimal part of the number is 3.45. Because the decimal point needs to be moved 8 places to produce 3.45, the exponent on the 10 is 8. Therefore, $345,000,000 = 3.45 \times 10^8$ in scientific notation.

6. Convert the following decimal numbers into scientific notation.

    a. 7,605,000,000,000   **b.** $-98,300,000$

Division Property of Exponents

You know that light travels at a speed of approximately 300,000 kilometers per second or approximately 25,920,000,000 kilometers per day. The nearest star to the earth other than the Sun is Proxima Centauri, which is approximately 39,740,000,000,000 kilometers from Earth. How long does it take light from Proxima Centauri to reach Earth?

The answer to the question is determined by dividing 39,740,000,000,000 kilometers by 25,920,000,000 kilometers per day. If you do this with your calculator, it produces the following screen.

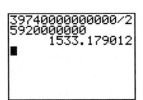

7. Change the mode of the calculator to Scientific Notation mode (Sci).

Now divide 39,740,000,000,000 kilometers per day by 25,920,000,000 kilometers by entering both numbers into your calculator in scientific notation. How does your result compare to the result performed in Normal mode?

8. Look at this calculation in scientific notation more closely:

$$\frac{3.974 \times 10^{13}}{2.592 \times 10^{10}} = 1.533179012 \times 10^3.$$

a. The 1.533179012 is a result of dividing 3.974 by 2.592. Check this result with your calculator.

b. The 10^3 is a result of dividing 10^{13} by 10^{10}. This means that $\frac{10^{13}}{10^{10}} = 10^3$. Note that

$$\frac{10^{13}}{10^{10}} = \frac{10 \times 10 \times 10 \times 10 \times 10 \times 10 \times 10 \times 10 \times 10 \times 10 \times 10 \times 10 \times 10}{10 \times 10 \times 10 \times 10 \times 10 \times 10 \times 10 \times 10 \times 10 \times 10} = 10^3$$

Rather than expanding the powers and dividing out the common factors, how could you obtain the exponent 3 from the exponents of the powers you are dividing?

Problem 8b demonstrates another important property of exponents.

Division Property of Exponents

To divide powers of the same base, keep the base and subtract the exponents $\frac{a^m}{a^n} = a^{m-n}$, where $a \neq 0$.

Example 3 **a.** $\frac{3^5}{3^2} = 3^{5-2} = 3^3$ **b.** $\frac{x^{15}}{x^9} = x^{15-9} = x^6$

c. $\frac{15t^6}{3t^2} = \frac{15}{3} \times \frac{t^6}{t^2} = 5 \times t^{6-2} = 5 \times t^4$

d. $\frac{-4a^7}{2a^5} = -2a^2$

e. $\frac{x^9}{y^5}$ *cannot be simplified because the bases x and y are different.*

9. Use the division property to simplify the following expressions.

a. $\dfrac{8^9}{8^4} =$

b. $\dfrac{x^6}{x} =$

c. $\dfrac{10w^8}{4w^5} =$

d. $\dfrac{6t^{13}}{2t^7} =$

e. $\dfrac{5^2}{5^2} =$

The result of Problem 9e using the division property is $\frac{5^2}{5^2} = 5^{2-2} = 5^0$. If you did the same problem by first writing 5^2 as 25, the result would be $\frac{5^2}{5^2} = \frac{25}{25} = 1$. Therefore, it must be true that $5^0 = 1$. In the same way it can be shown that $2^0 = 1$, $10^0 = 1$, etc.

Zero Exponents

$a^0 = 1$ if $a \neq 0$.

Example 4   **a.** $16^0 = 1$ **b.** $\left(\frac{3}{x}\right)^0 = 1$, $x \neq 0$

c. $(3x)^0 = 1$ provided that $x \neq 0$. Note that $3x^0 = 3 \cdot 1 = 3$.

d. $5(x + 3)^0 = 5$

10. Simplify the following expressions. Assume $x \neq 0$.

a. 7^0 **b.** $2x^0$ **c.** $(5x)^0$

d. $\left(\dfrac{4}{x}\right)^0$ **e.** $-3(x^2 + 4)^0$

Negative Integer Exponents

You know that light travels at a rate of 25,920,000,000 kilometers per day. The Sun is 149,600,000 kilometers from Earth. You have determined that it takes 1533 days for light to travel from the second nearest star, Proxima Centauri, to Earth. How many days does it take for light from the nearest star, the Sun, to travel to Earth?

To answer this question, divide 149,600,000 kilometers by 25,920,000,000 kilometers per day. If you convert both numbers to scientific notation and work in Sci mode on your calculator, your results should resemble the following.

```
1.496E8/2.592E10
     5.771604938E-3
```

This says that light travels from the Sun to Earth in 5.772×10^{-3} days. If you perform the same calculation in Normal mode, your results should resemble the following.

```
1.496E8/2.592E10
     5.771604938E-3
1.496E8/2.592E10
       .0057716049
■
```

Therefore, you know that 5.772×10^{-3} days $= 0.005772$ days.

11. Describe in your own words how to convert a number such as 5.772×10^{-3} (written in scientific notation) to its equivalent representation 0.005772 (written in decimal notation).

Therefore, to convert a number in scientific notation with a negative exponent, n, to decimal notation, move the decimal point $|n|$ places to the left.

Example 5   *Convert* 6.3×10^{-3} *and* -17.7×10^{-4} *to decimal notation.*

SOLUTION

a. $6.3 \times 10^{-3} = 0.0063$

b. $-17.7 \times 10^{4} = -0.00177$

You can check these results by entering the numbers into your calculator in scientific notation as long as your calculator is in Normal mode.

12. Convert the following to decimal notation.

a. $5.61 \times 10^{-5} =$

b. $9.071 \times 10^{-7} =$

Look again at 5.772×10^{-3} days $= 0.005772$ days. You know that

$$0.005772 = 5.772 \times 0.001$$

$$= 5.772 \times \frac{1}{1000}$$

$$= 5.772 \times \frac{1}{10^3}$$

Therefore, $5.772 \times 10^{-3} = 5.772 \times \frac{1}{10^3}$. It follows that $10^{-3} = \frac{1}{10^3}$. This is another specific case of a more general property of exponents.

Negative Exponents

If $a \neq 0$ and n is a real number, then $a^{-n} = \dfrac{1}{a^n}$.

Example 6   *Rewrite the following expressions using only positive exponents.*

a. $3^{-4} = \frac{1}{3^4} = \frac{1}{81}$

b. $(2x)^{-3} = \frac{1}{(2x)^3} = \frac{1}{8x^3}$

c. $x^{-1} = \frac{1}{x}$

d. $\frac{1}{x^{-4}} = x^4$

e. $3y^{-2} = \frac{3}{y^2}$

f. $\dfrac{-2a^{-3}}{b^{-2}} = \dfrac{-2b^2}{a^3}$

g. $x^{-3} \cdot x^{-5}$

Method 1. Apply the multiplication property of exponents first.

$$x^{-3} \cdot x^{-5} = x^{-8} = \frac{1}{x^8}$$

Method 2. Apply the definition of negative exponents first.

$$x^{-3} \cdot x^{-5} = \frac{1}{x^3} \cdot \frac{1}{x^5} = \frac{1}{x^8}$$

h. $\frac{x^{-3}}{x^4}$

Method 1. Apply the division property of exponents first.

$$\frac{x^{-3}}{x^4} = x^{-3-4} = x^{-7} = \frac{1}{x^7}$$

Method 2. Apply the definition of negative exponents first.

$$\frac{x^{-3}}{x^4} = \frac{1}{x^3 \cdot x^4} = \frac{1}{x^7}$$

13. Rewrite the following expressions using positive exponents only.

a. 5^{-3}

b. $(2z)^{-4}$

c. $6y^{-5}$

d. $\frac{4}{x^{-1}}$

e. $\left(\frac{x}{y}\right)^{-3}$

f. $\frac{x^3}{y^{-4}}$

g. $x^{-4} \cdot x^{-2}$

h. $\frac{a^{-2}}{a^{-5}}$

SUMMARY
Activity 2.4

1. In scientific notation, a positive number is written as a number (the base) between 1 and 10 times a power of 10. A negative number is written as a number (the base) between -10 and -1 times a power of 10.

2. To convert a number whose absolute value is greater than 1 from scientific notation to decimal notation, you move the decimal point of the base to the right, the number of decimal places indicated by the exponent of the power of 10.

3. To convert a number whose absolute value is greater than 1 from decimal notation into scientific notation, move the decimal point to the immediate right of the first nonzero digit. Count the decimal places moved. This number is the exponent on the power of 10.

4. To divide powers of the same base, keep the base and subtract the exponents $\frac{a^m}{a^n} = a^{m-n}$, where $a \neq 0$.

5. $a^0 = 1$, where $a \neq 0$.

6. To convert a number in scientific notation with a negative exponent, n, to decimal notation, move the decimal point $|n|$ places to the left.

7. If $a \neq 0$ and n is a real number, then $a^{-n} = \frac{1}{a^n}$.

EXERCISES
Activity 2.4

1. The total currency in circulation in the United States is approximately $566,075,000,000. Write this number using scientific notation.

   (*Source*: Financial Management Service; U.S. Dept. of the Treasury, June 29, 2001)

2. According to the International Telecommunication Union, the estimated number of cell phone subscribers in the world was 485,040,000 in 2000.

   a. Write the number of cell phone subscribers in the world in scientific notation.

   b. In Estonia there are 6×10^5 subscribers. Write this number of cell phone subscribers in standard notation.

   c. The number of cell phone users in the U.S. approximately doubled from 1997 to 2000. The information was given in this form: 2000: 1.09×10^8 and 1997: 5.5×10^7. Using the rules of exponents, show, algebraically, how you determine that the number has approximately doubled.

3. **a.** A sextillion has 21 zeros. Write three sextillion in scientific notation.

   b. The number 45,000,000,000,000,000 is read 45 quadrillion. Write this number in scientific notation.

   c. Write the number 9×10^{27} in standard notation. The number will be read 9 octillion.

   d. Use scientific notation to divide 9 octillion by 45 quadrillion. Use the rules of exponents.

4. **a.** One square inch is equivalent to approximately 0.000000159423 acre. Write this number in scientific notation.

b. 5.78704×10^{-4} cubic feet is equivalent to 1 cubic inch. Write this number in standard notation.

5. The amount of federal acreage in the U.S. is approximately 6.35×10^8. The total acreage in the U.S. is approximately 2.27×10^9.

  a. Use scientific notation to write the ratio of the federal acreage to the total acreage.

  b. Use the rules of exponents to simplify your answer.

In Exercises 6–20, use the properties of exponents to simplify the following, where $x \neq 0$. Write your results with positive exponents only.

6. $\dfrac{3^5}{3^2}$

7. $\left(\dfrac{6}{x}\right)^0$

8. 2^{-5}

9. $10x^0$

10. $\left(\dfrac{1}{x}\right)^{-2}$

11. $4x^{-4}$

12. $(2x)^{-3}$

13. $\dfrac{6x^8}{3x}$

14. $\dfrac{9x^8}{3x^{12}}$

15. $\dfrac{6x^3y^5z^2}{10x^7yz^2}$

16. $x^{-3} \cdot x$

17. $(3x^{-2})(x^{-3})x^2$

18. $\dfrac{10x^4}{5x^{-3}}$

19. $\dfrac{4a^0b^{-4}}{-8a^2b^{-1}}$

20. $a^{-3}(4a^{-1})(-5a^7)$

CLUSTER 1   ## What Have I Learned?

1. The graphs of functions f, g, and h appear in the accompanying figure.

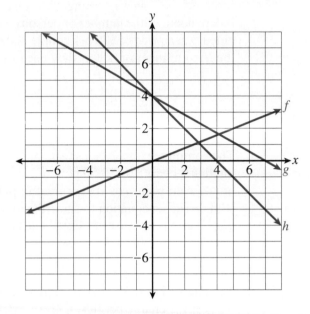

   a. Which function represents the sum of the other two functions?

   b. Which function represents the difference between the other two functions?

2. If the output of the function $(f - g)$ is constant, what must be true about the graphs of the functions f and g?

3. Given $f(x) = 2x - 3$ and $g(x) = 4$, for what values of x is $(f + g)(x) = (f \cdot g)(x)$?

4. Given the defining equations for two functions, describe how to find the output of the product function for a particular input value.

5. Explain the difference between 3^4 and 3^{-4}.

6. What will be the sign of the answer if you raise a positive base to a negative exponent?

How Can I Practice?

1. a. The Snowbelt Quilters Guild is planning a holiday dinner. A banquet room has been reserved, and catering arrangements have been made for a total of $600, a fixed fee independent of the number of persons who attend. The planning committee has decided that a price of $20 per couple is the most they will charge for tickets. What is the *least* number of tickets they need to sell to break even?

b. The committee also decides that once they have met their expenses, they will reduce the ticket charge by $0.50 per couple for each additional ticket (couple) above the break-even point you determined in part a. Let t represent the *additional* tickets sold. Write an equation to show the total number, N, of couples attending the banquet (i.e., the number of tickets sold) as a function of t. Call this function f.

c. The charge per ticket can be represented by the function $C = g(t)$. Write an equation for the charge per ticket, C.

d. Complete the following table for each of the functions $N = f(t)$ and $C = g(t)$.

t	0	2	4	6	8	10
$N = f(t)$						
$C = g(t)$						

e. The total revenue obtained from the ticket sales is the total number of tickets sold multiplied by the charge per ticket. Use the output values from part dc to complete the following table for the total revenue function, $R(t) = N \cdot C = f(t) \cdot g(t)$.

t	0	2	4	6	8	10
$R(t)$						

f. Use the results from parts b and c to determine a symbolic rule for $R(t)$.

g. Use your grapher to graph the total revenue function, $R(t)$, on the accompanying grid. What window values, Xmin, Xmax, Ymin, Ymax, do you use?

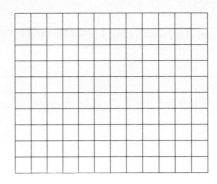

h. Determine the maximum revenue that can be obtained.

i. What is the total number of tickets that must be sold to obtain this maximum revenue? (Be careful.)

2. If $f(x) = x + 2$ and $g(x) = 2x - 3$, determine the following.

a. $f(x) + g(x)$ **b.** $f(x) - g(x)$

c. $f(x) \cdot g(x)$ **d.** $f(3) - g(3)$

e. $f(-2) \cdot g(-2)$ **f.** $3 \cdot f(x)$

3. If $f(x) = x^2 - 2x + 1$ and $g(x) = x^2 + x - 4$, determine each of the following.

a. $f(x) - g(x)$ **b.** $f(x) \cdot g(x)$

c. $f(-1) + g(-1)$ **d.** $2f(x) - 3g(x)$

4. Perform the indicated operations and simplify.

a. $(4x + 5) + (x - 7)$ **b.** $(x^2 - 3x + 1) + (x^2 + x - 9)$

c. $(x + 4) - (3x - 8)$ **d.** $(3x^2 - 4x - 5) - (x^2 + 9x + 3)$

e. $(5x^2 + 6x - 1) - (7x - 3)$

5. Perform the indicated operations and simplify.

a. $x \cdot x^3$ **b.** $x^4 \cdot x^5$

c. $(2x^6)(3x^2)$ **d.** $(xy^3)(x^4y^3z)$

e. $(2x^4y^5z^7)(5x^2z)$ **f.** $(-3a^2b)(-2a)(-5a^2b^2)$

6. Perform the indicated operations and simplify.

a. $(x - 2)(x - 5)$ **b.** $(4x - 3)(x + 7)$

c. $(2x - 3)(2x + 3)$ **d.** $(x - 2)(x^2 + 3x - 5)$

e. $(2x + 1)(x^2 - x + 2)$ **f.** $3(x - 7) - 2(x^2 + 4x)$

g. $2x(x + 5) - 3x(4 - 3x)$ **h.** $3x^2 - (x^3 + 1) - x(x^4 - 2)$

i. $(3x + 5)^2$ **j.** $(2x - 7)^2$

k. $(x + 4)^3$ **l.** $(5x - 7)(5x + 7)$

7. The tuition at a local state college was $2650 in 1990. Since then, tuition has increased each year approximately $75 per year. Let t represent the years since 1990 and $f(t)$ represent the tuition for any given year. At this same college, the cost for room and board was $1500 in 1990, and it has increased at rate of $125 per year since 1990. Let $g(t)$ represent the cost of room and board for any given year since 1990. The college fees have not changed over the years. They remain $50. Let $h(t)$ represent the college fees.

a. Complete the following input/output table for the tuition function, $f(t)$.

t, YEARS SINCE 1990	0	5	10	15	20
$f(t)$, COST OF TUITION ($)					

b. Write an equation that will give the cost of tuition for any year beginning in 1990.

c. Complete the following input/output table for the room and board, $g(t)$.

t, YEARS SINCE 1990	0	5	10	15	20
$g(t)$, COST OF ROOM AND BOARD ($)					

d. Write an equation that will give the cost of room and board for any year beginning in 1990.

e. Complete the following input/output table for the fees, $h(t)$.

t, YEARS SINCE 1990	0	5	10	15	20
$h(t)$, COST OF FEES ($)					

f. Write an equation that will give the cost of fees for any year beginning in 1990.

g. Use the equations in parts b, d, and f to determine a function, k, that will give the total cost of attending this college for any year after 1990. Write the function in simplest form.

h. Use your grapher to graph functions f, g, h, and k. Use the trace or table feature to complete the following table for four input values.

t, YEARS SINCE 1990	$f(t)$	$g(t)$	$h(t)$	$k(t)$
3				
12				
18				
25				

i. If the increase continues at the same rate, in what year will the tuition first equal or exceed $10,000?

8. Suppose f and g are defined by the following tables.

x	-4	-2	0	2	4	6
$f(x)$	49	9	-7	1	33	89

x	-4	-2	0	2	4	6
$g(x)$	32	12	0	-4	0	12

a. Complete the following table for $f - g$.

x	-4	-2	0	2	4	6
$(f - g)(x)$						

b. If $f(x) = 3x^2 - 2x - 7$ and $g(x) = x^2 - 4x$, determine an algebraic expression for $(f - g)(x)$.

c. Check your answers in the table in part a by using the function you found in part b.

9. Use the properties of exponents to simplify the following, where $x \neq 0$. Write your results with positive exponents only.

a. $2x^{-3}$

b. $(-3x)^2$

c. 3^{-4}

d. $\left(\dfrac{x}{5}\right)^0$

e. $\dfrac{4x^9}{2x^2}$

f. $(-2x^3y^6)(3xy^2)$

g. $4x^0$

h. $\dfrac{8x}{12x^5}$

i. $\dfrac{10xy^3z^4}{2x^5yz^4}$

j. $(3x^{-2})(-5x^{-4})(x)$

k. $\dfrac{-4x^3}{2x^{-3}}$

l. $\dfrac{a^4b^{-2}c^{-5}}{a^0b^2}$

10. a. The Exxon Valdez spilled oil in Prince William Sound, Alaska, when it ran aground in March 1989. It is estimated that 1.008×10^7 gallons of oil spilled. Write this number in standard form.

b. For insurance purposes you need to determine how many square miles of hunting land you own. One square foot is equivalent to 0.00000003587006 square miles. If your hunting land measures approximately 6,000,000 square feet, how many square miles is this? Use scientific notation and the rules of exponents to determine how many square miles of land you own.

c. The mean radius of the largest planet, Jupiter, is 43,441 miles. The formula for determining the volume of a sphere, V, is

$$v + \frac{4}{3}\pi r^3$$

Determine the approximate volume, in cubic miles, of Jupiter. Express your answer in scientific notation rounded to four decimal places.

11. In the year 2000, there were approximately 275,000,000 people living in the United States. In the same year, according to Gartner Group's Dataquest research, there were 86.1 million subscribers of cellular phones.

a. Write each of these numbers in scientific notation.

b. Use the numbers in scientific notation to estimate how many (say, one in approximately _____) of the people living in the U.S. are cell phone subscribers.

12. In 1999 Yale University's endowment was approximately $7,198,000 and Columbia University's was $3,636,000. By using scientific notation and rounding to two places, show that Yale's endowment is double the endowment at Columbia. Which property of exponents did you use in your explanation?

13. You have planned to put in a rectangular patio that measures 5 feet by 7 feet. However, you neglected to include enough seating room around your patio table. Let x be the number of additional feet you will extend your plan in each direction.

a. Determine a formula for the area of the extended patio.

b. If x is 4 feet, by how much have you increased the area of the patio from that of the original plan?

c. Your patio table is round with a radius of r feet. You need to purchase an umbrella for the table with an overhang of 2 feet all around. Write an expression for the area that the umbrella will cover.

CLUSTER 2 **Composition and Inverse of Functions**

ACTIVITY 2.5
Inflated Balloons

OBJECTIVES

1. Determine the composition of two functions.

2. Explore the relationship between $f(g(x))$ and $g(f(x))$.

The volume of an inflated balloon increases as the air temperature rises. The following table shows the data from experimental measurements for a particular balloon.

TEMPERATURE (in °F)	32	39	42	45	50	58	63	68
VOLUME (in cu in.)	35.1	36.5	37.1	37.7	38.7	40.3	41.3	42.3

1. Treating volume as a function of temperature, is this relationship a linear function? Describe how you determined your answer.

Appendix

2. Use your graphing calculator to sketch a scatterplot of these data points. Refer to Appendix C for procedures to plot a set of data on the TI-83 Plus. Your screen should resemble the following.

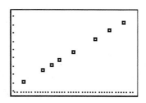

Does your plot verify your answer to Problem 1?

3. **a.** Use the table of values and/or your grapher to determine the equation for this function. Use V for the output variable and F for the input variable. Call the function g so that $V = g(F)$.

   b. Use your graphing calculator to sketch the graph of the volume function, g. Explain how you can be reasonably sure that your function is correct.

   c. Use the volume function, g, to determine the volume of the balloon when the temperature is 55°F.

4. Suppose you only have a Celsius thermometer and you want to know the volume of the balloon when the temperature is 10°C. To use the volume function from Problem 3, you must first convert degrees Celsius to degrees Fahrenheit.

a. The formula $F = 32 + 1.8C$ is used to convert from degrees Celsius, C, to degrees Fahrenheit, F. Note that the formula defines F as a function of C. Therefore, F can be written as $h(C)$, where h is the name of the function. Use the given formula to determine the Fahrenheit temperature equivalent to 10°C. That is, determine $h(10)$.

b. Now, use the result from part a to determine the volume for 10°C.

5. You have two functions. For the first function defined by $V = g(F) = 0.2F + 28.7$, F is the input, and $V = g(F)$ is the output. In the second function defined by $F = h(C) = 32 + 1.8C$, C is the input and $F = h(C)$ is the output.

Complete the following table using a combination of these functions. (Note that the temperature is given in Celsius units)

TEMPERATURE (in °C)	0	10	20	30	40
VOLUME (in cu in.)					

In calculating the volumes in Problem 5, you followed a two-step calculation. First you used the degrees Celsius, C, as the input to $F = g(C) = 32 + 1.8C$, to convert the temperature to degrees Fahrenheit. Then you used this output, degrees Fahrenheit, as the input to the function, $V = f(F) = 0.2F + 28.7$, to obtain the volume, V. To shorten this calculation, you can combine these two functions in a special way as described in Problem 6.

6. a. Substitute the expression $32 + 1.8C$ for F in $V = 0.2F + 28.7$, and simplify. You have just determined an equation for V as a function of C.

b. Use the equation in part a and your graphing calculator to verify the table of values in Problem 5.

Using function notation to describe the procedures in Problem 6a, you started with $F = h(C)$. Then, substituting $h(C)$ for F in the second function, $V = g(F)$, you have

$$V = g(F) = g(h(C))$$

This function of a function is called the **composition of g and h**.

Example 1   *Given $f(x) = 2x + 3$ and $g(x) = 4x - 1$, determine $g(f(x))$.*

SOLUTION

Substitute $2x + 3$ for $f(x)$ in $g(f(x))$.

$g(f(x)) = g(2x + 3)$ Replace x in the function rule for $g(x)$ with the expression $2x + 3$.

$\quad\quad\quad = 4(2x + 3) - 1$

$\quad\quad\quad = 8x + 12 - 1$ Simplify.

$\quad\quad\quad = 8x + 11$

Therefore, $g(f(x)) = 8x + 11$.

7. a. If $f(x) = 2x + 1$ and $g(x) = 4x - 3$, determine an equation for $f(g(x))$.

b. Use the result from part a to determine $f(g(3))$.

c. Determine an equation for $g(f(x))$.

d. Does $f(g(x)) = g(f(x))$?

SUMMARY
Activity 2.5

1. If x is the input of a function g, the output is $g(x)$. If $g(x)$ is then used as the input of a function f, the output is $f(g(x))$. The result is a function h, defined by $h(x) = f(g(x))$. The function h is the **composition** of the functions f and g.

2. In general, $f(g(x)) \neq g(f(x))$.

EXERCISES
Activity 2.5

1. Oil is leaking from a tanker and is spreading outward in the shape of a circle. The area, A, of the oil slick is a function of radius, r (in feet), and is given by $A = f(r) = \pi r^2$. The input, r, for the area function is itself a function of time, t (in hours), since the oil began leaking.

   a. If $r = g(t) = 100t$, determine $g(2)$, and interpret its meaning in this situation.

   b. Determine $f(g(2))$, and interpret its meaning in this situation.

   c. Determine $f(g(10))$, and interpret its meaning in this situation.

   d. Determine a general expression for $f(g(t))$.

   e. Determine the area of the circular oil spill after 10 hours using the new composite area function found in part d, and compare the result with your answer to part c.

2. If $V = g(r) = \frac{4}{3}\pi r^2$, and $r = f(t) = t + 1$, find $g(f(t))$.

3. If $s = u(t) = -2t^2 + 2t + 1$, and $t = v(x) = 3x - 1$, determine each of the following.

   a. $u(v(x))$

   b. $v(u(t))$

4. Use the first two tables to complete the third.

x	1	2	3	4	5	6
$f(x)$	2	-1	5	7	19	4

x	-1	2	4	5	7	19
$g(x)$	0	-3	4	1	5	12

x	1	2	3	4	5	6
$g(f(x))$						

5. You read about the safety features on your brand new car. In a 30-mile-per-hour collision, according to *Consumer Reports*, the seat belt locks properly about 99% of the time. In approximately 90% of such collisions, the air bag will successfully deploy. Let x represent the number of 30-mile-per-hour collisions in cars of the same year and model that you purchased.

   a. Write an equation for $L(x)$ that represents the number of collisions in which the seat belt locks.

   b. Write an equation for $D(x)$ that represents the number of collisions in which the air bag deploys.

   c. If we suppose that every occupant survives only when everything works as it should in the 30-mile-per-hour collision—that is, the seat belts lock and air bags deploy— write an equation for $S(x)$ that represents the number of survivors of x collisions. *Hint:* Use the results of parts a and b to write a composite function.

d. Evaluate $S(500)$.

e. Examine $L(x)$ and $D(x)$, and decide which of the safety features should be improved immediately to increase the number of survivors, $S(x)$.

6. A quality-control inspector at a bottler of carbon-filtered drinking water notes that the first six bottles processed each day are not acceptable because they are not properly labeled. After those first six bottles, the labeler is warmed up and works just fine for the rest of the shifts. After labeling, the bottles are filled, and caps are put in place. Caps are properly applied approximately 99% of the time. After capping, the bottles are inspected. Let x represent the number of bottles processed in a day.

a. Write an equation for $f(x)$ that represents the number of bottles that are properly labeled.

b. Write an equation for $g(x)$ that represents the number of bottles that are appropriately capped.

c. Now find $f(g(10,000))$, and interpret its meaning in this situation.

ACTIVITY 2.6

Finding a
Bargain

OBJECTIVE

1. Solve problems using the
   composition of functions.

You have been waiting for the best price for a winter coat. You see the following advertisement.

SUPER SUNDAY
60–70% OFF

original price
when you take
an additional 40% off
already reduced prices

This is it! The time is right! Last week the coat was on sale for 25% off the original price, and now you can get the coat for 40% less than last week's sale price.

1. a. Complete the following table.

ORIGINAL COST ($) OF THE COAT, x	LAST WEEK'S PRICE ($)	TODAY'S PRICE ($)
80	60	36
100		
120		
140		

   b. What percent of the original price would you have paid for the coat during last week's sale?

   c. Write an equation that gives last week's sale price, y, as a function of the original price, x. Therefore, y represents the sale price after the first reduction.

2. The ad indicates that you can now save an additional 40%.

   a. What percent of last week's sale price would you now have to pay?

   b. Write an equation that gives this week's sale price, z, as a function of last week's sale price, y. Therefore, z represents the sale price after the second reduction.

3. a. The original price of the coat was $160. Use the result from Problem 1c to determine the price y after the first reduction.

b. Use the result from Problem 2b to determine the price z after the second reduction.

c. What percent of the original $160 price did you pay?

d. What is the total discount as a percent off the original price?

4. The salesclerk couldn't get her register to accept your credit card. The manager told her to use a different register. In her frustration, the sales clerk entered the discount as a 65% reduction.

a. Why did she take a 65% reduction?

b. Is this okay with you? Explain.

c. Would this be okay with the manager of the department? Explain.

Determining the final sale price of the coat requires a sequence of two calculations:

First: $y = f(x) = 0.75x$ (With 25% off, you pay 75% of the cost.)

Second: $z = g(y) = 0.60y$ (With 40% off, you now pay 60% of the reduced price.)

5. a. Substitute the expression $0.75x$ for y into the second equation to determine the single-step equation for this composition function.

b. Your answer to part a should imply a savings of 55% from the original price. Does it? Explain.

c. Using function notation, you can determine the equation of the composition function as follows:

$$z = g(y) = g(f(x))$$

Determine $g(f(x))$. How does this compare with the final equation in part a?

SUMMARY
Activity 2.6

Taking a discount of a discount is an example of composition of functions denoted by $g(f(x))$, where

1. f is the function that calculates the first reduction.

2. x is the original price and $f(x)$ is the price after the first reduction.

3. g is the function that calculates the second reduction.

4. $g(f(x))$ is the price after the second reduction has been applied to the reduced price, $f(x)$.

EXERCISES
Activity 2.6

1. **a.** Let x represent the number of miles traveled. Write a function rule that converts x miles to an equivalent number of feet (1 mile = 5280 feet). Call this function f.

   b. Use the function in part a to convert 5 miles to an equivalent number of feet.

   c. Write a function rule to convert w feet to an equivalent number of inches. Call this function g.

   d. Use the function in part c to convert the number of feet found in part b to inches.

   e. Determine $g(f(x))$. This composition will combine the two-step process in parts b and d into one.

   f. Determine $g(f(5))$. Interpret your answer.

2. Function f gives the approximate percent increase in harmful ultraviolet rays for an x percent decrease in the thickness of the ozone layer.

x	0	1	2	3	4	5	6
f(x)	0	1.5	3.0	4.5	6.0	7.5	9.0

Function g gives the expected percent increase in cases of skin cancer for a p percent increase in ultraviolet radiation.

p	0	1.5	3.0	4.5	6.0	7.5	9.0
$g(p)$	0	5.25	10.5	15.75	21.0	26.75	31.5

In Exercises a–d, determine the output value, and interpret your answer.

a. $f(3)$

b. $g(4.5)$

c. $g(f(3))$

d. $g(f(6))$

e. What does $g(f(x))$ determine? That is, what does the output for $g(f(x))$ represent in this situation?

f. Complete the following table.

x	0	1	2	3	4	5	6
$g(f(x))$							

3. A car dealership advertises a factory rebate of $1500 and a 10% discount.

a. Let x represent the price of the car. Let $f(x)$ represent the price of the car after the rebate. Determine a rule for $f(x)$.

b. If $g(x)$ represents the price of the car after the 10% discount, determine a rule for $g(x)$.

c. Determine $g(f(20{,}000))$, and interpret your answer.

d. Suppose the price of a car is $20,000. Determine $f(g(\$20{,}000))$, and interpret your answer.

e. Compare the sale price obtained by subtracting the rebate first and then taking the discount, with the sale price obtained by taking the discount first and then subtracting the rebate.

4. You drop a pebble off a bridge. Ripples move out from the point of impact as concentric circles. The radius (in feet) of the outer ripple is given by

$$R = f(t) = 0.5t,$$

where t is the number of seconds after the pebble hits the water. The area, A, of a circle is a function of its radius and is given by

$$A = g(r) = \pi r^2.$$

a. Determine a formula for $g(f(t))$.

b. What are the input and output for the function defined in part a?

5. The functions f and g are defined by the following tables.

x	-3	-2	0	1	4	5	8	10	12
$f(x)$	8	6	3	2	5	8	11	15	20

x	0	2	3	4	5	8	9	11	15
$g(x)$	1	3	5	10	4	2	0	-2	-5

Determine the values for each of the following.

a. $f(g(3))$

b. $g(f(4))$

c. $f(g(4))$

d. $g(f(-3))$

ACTIVITY 2.7
The Square of a Cube

OBJECTIVES

1. Apply the property of exponents to simplify an expression involving a power to a power.

2. Apply the property of exponents to expand the power of a product.

3. Determine the nth root of a real number.

4. Write a radical as a power having a rational exponent and as a rational exponent having a power.

Suppose that one function squares the input x: $f(x) = x^2$. Further suppose that a second function cubes the input x: $g(x) = x^3$.

1. a. Determine each of the following.

    i. $f(3)$ **ii.** $g(4)$

    iii. $g(f(2))$

b. The general composition function in part a iii is written $g(f(x))$. Write the expression for $g(f(x))$ as a single power of x.

c. Use the expression from part b to determine $g(f(2))$, and compare the result to your answer from part a.

d. Now determine $f(g(x))$.

e. How does $f(g(x))$ compare to $g(f(x))$?

The composition functions you determined in Problems 1b and 1d illustrate an important property of exponents.

> **Property of Exponents (Power to a Power)**
>
> If a is a real number and m and n are integers, then
>
> $$(a^n)^m = (a^m)^n = a^{mn}.$$

Example 1   **a.** $(3^2)^3 = 3^{2 \cdot 3} = 3^6 = 729$; $(3^3)^2 = 3^{2 \cdot 3} = 3^6 = 729$

                 b. $(x^3)^5 = x^{3 \cdot 5} = x^{15}$; $(x^5)^3 = x^{5 \cdot 3} = x^{15}$

                 c. $(a^{-2})^3 = a^{-6} = \dfrac{1}{a^6}$

2. If $f(x) = x^4$ and $g(x) = x^5$, determine each of the following.

    a. $f(g(x))$ **b.** $g(f(x))$

The Power of a Product

3. If $h(x) = 4x^3$ and $k(x) = x^2$, determine each of the following.

   a. $h(k(2))$ **b.** $k(h(x))$

In Problem 3b, you obtain the expression $(4x^3)^2$. Since the base for the squaring is $4x^3$, you can write $(4x^3)^2$ as

$$(4x^3)^2 = 4x^3 \cdot 4x^3 = 16x^6 \qquad (1)$$
$$\uparrow$$
$$\text{base}$$

Note that $4x^3 \cdot 4x^3$ can be written equivalently as

$$4x^3 \cdot 4x^3 = 4 \cdot 4 \cdot x^3 \cdot x^3 = 4^2(x^3)^2. \quad (2)$$

Comparing the results on lines 1 and 2, the expression $(4x^3)^2$ can be written as

$$(4x^3)^2 = 4^2 \cdot (x^3)^2$$

Note that each factor in the base $4x^3$ (namely 4 and x^3) can be raised to the second power. This illustrates another important property of exponents.

Property of Exponents (Power of a Product)

If a and b are real numbers, and n is an integer, then

$$(ab)^n = a^n b^n.$$

Example 2 **a.** $(x^2 y^3)^4 = (x^2)^4 (y^3)^4 = x^8 y^{12}$ **b.** $(-2a^5)^3 = (-2)^3 (a^5)^3 = -8a^{15}$

   c. $(3x^{-3})^4 = 3^4 (x^{-3})^4 = 81x^{-12} = \dfrac{81}{x^{12}}$

4. a. If $h(x) = 3x^2$ and $k(x) = x^3$, determine each of the following.

   i. $h(k(x))$ **ii.** $k(h(x))$

   b. Does changing the order of the composition result in the same output? That is, does $h(k(x)) = k(h(x))$?

As you have observed several times in this chapter, the composition of functions is not commutative. That is, changing the order of the compositions may change the result. In general,

$$f(g(x)) \neq g(f(x)).$$

5. a. Apply the properties of exponents to simplify the following.

   i. $y = (5x^3)^2$ **ii.** $y = (-x^4)^3$

   b. Determine $f(g(x))$ if $f(x) = x^4$ and $g(x) = -3x^2$.

Fractional Exponents

What does $a^{\frac{1}{2}}$ represent? The properties of exponents allow you to adopt a reasonable definition for rational (fractional) exponents such as $\frac{1}{2}$. Let us begin by reviewing the definition of square root.

> **DEFINITION**
>
> Let a represent a nonnegative number, symbolically written as $a \geq 0$. The principal **square root** of a, denoted by $\sqrt{a}$, is defined as the nonnegative number that when squared produces a.

Example 3   **a.** $\sqrt{9} = 3$, *because* $3^2 = 9$

b. $\sqrt{100} = 10$, *because* $10^2 = 100$

Note that because $\sqrt{9} = 3$ *and* $3^2 = 9$, *it follows that* $(\sqrt{9})^2 = 9$.

> In general, $(\sqrt{a})^2 = a$ if $a \geq 0$.

Is there a relationship between $\sqrt{a}$ and $a^{\frac{1}{2}}$? To answer this question, you need to determine an exponent, m, such that

$$a^m = \sqrt{a}, a \geq 0.$$

Squaring both sides of this equation, you have

$$(a^m)^2 = (\sqrt{a})^2$$
$$a^{2m} = a^1.$$

Because the bases are equal, the exponents must be equal. Therefore,

$$2m = 1$$
$$= \frac{1}{2}.$$

From this result, the following definition is obtained.

> **DEFINITION**
>
> It follows that $a^{\frac{1}{2}} = \sqrt{a}$, where $a \geq 0$.

6. Evaluate each of the following, if possible, and check the answer using your grapher.

   a. $36^{\frac{1}{2}}$ **b.** $-9^{\frac{1}{2}}$ **c.** $\left(-9\right)^{\frac{1}{2}}$ **d.** $0^{\frac{1}{2}}$

Cube Roots

7. The volume of the following cube is 64 cubic inches.

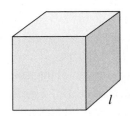

$V = 64$ cubic inches

   a. Determine the length, l, of one side (edge) of this cube.

   b. Explain how you obtained your answer.

The answer in Problem 7a is called the cube root of 64.

> **DEFINITION**
>
> The cube root of any real number a, denoted by $\sqrt[3]{a}$, is defined as the number that when cubed gives a.

Example 4   **a.** $\sqrt[3]{8} = 2$ *since* $2^3 = 8$

                b. $\sqrt[3]{125} = 5$ *since* $5^3 = 125$

                c. $\sqrt[3]{-1000} = -10$ *since* $(-10)^3 = -1000$

Numbers such as 8, 125, and -1000 that have exact cube roots are called perfect cubes.

Appendix

8. Evaluate each of the following, and check the answer using your grapher. See Appendix C for help in determining cube roots on the TI-83 Plus.

   a. $\sqrt[3]{1000}$ **b.** $\sqrt[3]{0}$ **c.** $\sqrt[5]{-32}$ **d.** $\sqrt[3]{100}$
                                                                              (nearest tenth)

Just as $\sqrt{a}$, where $a \geq 0$, can be written equivalently as $a^{\frac{1}{2}}$, the cube root of a real number a can be written as $a^{\frac{1}{3}}$.

$$\sqrt[3]{a} = a^{\frac{1}{3}}, \text{ the cube root of a real number } a$$

Similarly, $\sqrt[4]{a} = a^{\frac{1}{4}}$, the fourth root of $a \geq 0$ and

$$\sqrt[5]{a} = a^{\frac{1}{5}}, \text{ the fifth root of a real number } a$$

> In general, $\sqrt[n]{a} = a^{\frac{1}{n}}$, the nth root of a. The number a, called the **radicand**, must be nonnegative if n, called the **index**, is even.

Appendix

9. Calculate each of the following, and then verify your answer using your graphing calculator. See Appendix C for determining nth roots on the TI-83 Plus.

   a. $\sqrt[4]{81}$ **b.** $32^{\frac{1}{5}}$ **c.** $\sqrt[5]{-32}$ **d.** $-225^{\frac{1}{4}}$

10. a. Try to compute $\sqrt[4]{-81}$ using your graphing calculator. Explain what happens.

   b. Explain why the value of a in $\sqrt[n]{a}$, where n is even, cannot be negative.

11. Yachts that compete in the America's Cup must satisfy the International America's Cup Class rule that requires

$$L + 1.25\sqrt{S} - 9.8\sqrt[3]{D} = 16.296 \text{ meters,}$$

where L represents the yacht's length in meters,
           S represents the rated sail area, in square meters, and
           D represents the water displacement, in cubic meters.

   a. Is a yacht having length 21.85 meters, sail area 305.5 square meters, and displacement 21.85 cubic meters eligible to compete? Explain.

   b. Explain why the units of your numerical answer in part a is meters.

Rational Exponents

The properties of exponents can be expanded to include rational exponents where the numerator is different than 1. For example,

$$8^{\frac{2}{3}} = 8^{2 \cdot \frac{1}{3}} \qquad \text{Apply the property of exponents.}$$

$$= \left(8^2\right)^{\frac{1}{3}} \qquad \text{Apply the definition of the 1/3 exponent.}$$

$$= \sqrt[3]{8^2}$$

$$= \sqrt[3]{64}$$

$$= 4$$

In a similar fashion,

$$8^{\frac{2}{3}} = 8^{\frac{1}{3} \cdot 2} = \left(8^{\frac{1}{3}}\right)^2 = \left(\sqrt[3]{8}\right)^2 = 2^2 = 4$$

Therefore, $8^{\frac{2}{3}}$ can be written equivalently as $\sqrt[3]{8^2}$ or $\left(\sqrt[3]{8}\right)^2$. Note that 3, the denominator of the rational exponent 2/3, is the index. The numerator 2 indicates the power.

DEFINITION

$a^{\frac{p}{q}} = \sqrt[q]{a^p}$ or $a^{\frac{p}{q}} = \left(\sqrt[q]{a}\right)^p$, where $a \geq 0$ if q is even and p is a positive integer.

Example 5 **a.** $(-27)^{\frac{2}{3}} = \left(\sqrt[3]{-27}\right)^2 = (-3)^2 = 9$

b. $16^{\frac{3}{4}} = \left(\sqrt[4]{16}\right)^3 = 2^3 = 8$

c. $8^{\frac{-2}{3}} = \dfrac{1}{\left(8^{\frac{1}{3}}\right)^2} = \dfrac{1}{\left(\sqrt[3]{8}\right)^2} = \dfrac{1}{2^2} = \dfrac{1}{4}$

12. Compute each of the following, and then verify the answer using your grapher.

a. $25^{\frac{3}{2}}$ **b.** $(-8)^{\frac{2}{3}}$ **c.** $32^{\frac{4}{5}}$

d. $-16^{\frac{3}{4}}$ **e.** $243^{\frac{2}{5}}$ **f.** $(-16)^{\frac{3}{4}}$

13. Compute $7^{\frac{2}{3}}$ on your calculator, and explain why your answer is reasonable.

14. Write each of the following using fractional exponents.

a. $\sqrt[3]{x}$ **b.** $\sqrt[5]{x^3}$

c. $\sqrt[4]{x + 1}$ **d.** $\sqrt{xy}$

15. Perform the indicated operations by applying the appropriate property of exponents.

    a. $x^{\frac{1}{2}}x^{\frac{2}{3}}$
    b. $\left(x^3\right)^{\frac{3}{4}}$
    c. $\dfrac{x^3}{x^{\frac{1}{3}}}$

16. Determine the domain of each of the following.

    a. $f(x) = \sqrt{x}$
    b. $g(x) = \sqrt[3]{x}$

17. If $f(x) = \sqrt{x + 2}$, then determine each of the following.

    a. $f(-2)$
    b. $f(0)$

    c. $f(7)$
    d. $f(-6)$

18. If $g(x) = \sqrt[3]{x - 5}$, then determine each of the following.

    a. $g(5)$
    b. $g(13)$
    c. $g(-3)$

19. The area of the base of a cube is related to the volume of the cube by the formula

$$A = V^{\frac{2}{3}}.$$

Determine the area of a cube having volume 216 cubic inches.

 Appendix Additional examples and exercises involving properties of exponents are given in appendix A.

SUMMARY
Activity 2.7

1. If a is a real number and m and n are integers, then $(a^m)^n = a^{mn}$.

2. If a and b are real numbers and n is an integer, then $(ab)^n = a^n b^n$.

3. Let a represent a nonnegative number, symbolically written as $a \geq 0$. The principal square root of a, denoted by $\sqrt{a}$, is defined as the nonnegative number that, when squared, produces a.

4. $\left(\sqrt{a}\right)^2 = a$, if $a \geq 0$.

5. $a^{\frac{1}{2}} = \sqrt{a}$, where $a \geq 0$.

6. $\sqrt[n]{a} = a^{\frac{1}{n}}$ the nth root of a. The number a, called the **radicand**, must be nonnegative if n, called the **index**, is even.

7. $a^{\frac{p}{q}} = \sqrt[q]{a^p}$ or $a^{\frac{p}{q}} = \left(\sqrt[q]{a}\right)^p$, where $a \geq 0$ if q is even.

EXERCISES
Activity 2.7

1. Given $f(x) = 2x + 8$ and $g(x) = x - 3$, determine the following.

   a. $f(g(2))$ **b.** $g(f(2))$

   c. $f(g(x))$ **d.** $g(f(x))$

2. Simplify each expression by applying the properties of exponents. Write your results with positive exponents only.

   a. $(x^3)^6$ **b.** $(2x^5)^2$

   c. $(-3x^2)^3$ **d.** $(x^4)^{-3}$

   e. $(-4a^{-5})^2$ **f.** $(a^2b^{-2}c^3)^{-4}$

3. Determine $f(g(x))$ and $g(f(x))$ if $f(x) = x^6$ and $g(x) = -x^3$.

4. Using the accompanying graph, estimate each of the following.

   a. $f(g(4))$ **b.** $g(f(4))$ **c.** $f(g(0))$

   d. $g(f(0))$ **e.** $f(g(2))$ **f.** $g(f(2))$

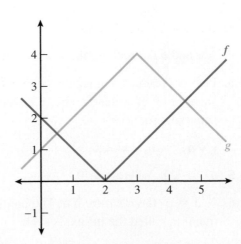

5. Compute each of the following quantities.

 a. $100^{\frac{1}{2}}$ **b.** $144^{\frac{1}{2}}$ **c.** $64^{\frac{1}{3}}$ **d.** $64^{\frac{4}{3}}$

 e. $5^{\frac{2}{5}}$ **f.** $(-8)^{\frac{2}{3}}$

 g. $25^{-\frac{1}{2}}$ **h.** $27^{-\frac{2}{3}}$

6. Simplify the following.

 a. $x^{\frac{1}{3}} \cdot x^{\frac{3}{4}}$ **b.** $\left(x^{-\frac{1}{3}}\right)^{-\frac{1}{2}}$ **c.** $\dfrac{x^{\frac{4}{5}}}{x^{\frac{1}{3}}}$

7. Write each of the following using fractional exponents.

 a. $\sqrt{x}$ **b.** $\sqrt[4]{x^3}$

 c. $\sqrt[3]{x+y}$ **d.** $\sqrt[5]{a^2 b^3}$

8. If $f(x) = \sqrt{x} - 3$, determine each of the following.

 a. $f(28)$ **b.** $f(3)$ **c.** $f(-1)$

9. If $g(x) = \sqrt[3]{x} + 10$, determine each of the following.

 a. $g(-74)$ **b.** $g(-10)$ **c.** $g(17)$

10. The length of time, t (in seconds), it takes the pendulum of a clock to swing through one complete cycle is a function of the length of the pendulum in feet. The relationship is defined by

$$t = f(L) = 2\pi \sqrt{\tfrac{L}{32}},$$

where L is the length of the pendulum.

 a. Rewrite the formula using fractional exponents.

 b. Determine the length of the cycle in time if the pendulum is 4 feet long.

ACTIVITY 2.8

Study Time

OBJECTIVES

1. Determine the inverse of a function represented by a table of values.

2. Use the notation f^{-1} to represent an inverse function.

3. Use the property $f(f^{-1}(x)) = f^{-1}(f(x)) = x$ to recognize inverse functions.

4. Determine the domain and range of a function and its inverse.

You are interested in taking sixteen credits this semester to complete your program, but you are concerned about the amount of time you will need for studying. The Academic Advising Center provides you with the information listed in the following table.

CREDITS TAKEN (INPUT)	HOURS OF STUDY PER WEEK (OUTPUT)
12	22
13	24
14	26
15	28
16	30
17	32
18	34

Notice that the number of hours of study time is a function of the number of credits taken. Call this function h.

1. Determine $h(14)$ and explain its meaning in practical terms.

2. Use the information from the preceding table to construct another table in which the number of hours of study time is the input and the number of credits taken is the output.

HOURS OF STUDY PER WEEK (INPUT)	CREDITS TAKEN (OUTPUT)

3. Using the input/output values from the second table, determine if the number of credits taken is a function of the number of hours of study time. Explain.

4. Call this new function c, and determine $c(30)$. Explain its meaning in practical terms.

5. a. What are the domain and the range of function h?

   b. What are the domain and the range of function c?

   c. How are the domain and range of function h related to the domain and range of function c?

6. Determine each of the following. Refer to the appropriate table.

   a. $h(15)$ **b.** $c(28)$

Notice that in Problem 6, the output of h—namely, 28—was used as the input of function c. Recall that this is the composition of h and c and can be written as $c(h(15))$.

7. Determine each of the following. Use the preceding tables given at the beginning of this activity.

   a. $c(h(18))$ **b.** $h(c(34))$

   c. $c(h(13))$ **d.** $h(c(32))$

8. a. Let x represent the input for the function h. What is the result of the composition $c(h(x))$?

   b. Let x represent the input for the function c. What is the result of the composition $h(c(x))$?

Inverse Functions

> **DEFINITION**
>
> When the input of the composition of two functions is the same as the output of the composition, the two functions are **inverses** of one another. One function "undoes" the other. Symbolically, if functions f and g are inverses, then $f(g(x)) = x$ and $g(f(x)) = x$.

Example 1 *Show that f and g, defined by $f(x) = 3x - 1$ and $g(x) = \frac{x+1}{3}$, are inverses.*

SOLUTION

First check the composition $f(g(x))$:

$$f(g(x)) = f\left(\tfrac{x+1}{3}\right) = 3\left(\tfrac{x+1}{3}\right) - 1 = (x + 1) - 1 = x$$

This verifies the first equation. Now check $g(f(x))$.

$$g(f(x)) = g(3x - 1) = \frac{(3x - 1) + 1}{3} = \frac{3x}{3} = x.$$

Since $f(g(x)) = g(f(x))$, f and g are inverses.

9. Are the functions h and c in Problem 8 inverses? Explain.

In general, the inverse of a function f is written f^{-1}. Using this notation, $h = c^{-1}$ and $c = h^{-1}$.

> **Inverse Functions**
>
> - The two functions f and g are inverses if $f(g(x)) = x$ and $g(f(x)) = x$.
> - The notation for the inverse of f is f^{-1}.
> - The domain of f is the range of f^{-1}, and the range of f is the domain of f^{-1}.
>
> **Important Note about Notation**
>
> The notation for the inverse function is potentially confusing. The -1 in $f^{-1}(x)$ is not an exponent! The notation is derived from the fact that the inverse function undoes the arithmetic operations of the original function. You may have seen -1 used as an exponent to denote the reciprocal of a number: $2^{-1} = \frac{1}{2}$; however, $f^{-1}(x) \neq \frac{1}{f(x)}$.

An important property of inverse functions is that the domain and range values are interchanged. For example, from the first table in this section,

$$h = \{(12, 22), (13, 24), (14, 26), \text{etc.}\}$$

If the input and output values of each ordered pair are interchanged, you have

$$\{(22, 12), (24, 13), (26, 14), \text{ etc.}\}$$

These ordered pairs match the (input, output) pairs of the function c, the inverse of h. See the second table in this section.

10. Given the function $p = \{(2, 4), (-5, 6), (0, 1), (7, 8)\}$, determine the following:

    a. The inverse function, p^{-1}.

    b. $p(2)$ **c.** $p^{-1}(4)$ **d.** $p(p^{-1}(4))$

    e. $p^{-1}(p(2))$ **f.** $p(p^{-1}(x))$ **g.** $p^{-1}(p(x))$

11. The function q is defined by the following table, where $s = q(t)$.

t (INPUT)	s (OUTPUT)
1	2
2	3
3	5
4	3

    a. Interchange the input and output values, and record the results in the following table.

s (INPUT)	t (OUTPUT)

    b. Does the table in part a represent a function? Explain.

    c. As a result of part b, the function q does not have an inverse. Could you have predicted that the function q does not have an inverse from the original table? Explain.

SUMMARY
Activity 2.8

1. The two functions f and g are inverses if $f(g(x)) = x$ and $g(f(x)) = x$.

2. The notation for the inverse of f is f^{-1}.

3. The domain of f is the range of f^{-1}, and the range of f is the domain of f^{-1}.

4. $f^{-1}(x) \neq \frac{1}{f(x)}$.

EXERCISES
Activity 2.8

1. The functions f and g are defined by the following tables.

x	y = f(x)
2	3
4	5
6	7
8	9

x	y = g(x)
3	2
5	4
7	6
9	8

Determine each of the following:

a. $f(g(7))$ **b.** $g(f(4))$ **c.** $f(g(x))$ **d.** $g(f(x))$

2. The function h is defined by the following set of ordered pairs.

$$\{(2, 3), (3, 4), (4, 5), (5, 6)\}$$

a. Write h^{-1} as a set of ordered pairs.

b. Determine $h(3)$ and $h^{-1}(h(3))$.

c. Determine $h^{-1}(5)$ and $h(h^{-1}(5))$.

3. The function r is defined by the following table.

x	r(x)
0	2
1	3
2	4
3	2

a. Determine the function r^{-1}.

x	r⁻¹(x)

b. Does the table in part a represent a function? Explain.

c. Does the function $r(x)$ have an inverse? Explain.

4. You are planning a trip to Canada, and you want to exchange some U.S. currency for Canadian money. The following table will help you with this conversion.

AMOUNT IN U.S. DOLLARS (input)	50	100	188.82	500	1000
AMOUNT IN CANADIAN DOLLARS (output)	79.44	158.88	300	794.40	1588.80

Source: Exchange rate, Bank of Canada, October 11, 2002.

a. Determine $f(100)$, and determine its meaning in practical terms.

b. Use the information from the preceding table to construct another table in which the amount in Canadian money is the input and the U.S. dollar amount is the output.

AMOUNT IN CANADIAN DOLLARS (input)					
AMOUNT IN U.S. DOLLARS (output)					

c. Use the information from the table in part b to determine if the amount in U.S. dollars is a function of the amount in Canadian dollars. Explain.

d. Let g represent the function in the table in part b. Determine $g(300)$. Explain its meaning in this practical situation.

e. Determine the domain of f and the range of f.

f. Determine the domain of g and the range of g.

g. How are the domain and range of f related to the domain and range of g?

h. Determine $g(f(500))$ by using the two preceding tables. Explain your answer.

i. Determine $f(g(79.44))$ by using the two preceding tables.

5. You are now familiar with the conversion tables in Exercise 4, but you would like to be able to take any amount of money and convert between U.S. and Canadian currencies.

Check to see if the functions f and g are linear, and refer to the data in the preceding tables to answer the following.

a. Determine the rate of change (slope) of function f. Round to four decimal places.

b. What is the practical meaning of the slope in this situation?

c. Using x to represent the input, write the linear equation of the function f.

d. Use the equation to determine $f(3000)$. Explain the practical meaning in this situation.

e. Determine the rate of change (slope) for function g. Round to four decimal places.

f. What is the practical meaning of the slope in function g?

g. Using x to represent the input, write the linear equation of the function g.

h. Use the equation to determine $g(6000)$. Explain the meaning in this situation.

i. Prove that the two functions are inverse functions by showing that $f(g(x)) = x$ and $g(f(x)) = x$. Round your answer to the nearest whole number.

ACTIVITY 2.9
Temperature
Conversions

OBJECTIVES

1. Determine the equation of the inverse of a function represented by an equation.

2. Describe the relationship between the graphs of inverse functions.

3. Determine the graph of the inverse of a function represented by a graph.

4. Use the graphing calculator to produce graphs of an inverse function.

In Activity 2.5, Inflated Balloons, you used the function defined by $F = 32 + 1.8C$ to convert from a temperature measured in degrees Celsius to a temperature measured in degrees Fahrenheit. Call this function T.

1. a. Identify the input and output variables for the conversion function T.

b. Determine $T(-5)$ and explain its meaning in this situation.

2. When the temperature is $70°$ F; what is the temperature in degrees Celsius?

3. If you need to determine the temperature in Celsius for several Fahrenheit temperatures, it is easier to have a single formula (in fact, a new function) in which Celsius is the output and Fahrenheit is the input. Solve $F = 32 + 1.8C$ for C to determine this new function. Call this function H.

4. a. Identify the input and output variables for the function H.

b. Determine $H(62)$, and explain its meaning in this situation.

5. Determine each of the following.

   a. $T(10)$ **b.** $H(T(10))$

6. Determine each of the following.

   a. $T(H(95))$ **b.** $H(T(0))$ **c.** $T(H(212))$

7. a. Write a general expression for $H(T(x))$ and simplify the result.

b. Write a general expression for $T(H(x))$ and simplify the result.

c. Based on the results of a and b, what can you conclude about the functions H and T?

Inverse Function Algorithm

Given an equation for a function, f, whose input is x and whose output is y, you can determine the equation for the inverse function, f^{-1}, as follows:

Step 1. Exchange x and y in the given equation. (This switches input and output.)

Step 2. Solve the resulting equation for y.

Step 3. Write y as $f^{-1}(x)$.

Example 1 *Given $F = T(C) = 32 + 1.8c$, determine the equation for the inverse T^{-1}.*

SOLUTION

Step 1. Exchange c and F: $F = 32 + 1.8c$ becomes $c = 32 + 1.8F$

Step 2. Solve for F:

$1.8F = C - 32$

$F = \dfrac{c - 32}{1.8}$

Step 3. Write F as $T^{-1}(c)$: $T^{-1}(c) = \dfrac{c - 32}{1.8}$

8. Given $y = f(x) = 5 - 3x$, determine an equation of the inverse function f^{-1}.

Graphs of Inverse Functions

9. a. Consider the function T in Example 1 defined by $y = T(x) = 32 + 1.8x$ and its inverse defined by $y = T^{-1}(x) = \frac{x - 32}{1.8}$. Then graph the two functions on the same coordinate system with domain $-50 \le x \le 250$ and range $-50 \le y \le 250$.

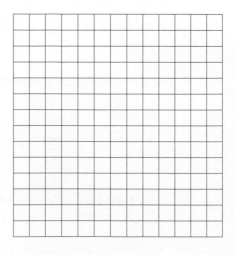

b. Draw the graph of the line $y = x$ on the coordinate system in part a. Describe the relationship between the graphs of T and T^{-1} and the line $y = x$.

You can draw the inverse of a function with your TI-83 Plus without determining the equation. Turn off the graph of T^{-1} from Problem 9, and keep the graphs of T and $y = x$. Your screen should resemble the following.

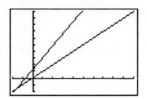

Access the Draw menu by pressing ⟮2nd⟯ ⟮PRGM⟯. Toggle down to option 8, Draw-Inv.

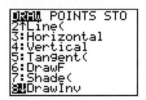

Pressing ⟮ENTER⟯ will place the DrawInv command into the Home screen. Now press ⟮VARS⟯, then ⟮Y-VARS⟯ and ⟮1: Functions⟯ to access the following screen.

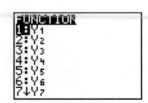

Choose the variable that corresponds to your function T, most likely Y1, and press ⟮ENTER⟯ again.

Your screen should resemble the following.

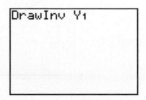

Press ⟮ENTER⟯. This adds the graph of the inverse to the graph screen.

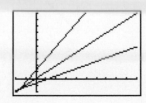

How does this compare with the graph of T^{-1} in Problem 9? Note that the graph of T^{-1} in the graph screen of the TI-83 Plus is not active. It cannot be traced, nor can values be viewed in the table.

10. Determine the point of intersection of the graphs of T and T^{-1}. What is the significance of this point in this situation?

Graphs of Inverse Functions

The graphs of inverse functions are reflections about the line $y = x$.

11. Sketch a graph of f^{-1} on the same coordinate system as the graph of f for the functions that follow.

a.

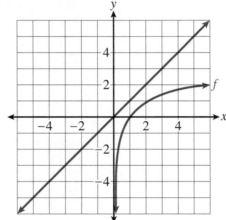

b.

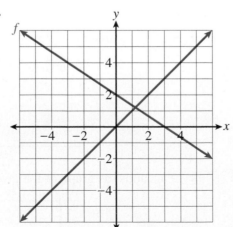

c.

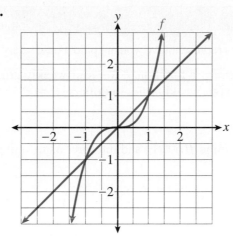

12. a. Determine the horizontal and vertical intercepts of the graphs of T and T^{-1} given in Problem 9.

b. Write a sentence that describes the relationship of the intercepts of T and the intercepts of T^{-1}.

13. a. Determine the slopes of the graphs of T and T^{-1} given in Problem 9?

b. Write a sentence that describes the relationship of the slope of T and the slope of T^{-1}.

SUMMARY
Activity 2.9

1. To determine the equation of the inverse of a function defined by $y = f(x)$,

Step 1. Exchange x and y in the given equation. (This switches input and output.)

Step 2. Solve the resulting equation for y.

Step 3. Write y as $f^{-1}(x)$.

2. The graphs of inverse functions are reflections about the line $y = x$.

3. If two linear functions are inverses, the slopes of the graphs of the lines are reciprocals.

EXERCISES
Activity 2.9

1. As a sales representative, you are paid a base salary of $250 a week, plus a 5% commission.

   a. Determine a function, f, for your weekly gross pay, P (before taxes), as a function of S, your weekly sales in dollars, $P = f(S)$.

   b. Determine $f(6000)$, and interpret its meaning in this situation.

   c. Solve the equation $P = 0.05S + 250$ for S to determine the equation for a new function, $g(P)$, whose input is P and whose output is S.

   d. Determine $g(400)$, and interpret its meaning in this situation.

   e. Determine $g(f(8000))$.

2. a. Complete the following table for the function $y = f(x) = 3x^2 - 2$.

x	-2	-1	0	1	2
y					

   b. Describe how you know that f is a function by examining the table in part a.

   c. Use the table of values from part a to determine f^{-1} if possible. If it is not possible, explain why.

3. Determine the equation of the inverse of the given function.

   a. $y = f(x) = 3x - 4$

   b. $w = g(z) = \frac{z - 4}{2}$ c. $s(t) = \frac{5}{t}$

4. a. Given $y = f(x) = 2x + 6$, determine f^{-1}.

Exercise numbers appearing in color are answered in the Selected Answers appendix.

b. Sketch the graphs f and f^{-1} on the same coordinate system.

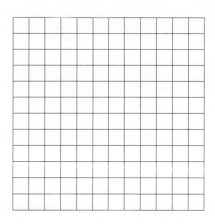

c. Add the sketch of the graph of $y = x$ to the sketches of f and f^{-1} in part b. Describe any symmetry in the graphs of f and f^{-1}.

d. Determine the horizontal and vertical intercepts of the graphs of f and of f^{-1}. Explain the relationship between the intercepts of f and the intercepts of f^{-1}.

e. Determine the slope of the graph of f and the slope of the graph of f^{-1}. What is the relationship between the slope of f and the slope of f^{-1}?

5. Consider the graphs of functions g and h. Are the functions inverses of each other? Explain, using a symmetry argument.

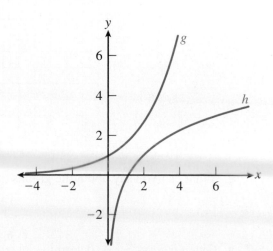

6. Consider the functions $f(x) = 3x + 6$ and $g(x) = \frac{1}{3}x - 2$.

   a. Determine $f(g(x))$.

   b. Determine $g(f(x))$.

   c. Are f and g inverse functions? Explain.

   d. Complete the following tables for the given functions.

x	f(x)
−2	
0	
2	
4	

x	g(x)
0	
6	
12	
18	

   e. Do you notice anything about the ordered pairs of f and g?

   f. Graph f and g on the following grid.

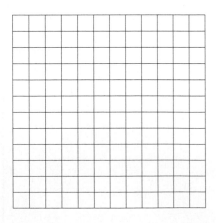

   g. What can you say about the graphs of f and g with respect to the graph of $y = x$?

7. a. Given $g(x) = \frac{6 + 4x}{3}$, determine an equation for $g^{-1}(x)$.

b. Sketch the graphs of g and g^{-1} on the following grid.

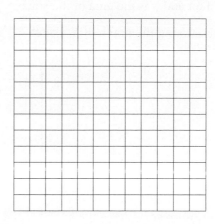

c. Do you believe that your equation for g^{-1} in part a is correct? Explain.

d. Determine $g^{-1}(g(x))$. Does your result support your answer in part c? Explain.

8. One inch measures 2.54 centimeters. To convert inches to centimeters, use the equation $C = 2.54x$, where x is the number of inches.

a. Using the graph of C, draw the graph of the inverse function C^{-1}. *Hint:* First draw the graph of $y = x$ on the same axes.

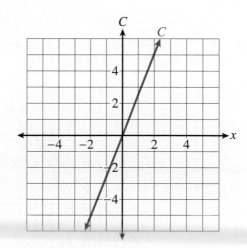

b. Locate $C^{-1}(5)$ on the graph, and interpret its meaning in this situation.

9. Your new kitchen has a square island that needs a granite top. The side of the largest square top you can purchase measures $S = \sqrt{a}$, where s is the length of the square in feet and a is the area of the square.

   a. Using the graph of S, sketch S^{-1}.

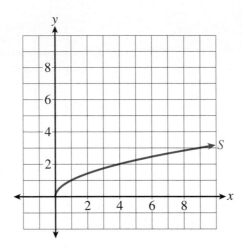

   b. Check your graph of S^{-1} by using the Draw command.

   c. What are input and output variables for S?

   d. What are the input and output variables for S^{-1}?

   e. Would S^{-1} ever be useful in this situation? Explain.

10. Sketch a graph of f^{-1} on the same coordinate system as the graph of f for the functions that follow.

   a.

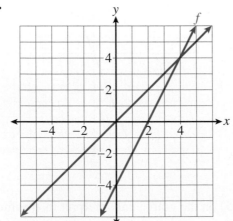

b.

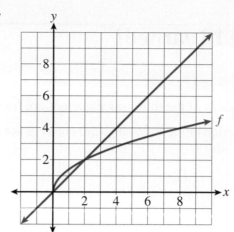

c.

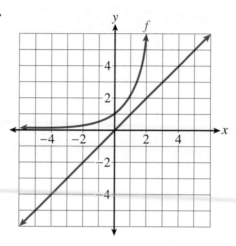

CLUSTER 2 **What Have I Learned?**

1. For any two functions f and g,

   i. determine which of the following equations are true and explain why.

   ii. determine which of the following equations are not always true and give an example to show why not.

   a. $(f + g)(x) = (g + f)(x)$

   b. $(f - g)(x) = (g - f)(x)$

   c. $(f \cdot g)(x) = (g \cdot f)(x)$

   d. $f(g(x)) = g(f(x))$

   e. $f(f^{-1}(x)) = f^{-1}(f(x))$

2. Describe how you would determine whether two functions are inverses of each other.

3. Suppose f is a nonconstant linear function.

   a. Will f always have an inverse function?

    b. What will always be true about the slopes of the lines representing the graphs of f and f^{-1}?

4. If $f(g(x)) = x^6$, determine at least three different ways to define f and g.

5. It is possible to compose more than two functions. If $f(x) = 2x - 1$, $g(x) = 5 - 3x$, and $h(x) = 2x^2$, determine each of the following.

    a. $f(g(h(1)))$ **b.** $g(f(h(1)))$

    c. $h(g(f(1)))$ **d.** $h(f(g(1)))$

How Can I Practice?

1. Given $f(x) = x^2 - 4$ and $g(x) = x + 2$, determine each of the following.

    a. $f(g(-3))$ **b.** $g(f(-3))$

    c. $f(g(x))$ **d.** $g(f(x))$

    e. $f(f(x))$

    f. $g^{-1}(x)$

2. Given $f(x) = x - 4$ and $g(x) = 4 + x - x^2$, determine each of the following.

    a. $f(g(3))$ **b.** $g(f(3))$

    c. $g(g(3))$ **d.** $f(g(x))$

    e. $g(f(x))$

    f. $f^{-1}(x)$

Answers to all How Can I Practice exercises are included in the Selected Answers appendix.

3. Given the following two tables, complete the third table.

x	0	1	2	3	4
f(x)	6	−1	−3	0	2

x	−3	−1	0	2	6
g(x)	2	3	1	0	−1

x	0	1	2	3	4
g(f(x))					

4. Simplify the following by applying the appropriate properties of exponents.

a. $(x^2)^4$

b. $(xy)^3$

c. $(2x^4y)^5$

d. $(3x^2)^3(2xy)^4$

e. $(-5)^2$

f. -5^2

g. $(-2x)^2$

h. $(-2x)^3$

i. $(x^2)^5(-2x^4)^3$

j. $25^{\frac{3}{2}}$

k. $64^{\frac{2}{3}}$

l. $(-27)^{\frac{1}{3}}$

m. $\left(\frac{9}{16}\right)^{\frac{3}{2}}$

n. $4^{\frac{1}{3}}$

o. $\left(x^3y^6\right)^{\frac{2}{3}}$

p. $x^{\frac{2}{3}}x^{\frac{1}{2}}$

q. $\left(8xy\right)^{\frac{1}{3}}$

5. Given $f(x) = 3x^2$ and $g(x) = -2x^3$, determine each of the following.

a. $g(f(x))$

b. $f(g(x))$

c. $g(f(-4))$

6. Given $s(x) = x^2 + 4x - 1$ and $t(x) = 4x - 1$, determine each of the following.

a. $s(t(x))$

b. $t(s(x))$

c. $t^{-1}(x)$

7. Given $p(x) = \frac{1}{x}$ and $c(x) = \sqrt{x + 2}$, determine each of the following.

a. $p(c(x))$　　　　　　　　**b.** $c(p(x))$　　　　　　　　**c.** $p^{-1}(x)$

8. The function q is defined by the following set of ordered pairs:

$$\{(4, 6), (7, -9)(-2, 1), (0, 0)\}$$

Determine q^{-1} as a set of ordered pairs.

9. Show that $f(x) = 2x - 3$ and $g(x) = \frac{x + 3}{2}$ are inverses of each other.

10. a. Determine the equations of the inverse function of $f(x) = 4x + 3$.

b. Sketch the graph of the function f and its inverse, f^{-1}, on the same coordinate system.

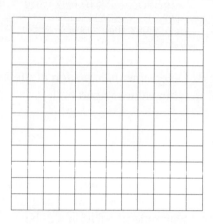

c. Determine the horizontal and vertical intercepts of the graph of f and of f^{-1}.

d. Determine the slopes of the graph of f and of f^{-1}.

e. Sketch a graph of the line $y = x$ on the same coordinate system.

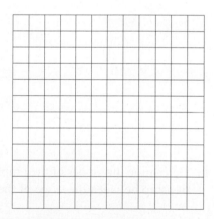

11. The 1990–2000 population boom in the United States public schools from kindergarten through grade 12 can be modeled by the function $f(x) = 0.6x + 41$, where x represents the years after 1990 and $f(x)$ represents the population in millions. (These figures are based on fall enrollments.) *Source:* U.S. Department of Education.

a. Complete the table using the given function model.

x, YEAR SINCE 1990	0	3	6	10
f(x), POPULATION IN MILLIONS				

b. Interchange the input and output data.

x, **POPULATION IN MILLIONS**				
f(x), **YEAR SINCE 1990**				

c. Determine the equation of the inverse function, and use this equation to check the table values in part b.

d. Graph the function and its inverse using your grapher. Use a window Xmin $= -20$, Xmax $= 20$, Ymin $= -100$, and Ymax $= 100$. Use Zoom, ZSQR to see the graphs in a square window. Using the trace feature and the appropriate graph, determine when the population will be 46 million. Round your answer to the nearest year.

e. Add the graph of $y = x$ to the graphs of the function and its inverse. Describe the symmetry in the graphs.

f. Determine the horizontal and vertical intercepts of the graphs of the function and its inverse. Explain the relationship between the intercepts of the two functions.

	HORIZONTAL INTERCEPT	VERTICAL INTERCEPT
$f(x) = 0.6x + 41$		
$f^{-1}(x) = \dfrac{x - 41}{0.6}$		

g. Determine the slopes of the graphs of the function and its inverse. What is the relationship between the slope of the function and the slope of its inverse?

h. Predict the year in which the school population will be 50 million.

12. A Japanese student is coming to the United States as an exchange student. She has saved 60,000 yen for the trip. She realizes that because she is going to stop briefly in Europe, she can convert her money to the euro. After looking up the currency conversion factor on the Internet for October 4, 2002, she determines that the function that will convert her Japanese money is $f(x) = 0.00826x$, with x representing the amount in Japanese currency and $f(x)$ representing the amount in euros.

a. Use the function to determine how many euros she will receive for her 60,000 yen.

b. She then comes to the U.S. with her euros. To convert her money to U.S. dollars, she finds another conversion factor on the Internet. The function is $g(E) = 0.9865\,E$, with E representing the amount in euros and $g(E)$ representing the U.S. dollar amount. How many dollars will she have if she has 350 euros?

c. If the Japanese student does not go to Europe, but flies directly to the U.S.A., write a new function that would tell her how much her 60,000 yen would be in U.S. dollars. *Hint:* Determine $g(f(x))$.

d. Use the function you found in part c to determine what her 60,000 yen would be worth in U.S. dollars.

13. Given $f(x) = 4x - 9$ and $g(x) = 10 - 3x$, what input to the composition function $f(g(x))$ will result in an output of 15?

14. The volume of a rectangular box is equal to the area of the base times the height. Suppose a particular box has a height of 10 in. and a square base.

a. Using V for the volume of the box, b for the area of the base, and x for the length of one side of the square base, express b as a function of x and V as a function of b.

b. Show how the volume is the composition of these two functions.

Summary

The bracketed numbers following each concept indicate the activity in which the concept is discussed.

CONCEPT / SKILL	DESCRIPTION	EXAMPLE
The sum function $f + g$ [2.1]	Given two functions, f and g, the sum function, $f + g$, is defined by $(f + g)(x) = f(x) + g(x)$.	See Example 1, Activity 2.1.
The difference function, $f - g$ [2.1]	Given two functions, f and g, the difference function, $f - g$, is defined by $(f - g)(x) = f(x) - g(x)$.	See Example 2, Activity 2.1.
Polynomial expression [2.1]	Any expression that is formed by adding or subtracting terms of the form ax^n, where a is a real number and n is a nonnegative integer, is called a **polynomial expression**.	$3x^3 - 2x^2 + 6x - 7$
Monomial [2.1]	A **monomial** is a polynomial with one term.	$13x^5$
Binomial [2.1]	A **binomial** is a polynomial with two terms.	$14x^4 - 3x$
Trinomial [2.1]	A **trinomial** is a polynomial with three terms.	$5x^4 - 7x + 13$
Polynomial function [2.1]	A polynomial function is any function defined by an equation of the form $y = f(x)$, where $f(x)$ is a polynomial expression.	$f(x) = 3x^3 - 2x - 7$
Adding or subtracting functions graphically [2.2]	When adding or subtracting two functions, f and g, graphically, for each input value x, add or subtract the output values (vertical displacements) represented by $f(x)$ and $g(x)$ to obtain the value of $(f + g)(x)$ or $(f - g)(x)$.	If $f(2) = 3$ and $g(2) = 5$, then $(f + g)(2) = 3 + 5 = 8$. That is, for the input value 2, the output of the sum function $f + g$ is 8, the sum of the vertical displacements for f and g.
FOIL [2.3]	FOIL is a common method used to multiply two binomials.	See Example 1, Activity 2.3
The product of two polynomials [2.3]	To multiply any two polynomials, multiply each term of the first by each term of the second.	$(x^2 + 1)(3x^2 + 6x - 2) =$ $3x^4 + 6x^3 - 2x^2 + 3x^2 + 6x - 2$ $= 3x^4 + 6x^3 + x^2 + 6x - 2$

The product of two functions [2.3]	Given two functions, f and g, the product function is defined by $y = (f \cdot g))(x) = f(x) \cdot g(x)$.	See Exercise 8, Activity 2.3.
$a^m \cdot a^n$ [2.3]	To multiply powers of the same base, keep the base and add the exponents. Symbolically, this property of exponents is written as $a^m \cdot a^n = a^{m+n}$.	$x^4 \cdot x^3 = x^{4+3} = x^7$
Composition of functions [2.5]	The **composition** of the functions f and g is a function, h, defined by $h(x) = f(g(x))$.	If $f(x) = 2x + 1$ and $g(x) = 3x - 2$, $f(g(x)) = f(3x - 2)$ $= 2(3x - 2) + 1 = 6x - 3$
Commutativity of composition [2.5]	In general, $f(g(x)) \neq g(f(x))$.	If $f(x) = 2x + 1$ and $g(x) = 3x - 2$, $f(g(x)) = 6x - 3$, $g(f(x)) = g(2x + 1)$ $= 3(2x + 1) - 2 = 6x + 1$
$(a^m)^n$ [2.7]	If a is a real numbers and m and n are integers, then $(a^m)^n = a^{mn}$.	$(x^2)^5 = x^{2 \cdot 5} = x^{10}$
$(ab)^n$ [2.7]	If a and b are real numbers and n is an integer, then $(ab)^n = a^n b^n$.	$(2x)^4 = 2^4 x^4 = 16x^4$
The principal square root of a, $\sqrt{a}$ [2.7]	Let a represent a nonnegative number, symbolically written as $a \geq 0$. The principal square root of a, denoted by $\sqrt{a}$, is defined as the nonnegative number that, when squared, produces a.	$\sqrt{16} = 4$ because $4^2 = 16$
Fractional exponents [2.7]	$a^{\frac{1}{2}} = \sqrt{a}$, where $a \geq 0$	$13^{\frac{1}{2}} = \sqrt{13}$
$\sqrt[n]{a}$ [2.7]	$\sqrt[n]{a} = a^{\frac{1}{n}}$, the nth root of a. The number a, called the **radicand**, must be nonnegative if n, called the **index**, is even.	$\sqrt[3]{36} = 36^{\frac{1}{3}}$
$a^{\frac{p}{8}}$ [2.7]	$a^{\frac{p}{q}} = \sqrt[q]{a^p}$ or $a^{\frac{p}{q}} = \left(\sqrt[q]{a}\right)^p$, where $a \geq 0$ if q is even, and p is a positive integer.	$16^{\frac{3}{4}} = \left(\sqrt[4]{16}\right)^3 = 2^3 = 8$
Inverse functions [2.8]	The two functions f and g are inverses if $f(g(x)) = x$ and $g(f(x)) = x$.	See Example 1, Activity 2.8.

Domain and range of inverse functions [2.8]	The domain of f is the range of f^{-1} and the range of f is the domain of f^{-1}.	See the tables at Problems 1 and 2, Activity 2.8.
Graphs of inverse functions [2.9]	The graphs of inverse functions are reflections about the line $y = x$.	See the graph before Problem 10, Activity 2.9.
Slopes of inverse linear functions [2.9]	If two linear functions are inverses, the slopes of the graphs of the lines are reciprocals.	The slope of $f(x) = 3x + 1$ is 3. The inverse, defined by $f^{-1}(x) = \frac{1}{3}x - \frac{1}{3}$, has a slope of $\frac{1}{3}$.

Gateway Review

1. Simplify the following.

   a. $(x + 6) + (2x^2 - 3x - 7)$ **b.** $(x^2 + 4x - 3) - (2x^2 - x + 1)$

   c. $(x - 3)(4x - 1)$ **d.** $(x - 5)(x^2 - 2x + 3)$

   e. $4(x + 2) - 3(5x - 1)$ **f.** $(2x^2 + x - 1)(x^2 - 3x + 4)$

2. Simplify the following. Write all of your results with positive exponents only.

   a. $(3x^3)(2x^5)$ **b.** $(4x^3y)^2$ **c.** $(xy)^2(-2x^3y)$

   d. $(5x^3y^4z)(-2x^2yz^3)$ **e.** $(3x^2y)^0(3x^3y)^2$ **f.** $(-5xy)^3$

   g. $\dfrac{6x^4}{3x}$ **h.** $2x^0$ **i.** $\dfrac{3^3}{3^3}$

   j. $\dfrac{6xy^4z^2}{4xyz^5}$ **k.** $(-5x^{-3})(x^{-5})$ **l.** $\dfrac{8x^{-4}}{-2x^{-6}}$

   m. $(-5x^{-3})^3$ **n.** $x^{\frac{4}{5}} \cdot x^{\frac{1}{2}}$

   o. $\left(x^{\frac{2}{3}}\right)^3$

3. Given $f(x) = 6x - 2$ and $g(x) = -2x + 3$, determine each of the following.

   a. $f(-3)$ **b.** $(f + g)(x)$

   c. $(f - g)(3)$ **d.** $(f \cdot g)(x)$

Answers to all Gateway exercises are included in the Selected Answers appendix.

e. $f(g(x))$

f. $g(f(2))$

g. f^{-1} (Find the inverse of f.)

4. Given $f(x) = x^2 - x + 3$ and $g(x) = 3x - 2$, determine each of the following.

   a. $(f - g)(x)$ **b.** $(f \cdot g)(x)$

   c. $(f(g(x)))$

   d. $(g(f(2)))$

5. Determine the value of each of the following.

   a. $49^{\frac{1}{2}}$ **b.** $32^{\frac{2}{5}}$ **c.** $(-27)^{\frac{4}{3}}$ **d.** $7^{\frac{3}{5}}$

   e. $\sqrt[3]{27^2}$ **f.** $\sqrt[4]{16^5}$ **g.** 4^{-2}

6. To ship the mail-order ceramic figures that you produce, you need to make square-bottomed boxes. For the size of the box to be proportional to the figurines, the height of the box must always be three times longer than the width. The cost of the material to make the top and bottom of the box molded to fit the figurine is $0.01 per square inch, and the cost of the material for the sides of the box sells for $0.004 per square inch.

   a. Write a function, $f(x)$, to represent the cost of producing the top and the bottom of a box. Use x to represent the width of the bottom of the box in inches.

   b. Write a function, $g(x)$, to represent the cost of producing the sides of the box.

c. Combine the functions in parts a and b to write one function that represents the total cost of making the box, $(f + g)(x)$. Write the equation in simplest form.

d. Using f, g, and $f + g$ as defined in parts a, b, and c, complete the following table.

x (in.)	f(x)	g	(f + g)(x)
2			
4			
6			
8			
10			

e. Use your grapher to graph all three functions. Determine $f(5)$, $g(5)$, and $(f + g)(5)$ from the graph. Explain the practical meaning of the values that you find.

7. You have a knitting machine in your home and your business is making ski hats. The fixed cost to run your knitting company is $300 per month, and the cost to produce each hat averages approximately $12. The hats will sell for $25.95.

a. Write a function f to represent the cost of making the hats. Use x to represent the number of hats made per month.

b. Write a function g to represent the revenue from the sale of the hats.

c. Write a function h to represent the profit for the month. Express this function in simplest form.

d. Graph the three functions on your grapher. How many hats must be sold in one month to break even?

e. Determine the value of $f(50)$, $g(50)$, and $h(f - g)(50)$. Explain the practical meaning of the values that you find.

f. Explain how the difference function $f(x) - g(x) = (f - g)(x)$ pertains to this situation.

8. The manufacturer of a certain brand of computer printer sells her printers at a wholesale price of $110 per printer, based on selling 60 printers. Because the warehouse is overstocked and new high-speed printers are arriving, the manufacturer is offering a one-time-only order of $2 less per printer on every printer you order over the usual 60. You may not, however, purchase more than 90 printers.

a. Let x represent the number of printers in excess of 60 that you will purchase. Write a function for the total cost, C, of the printers as a function of the number of printers x. Call this function f.

b. Rewrite the cost function by multiplying the factors and combining like terms.

c. What is the domain of the cost function?

d. If you decide that it is to your advantage to purchase 75 printers rather than 60 printers, explain the cost savings to you and your business.

9. Functions f and g are defined by the following tables.

x	−1	0	1	2	4	5	8
f(x)	8	5	2	−1	−7	−10	−19

x	−1	0	1	2	4	5	8
g(x)	5	1	−1	−1	5	11	41

Determine the values for each of the following.

a. $f(g(4))$　　　　　　　　　**b.** $g(f(-1))$

c. $f(g(0))$　　　　　　　　　**d.** $g(f(2))$

10. You manufacture snowboards. You cannot produce more than 30 boards per day. The cost, C, of producing x boards is represented by the function

$$C = f(x) = 150x - 0.9x^2.$$

a. What is the practical domain of the function?

b. What is the cost if 22 boards are produced?

c. The number of snowboards that can be produced in t hours is represented by the function $g(t) = 3.75t$. Determine $f(g(t))$.

d. What is the input variable in the composition of the functions in part c?

e. Because of a blizzard, your employees work only four hours on a certain day. Determine the production cost for that day.

f. The company prefers to keep production costs at approximately $3500. How many hours each day does the company have to operate to maintain this production cost? Your employees do not work more than nine hours per day.

11. a. Determine the equation of the inverse of the function $f(x) = \dfrac{2x - 3}{5}$.

b. What is the slope of the line for each function? What is the relationship between the slopes of the two functions?

12. a. Show that $f(x) = -2x + 1$ and $g(x) = \dfrac{1 - x}{2}$ are inverse functions of each other.

b. Sketch the graphs of the functions in part a on the same axis, and check the result with your grapher.

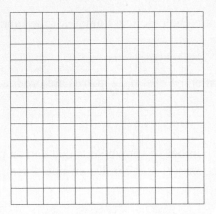

c. What can you say about the graphs of f and g with respect to the graph of $y = x$?

13. You work in the box office of a movie theater. The system for purchasing tickets is automated. You enter the number of tickets you need for each category (adult, child, senior), and the output is the total cost. The following table is a sample from a computer screen that shows the cost for adult tickets only. The number of tickets represents the input, and the total cost represents the output.

NUMBER OF ADULT TICKETS	TOTAL COST
2	$11.00
5	$27.50
7	$38.50
12	$66.00

a. Does this table represent a linear function? Explain.

b. Write a function, f, to represent the total cost as a function of the number of tickets, x, that are purchased.

c. Determine the cost of one ticket. What does this value represent in your function?

Often the customer approaches the window with the exact amount of cash for the tickets. In that case, you enter the total amount into the computer and press adult ticket, and out comes the number of tickets. In this case, the total cost is the input, and the number of tickets is the output.

d. Fill in the table showing this situation.

TOTAL COST	NUMBER OF TICKETS

e. Write a function g that represents the total number of tickets purchased as a function of the total cost, x.

f. What is the slope of this line? What is the relationship between the slopes of the two functions f and g?

g. Determine $f(g(x))$ and $g(f(x))$. Are the functions inverses of each other? Explain.

Exponential and Logarithmic Functions

CLUSTER 1 **Exponential Functions**

ACTIVITY 3.1
The Summer Job

OBJECTIVES

1. Determine the growth or decay factor of an exponential function.

2. Identify the properties of the graph of an exponential function defined by $y = b^x$, where $b > 0$ and $b \neq 1$.

3. Graph an exponential function.

Your neighbor's son will be attending college in the fall, majoring in mathematics. On July 1, he comes to your house looking for summer work to help pay for college expenses. You are interested since you need some odd jobs done, but you don't have a lot of extra money to pay him. He can start right away and will work all day July 1 for two cents. This gets your attention, but you wonder if there is a catch. He says that he will work July 2 for four cents, July 3 for eight cents, July 4 for sixteen cents, and so on for *every* day of the month of July.

1. Do you hire him?

For Problems 2–8, assume that you do hire him.

2. How much will he earn on July 5? July 6?

3. What will be his total pay for the first week of July (July 1 through July 7)?

4. **a.** Complete the following table.

DAY IN JULY (Input)	PAY IN CENTS (Output)
1	
2	
3	
4	
5	
6	
7	
8	

b. Do you notice a pattern in the output values? Describe how you can obtain the pay on a given day knowing the pay on the previous day.

c. Use what you discovered in part b to determine the pay on July 9.

5. a. The pay on any given day can be written as a power of two. Write each pay entry in the output column of the table in Problem 4 as a power of two. For example, $2 = 2^1, 4 = 2^2$.

b. Let n represent the number of days worked. Write an equation for the daily pay, $P(n)$, (in cents) as a function of n, the number of days worked. Note that the number of days worked is the same as the July date.

c. Use the rule from part b to determine how much your neighbor's son will earn on July 20. That is, determine the value of $P(n)$ when $n = 20$. What are the units of measurement of your answer?

d. How much will he earn on July 31? Be sure to indicate the units of your answer.

e. Was it a good idea to hire him?

6. a. Determine the rate of change of $P(n)$ as n increases from $n = 3$ to $n = 4$. What are the units of measurement of your answer?

b. Determine the rate of change of $P(n)$ as n increases from $n = 7$ to $n = 8$. Include units in your answer.

c. Is the function linear? Explain.

7. a. What is the practical domain of the function defined by $P(n) = 2^n$?

b. Sketch a scatterplot of ordered pairs of the form $(n, P(n))$ from July 1 to July 10 on appropriately scaled and labeled axes.

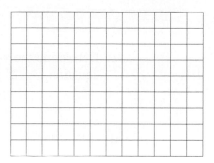

The function rule $P(n) = 2^n$ gives the relationship between the pay $P(n)$ (in cents) and the given July date, n, worked. This function belongs to a family of functions called **exponential functions**.

DEFINITION

Exponential functions are defined by equations of the form $y = b^x$, where the base b is a constant such that b is a positive number not equal to 1 ($b > 0$ and $b \neq 1$). Such functions are called **exponential** functions because the independent variable (input) x is the exponent.

Example 1   *Some examples of exponential functions are*

$$g(x) = 10^x, \; h(x) = (1.08)^x, \; V(x) = \left(\frac{1}{2}\right)^x, \text{ and } T(x) = (0.75)^x.$$

Graphs of Exponential Functions

Because n in $P(n) = 2^n$ (the summer job situation) represents a given day, the practical domain (whole numbers from 1 to 31) limits the investigation of the exponential function defined by $P(n) = 2^n$.

8. a. Consider the general function defined by $f(x) = 2^x$. Use your grapher to sketch a graph of this function. Use the window Xmin $= -10$, Xmax $= 10$, Ymin $= -2$, and Ymax $= 10$.

b. Because the graph of the general function $f(x) = 2^x$ is continuous (it has no holes or breaks), what appears to be the domain of the function f? What is the range of the function f?

c. Determine the y-intercept of the graph of f by substituting 0 for x in the equation $y = 2^x$ and solving for y.

d. Is the function f increasing or decreasing?

> **DEFINITION**
>
> If the base b of an exponential function defined by $y = b^x$ is greater than 1, then b is the **growth factor**. The graph of $y = b^x$ is increasing if $b > 1$.

Example 2 *The base 2 of $f(x) = 2^x$ is the growth factor because each time the input, x, is increased by 1, the output is multiplied by 2.*

9. Identify the growth factor, if any, for the given function.

a. $y = 1.08^x$

b. $h(x) = 0.8^x$

c. $y = 8x$

d. $g(x) = 10^x$

10. Return to the graph of $f(x) = 2^x$.

a. Does the graph of $f(x) = 2^x$ appear to have an x-intercept?

b. Use your calculator to complete the following table.

x	−10	−8	−6	−4	−2	−1
$f(x) = 2^x$						

Note: 2^{-10} is equivalent to $\frac{1}{2^{10}} \approx 0.0010$.

c. As the values of the input variable x decrease, what happens to the output values?

d. Use the trace feature of your grapher to trace the graph of $f(x) = 2^x$ for $x < 0$. What appears to be the relationship between the graph of $y = 2^x$ and the x-axis when x becomes smaller?

DEFINITION

A horizontal axis having equation $y = 0$ is called a **horizontal asymptote** of the graph of a function defined by $y = b^x$, where $b > 0$ and $b \neq 1$, because the graph of the function gets closer and closer to the x-axis $(y = 0)$ as the input gets farther from the origin, in the negative direction.

Example 3 *The x-axis is the horizontal asymptote of $y = 3^x$ and $y = 7^x$ because, as x gets smaller, the graph gets closer and closer to the x-axis. See the graph that follows.*

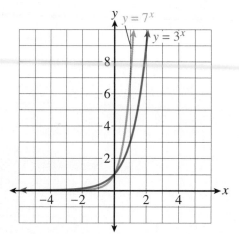

11. a. Complete the following table.

x	−3	−2	−1	0	1	2	3	4	5
$f(x) = 2^x$									
$g(x) = 10^x$									

b. Sketch the graph of the functions f and g on your grapher. Use the window Xmin $= -5$, Xmax $= 5$, Ymin $= -2$, and Ymax $= 9$.

c. Use the results from parts a and b to describe how the graphs of $f(x) = 2^x$ and $g(x) = 10^x$ are similar and how they are different. Be sure to include domain, growth factor, x- and y-intercepts, and horizontal asymptotes. Also discuss whether the graph of g increases faster or slower than the graph of f.

12. a. Complete the following table.

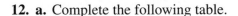

x	−3	−2	−1	0	1	2	3	4	5
$V(x) = \left(\frac{1}{2}\right)^x$									

b. Describe how you can obtain the output value for $x = 6$, using the output value for $x = 5$.

c. Sketch the graph of $V(x) = \left(\frac{1}{2}\right)^x$. Verify your sketch using your graphing calculator.

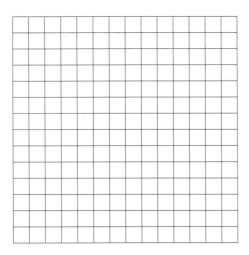

d. What are the domain and range of the function V?

e. Determine the vertical intercept of the graph of V.

f. Is the function V increasing or decreasing?

DEFINITION
If the base b of an exponential function $y = b^x$ is between 0 and 1, then b is the **decay factor**. The graph of $y = b^x$ is decreasing if $0 < b < 1$.

Example 4   *The base $\frac{1}{2}$ in the function $V(x) = \left(\frac{1}{2}\right)^x$ is the **decay** factor because each time x is increased by 1, the output value is multiplied by $\frac{1}{2}$.*

13. Identify the decay factor, if any, for the given function.

    a. $y = 0.98^x$

    b. $h(x) = 1.8^x$

    c. $y = 0.8x$

    d. $g(x) = \left(\frac{2}{7}\right)^x$

14. Return to the graph of $V(x) = \left(\frac{1}{2}\right)^x$.

    a. Does the graph of $V(x) = \left(\frac{1}{2}\right)^x$ have an x-intercept?

    b. Complete the following table.

x	1	3	5	7	10
$V(x) = \left(\frac{1}{2}\right)^x$					

    c. As the values of the input variable x get larger, what happens to the output values?

    d. Does the graph of V have a horizontal asymptote? Explain.

15. **a.** For each of the following exponential functions, identify the base, b, and determine whether the base is a growth or decay factor. Graph each function on your grapher, and complete the table below.

FUNCTION	BASE, b	GROWTH OR DECAY FACTOR	x-INTERCEPT	y-INTERCEPT	HORIZONTAL ASYMPTOTE	INCREASING OR DECREASING
$h(x) = (1.08)^x$						
$T(x) = (0.75)^x$						
$f(x) = (3.2)^x$						
$r(x) = \left(\frac{1}{4}\right)^x$						

b. Without graphing, how might you determine which of the functions in part a increase and which decrease? Explain.

16. Examine the output pattern to determine which of the following data sets is linear and which is exponential. For the linear set, determine the slope. For the exponential set, determine the growth or decay factor.

a.

x	−2	−1	0	1	2	3	4
y	−8	−4	0	4	8	12	16

b.

x	−2	−1	0	1	2	3	4
y	$\frac{1}{16}$	$\frac{1}{4}$	1	4	16	64	256

17. Determine the decay factor of the function represented by the data, and complete the table.

x	−2	−1	0	1	2
f(x)	16	4			

18. a. Complete the following table.

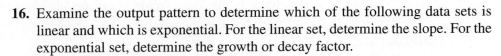

x	−3	−2	−1	0	1	2	3	4
$f(x) = 3^x$								
$g(x) = 3^x + 1$								
$h(x) = 3^{x+1}$								

b. Sketch a graph of the functions f, g, and h. Verify using your grapher.

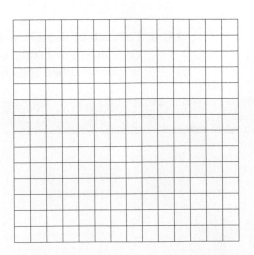

c. Describe any similarities or differences that you observe in the graphs, especially vertical and horizontal asymptotes.

SUMMARY
Activity 3.1

Functions defined by equations of the form $y = b^x$, where $b > 0$, and $b \neq 1$, are called **exponential functions** and have the following properties:

1. The domain is all real numbers.

2. The range is $y > 0$.

3. If $0 < b < 1$, the function is decreasing and has the following general shape.

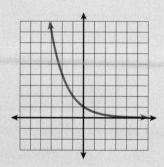

In this case, b is called the **decay factor**.

4. If $b > 1$, the function is increasing and has the following general shape.

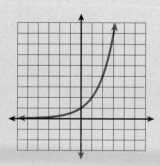

In this case, b is called the **growth factor**.

5. The vertical intercept (y-intercept) is $(0, 1)$.

6. The graph does not intersect the horizontal axis.

7. The line $y = 0$ (the x-axis) is a **horizontal asymptote**.

1. a. Complete the following tables.

x	-3	-2	-1	0	1	2	3
$h(x) = 5^x$							

x	-3	-2	-1	0	1	2	3
$g(x) = \left(\frac{1}{5}\right)^x$							

b. Sketch graphs of h and g on the following grid.

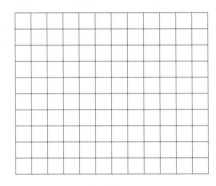

c. Use the tables and graphs in part a and b to complete the following table.

FUNCTION	BASE, b	GROWTH OR DECAY FACTOR	x-INTERCEPT	y-INTERCEPT	HORIZONTAL ASYMPTOTE	INCREASING OR DECREASING
$h(x) = 5^x$						
$g(x) = \left(\frac{1}{5}\right)^x$						

2. a. Complete the following table.

x	-3	-2	-1	0	1	2	3
$f(x) = 3^x$							
$g(x) = x^3$							
$h(x) = 3x$							

b. Sketch a graph of each of the given functions f, g, and h.

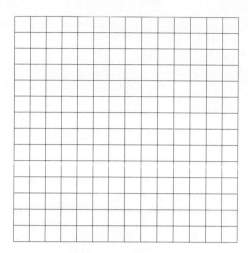

c. Describe any similarities or differences that you observe in the graphs.

3. Using your grapher, investigate the graphs of the following families (groups) of functions. Describe any relationships within each family, including domain and range, growth or decay factors, vertical and horizontal intercepts, and asymptotes. Identify the functions as increasing or decreasing.

a. $f(x) = 2^x$, $g(x) = 2^x + 1$, $h(x) = 2^x - 3$

b. $f(x) = \left(\dfrac{3}{4}\right)^x$, $g(x) = \left(\dfrac{4}{3}\right)^x$

c. $f(x) = 10^x$, $g(x) = -10^x$

d. $f(x) = 3^x$, $g(x) = \left(\frac{1}{3}\right)^x$

4. Determine which of the following data sets are linear and which are exponential. For the linear sets, determine the slope. For the exponential sets, determine the growth factor or the decay factor.

a.

x	−2	−1	0	1	2	3	4
y	$\frac{1}{9}$	$\frac{1}{3}$	1	3	9	27	81

b.

x	−2	−1	0	1	2	3	4
y	2	2.5	3	3.5	4	4.5	5

c.

x	−2	−1	0	1	2	3	4
y	0.75	1.5	3	6	12	24	48

d.

x	−2	−1	0	1	2	3	4
y	6.25	2.5	1	0.4	0.16	.064	.0256

5. Assume that y is an exponential function of x.

a. If the growth factor is 1.08, then complete the following table.

x	0	1	2	3
y	23.1			

b. If the decay factor is 0.75, then complete the following table.

x	0	1	2	3
y	10			

6. a. Would you expect $f(x) = 3^x$ to increase faster or slower than $g(x) = 2.5^x$ for $x > 0$? Explain. (*Hint:* You may want to use your grapher for help.)

b. Would you expect $f(x) = \left(\frac{1}{2}\right)^x$ to decrease faster or slower than $g(x) = (0.70)^x$ for $x > 0$? Explain.

7. Determine the domain and range of each of the following functions.

a.

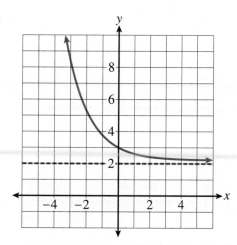

b.

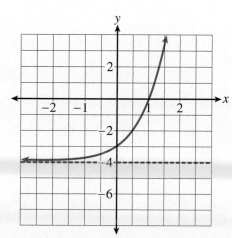

c.

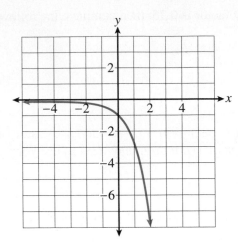

8. Use your graphing calculator to determine the domain and range of the following functions.

   a. $f(x) = \left(\frac{1}{2}\right)^x - 2$

   b. $f(x) = -4^x + 3$

9. Take a piece of paper from your notebook. Let x represent the number of times you fold the paper and $f(x)$ represent the number of sections the paper is divided into after the folding.

   a. Complete the table of values.

x	0	1	2	3	4	5
$f(x)$	1					

   b. If you could fold the paper eight times, how many individual sections will there be on the paper?

   c. Does this data represent an exponential function? Explain.

   d. What is the practical domain and range in this situation?

ACTIVITY 3.2
Cellular Phones

OBJECTIVES

1. Determine the growth and decay factor for an exponential function represented by a table of values or an equation.

2. Graph exponential functions defined by $y = ab^x$, where $b > 0$ and $b \neq 1$.

3. Determine the doubling and halving time.

During a meeting, you hear the familiar ring of a cell phone. Without hesitation, several of your colleagues reach into their jacket pockets, brief cases, and purses to receive the anticipated call. Although sometimes annoying, cell phones have become part of our way of life.

The following table shows the rapid increase in the number of cellular phones (figures are approximate) in recent years. Note that the input variable (year) increases in steps of one unit (year).

Calling All Cells

YEAR	NUMBER OF CELLULAR PHONES AS OF JAN. I (in millions)
1996	44.248
1997	55.312
1998	69.140
1999	86.425
2000	108.031

1. Is this a linear function? How do you know?

2. **a.** Evaluate the indicated ratios to complete the following table.

No. of phones in 1997 / No. of phones in 1996	No. of phones in 1998 / No. of phones in 1997	No. of phones in 1999 / No. of phones in 1998	No. of phones in 2000 / No. of phones in 1999

b. What do you notice about the values of the table?

In an exponential function with base b, equally spaced input values yield output values whose successive ratios are constant. If the input values increase by increments of 1, the common ratio is the base b. If $b > 1$, b is the growth factor; if $0 < b < 1$, b is the decay factor.

3. **a.** Does the relationship in the table preceding Problem 1 represent an exponential function? Explain.

b. What is the growth factor?

c. As a consequence of the result found in part b, you can start with 44.248, the number of cellular phones in 1996, and obtain the number of cellular phones in 1997 by multiplying by the growth factor, $b = 1.25$. You can then determine the number of cellular phones in 1998 by multiplying the number of cellular phones in 1997 by b, and so on. Verify this with your calculator. Note that because the exponential function is a mathematical model, the results will vary slightly from the actual number of phones given in the table preceding Problem 1.

Once you know the growth factor ($b = 1.25$), you can determine the equation that gives the number of phones as a function of t, the number of years since 1996. Note that $t = 0$ corresponds to 1996, $t = 1$ to 1997, and so on.

4. a. Complete the table.

t	CALCULATION FOR THE NUMBER OF CELL PHONES	EXPONENTIAL FORM	NUMBER OF CELL PHONES
0	44.248	$44.248(1.25)^0$	44.248
1	$(44.248)1.25$	$44.248(1.25)^1$	
2			
3			

b. Use the pattern in the preceding table to help you write the equation of the form $N(t) = a \cdot b^t$, where $N(t)$ represents the number of cell phones (in millions) in use at time t, the number of years since 1996.

c. What is the practical domain of the function N?

d. Graph the function N on your graphing calculator, and then sketch the result below on an appropriately scaled and labeled axis.

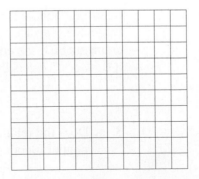

e. Determine the vertical intercept of the graph of N by substituting 0 for the input, t. What is the practical meaning of the vertical intercept in this situation?

DEFINITION

Many exponential functions can be represented symbolically by $f(t) = a \cdot b^t$, where a is the value of f when $t = 0$ and b is the growth or decay factor. If the input, t, of $y = a \cdot b^t$ represents time, then the coefficient a is called the initial value.

Example 1   *The exponential function defined by $f(x) = 5 \cdot 2^x$ has y-intercept $(0, 5)$ and growth factor $b = 2$. The exponential function defined by $h(x) = \frac{1}{2}(0.75)^x$ has y-intercept $\left(0, \frac{1}{2}\right)$ and decay factor $b = 0.75$.*

---

5. Use the function defined by $N(t) = 44.248(1.25)^t$ to estimate the number of cell phone users in 2005. Do you think this is a good estimate? Explain.

6. a. Use the graph of the exponential function $N(t) = 44.248(1.25)^t$ and the trace or table feature of your graphing calculator to estimate the number of years it takes for the number of cell phone users to double from 44.248 million to 88.496 million.

b. Estimate the time necessary for the number of cell phone users to double from 88.496 million to 176.992 million. Verify your estimate using your calculator.

c. How long will it take for any given number of cell phone users to double?

DEFINITION

The **doubling time** of an exponential function is the time it takes for an output to double. The doubling time is determined by the growth factor and remains the same for all output values.

Example 2   *The balance B(t), in dollars, of an investment account is defined by $B(t) = 5500(1.12)^t$, where t is the number of years. The initial value for this function is $5500. Determine the value of t when the balance is doubled or equal to $11,000.*

SOLUTION

If you use the table feature of your calculator, the doubling time is estimated at 6.1 years (see the following calculator graphic). The intersect feature on the graphing calculator shows the doubling time to be 6.12 years to the nearest hundredth.

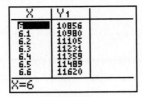

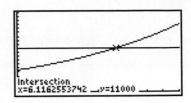

Decreasing Exponential Functions, Decay Factor, and Halving Time

You have just purchased a new automobile for $16,000. Much to your dismay, you have just learned that you should expect the value of your car to depreciate by 30% per year! The following table shows the retail value of the car for the next several years, where V is the value in thousands of dollars.

DEPRECIATION: TAKING ITS TOLL					
t (year)	0	1	2	3	4
V(t) (in thousands of dollars)	16	11.2	7.8	5.5	3.8

The values of the input, t, are incremented by 1, and a value of the output is obtained by multiplying the previous value by a constant factor. This is another example of an exponential function. However, because the value of the car is decreasing, the constant factor is a *decay factor* and its value will be between 0 and 1.

7. **a.** Calculate the decay factor, b, in this situation. Show the calculations you perform to determine the factor.

b. You can now start with 16, the initial value of the car (in thousands of dollars), and obtain the value after one year by multiplying by the decay factor, $b = 0.70$. The value of the car after two years is the value of the car after one year times the decay factor. Verify this on your calculator, and compare your results with the entries in the graphic prior to Problem 7.

8. a. Complete the following table.

T	CALCULATION OF THE VALUE OF THE CAR	EXPONENTIAL FORM	VALUE, $V(t) = a \cdot b^t$ (in thousands of dollars)
0	16	$16(0.70)^0$	16
1	$16 \cdot 0.70$	$16(0.70)^1$	
2			
3			

b. Use the pattern in the preceding table to help you write an equation of the form $V(t) = a \cdot b^t$, where $V(t)$ represents the value of the car as a function of time, t.

c. What is the practical domain of this function?

d. Graph this function on your grapher, and sketch the result below on an appropriate scaled and labeled set of axes.

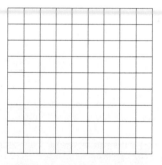

e. Determine the vertical intercept of the graph by substituting 0 for the input t. What is the practical meaning of the intercept in this situation?

9. a. Use your grapher to estimate the number of years it takes to halve the value of the automobile from $16,000 to $8000.

b. Estimate the time necessary to halve the value from $8000 to $4000. Verify your answer using your grapher.

c. How long will it take for any specific value of the car to halve?

> **DEFINITION**
>
> The **half-life** of an exponential function is the time it takes for an output to decay by one-half. The half-life is determined by the decay factor and remains the same for all output values.

10. Homemade chocolate chip cookies lose their freshness over time. Let the taste quality be 1 when the cookies are fresh. The taste quality decreases according to the function:

$$Q = 0.8^x,$$

where x is the number of days since the cookies were baked.

Determine when the taste quality will be one-half of its value. Use the intersect feature of your calculator to determine when $f(x)$ is $\frac{1}{2}$ of 1 or 0.5.

SUMMARY
Activity 3.2

1. For **exponential functions** defined by $f(x) = ab^x$, a is the value of f when $x = 0$ (sometimes called the initial value), and b is the growth or decay factor.

2. The vertical intercept of these functions is $(0, a)$.

3. In an exponential function, equally spaced input values yield output values whose successive ratios are constant. If the input values increase by one unit, then
   a. the constant ratio is the **growth factor** if the output values are increasing.
   b. the constant ratio is the **decay factor** if the output values are decreasing.

4. The **doubling time** of an increasing exponential function is the time it takes for an output to double. The doubling time is set by the growth factor and remains the same for all output values.

5. The **half-life** of a decreasing exponential function is the time it takes for an output to decay by one-half. The half-life is determined by the decay factor and remains the same for all output values.

EXERCISES
Activity 3.2

1. The population of Russia in selected years can be approximated by the following table.

YEAR	1995	1996	1997	2000
POPULATION (in millions)	148.0	147.6	146.9	146.0

a. Let 1995 correspond to $t = 0$. Let b be the ratio between the population of Russia in 1996 and 1995. Determine an exponential function of the form $y = a \cdot b^t$ to represent the population of Russia symbolically. Round to four decimal places.

b. Does the function in part a give an accurate value of the population of Russia in 2000? Explain.

c. Use your model in part b to predict the population of Russia in 2007.

2. Without using your graphing calculator, match each graph with its equation. Then check your answer using your graphing calculator.

a. $f(x) = 0.5(0.73)^x$ **b.** $g(x) = 3(1.73)^x$ **c.** $h(x) = -2(1.73)^x$

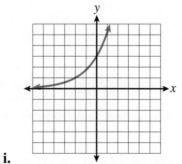

i.

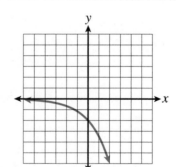

ii.

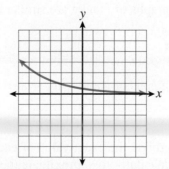

iii.

3. Which of the following tables represent exponential functions? Indicate the growth or decay factor for the data that is exponential.

a.

x	0	1	2	3	4
y	0	2	16	54	128

b.

x	0	1	2	3	4
y	1	4	16	64	256

c.

x	1	2	3	4	5
y	1750	858	420	206	101

4. **a.** Sketch a graph of $f(x) = 2^x$, $g(x) = 3 \cdot 2^x$, and $h(x) = -3 \cdot 2^x$ on the same coordinate axis.

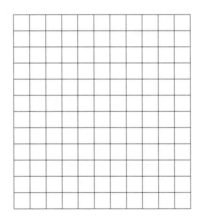

b. Describe how the graphs of f and g are similar and how they are different.

c. Describe how the graphs of g and h are related to each other.

5. If $f(x) = 3 \cdot 4^x$, determine the exact value of each of the following, when possible. Otherwise, use your calculator to approximate the value to the nearest hundredth.

a. $f(-2)$ **b.** $f\left(\frac{1}{2}\right)$

c. $f(2)$ d. $f(1.3)$

6. In 1995, the U.S. emitted approximately 1400 million tons of carbon into the atmosphere. This represented about one-fourth of the world total. The U.S. emissions were increasing at about 1.3% per year. If t represents the number of years since 1995 and $A(t)$ represents the amount of carbon (in millions of tons) emitted in a given year, then $A(t) = 1400(1.013)^t$.

   a. Complete the following table.

t, NUMBER OF YEARS SINCE 1995	0	1	2	3	4	5
A(t), AMOUNT OF U.S. CARBON EMISSIONS (in millions of tons)						

   b. Determine the growth factor for carbon emissions.

   c. Sketch a graph of this exponential equation. Use $0 \le t \le 25$ and $0 \le A(t) \le 2500$.

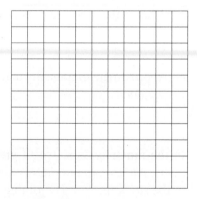

   d. Use the equation $A(t) = 1400(1.013)^t$ to determine the amount of carbon emission in 2010. Include the units of measurement in your answer.

   e. Use the graph and trace features of your grapher to approximate the year in which carbon emissions in the U.S. will exceed 2000 million tons.

7. Chlorine is used to disinfect swimming pools. The chlorine concentration should be between 1.5 and 2.5 parts per million (ppm). On sunny, hot days, 30% of the chlorine dissipates into the air or combines with other chemicals. The chlorine concentration, $A(x)$, (in parts per million) in a pool after x sunny days can be modeled by

$$A(x) = 2.5(0.7)^x.$$

a. What is the initial concentration of chlorine in the pool?

b. Complete the following table.

x	0	1	2	3	4	5
A(x)						

c. Sketch a graph of the chlorine function.

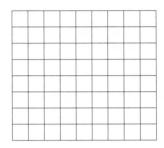

d. What is the chlorine concentration in the pool after three days?

e. Approximate graphically and numerically the number of days before chlorine should be added.

8. In the nineties, there was a growing national increase in the number of investment clubs. An investment club is a group of people who meet on a regular basis to invest in the stock market. By joining a club, members are able to share in a diverse portfolio and therefore reduce the risk of losing money.

The following table shows the rapid growth in the number of clubs from 1990 to 1996 (figures for the number of clubs are approximate). Note that the input variable (year) increases in steps of one unit.

YEAR	NUMBER OF CLUBS AS OF JAN. I
1990	5820
1991	7180
1992	8860
1993	10,930
1994	13,480
1995	16,630
1996	20,510

a. Does the relationship in the table represent an exponential function? Explain.

b. What is the growth factor?

c. Determine the equation that gives the number of clubs $N(t)$ as a function of t, the number of years since 1990. Note that $t = 0$ corresponds to 1990.

d. Graph the function.

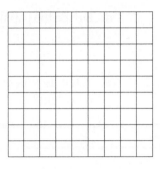

e. What is the vertical intercept? What is the practical meaning of this intercept in this situation?

f. Use the equation to estimate the number of clubs in 2000. Do you think this is a good estimate? Explain.

g. Use the graph of the exponential function and the trace or table feature of your grapher to estimate the number of years it takes for the number of clubs to double from 5820 to 11,640.

h. How long will it take for any given number of clubs to double?

ACTIVITY 3.3

Population
Growth

OBJECTIVES

1. Determine annual growth
   or decay rate of an
   exponential function
   represented by a table of
   values or an equation.

2. Graph an exponential
   function having equation
   $y = a(1 + r)^x$.

According to the 2000 U.S. census, the city of Charlotte, North Carolina, had a population of approximately 541,000.

1. a. Assuming that the population increases at a constant rate of 3.2%, determine the population of Charlotte (in thousands) in 2001.

b. Determine the population of Charlotte (in thousands) in 2002.

c. Divide the population in 2001 by the population in 2000, and record this ratio.

d. Divide the population in 2002 by the population in 2001, and record this ratio.

e. What do you notice about the ratios in parts c and d? What do these ratios represent?

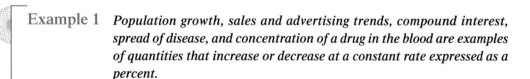

Linear functions represent quantities that change at a constant rate (slope). Exponential functions represent quantities that change at a constant rate, expressed as a percent.

Example 1 *Population growth, sales and advertising trends, compound interest, spread of disease, and concentration of a drug in the blood are examples of quantities that increase or decrease at a constant rate expressed as a percent.*

2. Let t represent the number of years since 2000 ($t = 0$ corresponds to 2000). Use the results from Problem 1 to complete the following table.

t, Years (since 2000)	0	1	2	3	4	5
P, population (in thousands)	541					

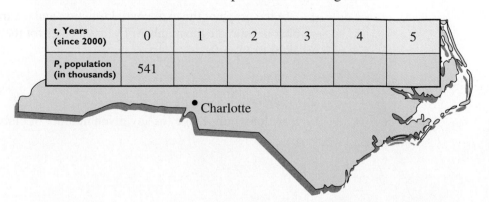

Charlotte

Once you know the growth factor ($b = 1.032$), you can determine the exponential model that describes the population of Charlotte as a function of t, where $t = 0$ corresponds to the year 2000.

3. a. Complete the following table.

t	CALCULATION FOR POPULATION (in thousands)	EXPONENTIAL FORM	$P(t)$, POPULATION (in thousands)
0	541	$541(1.032)^0$	541
1	$(541)1.032$	$541(1.032)^1$	
2	$(541)(1.032)(1.032)$		
3			

b. Use the pattern in the table in part a to help you write the symbolic rule for $P(t)$, the mathematical model of the population of Charlotte (in thousands), using t, the number of years since 2000, as the input value.

The equation $P(t) = 541(1.032)^t$ has the general form $P = P_0(1 + r)^t$, where r is the annual **growth rate**, $(1 + r)$ is the **growth factor** or the base, b, of the exponential function, t is the time in years, and P_0 is the initial value, the population when $t = 0$.

Example 2

a. *Determine the growth factor and the growth rate of the function defined by* $f(x) = 250(1.7)^x$.

SOLUTION

The growth factor $1 + r$ is the base 1.7. To determine the growth rate, solve the equation $1 + r = 1.7$ for r.

$$r = 0.7 \text{ or } 70\%$$

b. *If the growth rate of a function is 5%, determine the growth factor.*

SOLUTION

If $r = 5\%$ or 0.05, the growth factor is $1 + r = 1 + 0.05 = 1.05$.

4. a. Determine the growth factor in the Charlotte population function $P(t) = 541(1.032)^t$.

b. Determine the growth rate. Express your answer as a percent.

5. a. Using the function defined by $P(t) = 541(1.032)^t$, determine the population of Charlotte in 2006. That is, determine $P(t)$ when $t = 6$.

b. Graph the population function with your graphing calculator, adjusting the window to show the population between 2000 and 2010. Use the window Xmin = 0, Xmax = 10, Ymin = 540, and Ymax = 740. What type of function does the graph resemble?

c. Reset the window to Xmin = -50, Xmax = 100, Ymin = 0, and Ymax = 13,000, and display the graph. Now what type of function does the graph look like?

d. Determine $P(0)$. What is the graphical and the practical meaning of $P(0)$?

6. a. Use the model to predict Charlotte's population in 2010.

b. Verify your prediction on the graph.

7. a. Use the graph to estimate when Charlotte's population will reach 700,000, assuming it continues to grow at the same rate. Remember, $P(t)$ is the number of thousands.

b. Evaluate $P(32)$ and describe what it means.

8. Use the model to estimate the population of Charlotte in 2002 and in 2020. In which prediction are you more confident? Why?

9. a. Assuming the growth rate remains constant, how long will it take for the population of Charlotte to double its 2000 population?

b. Explain how you reached your conclusion in part a.

Waste-Water Treatment Facility

You are working at a waste-water treatment facility. You are presently treating water contaminated with 18 micrograms of pollutant per liter. Your process is designed to remove 20% of the pollutant during each treatment. Your goal is to reduce the pollutant to less than 3 micrograms per liter.

10. a. What percent of pollutant present at the start of a treatment remains at the end of the treatment?

b. The concentration of pollutant is 18 micrograms per liter at the start of the first treatment. Use the result of part a to determine the concentration of pollutant at the end of the first treatment.

c. Complete the following table. Round the results to the nearest hundredth.

n, NUMBER OF TREATMENTS	0	1	2	3	4	5
$C(n)$, CONCENTRATION OF POLLUTANT, IN $\mu g/l$, AT THE END OF THE nTH TREATMENT	18	14.4				

d. Write an equation for the concentration, $C(n)$, of the pollutant as a function of the number of treatments, n.

The equation $C(n) = 18(0.80)^n$ has the general form $C = C_0(1 - r)^n$, where r is the **decay rate**, $(1 - r)$ is the **decay factor** or the base of the exponential function, n is the number of treatments, and C_0 is the initial value, the concentration when $n = 0$.

Example 3   **a.** *Determine the decay factor and the decay rate of the function defined by $h(x) = 123(0.43)^x$.*

SOLUTION

The decay factor $1 - r$ is the base, 0.43. To determine the decay rate, solve the equation $1 - r = 0.43$ for r.

$$r = 0.57 \text{ or } 57\%$$

b. *If the decay rate of a function is 5%, determine the decay factor.*

SOLUTION

If $r = 5\%$ or 0.05, the decay factor is $1 - r = 1 - 0.05 = 0.95$.

11. a. If the decay rate is 2.5%, what is the decay factor?

b. If the decay factor is 0.76, what is the decay rate?

12. a. Use the function defined by $C(n) = 18(0.8)^n$ to predict the concentration of contaminants at the waste-water treatment facility after seven treatments.

b. Sketch a graph of the concentration function on your grapher. Use the table in Problem 10c to set a window. Does the graph look like you expected it would? Explain.

c. What is the vertical intercept? What is the practical meaning of the intercept in this situation?

d. Reset the window of your grapher to Xmin = −5, Xmax = 15, Ymin = −10, and Ymax = 50. Does the graph have a horizontal asymptote? Explain what this means in this situation.

13. Use the table or trace feature of your grapher to estimate the number of treatments necessary to bring the concentration of pollutant below 3 micrograms per liter.

SUMMARY
Activity 3.3

1. **Exponential functions** are used to describe phenomena that grow or decay by a constant percent rate over time.

2. If r represents the **annual growth rate**, the exponential function that models the quantity, P, can be written as

$$P(t) = P_0(1 + r)^t,$$

where P_0 is the initial amount, t represents the number of elapsed years, and $1 + r$ is the growth factor.

3. If r represents the **annual decay rate**, the exponential function that models the amount remaining can be written as

$$P(t) = P_0(1 - r)^t,$$

where $1 - r$ is the decay factor.

EXERCISES
Activity 3.3

1. Determine the growth and decay factors and growth and decay rates in the following table.

GROWTH FACTOR	GROWTH RATE		DECAY FACTOR	DECAY RATE
1.02			0.77	
	2.9%			68%
2.23			0.953	
	34%			19.7%
1.0002			0.9948	

2. The 2000 U.S. Census reports the populations of Bozeman, Montana, as 27,509 and Butte, Montana, as 32,370. Since the 1990 census, Bozeman's population has been increasing at approximately 1.96% per year. Butte's population has been decreasing at approximately 0.29% per year. Assume that the growth and decay rates stay constant.

   a. Let $P(t)$ represent the population t years after 2000. Determine the exponential functions that model the populations of both cities.

   b. Use your models to predict the populations of both cities in 2005.

c. Estimate the number of years necessary for the population of Bozeman, Montana, to double.

d. Using the table and/or graphs of these functions, predict when the populations will be equal.

3. You have just taken over as the city manager of a small city. The personnel expenses were \$8,500,000 in 2002. Over the past five years, the personnel expenses have increased at a rate of 3.2% annually.

a. Assuming that this rate continues, write an equation describing personnel costs, $C(t)$, in millions of dollars, where $t = 0$ corresponds to 2002.

b. Sketch a graph of this function up to the year 2012 ($t = 10$).

c. What are your projected personnel costs in the year 2007?

d. What is the vertical intercept? What is the practical meaning of the intercept in this situation?

e. In what year will the personnel expenses be double the 2002 personnel expenses?

4. According to the U.S. Bureau of the Census, the population of the United States from 1930 to 2000 can be modeled by $P(t) = 120.6 \cdot 1.0125^t$, where t represents the number of years since 1930.

a. Sketch a graph of the U.S. population model from 1930 to 2000.

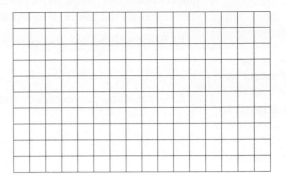

b. Determine the annual growth rate and the growth factor from the equation.

c. Use the population equation to determine the population (in millions) of the U.S. in 2000. How does your answer compare to the actual population of 281.4 million?

5. You have recently purchased a new car for $20,000, by arranging financing for the next five years. You are curious to know what your new car will be worth when the loan is completely paid off.

a. Assuming that the value depreciates at a constant rate of 15%, write an equation that represents the value $V(t)$ of the car t years from now.

b. What is the decay rate in this situation?

c. What is the decay factor in this situation?

d. Use the equation from part a to estimate the value of your car five years from now.

e. Use the trace and table features of your graphing calculator to check your results in part d.

f. Use the trace or table features of you graphing calculator to determine when your car will be worth $10,000.

**PROJECT
ACTIVITY 3.4**
Photocopying
Machines

OBJECTIVES

1. Generate data given the growth or decay rate of an exponential function.
2. Write exponential functions given the growth or decay rate.
3. Graph exponential functions from data.
4. Determine doubling and halving times from exponential functions.

Most photocopy machines allow for enlarging or reducing the size of the original. Suppose you have a chart you wish to photocopy for a report you need to submit to your supervisor. The chart is 10 inches wide by 7 inches high. To reduce the size of the copy 20%, you set the machine to reduce by taking 80% of the original dimensions.

1. What will be the dimensions of your photocopy?

2. What is the percentage reduction in *area* of the photocopy?

3. If your chart must fit in a space only 4 inches high in your report, how many times would you need to reduce the original? (Assume the photocopier is set at 80% reduction.)

4. **a.** Complete the following table showing the dimensions of the chart after *x* 80% reductions (photocopies of photocopies). Record each length to the nearest hundredth of an inch.

Smaller and Smaller

x, THE NUMBER OF 80% REDUCTIONS	0	1	2	3	4	5	6	7	8	9	10
h(x), HEIGHT (in.)	7	5.6									
w(x), WIDTH (in.)	10	8									

b. If you were actually to perform ten 80% reductions (photocopies of photocopies), what do you think your results would look like?

5. Use the data from your table to plot points for two curves on separate axes. Use the number of reductions, *x*, versus the heights of the reduced copies for one curve, and the number of reductions versus the widths of the reduced copies for a second curve. Describe how the two graphs are similar and how they are different.

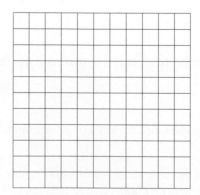

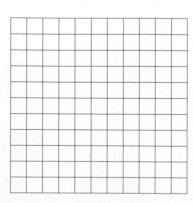

6. a. Notice that each entry in the table for $h(x)$ or $w(x)$ can be obtained by multiplying the previous entry by a constant factor. What is the constant factor in this situation?

b. What is the practical meaning of the constant factor in this situation?

7. Determine the equations that define the functions h and w, where x is the number of 80% reductions. Enter these functions on your grapher to verify your work in Problems 4 and 5.

8. The 80% reduction resulted in a decreasing function. What kind of percentage would you need to enter on the photocopier to result in an increasing function?

9. a. Complete the following table showing the width of a 10-inch-wide chart in which x represents the number of 20% enlargements.

x, NUMBER OF 20% ENLARGEMENTS	0	1	2	3	4	5
$w(x)$, WIDTH (in.)						

b. What is the growth factor? What is the practical meaning of the growth factor in this situation?

c. Write an equation for w in terms of x, where x is the number of 20% enlargements.

d. Use your graphing calculator to graph the function.

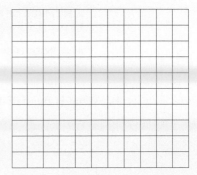

10. a. If the machine you are using will enlarge only at 20%, how many times would you need to copy the 10-inch width to make it at least 20 inches wide?

b. Will the chart ever be exactly 20 inches wide? Explain.

c. Describe in detail how you found your answers.

11. Assuming a constant 20% enlargement, how many copies would it take to get your original 10-inch width to grow to at least 40 inches? At least 80 inches?

12. Complete the following table with your results from Problems 10 and 11.

x, NUMBER OF 20% ENLARGEMENTS	0	4	8	12
WIDTH (in.)				

13. Use your grapher to determine how many copies are needed to double the output.

14. Examine the height and width functions given in Problem 7, $h(x) = 7(0.8)^x$ and $w(x) = 10(0.8)^x$.

a. Complete the following table

x, NUMBER OF 80% REDUCTIONS	0	3	6	9
$h(x)$, HEIGHT (in.)				

x, NUMBER OF 80% REDUCTIONS	0	3	6	9
$w(x)$, WIDTH (in.)				

b. What is the half-life of each function?

ACTIVITY 3.5

Compound
Interest

OBJECTIVES

1. Apply the compound interest and continuous compounding formulas to a given situation.

2. Graph base e exponential functions.

3. Solve problems involving continuous growth and decay models.

Congratulations, you have inherited $20,000! Your grandparents suggest that you use half of the inheritance to start a retirement fund. Your grandfather claims that an investment of $10,000 could grow to over half a million dollars by the time of retirement. You are intrigued by this statement and decide to investigate if this could possibly happen.

Suppose you deposit $10,000 in the bank at a 6.5% annual interest rate. After one year, your balance is

$$10,000 + 0.065(10,000) = 10,000 + 650 = 10,650$$

The interest, $650, earned during the year becomes part of the new balance. At the end of the second year, your balance is

$$10,650 + 0.065(10,650) = 10,650 + 692.25 = 11,342.25$$

Note that you made interest on the original deposit, plus interest on the first year's interest. In this situation, we say that interest is compounded. Usually, the compounding occurs at fixed intervals (typically at the end of every year, quarter, month, or day). In the preceding situation, interest is compounded annually.

If interest is compounded, then the current balance is given by the formula

$$A = P\left(1 + \tfrac{r}{n}\right)^{nt},$$

where A is the current amount, or balance, in the account;

P is the principal (the original amount deposited);

r is the annual interest rate (annual percentage rate in decimal form);

n is the number of times per year that interest is compounded; and

t is the time in years the money has been invested.

> **Example 1**   *You invest $100 at 4% compounded quarterly. How much money do you have after five years?*
>
> SOLUTION
>
> The principal is $100, so $P = 100$. The annual interest rate is 4%, so $r = 0.04$. Interest is compounded quarterly, that is, four times per year, so $n = 4$. The money is invested for five years, so $t = 5$. Substituting numbers for the variables in the preceding formula, you have
>
> $$A = 100\left(1 + \tfrac{0.04}{4}\right)^{4 \cdot 5} = \$122.02.$$

1. **a.** Suppose you deposit $10,000 in an account that has a 6.5% annual interest rate (usually referred to as APR, for annual percentage rate), and whose interest is compounded annually ($n = 1$). Substitute the appropriate values for P, n and r into the preceding formula to get the balance, A, as a function of time, t.

b. Use the formula from part a to determine your balance, A, at the end of the first year $(t = 1)$?

c. What will be the amount of interest earned in the first year?

d. Use the formula developed in part a and complete the following table.

t, YEAR	0	1	2	3	4
A, BALANCE	10,000.00				

e. The formula in part a defines A as an exponential function of t. Identify the base.

f. Is the base a growth or decay factor? Explain.

2. a. Suppose you deposit the $10,000 into an account that has the same interest rate (APR) of 6.5%, with compounding quarterly $(n = 4)$ rather than annually $(n = 1)$. Write a new formula for your balance, A, as a function of time.

b. What would be your balance after the first year?

c. Use the table feature of your calculator to determine the balance at the end of each year for ten years, and record the values in the table in Problem 3 under $n = 4$ (compounded quarterly).

d. What is the base of this exponential function?

3. Now deposit your $10,000 into a 6.5% APR account with *monthly* compounding $(n = 12)$ and then in an account with *daily* compounding $(n = 365)$. Use your grapher and the appropriate formula to complete the following table.

COMPARISON OF $10,000 PRINCIPAL IN 6.5% APR ACCOUNTS WITH VARYING COMPOUNDING PERIODS			
T	n = 4	n = 12	n = 365
0			
1			
2			
3			
4			
5			
6			
7			
8			
9			
10			

4. In Problem 3, you calculated the balance on a deposit of $10,000 at an annual interest rate of 6.5% that was compounded at different intervals. After ten years, which account has the higher balance? Does this seem reasonable? Explain.

Continuous Compounding

You could extend this problem so that interest is compounded every hour or every minute or even every second. However, compounding more frequently than every hour does not increase the balance very much.

To discover why this happens, take a closer look at the exponential functions from Problems 1–3.

$$A = 10{,}000\left(1 + \tfrac{0.065}{1}\right)^{\cdot\, t} = 10{,}000[\underline{(1 + 0.065)^1}]^t$$

$$A = 10{,}000\left(1 + \tfrac{0.065}{4}\right)^{4\,\cdot\, t} = 10{,}000\left[\underline{\left(1 + \tfrac{0.065}{4}\right)^4}\right]^t$$

$$A = 10{,}000\left(1 + \tfrac{0.065}{12}\right)^{12\,\cdot\, t} = 10{,}000\left[\underline{\left(1 + \tfrac{0.065}{12}\right)^{12}}\right]^t$$

$$A = 10{,}000\left(1 + \tfrac{0.065}{365}\right)^{365\,\cdot\, t} = 10{,}000\left[\underline{\left(1 + \tfrac{0.065}{365}\right)^{365}}\right]^t$$

Can you discover a pattern in the form of the underlined expressions?

Each formula can be expressed as $A = 10{,}000b^t$, where $b = \left(1 + \tfrac{0.065}{n}\right)^n$ for $n = 1, 4, 12,$ and 365. The number b is called the **growth** factor, and n is the number of compounding periods per year.

> **Example 2** *If $n = 4$ in the formula $b = \left(1 + \frac{0.065}{n}\right)^n$, then $b = \left(1 + \frac{0.065}{4}\right)^4 = 1.06660$. The number 1.06660 is the growth factor.*

5. a. Determine the value of b in the following table, where $b = \left(1 + \frac{0.065}{n}\right)^n$. Round to five decimal places.

n, NUMBER OF COMPOUNDING PERIODS	1	4	12	365
b, GROWTH FACTOR				

b. If interest is compounded hourly, then $n = 365 \cdot 24 = 8760$. Compute the growth factor b for compounding hourly.

c. Write a sentence comparing the growth factor b for compounding hourly, $n = 8760$, to that for daily compounding, $n = 365$.

If the compounding periods become shorter and shorter (compounding every hour, every minute, every second), n gets larger and larger. If you consider the period to be so short that it's essentially an instant in time, you have what is called **continuous compounding**. Some banks use this method for compounding interest.

The formula $A = P\left(1 + \frac{r}{n}\right)^{nt}$ is no longer used when interest is compounded continuously. The following develops a formula for continuous compounding.

Step 1. Rewrite the given formula as indicated using properties of exponents.

$$A = P\left(1 + \frac{r}{n}\right)^{nt} = P\left[\left(1 + \frac{r}{n}\right)^{\frac{n}{r}}\right]^{rt}, \text{ since } \frac{n}{r} \cdot rt = nt$$

Step 2. Let $\frac{n}{r} = x$. It follows that $\frac{r}{n} = \frac{1}{x}$. Note that as n gets very large, the value of x also gets very large.

Step 3. Substituting $\frac{1}{x}$ for $\frac{r}{n}$ and x for $\frac{n}{r}$, in the rewritten formula in step 1, you have

$$A = P\left[\left(1 + \frac{r}{n}\right)^{\frac{n}{r}}\right]^{rt} = P\left[\left(1 + \frac{1}{x}\right)^x\right]^{rt}.$$

6. a. Now take a closer look at the expression $\left(1 + \frac{1}{x}\right)^x$. Enter $\left(1 + \frac{1}{x}\right)^x$ into your calculator as a function of x. Display a table that starts at 0 and is incremented by 100. The results are displayed below.

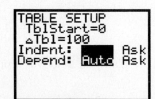
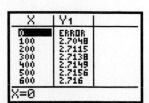

b. In the table of values, why is there an error at $x = 0$?

c. Scroll down in the table and describe what happens to the output, $\left(1 + \frac{1}{x}\right)^x$, as the input, x, gets very large.

The letter e is used to represent the number that $\left(1 + \frac{1}{x}\right)^x$ approaches as x gets very large. This notation was devised by mathematician Leonhard Euler (1707-1783). Euler used the letter e to denote this number because the number is irrational and its decimal representation never ends and never repeats.

7. The number e is a very important number in mathematics. Find it on your calculator, and write its decimal approximation below. How does this approximation compare to the result in Problem 6a?

You are now ready to complete the formula for continuous compounding. Substituting e for $\left(1 + \frac{1}{x}\right)^x$ in $A = P\left[\left(1 + \frac{1}{x}\right)^x\right]^{rt}$, you obtain the continuous compounding formula

$$A = Pe^{rt},$$

where A is the current amount, or balance, in the account;

P is the principal;

r is the annual interest rate (annual percentage rate in decimal form);

t is the time in years that your money has been invested; and

e is the base of the continuously compounded exponential function.

Example 3   *You invest $100 at a rate of 4% compounded continuously. How much money will you have after five years?*

SOLUTION

The principal is $100, so $P = 100$. The annual interest rate is 4%, so $r = 0.04$. The money is invested for five years, so $t = 5$. Because interest is compounded continuously, you use the formula for continuous compounding as follows.

$$A = 100e^{0.04 \cdot 5} = \$122.14$$

8. Calculate the balance of your $10,000 investment in ten years with an annual interest rate of 6.5% compounded continuously.

9. a. Historically, investments in the stock market have yielded an average rate of 11.7% per year. Suppose you invest $10,000 in an account at an 11% annual interest rate that compounds continuously. Use the formula $A = Pe^{rt}$ to determine the balance after 35 years.

b. What is the balance after 40 years?

c. Your grandfather claimed that $10,000 could grow to more than half a million dollars by retirement time (40 years). Is your grandfather correct in his claim?

Exponential Functions with Base *e*

10. a. Enter the function $y = e^x$ into your graphing calculator. Use the table feature to complete the following table.

x	-2	-1	0	2	4	6
$y = e^x$						

a. Sketch a graph of $y = e^x$. Verify your sketch using your graphing calculator.

b. Is this function increasing or decreasing?

c. Determine the vertical and horizontal intercepts (if any).

d. Does the graph of $y = e^x$ have a horizontal asymptote? Explain.

11. a. Enter the function $y = e^{-x}$ into your graphing calculator. Use the table feature to complete the following table.

x	-2	-1	0	2	4	6
$y = e^{-x}$						

b. Sketch a graph of $y = e^{-x}$. Verify your sketch using your graphing calculator.

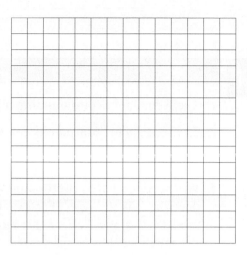

c. Is this function increasing or decreasing?

d. Determine the vertical and horizontal intercepts (if any).

e. Does the graph of $y = e^{-x}$ have a horizontal asymptote? Explain.

Continuous Growth and Decay

Whenever growth is continuous at a constant rate, the exponential model used to describe it is $y = y_0 e^{rt}$. For exponential decay at a continuous, constant rate, the model is $y = y_0 e^{-rt}$. For either model, r is the constant growth (or decay) rate, y_0 is the amount present initially (when $t = 0$), and e is the constant irrational number approximately equal to 2.718.

12. *E. coli* bacteria is capable of very rapid growth, doubling in number approximately every 49.5 minutes. The number, N, of *E. coli* bacteria per milliliter after x minutes can be modeled by the equation

$$N = 500,000 e^{0.014x}.$$

a. What is the initial number of bacteria?

b. How many *E. coli* bacteria would you expect after 99 minutes? (*Hint:* There will be two doublings.) Verify your estimate using the equation.

    c. Use a graphing or numerical approach to determine the elapsed time when there would be 20,000,000 *E. coli* bacteria.

SUMMARY
Activity 3.5

1. The formula for **compounding interest** is $A = P\left(1 + \frac{r}{n}\right)^{nt}$.

2. The formula for **continuous compounding** is $A = Pe^{rt}$.

3. If the number of **compounding periods** is large, $A = P\left(1 + \frac{r}{n}\right)^{nt} \approx A = Pe^{rt}$.

4. Whenever growth is **continuous at a constant rate**, the exponential model used to describe it is $y = y_0 e^{rt}$. For exponential decay at a continuous constant rate, the model is $y = y_0 e^{-rt}$. For either model, r is the constant growth (or decay) rate, y_0 is the amount present initially (when $t = 0$), and e is the constant irrational number approximately equal to 2.718.

EXERCISES
Activity 3.5

1. You inherit $25,000 and deposit it into an account that earns 4.5% compounded quarterly.

    a. Write an equation that gives the amount of money in the account after *t* years.

    b. How much money will be in the account after ten years?

    c. You want to have approximately $65,000 in the bank when your first child begins college. Use your grapher to determine in how many years you will reach this goal.

    d. If the interest were to be compounded continuously at 4.5%, how much money would be in the account after ten years?

    e. Compounding continuously, use your grapher to determine in how many years you would reach your goal of $65,000.

    f. Should you look for an investment account that will be compounded continuously?

2. You deposit $2000 in an account that earns 5% compounded monthly.

   a. What will be your balance after two years?

   b. Estimate how long it would take for your investment to double.

   c. Identify the growth rate and the growth factor.

3. Your friend deposits $1900 in an account that earns 6% compounded continuously.

   a. What will be her balance after two years?

   b. Estimate how long it will take for your friend's investment to double.

4. You are 25 years old and begin to work for a large company that offers you two different retirement options.

   Option 1. You will be paid a lump sum of $20,000 for each year you work for the company.

   Option 2. The company will deposit $10,000 into an account that will pay you 12% compounded monthly. When you retire, the money will be given to you.

   Let A represent the amount of money you will have for retirement after t years.

   a. Write an equation that represents option 1.

   b. Write an equation that represents option 2.

   c. Use your graphing calculator to sketch a graph of the two options on the same axis.

   d. If you plan to retire at age 65, which would be the better plan? Explain.

    e. If you decide to retire at age 55, which would be the better plan? Explain.

    f. Use your grapher to determine at what age it would not make a difference which plan you choose.

5. Strontium 90 is a radioactive material that decays according to the function defined by $y = y_0 e^{-0.0244t}$, where y_0 is the amount present initially and t is time in years.

    a. If there are 20 grams of strontium 90 present today, how much will be present in 20 years?

    b. Use the graph of the function to approximate how long it will take for 20 grams to decay to 10 grams, 10 grams to decay to 5 grams, and 5 grams to decay to 2.5 grams. The length of time is called the *half-life*. In general, a half-life is the time required for half of a radioactive substance to decay.

    c. Identify the annual decay rate and the decay factor.

6. When drugs are administered into the bloodstream, the amount present decreases continuously at a constant rate. The amount of a certain drug in the bloodstream is modeled by the function $y = y_0 e^{-0.35t}$, where y_0 is the amount of the drug injected (in milligrams) and t is time (in hours).

    a. Suppose that 10 milligrams are injected at 10:30 A.M., how much of the drug is still in the bloodstream at 2:00 P.M.?

    b. If another dose needs to be administered when there is 1 milligram of the drug present in the bloodstream, approximately when should the next dose be given (to the nearest quarter hour)?

7. The amount of credit-card spending from Thanksgiving to Christmas has increased by 14% per year since 1987. The amount, A, in billions of dollars of credit-card spending during the holiday period in a given year can be modeled by

$$A = f(x) = 36.2e^{0.14x},$$

where x represents the number of years since 1987.

    a. How much was spent using credit cards from Thanksgiving to Christmas in 1996?

b. Sketch a graph of the credit-card function.

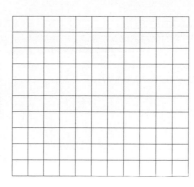

c. What is the vertical intercept of the graph? What is the practical meaning of the intercept in this situation?

d. Determine, graphically and numerically, the year when credit-card spending reached 75 billion dollars.

e. What is the doubling time?

8. Sketch a graph of each of the following and then verify using your graphing calculator.

    a. $f(x) = e^x + 2$ **b.** $g(x) = e^{x^2}$

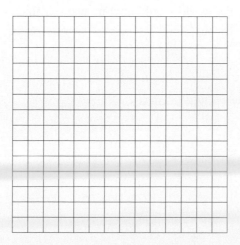
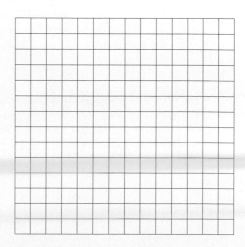

c. $h(x) = e^{2x}$ **d.** $F(x) = 3e^{x+1}$

 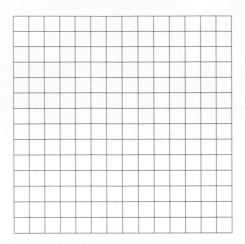

9. **Lab Exercise:** Being irrational, the number e cannot be expressed as the ratio of two integers. Its significance in higher mathematics is unmatched by any other real number, with the possible exception of π, another irrational number. Research the following questions.

     a. Why is the letter e used to represent this important number?

     b. Give examples of additional numbers that are irrational.

     c. How many decimal places do mathematicians and scientists know for either e or π?

     d. Since it has been proven that irrational numbers never have a repeating pattern of decimals, why in the world would anyone want to find so many digits for either e or π?

ACTIVITY 3.6

College Graduates

OBJECTIVES

1. Determine the equation of an exponential function that best fits the given data.

2. Make predictions using an exponential regression equation.

3. Determine whether a linear or exponential model best fits the data.

According to the U.S. Department of Education, the number of college graduates increased significantly during the twentieth century. The following table gives the number (in thousands) of college degrees awarded from 1900 through 2000.

College Bound

YEAR	NUMBER OF COLLEGE GRADUATES (thousands)
1900	30
1910	54
1920	73
1930	127
1940	223
1950	432
1960	530
1970	878
1980	935
1990	1017
2000	1180

1. Let t represent the number of years since 1900 ($t = 0$ corresponds to 1900, $t = 10$ to 1910, etc.). Let N represent the number of college graduates (in thousands) at time t. Sketch a scatterplot of the given data on your graphing calculator. Your scatterplot should appear as follows:

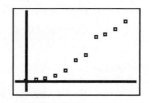

In Chapter 1, you learned how to use your calculator to determine a linear regression equation to model data with a linear function (see Activity 1.9 and Appendix C). You can use your calculator to model data with an exponential function just as easily. Use option 0, ExpReg, instead of option 4, LinReg in the STAT CALC menu.

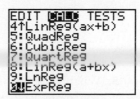

2. **a.** Use your graphing calculator to determine the equation of an exponential function that best fits the given data.

b. Sketch a graph of the exponential model using your graphing calculator.

c. What is the practical domain of this exponential function?

d. What is the vertical intercept of the graph? How does it compare to the actual initial value ($t = 0$) from the table?

3. a. What is the base of the exponential model? Is the base a growth or decay factor? How do you know?

b. What is the annual growth rate?

4. a. Use the exponential model to determine the number of college graduates in 2010 ($t = 110$).

b. Use the exponential model to write an equation that can be used to determine the year in which there will be 2 million college graduates. Remember that the number of college graduates is measured in thousands.

c. Solve the equation in part a using a graphing approach. Use the intersect feature of your graphing calculator; the screen containing the solution should appear as follows.

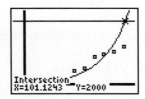

5. What is the doubling time for your exponential model? That is, approximately how many years will it take for a given number of college graduates to double?

Decreasing Exponential Model

Students in U.S. public schools have had much greater access to computers in recent years. The following table shows the number of students per computer in selected years.

YEAR	1983	1984	1985	1987	1989	1992	1995	1999
NUMBER OF STUDENTS PER COMPUTER	125	75	50	32	22	16	10	5.7

6. **a.** Use your grapher to determine the equation of an exponential function that models the given data. Let your input, t, represent the number of years since 1983.

b. Sketch a graph of the exponential model using your graphing calculator.

c. What is the base of the exponential model? Is the base a growth or decay factor? How do you know?

d. What is the annual decay rate?

e. Does the graph have a horizontal asymptote? What is the practical meaning of this asymptote in the context of the situation?

EXERCISES
Activity 3.6

1. The total amount of money spent on health care in the United States is increasing at an alarming rate. The following table gives the total national health care expenditures in billions of dollars in selected years from 1970 through 1999.

YEAR	1970	1975	1980	1985	1990	1995	1999
TOTAL SPENT (BILLIONS OF DOLLARS)	73.1	129.8	245.8	426.5	695.6	987	1210.7

Source: National Center for Health Statistics.

Exercise numbers appearing in color are answered in the Selected Answers appendix.

a. Would the data in the preceding table be better modeled by a linear model, $y = mx + b$, or an exponential model, $y = a \cdot b^x$? Explain.

b. Sketch a scatterplot of this data.

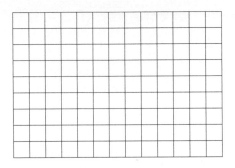

c. Does the graph reinforce your conclusion in part a? Explain.

d. Use your grapher to determine the exponential regression equation that best fits the health care data in the preceding table. Let your input, t, represent the number of years since 1970.

e. Using the regression equation from part d, determine the predicted total health care expenditures for the year 1990.

f. According to the exponential model, what is the growth factor for the total health care costs per year?

g. What is the growth rate?

h. According to the exponential model, in what year did the total heath care costs first exceed $1 trillion?

i. What is the doubling time for your exponential model?

2. a. Consider the following data set for the variables x and y.

x	5	8	11	15	20
y	70.2	50.7	35.1	22.6	9.5

Plot these points on the following grid.

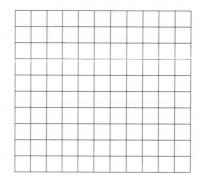

b. Use your grapher to determine both a linear regression and an exponential regression model of the data. Record the equations for these models here.

c. Which model appears to fit the data better? Explain.

d. Use the better model to determine y when $x = 13$ and y when $x = 25$.

e. For the exponential model, what is the decay factor?

f. What does it mean that the decay factor is between 0 and 1?

g. What is the half-life for the exponential model?

3. Use the graph of $y = 5 \cdot 2^x$ as a check, and summarize the properties of the exponential function $y = a \cdot b^x$, where $a > 0$.

   a. What is the domain?

   b. What is the range?

   c. When is $y = a \cdot b^x$ positive?

   d. When is $y = a \cdot b^x$ negative?

   e. What is the vertical intercept of the graph of $y = a \cdot b^x$?

CLUSTER 1

What Have I Learned?

1. Consider a linear function defined by $g(x) = mx + b, m \neq 0$, and an exponential function defined by $f(x) = a \cdot b^x$. Explain how you can determine from the equation whether the function is increasing or decreasing.

2. Suppose you have an exponential function of the form $f(x) = a \cdot b^x$, where $a > 0$ and $b > 0$ and $b \neq 1$. By inspecting the graph of f, can you determine if $b > 1$ or if $0 < b < 1$? Explain.

3. You are given a function defined by a table, and the input values are in increments of 1. By looking at the table, can you determine whether or not the function can be approximated by an exponential model? Explain.

4. Explain the difference between growth rate and growth factor.

5. An exponential function $y = a \cdot b^x$ passes through the point (0, 2.6). What can you conclude about the values of a and b?

6. You have just received a substantial tax refund of P dollars. You decide to invest the money in a CD for two years. You have narrowed your choices to two banks. Bank A will give you 6.75% interest compounded quarterly. Bank B offers you 6.50% compounded continuously. Where do you deposit your money? Explain.

7. Explain why the base in an exponential function cannot equal 1.

CLUSTER 1 How Can I Practice?

1. You are planning to purchase a new car and have your eye on a specific model. You know that new car prices are projected to increase at a rate of 4% per year for the next few years.

    a. Write an equation that represents the projected cost, C, of your dream car t years in the future, given that it costs $17,000 today.

    b. Identify the growth rate and the growth factor.

    c. Use your equation in part a to project the cost of your car three years from now.

    d. Use your graphing calculator to approximate how long it will take for your dream car to cost $30,000, if the price continues to increase at 4% per year.

2. Without using your graphing calculator, match the graph with its equation.

    a. $f(x) = -3(1.47)^x$ **b.** $g(x) = 2.5(0.47)^x$ **c.** $h(x) = 1.5(1.47)^x$

    i. 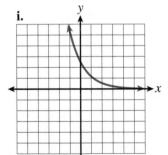   ii. 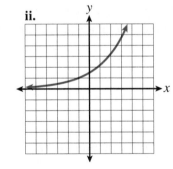   iii.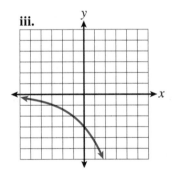

3. Explain the reasons for your choices in Problem 2.

4. Complete the following tables representing exponential functions. Round calculations to two decimal places whenever necessary.

a.

x	0	1	2	3	4
y	2.00	5.10			

b.

x	0	1	2	3	4
y	3.50	2.10			

c.

x	0	1	2	3	4
y	$\frac{1}{6}$	6			

5. Write the equation of the exponential function that represents the data in each table in Problem 4.

a.   b.   c.

6. Without graphing, classify each of the following functions as increasing or decreasing, and determine $f(0)$. (Use your graphing calculator to verify.)

   a. $f(x) = 1.3(0.75)^x$ b. $f(x) = 0.6(1.03)^x$

   c. $f(x) = 3(1/5)^x$

7. **a.** Given the following table, do you believe that it can be approximately modeled by an exponential function?

x	0	1	2	3	4	5	6
y	2	5	12.5	31.3	78.1	195.3	488.3

   b. If you answered yes to part a, what is the constant ratio of successive output values?

   c. Determine an exponential equation that models this data.

8. Your starting salary for a new job is $22,000 per year. You are offered two options for salary increases:

   Plan 1: an annual increase of $1000 per year or
   Plan 2: an annual percentage increase of 4% of your salary.

   Your salary is a function of the number of years of employment at your job.

a. Write an equation to determine the salary S after x years on the job using plan 1; using plan 2;

b. Complete the following table using the equations from part a.

x	0	1	3	5	10	15
S, PLAN 1						
S, PLAN 2						

c. Which plan would you choose? Explain.

9. The number of victims of a flu epidemic is increasing at a continuous rate of 7.5% per week.

   a. If 2000 people are currently infected, write an exponential model of the form $N = f(t) = N_0 e^{rt}$, where

   N is the number of victims in thousands,

   N_0 is the initial number infected in thousands,

   r is the weekly percent rate expressed as a decimal, and

   t is the number of weeks.

   b. Use the exponential model to predict the number of people infected after eight weeks.

   c. Sketch the graph of the flu function using your graphing calculator.

   d. Use a graphing approach to predict when the number of victims of the flu will triple.

10. a. Complete the following tables.

x	−3	−2	−1	0	1	2	3
$h(x) = 4^x$							

x	−3	−2	−1	0	1	2	3
$g(x) = \left(\frac{1}{4}\right)^x$							

b. Sketch graphs of *h* and *g* on the following grid.

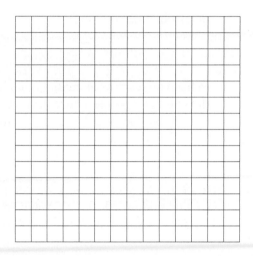

c. Use the tables and graphs in part a and b to complete this table.

FUNCTION	BASE, b	GROWTH OR DECAY FACTOR	x-INTERCEPT	y-INTERCEPT	HORIZONTAL ASYMPTOTE	INCREASING OR DECREASING
$h(x) = 4^x$						
$g(x) = \left(\frac{1}{4}\right)^x$						

11. You are a college freshman, and have a credit card. You immediately purchase a stereo system for $415. Your credit limit is $500. Let's assume that you make no payments and purchase nothing more and there are no other fees. The monthly interest rate is 1.18%.

a. What is your initial credit card balance?

b. What is the growth rate of your credit card balance?

c. What is the growth factor of your credit card balance?

d. Write an exponential function to determine how much you will owe (represented by $f(x)$) after x months with no more purchases or payments.

e. Use your graphing calculator to graph this function. What is the vertical intercept?

f. What is the practical meaning of this intercept in this situation?

g. How much will you owe after ten months? Use the table feature on your grapher to determine the solution.

h. When you reach your credit limit of $500, the bank will expect a payment. How long do you have before you will have to start paying the money back? Use the trace feature on your grapher to approximate the solution.

12. You are working part-time for a computer company while going to college. The following table shows the hourly wage, $w(t)$, in dollars, that you earn as a function of time, t. Time is measured in years since the beginning of 1996 when you started working.

TIME, t, YEARS, SINCE 1996	0	1	2	3	4	5
HOURLY WAGE, w(t), ($)	12.50	12.75	13.01	13.27	13.53	13.81

a. Calculate the ratios of the outputs to determine if the data in the table is exponential. Round each ratio to the nearest hundredth.

b. What is the growth factor?

c. Write an exponential equation that models the data in the table.

d. What percent raise did you receive each year?

e. If you continue to work for this company, what can you expect your hourly wage to be in 2005?

f. For approximately how many years will you have to work for the company in order for your hourly wage to double? (Assume you will receive the same percentage increase each year.)

13. You deposited $10,000 in an account that pays 12% annual interest compounded monthly.

   a. Write an equation to determine the amount, A, you will have in t number of years.

   b. How much will you have in five years?

   c. Use your grapher to determine in how many years your investment will double.

   d. Write an equation to determine the amount, A, you will have in t number of years if the interest is compounded continuously.

   e. Use the equation in part d and determine how much you will have in five years. Compare your answer to your answer in part b.

14. The number of farms in the U.S. has declined from 1940 to 2000, as the data in the following table shows. The data is estimated from National Agricultural Statistics Service, U.S. Department of Agriculture.

YEAR	1940	1950	1960	1970	1980	1990	2000
NUMBER OF FARMS, IN MILLIONS	6.2	5.8	4	3	2.5	2.2	2

   a. Make a scatterplot of this data.

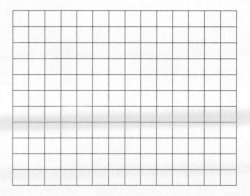

b. Does the scatterplot show that the data would be better modeled by a linear model or by an exponential model? Explain.

c. Use your graphing calculator to determine the exponential regression equation that best fits the U.S. farm data. Let x represent the number of years since 1940.

d. Use the regression equation to predict the total number of farms in the U.S. in 2010.

e. According to your exponential model, what is the decay factor for the total number of farms in the U.S.?

f. What is the decay rate?

g. Use your graphing calculator to determine the halving time for your exponential model.

CLUSTER 2   Logarithmic Functions

ACTIVITY 3.7
The Diameter of
Spheres

OBJECTIVES

1. Define logarithm.

2. Write an exponential statement in logarithmic form.

3. Write a logarithmic statement in exponential form.

4. Determine log and ln using the calculator.

Spheres are all around you (pardon the pun). You play sports with spheres like baseballs, basketballs, and golf balls. You live on a sphere. The Earth is a big ball in space, as are the other planets, the Sun, and the Moon. All spheres have properties in common. For example, the formula for the volume, V, of any sphere is $V = \frac{4}{3}\pi r^3$, and the formula for the surface area, S, of any sphere is $S = 4\pi r^2$, where r represents the radius of the sphere.

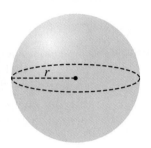

However, not all spheres are the same size. The following table gives the diameter, d, of some spheres you all know. Recall that the diameter, d, of a sphere is twice the radius, r.

SPHERE	DIAMETER, d, IN METERS
golf ball	0.043
baseball	0.075
basketball	0.239
Moon	3,476,000
Earth	12,756,000
Jupiter	142,984,000

If you want to determine either the volume or surface area of any of the spheres in the preceding table, the diameter of the given sphere would be the input value and would be referenced on the horizontal axis. But how would you scale this axis?

1. **a.** Plot the values in the first three rows of the table. Scale the axis starting at 0 and incrementing by 0.02 meters.

   b. Can you plot the values in the last three rows of the table on the same axis? Explain.

2. a. Plot the values in the last three rows of the table on a different axis. Scale the axis starting at 0 and incrementing by 10,000,000 meters.

b. Can you plot the values in the first three rows of the table on the axis in part a? Explain.

Logarithmic Scale

3. There is a way to scale the axis so that you can plot all the values in the table on the same axis.

a. Starting with the leftmost tick mark, give the first tick mark a value of 10^{-2} meters. Give the next tick mark a value of 10^{-1} meters. Continue in this way by giving each consecutive tick mark a value that is one power of ten greater than the preceding tick mark.

10^{-2} 10^{-1}

b. Complete the following table by writing all of the diameters from the preceding table in scientific notation.

SPHERE	DIAMETER, d, IN METERS	d, IN SCIENTIFIC NOTATION
golf ball	0.043	
baseball	0.075	
basketball	0.239	
Moon	3,476,000	
Earth	12,756,000	
Jupiter	142,984,000	

c. To plot the diameter of a golf ball, notice that 0.043 meter is between $10^{-2} = 0.01$ meter and $10^{-1} = 0.1$ meter. Now using the axis in part a, plot 0.043 meter between the tick mark labeled 10^{-2} and 10^{-1} meter, closer to the tick mark labeled 10^{-2} meter.

d. To plot the diameter of the Earth, notice that 12,756,000 meters is between $10^7 = 10,000,000$ meters and $10^8 = 100,000,000$ meters. Now plot 12,756,000 meters between the tick marks labeled 10^7 and 10^8 meters, closer to the tick mark labeled 10^7.

e. Plot the remaining data in the same way by first determining between which two powers of ten the number lies.

> The scale you used to plot the diameter values is a *logarithmic* or *log scale*. The tick marks on a logarithmic scale are usually labeled with just the exponent of the power of ten.

4. a. Rewrite the axis from Problem 3a by labeling the tick marks with just the exponents of the power of ten.

b. The axis looks like a standard axis with tick marks labeled $-2, -1, 0, 1$, etc. However, it is quite different. Describe the difference between this log scale and a standard axis labeled in the same way. Focus on the values between consecutive tick marks.

DEFINITION

The exponents used to label the tick marks of the preceding axis are **logarithms** or simply **logs**. Since these are exponents of powers of ten, the exponents are logs base ten, known as **common logarithms** or common logs.

Example 1

a. *The common logarithm of 10^3 is the exponent to which 10 must be raised to obtain a result of 10^3. Therefore, the common log of 10^3 is 3.*

b. *The common log of 10^{-2} is -2.*

c. *The common log of $100 = 10^2$ is 2.*

5. Determine the common log of each of the following.

    a. 10^{-1} **b.** 10^4 **c.** 1000

    d. 100,000   **e.** 0.0001

Logarithmic Notation

Remember that a logarithm is an exponent. The common log (base ten) is an exponent, y, to which the base, 10, must be raised to get result x.

Example 2

x, THE NUMBER	y, THE EXPONENT TO WHICH THE BASE, 10, MUST BE RAISED TO GET x	LOG NOTATION $\log_{10} x = y$
10^3	3	$\log_{10} 10^3 = 3$
10^{-2}	-2	$\log_{10} 10^{-2} = -2$
100	2	$\log_{10} 100 = 2$

If the base of a logarithm is 10, the notation $\log_{10}$ is written with the 10 omitted. Therefore,

$$\log_{10} 10^3 = \log 10^3 = 3; \log_{10} 100 = \log 100 = 2$$

6. Determine each of the following:

a. $\log 10^{-1}$ **b.** $\log 10^4$ **c.** $\log 1000$

d. $\log 100,000$ **e.** $\log 0.0001$

Bases for Logarithms

The logarithmic scale for the diameter of spheres situation was the exponents of powers of ten. Using 10 as the base for logarithms is common since the number 10 is the base of our number system. However, other numbers could be used as the base for logs. For example, you could use exponents of powers of five or exponents of powers of two.

Example 3

Base five logarithms: The log of a number, x, is the exponent to which the base, 5, must be raised to obtain x. For example,

a. $\log_5 5^4 = 4$

b. $\log_5 125 = \log_5 5^3 = 3$

c. $\log_5 \frac{1}{25} = \log_5 5^{-2} = -2$

Base two logarithms: The log of a number, x, is the exponent to which 2 must be raised to obtain x. For example,

a. $\log_2 2^5 = 5$

b. $\log_2 16 = \log_2 2^4 = 4$

c. $\log_2 \frac{1}{8} = \log_2 2^{-3} = -3$

In general, a statement in logarithmic form is $\log_b x = y$, where b is the base of the logarithm, x is a power of b, and y is the exponent. The base b for a logarithm is any positive number except 1.

7. Determine each of the following.

a. $\log_5 125$

b. $\log_2 \frac{1}{16}$

c. $\log_3 9$

d. $\log_3 \frac{1}{27}$

The examples and problems so far in this activity demonstrate the following property of logarithms:

Property of Logarithms

For appropriate values of b, $\log_b b^n = n$.

8. Determine each of the following:

a. $\log 1$

b. $\log_5 1$

c. $\log_{\frac{1}{2}} 1$

d. $\log 10$

e. $\log_5 5$

f. $\log_{\frac{1}{2}} \left(\frac{1}{2}\right)$

9. a. Referring to Problem 8 a–c, write a general rule for $\log_b 1$.

b. Referring to Problem 8 d–f, write a general rule for $\log_b b$.

Property of Logarithms

In general, $\log_b 1 = 0$ and $\log_b b = 1$, where $b > 0, b \neq 1$

Natural Logarithms

Because the base of a log can be any positive number except 1, the base can be the number e. Many applications involve the use of log base e. Log base e is called the natural log and has the following special notation:

$\log_e x$ is written as $\ln x$, read simply as l-n-x.

Example 4

a. $\ln e^2 = \log_e e^2 = 2$

b. $\ln \frac{1}{e^4} = \ln e^{-4} = -4$

10. Evaluate the following.

    a. $\ln e^7$ **b.** $\ln\left(\frac{1}{e^3}\right)$

    c. $\ln 1$ **d.** $\ln e$

Logarithmic and Exponential Forms

Because logarithms are exponents, logarithmic statements can be written as exponential statements, and exponential statements can be written as logarithmic statements.

For example, in the statement $3 = \log_5 125$, the base is 5, the exponent (logarithm) is 3, and the power is 125. This relationship can also be written as the equation $5^3 = 125$.

> In general, the logarithmic equation $y = \log_b x$ is equivalent to the exponential equation $b^y = x$.

Example 5   *Rewrite the exponential equation $e^{0.5} = x$ as an equivalent logarithmic equation.*

SOLUTION

In the equation $e^{0.5} = x$, the base is e, the power is x, and the exponent (logarithm) is 0.5. Therefore, the equivalent logarithmic equation is $0.5 = \log_e x$ or $0.5 = \ln x$.

11. Rewrite each exponential equation as a logarithmic equation and each log equation as an exponential equation.

    a. $3 = \log_2 8$ **b.** $\ln e^3 = 3$ **c.** $\log_2 \frac{1}{16} = -4$

    d. $6^3 = 216$ **e.** $e^1 = e$ **f.** $3^{-2} = \frac{1}{9}$

Logarithms and the Calculator

The arguments or powers of the logarithms you have been working with have been exact powers of the base. However, in many situations, you have to evaluate a logarithm whose argument is not an exact power of the base. For example, what is log 20 or ln 15? You can do this easily for the common log (base ten) and the natural log (base e) because these two logs are on your calculator.

12. Use your calculator to evaluate the following.

    a. log 20   **b.** ln 15

    c. $\ln \frac{1}{2}$ **d.** log 0.02

    e. Use your calculator to check your answers to Problems 6 and 10.

13. Use your calculator to complete the following table. Confirm the placement of the diameter values on the log-scaled axis.

SPHERE	DIAMETER, d, IN METERS	d, IN SCIENTIFIC NOTATION	$\log(d)$
golf ball	0.043	4.3×10^{-2}	
baseball	0.075	7.5×10^{-2}	
basketball	0.239	2.39×10^{-1}	
Moon	3,476,000	3.476×10^{6}	
Earth	12,756,000	1.2756×10^{7}	
Jupiter	142,984,000	1.4298×10^{8}	

SUMMARY Activity 3.7

1. The notation for logarithms is $\log_b x = y$, where b is the base of the log, x is the power of b, and y is the exponent. The base b can be any positive number except 1; x can be any positive number, y is any real number.

2. The notation for **common logarithm** or base ten logarithms is $\log_{10} x = \log x$.

3. The notation for the **natural logarithm** or base e logarithm is $\log_e x = \ln x$.

4. The **logarithmic equation** $y = \log_b x$ is equivalent to the **exponential equation** $b^y = x$.

5. If $b > 0$ and $b \neq 1$,
    a. $\log_b 1 = 0$
    b. $\log_b b = 1$
    c. $\log_b b^n = n$

EXERCISES
Activity 3.7

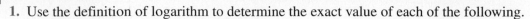

1. Use the definition of logarithm to determine the exact value of each of the following.

   a. $\log_2 32$ **b.** $\log_3 27$ **c.** $\log 0.1$

   d. $\log_2 \left(\frac{1}{64}\right)$ **e.** $\log_5 1$ **f.** $\log_{\frac{1}{2}} \left(\frac{1}{4}\right)$

   g. $\log_7 \sqrt{7}$ **h.** $\log_{100} 10$ **i.** $\log 1$
   Hint: $\sqrt{7} = 7^{\frac{1}{2}}$

   j. $\log_2 1$ **k.** $\ln e^5$ **l.** $\ln \left(\frac{1}{e^2}\right)$ **m.** $\ln 1$

2. Evaluate each common logarithm without the use of a calculator.

   a. $\log \left(\dfrac{1}{1000}\right) = $ _____   **b.** $\log \left(\dfrac{1}{100}\right) = $ _____

   c. $\log \left(\dfrac{1}{10}\right) = $ _____   **d.** $\log 1 = $ _____

   e. $\log 10 = $ _____   **f.** $\log 100 = $ _____

   g. $\log 1000 = $ _____

3. Rewrite the following equations in logarithmic form.

   a. $3^2 = 9$ **b.** $\sqrt{121} = 11$ *Hint:* First rewrite $\sqrt{121}$ in exponential form.

   c. $4^t = 27$ **d.** $b^3 = 19$

4. Rewrite the following equations in exponential form.

   a. $\log_3 81 = 4$ **b.** $\frac{1}{2} = \log_{100} 10$

c. $\log_9 N = 12$ **d.** $y = \log_7 x$

e. $\ln \sqrt{e} = \frac{1}{2}$ **f.** $\ln \left(\frac{1}{e^2}\right) = -2$

5. Estimate between what two integers the solutions for the following equations fall. Then solve each equation exactly after first changing it to log form. Use your calculator to approximate your answer to three decimal places.

   a. $10^x = 3.25$ **b.** $10^x = 590$ **c.** $10^x = 0.0000045$

ACTIVITY 3.8

Walking Speed of Pedestrians

OBJECTIVES

1. Determine the inverse of the exponential function.

2. Identify properties of the graph of $y = \log x$.

3. Identify the properties of the graph of a logarithmic function.

4. Graph the natural logarithmic function.

On a recent visit to Boston, you notice that people seem rushed as they move about the city. Upon returning to college, you mention this observation to your psychology instructor. The instructor refers you to a psychology study that investigates the relationship between the average walking speed of pedestrians and the population of the city. The study cites the following statistics presented graphically as follows.

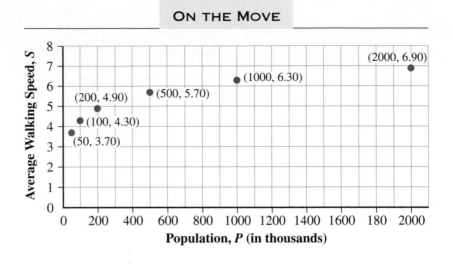

ON THE MOVE

1. a. Does the data appear to be linear? Explain.

b. Does the data appear to be exponential? Explain.

This data is actually logarithmic. Situations that can be modeled by logarithmic functions will be the focus of this and the following activity.

Introduction to the Logarithmic Function

The logarithmic function is defined by $y = \log_b x$, where

b represents the base of the log $(b > 0, b \neq 1)$,

x is the input and represents a power of the base b (x is also called the argument), and

y is the output and represents an exponent.

2. a. Evaluate $\log_{10}(-100)$ using your calculator. What do you observe? Does it seem reasonable? Explain.

b. Is it possible to determine log (0)?

c. What is the domain for the function defined by $y = \log x$?

d. What is the range? Remember, the output y is an exponent.

3. The exponential function defined by $f(x) = 10^x$ has a special relationship with the corresponding logarithmic function defined by $g(x) = \log_{10} x = \log x$.

a. Complete the following tables for $f(x) = 10^x$ and $g(x) = \log x$.

x	$f(x) = 10^x$
-2	
-1	
0	
1	
2	

x	$g(x) = \log x$
0.01	
0.1	
1	
10	
100	

b. Compare the input and output values for functions f and g.

c. Sketch the graphs of $Y1 = 10^x$ and $Y2 = \log_{10} x$ using your graphing calculator. Use the window Xmin $= -4$, Xmax $= 4$, Ymin $= -3$, and Ymax $= 3$. Your screen should appear as follows:

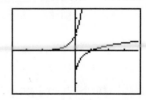

d. Graph $y = x$ on the same coordinate axes as functions f and g. Describe in a sentence or two the symmetry you observe in the graphs of f and g.

Recall the concept of an inverse function from Chapter 2. The inverse function interchanges the domain and range of the original function. Also, the graph of an inverse function is the reflection of the original function about the line $y = x$. Therefore, it appears from the results in Problem 3 that $f(x) = 10^x$ and $g(x) = \log x$ are inverse functions.

You can determine the equation of the inverse function by interchanging the input (x-values) and the output (y-values) in the given equation for the function and solving the new equation for y.

Example 1 *Determine the equation of the inverse of the function defined by $y = 5^x$.*

SOLUTION

Step 1. Interchange the x and y variables. $x = 5^y$

Step 2. Solve the resulting equation for y by $y = \log_5 x$
writing the statement in logarithmic notation.

4. Use the algebraic approach demonstrated in Example 1 to verify that $y = \log x$ is the inverse of $y = 10^x$.

Problems 2, 3, and 4 illustrate the following properties of the common logarithmic function.

> **Properties of the Common Logarithmic Function $f(x) = \log x$**
>
> **1.** The domain of f is the set of all positive real numbers $(x > 0)$.
> **2.** The range of f is all real numbers.
> **3.** f is the inverse of the function defined by $g(x) = 10^x$.

The Graph of the Natural Logarithmic Function

5. a. Using your calculator, complete the following table. Round your answers to three decimal places.

x	0.1	0.5	1	5	10	20	50
y = ln (x)							

b. Sketch a graph of $y = \ln x$.

c. Verify your graph in part b using your graphing calculator. Using the window Xmin = −1, Xmax = 4, Ymin = −2.5, and Ymax = 2.5, your screen should appear as follows.

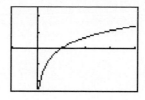

d. What are the domain and range of the function defined by $y = \ln x$?

e. Does the graph of $y = \ln x$ have a horizontal asymptote? Explain.

f. Does the graph of $y = \ln x$ have a vertical intercept?

g. Complete the following table using your calculator. Round your answers to the nearest tenth.

x	0.001	0.01	0.1	0.25	0.5	1
y = ln x						

h. As the input values take on values closer and closer to 0, what happens to the corresponding output values?

DEFINITION

The vertical axis, having equation $x = 0$, is called a vertical asymptote because as the input values get closer and closer to a value of 0, the graph gets closer and closer to the vertical line $x = 0$.

Example 2   *The vertical asymptote of the graphs of $y = \log x$ and $y = \ln x$ is the vertical line $x = 0$.*

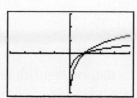

6. a. Graph $y = e^x$, $y = \ln x$, and $y = x$ on the same set of coordinate axes as using the window Xmin $= -7.5$, Xmax $= 7.5$, Ymin $= -5$, and Ymax $= 5$. Describe the symmetry that you observe.

b. Use an algebraic approach to determine the inverse of the exponential function defined by $y = e^x$.

SUMMARY
Activity 3.8

1. Properties of the log function defined by $y = \log x$.

a. The domain of f is $x > 0$.

b. The range of f is all real numbers.

c. f is the inverse of the function defined by $g(x) = 10^x$.

2. The graph of a logarithmic function defined by $y = \log_b x$, where $b > 1$, resembles the following graph, and the function

a. is increasing for all $x > 0$.

b. has an x-intercept of $(1, 0)$.

c. has no y-intercept.

d. has a vertical asymptote of $x = 0$, the y-axis.

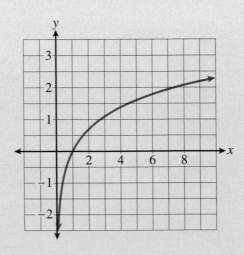

3. The **natural logarithmic function** is defined by

$$y = \ln x = \log_e x.$$

4. The graph of the natural logarithmic function $y = \ln x$
   a. is increasing for all $x > 0$.
   b. has an x-intercept of $(1, 0)$.
   c. has a vertical asymptote of $x = 0$, the y-axis.

5. You can determine the equation of the inverse of the function by interchanging the input (x-values) and the output (y-values) in the given equation for the function and solving the new equation for y.

EXERCISES
Activity 3.8

1. Using the graph of $y = \log x$ as a check, summarize the following properties of the common logarithmic function.

   a. What is the domain?

   b. What is the range?

   c. For what values of x is $\log x$ positive?

   d. For what values of x is $\log x$ negative?

   e. For what values of x is $\log x = 0$?

   f. For what values of x is $\log x = 1$?

2. a. Complete the following table using your calculator. Round your answers to the nearest tenth.

x	0.001	0.01	0.1	0.25	0.5	1
y = log x						

   b. As the positive input values take on values closer and closer to 0, what happens to the corresponding output values?

Exercise numbers appearing in color are answered in the Selected Answers appendix.

c. Determine the vertical asymptote of the graph of $y = \log x$.

3. The exponential function defined by $y = 2^x$ has an inverse. Determine the equation of the inverse function. Write your answer in logarithmic form.

4. Using the graph of $y = \ln x$ as a check, summarize the following properties of the natural logarithmic function.

   a. What is the domain?

   b. What is the range?

   c. For what values of x is $\ln x$ positive?

   d. For what values of x is $\ln x$ negative?

   e. For what values of x is $\ln x = 0$?

   f. For what values of x is $\ln x = 1$?

5. The life expectancy for a piece of equipment is the number, n, of years for the equipment to depreciate to a known salvage value, V. The life expectancy, n, is given by the formula

$$n = \frac{\log V - \log C}{\log (1 - r)},$$

   where C is the initial cost of the piece of equipment and r is the annual rate of depreciation expressed as a decimal. If a computer costs \$34,000 and has a salvage value of \$1000, what is the life expectancy if the annual rate of depreciation is 40%?

ACTIVITY 3.9

Walking Speed
of Pedestrians,
continued

OBJECTIVES

1. Compare the average rate
of change of increasing
logarithmic, linear, and
exponential functions.

2. Determine the regression
equation of a natural
logarithmic function that
best fits a set of data.

In Activity 3.8 you looked at a psychology study that investigated the relationship
between the average walking speed of pedestrians and the population of the city.
Graphically the data was presented as follows.

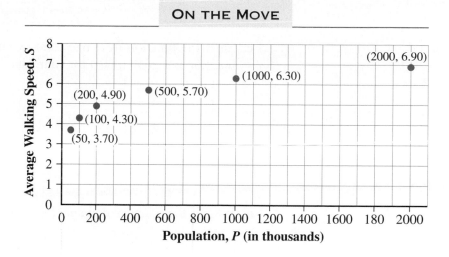

1. a. Does the data appear to be logarithmic? Explain.

b. Use the data in the graph to complete the following table.

POPULATION, P (in thousands)	50	100	200	500	1000	2000
AVERAGE WALKING SPEED, S						

The natural logarithmic function can be used to model a variety of scientific and natural phenomena. The natural logarithmic function is so prevalent that on most graphing calculators it has its own built-in regression finder.

2. a. Use your grapher and the table in Problem 1b to produce a scatterplot of the
average walking speed data.

b. Use the regression feature of your calculator to produce a natural logarithmic curve that approximates the data in the table. Use option 9 from the
STAT CALC menu.

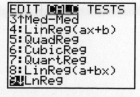

The LnReg option will generate a regression equation of the form $y = a + b \ln x$. Round a and b to the nearest thousandth, and record the function below.

c. Enter the function from part b into your grapher. Verify visually that this function is a good model for your data.

d. What is the practical domain of this function?

e. Use the function from part b to predict the average walking speed in Boston, population 589,121. *Note: P* is the number be in thousands (589.121 thousands).

f. Use the model to predict the average walking speed in New York City, population 8,008,278

3. a. If the average walking speed in a certain city is 5.2 feet per second, write an equation that can be used to determine the population P of the city.

b. Solve the equation using a graphical approach.

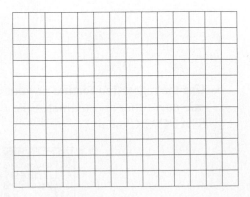

Comparing the Average Rate of Change of Logarithmic, Linear, and Exponential Functions

4. a. Complete the following table using the function defined by
$S = 0.3 + 0.868 \ln (P)$.

P, POPULATION, (thousands)	10	20	150	250
S, AVERAGE WALKING SPEED (ft/sec)				

b. Determine the average rate of change of S as the population increases from

    i. 10 to 20 thousand

    ii. 20 to 150 thousand

    iii. 150 to 250 thousand

c. What can you say in general about the average rate of change in the walking speed as the population increases?

You should have discovered that the average rate of change is always positive. This means that the walking speed increases as the population increases. Nevertheless, in general, the increase gets smaller as the population increases. This is characteristic of logarithmic functions.

> As the input of a logarithmic function with $b > 1$ increases, the output increases at a slower rate (the graph becomes less steep).

5. Complete the following statements by describing the rate at which the output values change.

    a. For an increasing linear function, as the input variable increases, the output

    _____

    b. For an increasing exponential function, as the input increases, the output

    _____

    c. For an increasing logarithmic function, as the input increases, the output

    _____

6. Consider the graphs of

   i. $f(x) = e^x$ **ii.** $h(x) = x$ **iii.** $g(x) = \ln x$

using the window Xmin $= -7.5$, Xmax $= 7.5$, Ymin $= -5$, and Ymax $= 5$.

   a. Which of the functions are increasing?

   b. Which of the functions are decreasing?

   c. As the input values get larger, which of the functions grows fastest?

   d. As the input values get larger, which of the functions grows slowest?

   e. Do any of these functions have a horizontal asymptote?

   f. Do any of these functions have a vertical asymptote?

   g. Compare the domains of these functions.

   h. Compare the ranges of these functions.

Problem 6 illustrates some of the relationships between $f(x) = b^x$, where $b > 1$, $g(x) = \log_b (x)$, where $b > 1$, and $y = mx + b$, where $m > 0$.

Application

7. You are working on the development of an "elastic" ball for the IBF Toy Company. The question you are investigating is, "If the ball is launched straight up, how far has it traveled vertically when it hits the ground for the tenth time?"

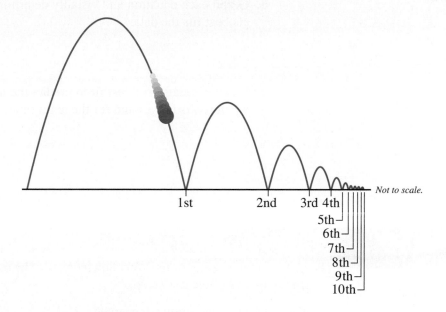

Not to scale.

1st   2nd   3rd 4th
5th
6th
7th
8th
9th
10th

Your launcher will project the ball 10 feet into the air. This means it will travel 20 feet (10 feet up and 10 down) before it hits the ground the first time. Assuming that the ball returns to 50% of its previous height, it will rebound 5 feet and travel 10 feet before it hits the ground again. The following table summarizes this situation.

N, times the ball hits the ground	1	2	3	4	5	6
Distance traveled since last time (ft)	20	10	5	2.5	1.25	0.625
T, Total distance traveled (ft)	20	30	35	37.5	38.75	39.375

a. Using the window Xmin = 0, Xmax = 7, Ymin = 0, and Ymax = 45, a plot of N versus T should resemble the following.

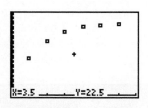

b. Do the table and scatterplot indicate the data is linear, exponential, or logarithmic?

c. Use your graphing calculator to produce linear, exponential, and natural log regression equations for the given data.

d. Graph each equation and visually determine which of the regression models best fits the data.

e. Use the equation of best fit to predict the total distance traveled by the ball when it hits the ground for the tenth time.

SUMMARY
Activity 3.9

1. As the input of a logarithmic function increases, the output increases at a slower rate (the graph becomes less steep).

2. The relationships among the graphs $f(x) = b^x$, where $b > 1$, $g(x) = \log_b (x)$, where $b > 1$, and $y = mx + b$, where $m > 0$, are identified in the following table.

GRAPHS	INCREASING OR DECREASING	GROWTH RATE	HORIZONTAL OR VERTICAL ASYMPTOTE	DOMAIN	RANGE
$f(x) = b^x$, $b > 1$	increasing	fastest	horizontal asymptote	all real numbers	$y > 0$
$g(x) = \log_b (x)$, $b > 1$	increasing	slowest	vertical asymptote	$x > 0$	all real numbers
$y = mx + b$, $m > 0$	increasing	constant	none	all real numbers	all real numbers

EXERCISES
Activity 3.9

1. *Chlamydia trachomatis* infections are the most commonly reported notifiable disease in the United States. These are among the most prevalent of all sexually transmitted diseases. The following data from the Centers for Disease Control and Prevention indicates the reported rates, R, in rates per 100,000 people from 1985 to 2000. Let t represent the number of years since 1985.

YEARS, t, SINCE 1985	1	3	5	9	13	15
REPORTED RATES: U.S. (rate per 100,000 population), R	40	90	160	200	250	260

Exercise numbers appearing in color are answered in the Selected Answers appendix.

a. Plot the points on the following grid.

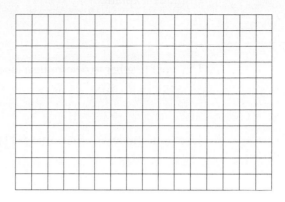

b. Does the scatterplot indicate that the data is logarithmic? Explain.

c. Determine the natural log regression equation. Record the regression equation below, and add a sketch of the regression curve to the scatterplot in part a.

d. Is the graph a good fit of the data?

e. Use your model to predict the reported rate of *Chlamydia trachomatis* infections per 100,000 people in 2004.

2. a. Consider a data set for the variables x and y.

x	1	4	7	10	13
f(x)	3.0	4.5	5.0	5.2	5.8

Plot these points on the following grid.

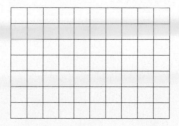

b. Does the scatterplot indicate the data is more likely linear, exponential, or logarithmic? Explain.

c. Use your grapher to determine a logarithmic regression model that represents this data.

d. Use your model to determine $f(11)$ and $f(20)$.

3. The barometric pressure, P, in inches of mercury at a distance x miles from the eye of a moderate hurricane can be modeled by

$$P = f(x) = 0.48 \ln (x + 1) + 27.$$

a. Determine $f(0)$. What is the practical meaning of the value in this situation?

b. Sketch a graph of this function.

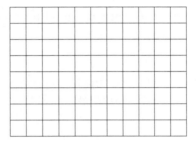

c. Describe how air pressure changes as you move away from the eye of the hurricane.

4. The formula $R = 80.4 - 11 \ln x$ is used to approximate the minimum required ventilation rate, R, as a function of the air space per child in a public school classroom. The rate R is measured in cubic feet per minute, and x is measured in cubic feet.

   a. Sketch a graph of the rate function for $100 \leq x \leq 1500$.

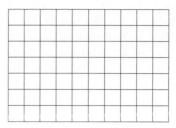

   b. Determine the required ventilation rate if the air space per child is 300 cubic feet.

5. You have recently accepted a job working in the coroner's office of a large city. Because of the large numbers of homicides, it has been very difficult for the coroners to complete all of their work. Your job is, in part, to assist them in the paperwork. On one particular day, you are working on a case in which you are attempting to establish the time of death.

   The coroner tells you that to establish the time of death, he uses the formula

   $$t = 4 \ln \frac{98.6 - T_s}{T_b - T_s},$$

   where t is the number of hours the victim has been dead,

   T_b represents the temperature of the body when discovered, and

   T_s represents the temperature of his surroundings.

   The coroner also tells you that the thermostat in the apartment in which the body was found was set at 68°F and that the victim's body temperature was 78°F.

   a. Using the preceding formula, determine the number of hours the victim has been deceased. Use your calculator to approximate your answer to one decimal place.

   b. If the body was discovered at 10:07 P.M., what do you estimate for the time of death?

ACTIVITY 3.10

The Elastic Ball

OBJECTIVES

1. Apply the log of a product property.

2. Apply the log of a quotient property.

3. Apply the log of a power property.

4. Discover change of base formula.

You are continuing your work on the development of the elastic ball. You are still investigating the question, "If the ball is launched straight up, how far has it traveled vertically when it hits the ground for the tenth time?" However, your supervisor tells you that you cannot count the initial launch distance. You must calculate only the rebound distance.

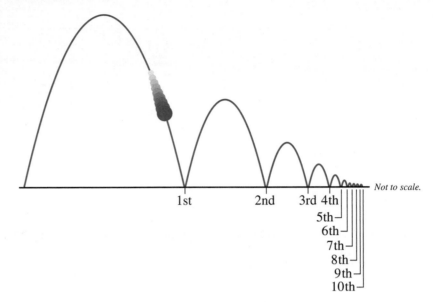

Using some physical properties, timers, and your calculator, you collect the following data:

N, NUMBER OF TIMES THE BALL HITS THE GROUND	1	2	3	4	5	6
T, TOTAL REBOUND DISTANCE (ft)	0	9.0	13.5	16.3	18.7	21.0

1. Does the data seem reasonable? Explain.

2. Use your graphing calculator to construct a scatterplot of the data with N as the input and T as the output. Using a window of Xmin = 0, Xmax = 7, Ymin = 0, and Ymax = 25, your graph should resemble the following.

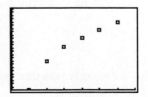

3. Do you believe the data can be modeled by a logarithmic function? Explain.

4. This data can be modeled by $T = 26.75 \log N$. Use your graphing calculator to verify visually that this is a reasonable model for the given data.

5. a. Using the log model, complete the following table. Round values to the nearest hundredth.

N	2	5	10
$T = 26.75 \log N$			

b. How are the outputs from 2 and 5 related to the output for 10?

c. Using the results from part b, how could you determine the total rebound distance after 10 bounces?

The results from Problem 5 can be written as follows:

$$26.75 = 8.05 + 18.70$$
$$26.75 \log 10 = 26.75 \log 2 + 26.75 \log 5$$
$$26.75 \log (2 \cdot 5) = 26.75 \log 2 + 26.75 \log 5.$$

Dividing both sides by 26.75, you have

$$\log (2 \cdot 5) = \log 2 + \log 5.$$

This result illustrates an important property of logarithms.

Property of the Logarithm of a Product

If $A > 0, B > 0$, then $\log_b (A \cdot B) = \log_b A + \log_b B$, where $b > 0, b \neq 1$

Expressed verbally, this property states that the logarithm of a product is the sum of the individual logarithms.

Example 1

a. Note that $\log_2 32 = \log_2 (4 \cdot 8) = \log_2 4 + \log_2 8 = 2 + 3 = 5$

b. $\log (5st) = \log (5) + \log (s) + \log (t)$

c. $\ln (xy) = \ln x + \ln y$

6. Use the property of the logarithm of a product to write the following as the sum of two or more logarithms.

a. $\log_b (7 \cdot 13)$ **b.** $\log_3 (xyz)$

c. $\log 15$ **d.** $\ln (3xy)$

7. Write the following as the logarithm of a single expression.

    a. $\ln a + \ln b + \ln c$ **b.** $\log_4 3 + \log_4 9$

Logarithm of a Quotient

Consider the following table from Problem 5.

N	2	5	10
T = 26.75 log N	8.05	18.70	26.75

This table also indicates that the rebound distance after this ball has hit the floor twice (8.05 ft) is the total rebound distance when the ball has hit the ground ten times (26.75 ft) minus the total rebound distance when the ball has hit the ground five times (18.70 ft).

This can be written as

$$8.05 = 26.75 - 18.70$$

$$26.75 \log 2 = 26.75 \log 10 - 26.75 \log 5$$

$$\log 2 = \log 10 - \log 5.$$

Substituting $\log \left(\frac{10}{5}\right)$ for $\log 2$, you have

$$\log \left(\tfrac{10}{5}\right) = \log 10 - \log 5.$$

This suggests another important property of logarithms. The property is demonstrated further in Problem 8.

  8. a. Complete the following table. Round your answers to the nearest thousandth.

x	$Y1 = \log\left(\frac{x}{4}\right)$	$Y2 = \log x - \log 4$
1		
5		
10		
23		

    b. Is the expression $\log \left(\frac{x}{4}\right)$ equivalent to $\log (x) - \log (4)$? Explain.

    c. Sketch the graph of $y = \log \left(\frac{x}{4}\right)$ and $y = \log x - \log 4$ using your graphing calculator. What do the graphs suggest about the relationship between $\log \left(\frac{x}{4}\right)$ and $\log x - \log 4$?

Property of the Logarithm of a Quotient

If $A > 0$, $B > 0$, then $\log_b \left(\frac{A}{B}\right) = \log_b A - \log_b B$, where $b > 0$, $b \neq 1$.

Expressed verbally, this property states that the logarithm of a quotient is the difference of the logarithm of the numerator and the logarithm of the denominator.

Example 2

a. $\log_3 \left(\frac{81}{27}\right) = \log_3 81 - \log_3 27 = 4 - 3 = 1$

Note that $\log_3 \left(\frac{81}{27}\right) = \log_3 (3) = 1$.

b. $\log \left(\frac{2x}{y}\right) = \log (2x) - \log (y) = \log 2 + \log x - \log y$

c. $\ln \left(\frac{x^2}{5}\right) = \ln (x^2) - \ln (5)$

9. Use the properties of logarithms to write the following as the sum or difference of logarithms.

a. $\log_6 \frac{17}{3}$

b. $\ln \frac{x}{23}$

c. $\log_3 \frac{2x}{y}$

d. $\log \frac{3}{2z}$

10. Write the following expressions as the logarithm of a single expression.

a. $\log x - \log 4 + \log z$

b. $\log x - (\log 4 + \log z)$

11. a. Use your graphing calculator to sketch the graphs of $y = \log x + \log 4$ and $y = \log (x + 4)$.

b. How do these graphs compare?

c. What do the graphs suggest about the relationship between $\log (A + B)$ and $\log A + \log B$?

Logarithm of a Power

Before calculators, logarithms were used to help in computing products and quotients of numbers. More important, logarithms were used to compute powers such as 734.21^3 and $\sqrt{0.0761} = (0.0761)^{\frac{1}{2}}$. In such a case, the first step was to take the logarithm of the power and rewrite the resulting expression. To determine how to rewrite $\log 734.21^3$, you can investigate the expression $\log x^3$.

12. a. Complete the following table. Round to the nearest thousandth.

x	$Y1 = \log x^3$	$Y2 = 3\log x$
2		
7		
15		

b. Sketch the graphs of $y = \log x^3$ and $y = 3 \log x$ using your graphing calculator.

c. What do the results of part a and part b demonstrate about the relationship between $\log x^3$ and $3 \log x$?

The results in Problem 12 illustrate another property of logarithms.

Property of the Logarithm of a Power

If $A > 0$ and p is any real number, then $\log_b A^p = p \cdot \log_b A$,

where $b > 0$, $b \neq 1$. In words, the property states that the logarithm of a power is equivalent to the exponent times the logarithm of the base.

Example 3

a. $\log_3 9^2 = 2 \log_3 9 = 2 \cdot 2 = 4$

b. $\log_5 x^4 = 4 \log_5 x$

c. $\ln (xy)^7 = 7 \ln (xy)$

d. $\ln x^{\frac{1}{4}} = \frac{1}{4} \ln x$

13. Use the properties of logarithms to write the given logarithms as the sum or difference of two or more logarithms, or as the product of a real number and a logarithm. All variables represent positive numbers.

a. $\log_3 x^{\frac{1}{2}}$

b. $\log_5 x^3$

c. $\ln t^2$

d. $\log \sqrt[3]{50}$ *Hint:* $\sqrt[3]{50} = 50^{\frac{1}{3}}$

e. $\log_5 \frac{x^2 y^3}{z}$

f. $\log_3 \frac{3x^2}{y^3}$

14. Write each of the following as the logarithm of a single expression with coefficient 1.

a. $2 \log_3 5 + 3\log_3 2$

b. $\frac{1}{2} \log x^4 - \frac{1}{2} \log y^5$

c. $3 \log_b 10 - 4 \log_b 5 + 2 \log_b 3$ d. $3 \ln 4 - (4 \ln 5 + 2 \ln 3)$

Change of Base Formula

Because the TI-83 Plus has only the log base ten and the log base e keys, you cannot graph a logarithmic function such as $y = \log_2 x$ directly. Consider the following argument to rewrite the expression $\log_2 x$ as an equivalent expression using log base ten.

By definition of inverse, $y = \log_2 x$ is the same as $x = 2^y$. Taking the log of both sides of the second equation, $x = 2^y$, you have

$$\log x = \log 2^y.$$

Using the property of the log of a power, $\log x = y \log 2$. Solving for y, you have

$$y = \frac{\log x}{\log 2}$$

Therefore, the equation $y = \log_2 x$ is equivalent to $y = \frac{\log x}{\log 2}$.

15. Using the formula $y = \log_2 x = \frac{\log x}{\log 2}$, use your graphing calculator to graph $y = \log_2 x$. Your graph should resemble the following.

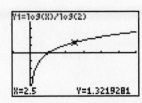

16. a. Write $y = \log_6 x$ as an equivalent equation using base ten.

b. Use the result from part a to graph $y = \log_6 x$.

c. What is the domain of the function?

d. What is the x-intercept of the graph?

The formula you used for graphing log functions of different bases is a special case of the formula

$$\log_b x = \frac{\log_a x}{\log_a b}.$$

This is often called the **change of base formula**, where $b > 0$ and $b \neq 1$.

The preceding formula is used to change from base b to base a. Because most calculators have log base ten (log) and log base e (ln) keys, you usually convert to one of those bases.

Therefore,

$$\log_b x = \frac{\log x}{\log b} \text{ or } \log_b x = \frac{\ln x}{\ln b}.$$

Example 4 *Change the equation $y = \log_5 x$ to an equivalent equation in base ten and/or base e.*

$$y = \log_5 x = \frac{\log x}{\log 5} \text{ or } y = \log_5 x = \frac{\ln x}{\ln 5}$$

17. Use each of the change of base formulas to determine $\log_4 1024$.

a. Using base ten:

b. Using base e:

c. How do the results in parts a and b compare?

SUMMARY
Activity 3.10

Properties of the Logarithmic Function

If $A > 0, B > 0, b > 0$, and $b \neq 1$, then

1. $\log_b (A \cdot B) = \log_b A + \log_b B$

2. $\log_b \left(\dfrac{A}{B} \right) = \log_b A - \log_b B$

3. $\log_b (x + y) \neq \log_b (x) + \log_b (y)$

4. $\log_b A^p = p \log_b A$

5. You can use the calculator to change logarithms in base b to common or natural logarithms by

$$\log_b x = \frac{\log x}{\log b} \text{ or } \log_b x = \frac{\ln x}{\ln b}.$$

EXERCISES
Activity 3.10

1. Use the preceding properties of logarithms to write the following as a sum or difference of two or more logarithms.

   a. $\log_b (3 \cdot 7)$ **b.** $\log_3 (3 \cdot 13)$

   c. $\log_7 \frac{13}{17}$ **d.** $\log_3 \frac{xy}{3}$

2. Write the following expressions as the logarithm of a single number.

   a. $\log_3 5 + \log_3 3$ **b.** $\log 25 - \log 17$

   c. $\log_5 x - \log_5 5 + \log_5 7$ **d.** $\ln (x + 7) - \ln x$

3. **a.** Sketch the graphs of $y = \log (2x)$ and $y = \log x + \log 2$ on your graphing calculator.

b. Are you surprised by the results? Explain.

4. a. Sketch the graphs of $y = \log \left(\frac{3}{x}\right)$ and $y = \log x - \log 3$ on your graphing calculator.

b. Are you surprised by your results? Explain.

c. If your graphs in part a are not identical, can you modify the second function to make the graphs identical? Explain.

5. You have been hired to handle the local newspaper advertising for a large used car dealership in your community. The owner tells you that your predecessor in this position used the formula

$$N(A) = 7.4 \log (A)$$

to decide how much to spend on newspaper advertising over a two-week period. The owner admitted that he didn't know much about the formula except that $N(A)$ represented the number of cars that the owner could expect to sell, and A was the amount of money that was spent on local newspaper advertising. He also indicated that the formula seemed to work well. You can purchase small ads in the local paper for $15 per day, larger ads for $50 per day, and giant ads for $750 per day.

a. How many cars do you expect to sell if you purchase one small ad?

b. To understand the relationship between the amount spent on advertising and the number of cars sold, you set up a table. Complete the following table.

AD COST, A	EXPECTED CAR SALES, $N(A)$
15	
50	
750	

c. How do the expected car sales from one small ad and one larger ad compare to the expected car sales from just one giant ad?

d. Are the results in the previous table above consistent with what you know about the properties of logarithms? Explain.

e. What are you going to advise the owner regarding the purchase of a giant ad?

6. Use the properties of logarithms to write the given logarithms as the sum or difference of two or more logarithms, or as the product of a real number and a logarithm. Simplify, if possible. All variables represent positive numbers.

a. $\log_3 3^5$

b. $\log_2 2^x$

c. $\log_b \dfrac{x^3}{y^4}$

d. $\ln \dfrac{\sqrt[3]{x}\sqrt[4]{y}}{z^2}$

e. $\log_3 (2x + y)$

7. Write each of the following as the logarithm of a single expression with coefficient 1.

a. $2 \log_2 7 + \log_2 5$

b. $\frac{1}{4} \log x^3 - \frac{1}{4} \log z^5$

c. $2 \ln 10 - 3 \ln 5 + 4 \ln z$

d. $\log_5 (x + 2) + \log_5 (x + 1) - 2 \log_5 (x + 3)$

8. Given that $\log_a x = 6$ and that $\log_a y = 25$, determine the numeric value of each of the following.

a. $\log_a \sqrt{y}$

b. $\log_a x^3$

c. $3 + \log_a x^2$

d. $\log_a \dfrac{x^2 y}{a}$

9. Use the change of base formula and your calculator to determine a decimal approximation of each of the following to the nearest ten thousandth.

  a. $\log_7 5$ **b.** $\log_6 \sqrt{15}$

  c. $\log_{13} 47$ **d.** $\log_5 \sqrt[3]{31}$

10. The formula

$$P = f(t) = 95 - 30\log_2 t$$

gives the percentage, P, of students who could recall the important content of a classroom presentation as a function of time, t, where t is the number of days that have passed since the presentation was given.

  a. Sketch a graph of the function.

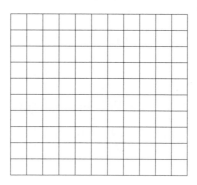

  b. After three days, what percentage of the students will remember the important content of the presentation?

  c. According to the model, after how many days do only half $(P = 50)$ of the students remember the important features of the presentation? Use a graphing approach.

ACTIVITY 3.11
Prison Growth

OBJECTIVE

1. Solve exponential equations both graphically and algebraically.

You are a criminal justice major at the local community college. The following statistics appeared in one of your required readings relating to the inmate population of U.S. federal prisons.

Year	1975	1979	1986	1990	1994	1998	2000
TOTAL SENTENCED POPULATION, P_T (in thousands)	20.1	21.5	31.8	47.8	76.2	95.5	112.3
TOTAL SENTENCED DRUG OFFENDERS, P_D (in thousands)	5.5	5.5	12.1	25.0	46.7	56.3	63.93

You decide to analyze the prison growth situation for a project in your criminology course.

1. Although the years in the table are not evenly spaced, you notice that each of the populations seems to grow rather slowly at first and more quickly later. Do you think the data will be better modeled by linear or exponential functions?

2. Let t represent the number of years since 1970. Use your graphing calculator to produce a scatterplot of the total inmate population, P_T. Your screen should appear as follows.

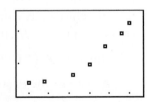

3. Use your graphing calculator to determine the regression equation of an exponential function that best represents the total inmate population, P_T. Remember that the input variable is t, the number of years since 1970. In your regression equation, $P_T = ab^t$, round the value for a to two decimal places and the value for b to three decimal places. Write your model below.

4. Use your graphing calculator to visually check how well the equation in Problem 3 fits the data. Using the window Xmin $= -3$, Xmax $= 40$, Ymin $= 0$, and Ymax $= 140$, your graph should resemble the following.

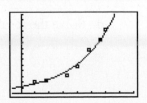

5. Use the exponential regression model from Problem 3 to determine the total federal prison inmate population in 2003.

6. a. Using your model from Problem 3, write an equation that can be used to determine the year in which the total federal inmate population, P_T, is 180,000. Remember, the population in the model is given in thousands.

b. Solve the equation in part a using a graphing approach. Your screen should resemble the following. What is the equation of the horizontal line in the graph?

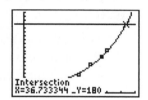

To solve the equation $11.80(1.077)^t = 180$ for t using an algebraic approach, you need to remove t as an exponent. The following problem guides you through this process. As you will discover, logarithms are essential in this algebraic approach.

7. Solve $11.80(1.077)^t = 180$ for t using an algebraic approach.

a. Isolate the exponential factor $(1.077)^t$ on one side of the equation.

b. Take the log (or ln) of each side of the equation in part a.

c. Apply the appropriate property of logarithms on the left side of the equation to remove t as an exponent.

d. Solve the resulting equation in part c for t.

e. How does your solution in part d compare to the estimate obtained graphically in Problem 6b?

8. You notice that over the years, the number of drug offenders seems to become a bigger percentage of the total population.

a. Determine an exponential regression equation to model the number of total sentenced drug offenders, P_D. Let the input variable t represent the number of years since 1970. In the regression equation $P_D = ab^t$, round the value for a to two decimal places and the value of b to three decimal places.

b. Use the exponential model to predict the total number of sentenced drug offenders in federal prisons in 2003.

c. Write an equation that can be used to determine the year in which the total number of sentenced drug offenders will reach 150,000.

d. Solve the equation in part c using an algebraic approach.

Radioactive Decay

Radioactive substances, such as uranium 235, strontium 90, iodine 131, and carbon 14, decay continuously with time. If P_0 represents the original amount of a radioactive substance, then the amount P present after a time t (usually measured in years) is modeled by

$$P = P_0 e^{kt},$$

where k represents the rate of continuous decay.

9. Strontium 90 decays continuously at a constant rate of 2.44% per year. Therefore, the equation for the amount P of strontium 90 after t years is

$$P = P_0 e^{-0.0244t}.$$

a. If 10 grams of strontium 90 are present initially, determine the number of grams present after 20 years.

b. How long will it take for the given quantity to decay to 2 grams?

c. How long would it take for the given amount of strontium 90 to decay to one-half of its original size (called its half-life)? Round to the nearest whole number.

d. Do you think that the half-life of strontium 90 is 28 years regardless of the initial amount? Answer part c using P_0 as the initial amount. (*Hint:* Find t when $P = \frac{1}{2}P_0$.)

SUMMARY
Activity 3.11

To solve exponential equations of the form $a \cdot b^x = c$, where $a > 0, b > 0, b \neq 1$, and $c > 0$:

1. Isolate the exponential factor on one side of the equation.

2. Take the log (or ln) of each side of the equation.

3. Apply the property $\log b^x = x \log b$ to remove the variable x as an exponent.

4. Solve the resulting equation for the variable.

EXERCISES
Activity 3.11

1. The number of arrests for possession of marijuana in New York City has increased dramatically. The number of arrests can be modeled by

$$N(t) = 0.37(1.77)^t,$$

where $N(t)$ represents the number of arrests in thousands and t represents the number of years since 1990.

a. How many arrests were made in 2000?

b. According to the model, in what year were there 50,000 arrests?

2. The U.S. Department of Transportation recommended that states adopt a 0.08% blood-alcohol concentration as the legal measure of drunk driving. Medical research has shown that as the concentration of alcohol in the blood increases, the risk of having a car accident increases exponentially. The risk, R, expressed as a percentage, is modeled by

$$R(x) = 6e^{12.77x},$$

where x is the blood-alcohol concentration, expressed as a percent.

a. What is the risk of having a car accident if your blood-alcohol concentration is 0.08% ($x = 0.08$)?

b. What blood-alcohol concentration has a corresponding 25% risk of a car accident?

3. In 1990, the International Panel on Climate Change projected the following future amounts of carbon dioxide (in parts per million or ppm) in the atmosphere.

YEAR	1990	2000	2075	2175	2275
AMOUNT OF CARBON DIOXIDE (ppm)	353	375	590	1090	2000

a. Use your graphing calculator to create a scatterplot of the data. Let t represent the number of years since 1990 and $A(t)$ represent the amount of carbon dioxide (in ppm) in the atmosphere. Do the carbon dioxide levels appear to be growing exponentially?

b. Use your graphing calculator to determine the regression equation of an exponential model that best fits the data.

   c. Use the model in part b to determine in what year the 1990 carbon dioxide level is
   expected to double.

   d. Verify your result in part c graphically.

*In Exercises 4–9, solve each equation using an algebraic approach. Verify your answers
graphically.*

4. $2^x = 14$

5. $3^{2x} = 8$

6. $1000 = 500(1.04)^t$

7. $e^{0.05t} = 2$ (*Hint:* Take the
natural log of both sides.)

8. $2^{3x+1} = 100$

9. $e^{-0.3t} = 2$

10. a. Iodine 131 disintegrates at a continuous constant rate of 8.6% per day. Determine
its half-life. Use the model

$$P = P_0 e^{-0.086t},$$

where t is measured in days. Round your answer to the nearest whole number.

b. If dairy cows eat hay containing too much iodine 131, their milk will be unsafe to drink. Suppose that hay contains five times the safe level of iodine 131. How many days should the hay be stored before it can be fed to dairy cows?

(*Hint:* Find t when $P = \frac{1}{5}P_0$.)

11. a. In 1969, a report written by the National Academy of Sciences (U.S.) estimated that the Earth could reasonably support a maximum world population of 10 billion. The world's population was approximately 3.6 billion, and growing continuously 2% per year. If this growth rate remained constant, in what year would the world population reach 10 billion, referred to as the Earth's carrying capacity? Use the model

$$P = P_0 e^{kt},$$

where P is the population (in billions), $P_0 = 3.6$, $k = 0.02$, and t is the number of years since 1969.

b. According to your growth model, when would this 1969 population double?

c. The world population in 1995 was approximately 5.7 billion. How does this compare with the population predicted by your growth model in part a?

d. The growth rate in 1995 was 1.5%. Assuming this growth rate remains constant, determine when the Earth's carrying capacity will be reached. Use the model $P = P_0 e^{kt}$.

ACTIVITY 3.12

Frequency
and Pitch

OBJECTIVE

1. Solve logarithmic equations both graphically and algebraically.

Raising a musical note one octave has the effect of doubling the pitch, or frequency, of the sound. However, you do not perceive the note to sound "twice as high," as you might predict. Perceived pitch is given by the function

$$P(f) = 2410 \log (0.0016f + 1),$$

where P is the perceived pitch in mels (units of pitch) and f is the frequency in hertz.

1. Let frequency (input) vary in value from 10 to 100,000 hertz, and let the perceived pitch (output) vary from 0 to 6000 mels. Graph this equation on your grapher, using the following window: Xmin = 0, Xmax = 100,000, Ymin = 0, and Ymax = 6000.

2. What is the perceived pitch, P, for the input value 10,000 hertz?

3. **a.** Write an equation that can be used to determine what frequency, f, gives an output value of 2000 mels.

   b. Solve the equation in part a using a graphing approach.

To determine the exact answer in Problem 3, you can use an algebraic approach. The following problem guides you through this process.

4. Solve the equation $2410 \log (0.0016x + 1) = 2000$ using an algebraic approach.

   a. Solve the equation for $\log (0.0016x + 1)$. That is, isolate the log on one side of the equation.

   b. The equation in part a is now in the form $\log_b N = E$, where $b = 10$, $N = 0.0016x + 1$ and $E = \frac{2000}{2410}$. Write the equation from part a in exponential form, $b^E = N$.

c. In exponential form, the equation in part a should be

$$0.0016x + 1 = 10^{\frac{2000}{2410}}.$$

Solve this equation for f. Of course, you will need to approximate a value of $10^{\frac{2000}{2410}}$ using your calculator.

d. How does your answer to part c compare to your answer to Problem 3b?

5. a. Use an algebraic approach to determine the frequency, f, that produces a perceived pitch of 3000 mels.

b. Verify your answer in part a using a graphing approach.

6. The formula $W = 0.35 \ln P + 2.74$ is a model for the average walking speed, W, in feet per second for a resident of a city with population P, measured in thousands.

a. Determine the walking speed of a resident of a small city having a population of 500,000.

b. If the average walking speed of a resident is 4.5 feet per second, what is the population of the city? Round your answer to the nearest thousand.

SUMMARY
Activity 3.12

To solve a logarithmic equation algebraically,

Step 1. Rewrite the equation in the form $\log_b (f(x)) = c$, where $b > 0$, $b \neq 1$, $c > 0$, and $f(x) > 0$.

Step 2. Rewrite the resulting equation from step 1 in exponential form $f(x) = b^c$.

Step 3. Solve the resulting equation from step 2 algebraically.

Step 4. Check the solutions in the original equation.

EXERCISES
Activity 3.12

In Exercises 1–6, solve each equation using an algebraic approach. Then verify your answer using a graphical approach.

1. $\log_2 x = 5$

2. $\ln x = 10$

3. $3 \log_5 (x + 2) = 5$

4. $\log_5 (x - 4) = 2$

5. $20 = 3.5 \ln x$

6. $4 + 1.75 \ln x = 31$

7. Stars have been classified into magnitude according to their brightness. Stars in the first six magnitudes are visible to the naked eye; those of higher magnitudes are visible only through a telescope. The magnitude, m, of the faintest star that is visible with a telescope having lens diameter d, in inches, is modeled by

$$m = 8.8 + 5.1 \log (d).$$

What is the highest magnitude of a star that is visible with the 200-inch telescope at Mount Palomar, California?

8. Coal consumption in the United States can be modeled by the equation

$$A(x) = 4.95 + 4.67 \ln x,$$

where x is the number of years since 1970, and $A(x)$ is the amount of coal consumed in quadrillions of British thermal units or quads. According to the model, in what year will the consumption of coal in the U.S. reach 30 quads?

9. The acidity or alkalinity of any solution is determined by the concentration of hydrogen ions, $[H^+]$, in the substance, measured in moles per liter. Acidity (or alkalinity) is measured on a pH scale, using the model

$$pH = -\log[H^+].$$

The pH scale ranges from 0 to 14. Values below 7 have progressively greater acidity; values greater than 7 are progressively more alkaline. Normal unpolluted rain has a pH of about 5.6. The acidity of rain over the northeastern United States, caused primarily by sulfur dioxide emissions, has had very damaging effects. One of the most acidic rainfalls on record had a pH of 2.4. What was the concentration of hydrogen ions?

10. The Richter scale is a well-known method of measuring the magnitude of an earthquake in terms of the amplitude, A (height), of its shock waves. The magnitude of any given earthquake is given by

$$m = \log\left(\frac{A}{A_0}\right),$$

where A_0 is a constant representing the amplitude of an average earthquake.

a. The magnitude of the 1906 San Francisco earthquake was 8.3 on the Richter scale. Write an equation that gives the amplitude, A, of the San Francisco earthquake in terms of A_0.

b. An earthquake with a magnitude of 5.5 will begin to cause serious damage. Write an equation that gives the amplitude, A, of a serious-damage earthquake in terms of A_0.

c. Determine the ratio of the amplitude of the San Francisco earthquake to the amplitude of a serious-damage (magnitude 5.5) earthquake. What is the significance of this number?

CLUSTER 2

What Have I Learned?

1. A logarithm is an exponent. Explain how this fact relates to the following properties of logarithms.

   a. $\log_b (x \cdot y) = \log_b x + \log_b y$

   b. $\log_b \frac{x}{y} = \log_b x - \log_b y$

   c. $\log_b x^n = n \cdot \log_b x$

2. You have \$20,000 to invest. Your broker tells you that mutual fund A has been growing exponentially for the past two years and that mutual fund B has been growing logarithmically over the same period. If you make your decision based solely on the past performances of the funds, in which fund would you choose to invest? Explain.

3. Study the following graphs showing various types of functions you have encountered in this course.

   a.

   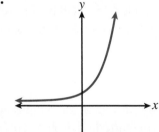

   b.

   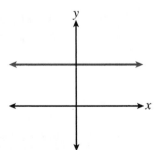

   c.

   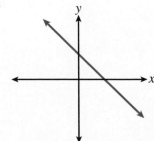

   d.

   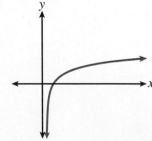

e.

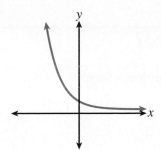

f.

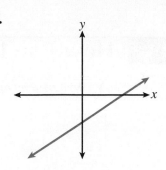

Complete the following table with respect to the preceding graphs.

DESCRIPTION	GRAPH NUMBER	GENERAL EQUATION
Constant function		
Linearly decreasing function		
Logarithmically increasing function		
Exponentially decreasing function		
Exponentially increasing function		
Linearly increasing function		

4. The graph of $y = \log_b x$ will never be located in the second or third quadrants. Explain.

5. What function would you enter into $Y1$ on your grapher to graph the function $y = \log_4 x$?

6. What values of x cannot be solutions to the equation $\log_b (3x - 2) = -1$?

7. What is the relationship between the functions $y = \log x$ and $y = 10^x$? How are the graphs related?

How Can I Practice?

1. Write each equation in logarithmic form.

   a. $4^2 = 16$ 　　　　　 **b.** $0.0001 = 10^{-4}$ 　　　　 **c.** $3^{-4} = \frac{1}{81}$

2. Write each equation in exponential form.

   a. $\log_2 32 = 5$ 　　　　　　　　　 **b.** $\log_5 1 = 0$

   c. $\log_{10} .001 = -3$ 　　　　　　 **d.** $\ln e = 1$

3. Solve each equation for the unknown variable.

   a. $\log_4 x = -3$ 　　　 **b.** $\log_b 32 = 5$ 　　　 **c.** $\log_5 125 = y$

4. **a.** Complete the table of values for the function $f(x) = \log_4 x$.

x	0.25	0.5	1	4	16	64
f(x)						

   b. Sketch a graph of the function, f.

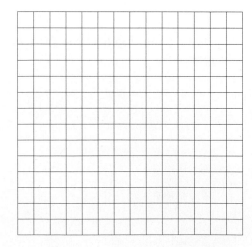

Exercise numbers appearing in color are answered in the Selected Answers appendix.

    c. Use your grapher to check your result in parts a and b.

    d. Determine the x-intercept.

    e. What is the domain of the function?

    f. What is the range?

    g. Does the graph have a vertical or horizontal asymptote? Explain.

    h. Use your grapher to determine $f(32)$.

    i. Use your grapher to determine x when $f(x) = 3.25$.

5. Write each of the following as a sum, difference, or multiple of logarithms. Assume that x, y, and z are all greater than 0.

    a. $\log_b \frac{xy^2}{z}$ **b.** $\log_3 \frac{\sqrt{x^3 y}}{z}$

    c. $\log_5 (x\sqrt{x^2 + 4})$ **d.** $\log_4 \sqrt[3]{\frac{xy^2}{z}}$

6. Rewrite the following as the logarithm of a single quantity.

    a. $\log x + \frac{1}{3}\log y - \frac{1}{2}\log z$ **b.** $3\log_3 (x + 3) + 2\log_3 z$

    c. $\frac{1}{3}\log_3 x - \frac{2}{3}\log_3 y - \frac{4}{3}\log z$

7. Use the change of base formula and your calculator to approximate the following.

    a. $\log_5 17$ **b.** $\log_{13} \sqrt[3]{41}$

8. Solve each of the following using an algebraic approach.

    a. $25 + 3 \ln(x) = 10$ **b.** $1.5 \log_4(x - 1) = 7$

9. Solve the following algebraically. Check your solutions using graphs or tables.

    a. $3^x = 17$ **b.** $42 = 3e^{1.7x}$

    c. $2 \cdot 3^x = 4^x$

10. The following table shows per capita health care expenditures (in dollars) in the United States from 1988 to 1993.

YEAR	1988	1989	1990	1991	1992	1993
EXPENDITURE ($)	2201	2422	2688	2902	3144	3331

    a. Plot these points on the following grid.

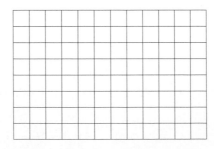

    b. The logarithmic model that fits this data is $E = f(x) = 2090 + 630 \ln x$, where E represents the per capita health care expenditures and x is the number of years since 1987. Add a sketch of this model to the grid in part a.

    c. Use this model to predict the per capita health care expenditures in 2005.

d. Use the model in part c to predict in what year health care expenditures will reach $3500. Use a graphing approach.

e. Write the equation that you would solve to determine the answer to part d.

f. Solve the equation in part e algebraically, and compare your result with your answer in part d.

Summary

The bracketed numbers following each concept indicate the activity in which the concept is discussed.

CONCEPT / SKILL	DESCRIPTION	EXAMPLE
The exponential function [3.1]	The exponential function is defined by $y = b^x, b > 0, b \neq 1$.	$y = 3^x$
Decay factor of an exponential function [3.1]	If $0 < b < 1$, the function is decreasing, and b is called the decay factor.	The exponential function $y = \left(\frac{1}{2}\right)^x$ has a decay factor of $\frac{1}{2}$.
Growth factor of an exponential function [3.1]	If $b > 1$, the function is increasing, and b is called the growth factor.	The exponential function $y = 3^x$ has a growth factor of 3.
Vertical intercept of an exponential function [3.1]	The vertical intercept (y-intercept) of an exponential function $y = b^x$ is $(0, 1)$.	The graph of $y = 2^x$ passes through the point $(0, 1)$.
Horizontal asymptote of an exponential function [3.1]	The line $y = 0$ is a horizontal asymptote of an exponential function $y = b^x$.	As x gets smaller, the output values of $y = 3^x$ approach 0.
Doubling time [3.2]	The doubling time of an exponential function is the time it takes for an output to double. The doubling time is set by the growth factor and remains the same for all output values.	Example 2, Activity 3.2
Half-life [3.2]	The half-life of an exponential function is the time it takes for an output to decay by one-half. The half-life is determined by the decay factor and remains the same for all output values.	Problem 10, Activity 3.2
Growth model [3.3]	If r represents the annual growth rate, the exponential function that models the quantity P can be written as $P(t) = P_0(1 + r)^t$, where P_0 is the initial amount, t represents the number of elapsed years, and $1 + r$ is the growth factor.	Example 2, Activity 3.3
Decay model [3.3]	If r represents the annual percent that decays, the exponential function that models the amount remaining can be written as $P(t) = P_0(1 - r)^t$, where $1 - r$ is the decay factor.	Example 3, Activity 3.3
Compound interest [3.5]	The formula for compounding interest is $A = P\left(1 + \frac{r}{n}\right)^{nt}$.	Example 1, Activity 3.5

Continuous compounding [3.5]	The formula for continuous compounding is $A = Pe^{rt}$.	Example 3, Activity 3.5
Continuous growth at a constant rate [3.5]	Whenever growth is continuous at a constant rate, the exponential model used is $y = y_0 e^{rt}$.	Problem 12, Activity 3.5
Continuous decay at a constant rate [3.5]	Whenever decay is continuous at a constant rate, the model used is $y = y_0 e^{-rt}$.	Exercise 6, Activity 3.5
Logarithm [3.7]	In the equation $y = b^x$, where $b > 0$ and $b \neq 1$, x is called a logarithm or log.	For the equation $3^4 = 81$, 4 is the logarithm of 81 to the base 3.
Notation for logarithms [3.7]	The notation for logarithms is $\log_b x = y$, where b is the base of the log, x (a positive number) is the power of b, and y is the exponent.	In the equation $\log_2 16 = 4$, 2 is the base, 4 is the log or exponent, and 16 is the power.
Common logarithm [3.7]	A common logarithm is a base ten logarithm. The notation is $\log_{10} x = \log x$.	$1000 = 10^3$. The common logarithm of 10^3 is 3; i.e., $\log 1000 = 3$.
Natural logarithm [3.7]	A natural logarithm is a base e logarithm. The notation is $\log_e x = \ln x$.	$\log_e e^3 = \ln e^3 = 3$
Logarithmic equation [3.7]	The logarithmic equation $y = \log_b x$ is equivalent to the exponential equation $b^y = x$.	The equations $6 = \log_4 x$ and $x = 4^6$ are equivalent.
Basic properties of logarithms [3.7]	If $b > 0$ and $b \neq 1$, $\log_b 1 = 0$ $\log_b b = 1$ and $\log_b b^n = n$.	$\log_4 1 = 0$, $\log_7 7 = 1$, $\log_6 6^4 = 4$
Logarithmic function [3.8]	If $b > 0$ and $b \neq 1$, the logarithmic function is defined by $y = \log_b x$.	$y = \log_4 x$
Graph of the logarithmic function [3.8]	The graph is increasing for all $x > 0$, has an x-intercept of $(1, 0)$, and has a vertical asymptote of $x = 0$, the y-axis.	
Comparison of the graphs of $f(x) = b^x$; **where** $b > 1$ **and** $g(x) = \log_b (x)$ **where** $b > 1$ [3.9]	Both graphs increase. The exponential function increases faster as x increases; the log function increases slower as x increases. The domain of the exponential function is the range of the log, which is all real numbers; the range of the exponential function is the domain of the log, which is the interval $(0, \infty)$.	Problem 6, Activity 3.9

If $A > 0$, $B > 0$, $b > 0$, and $b \neq 1$, then $\log_b (A \cdot B) = \log_b A + \log_b B$ [3.10]	The logarithm of a product is the sum of the logarithms.	$\log_2 (4 \cdot 8) = \log_2 (4) + \log_2 (8)$ $= 2 + 3 = 5$
If $A > 0$, $B > 0$, $b > 0$, and $b \neq 1$, then $\log_b \left(\frac{A}{B}\right) = \log_b A - \log_b B$ [3.10]	The logarithm of a quotient is the difference of the logarithms.	$\log_3 \left(\frac{81}{27}\right) = \log_3 81 - \log_3 27$ $= 4 - 3 = 1$
If $A > 0$, $B > 0$, $b > 0$, and $b \neq 1$, then $\log_b (A + B) \neq \log_b (A) + \log_b (B)$ [3.10]	The logarithm of a sum is not the sum of the logarithms.	$\log 2 + \log 3 =$ $0.3010 + 0.4771 = 0.7781$ $\log (2 + 3) = \log 5 = 0.6990$
If $A > 0$, p is a real number, $b > 0$, and $b \neq 1$, then $\log_b A^p = p \log_b A$ [3.10]	The logarithm of a power is the exponent times the logarithm of the base.	$\log_5 x^4 = 4 \log_5 x$
Change of base formula [3.10]	The logarithm of any positive number x to any base can be found using the formula $$\log_b x = \frac{\log x}{\log b} \text{ or } \log_b x = \frac{\ln x}{\ln b}.$$	$\log_2 (2.5) = \dfrac{\log (2.5)}{\log 2} = 1.3219$

Gateway Review

1. a. Determine the output values for the function $f(x) = 8^x$ by completing the following table.

x	−1	−$\frac{1}{3}$	0	1	$\frac{4}{3}$	2	3
$f(x) = 8^x$							

b. Sketch the graph of the function, f.

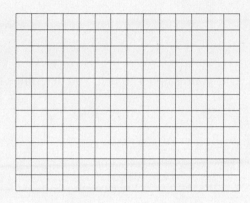

c. Is this function increasing or decreasing? Explain how you know this by looking at the equation of the function.

d. What is the domain?

e. What is the range?

f. What are the x-and y-intercepts?

g. Are there any asymptotes? If yes, write the equations of the asymptotes.

h. Compare the graph of f to the graph of $g(x) = \left(\frac{1}{8}\right)^x$. What are the similarities and the differences?

i. In what way does the graph of $h(x) = 8^x + 5$ differ from $f(x) = 8^x$?

Answers to all Gateway exercises are included in the Selected Answers appendix.

j. Write the equation of the function that is the inverse of the function $f(x)$.

2. Complete the table for each exponential function. Use your grapher to check your work.

FUNCTION	BASE, b	GROWTH OR DECAY FACTOR	x-INTERCEPT	y-INTERCEPT	HORIZONTAL ASYMPTOTE	INCREASING OR DECREASING
$h(x) = 6^x$						
$g(x) = (\frac{1}{3})^x$						
$p(x) = 5(2.34)^x$						
$q(x) = 3(0.78)^x$						
$r(x) = 2^x - 4$						

3. Use your grapher to help you determine the domain and range for each function.

FUNCTION	$f(x) = 0.8^x$	$h(x) = 6^x + 2$	$t(x) = 3^x - 5$	$q(x) = \log_4 x$	$r(x) = \ln(x - 3)$
DOMAIN					
RANGE					

4. a. Given the following table, determine whether the given data can be approximately modeled by an exponential function. If it can, what is the growth or decay factor?

x	0	1	2	3	4
y	10	15.5	24	36	55.5

b. Determine an exponential equation that models this data.

5. a. Your salary has increased at the rate of 1.5% annually for the past five years, and your boss projects this will remain unchanged for the next five years. You were making $15,000 annually in 2002. Complete the following table.

2002	2003	2004	2005	2006	2007

b. Write the exponential growth function that models your annual salary during this period of time. Let x represent the number of years since 2002.

c. If your increase in salary continues at this rate, how much will you make in 2010? Is this realistic?

d. You would like to double your salary. How many years will you have to work before your salary will be twice the salary you made in 2002?

6. a. You just inherited $5000. You can invest the money at a rate of 6.5% compounded continuously. In eight years, your oldest child will be going to college. How much will be in the bank for her education? Use the equation $A = A_0 e^{rt}$.

b. You actually need to have $12,000 for your child's first year of college. For how many years would you have to leave the money in the bank to have the $12,000?

7. The number of multiple births (triplets or higher) in the United States between 1990 and 1999 is listed in the following table, with 0 representing the year 1990.

NUMBER OF YEARS SINCE 1990	0	2	3	4	5	6	7	8	9
NUMBER OF MULTIPLE BIRTHS	3028	3883	4168	4594	4973	5939	6737	7625	7321

Source: National Center for Health Statistics.

a. Plot the data on a appropriately scaled and labeled coordinate axis.

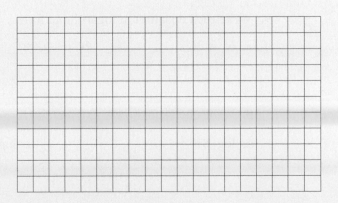

b. Does the scatterplot show that the data would be better modeled by a linear or an exponential model?

c. Use your grapher to determine the exponential regression equation that best fits the multiple births data.

d. According to your model, what is the growth factor for the multiple births data?

e. Estimate the growth rate (written as a percent) in multiple births each year.

f. Use the regression equation to determine the predicted total number of multiple births in 2012.

g. Use your grapher to determine the doubling time for your exponential model.

8. Determine the value of each of the following without using your calculator.

a. $25^{\frac{3}{2}}$ **b.** $81^{\frac{3}{4}}$ **c.** $64^{\frac{-5}{6}}$

d. $\sqrt[3]{125^2}$ **e.** $\log_3 \frac{1}{9}$ **f.** $\log_5 625$

g. $\log 0.001$

9. Write each equation in logarithmic form.

a. $6^2 = 36$ **b.** $0.000001 = 10^{-6}$ **c.** $2^{-5} = \frac{1}{32}$

10. Write each equation in exponential form.

a. $\log_3 81 = 4$ **b.** $\log_7 1 = 0$

c. $\log_{10} 0.0001 = -4$ **d.** $\ln e = 1$

11. Solve each equation for the unknown variable.

    a. $\log_5 x = -3$ **b.** $\log_b 256 = 4$ **c.** $\log_2 64 = y$

12. a. Complete the table of values for the function $f(x) = \log_5 x$.

x	0.008	0.04	0.2	1	5	25
$f(x)$						

    b. Sketch a graph of the function.

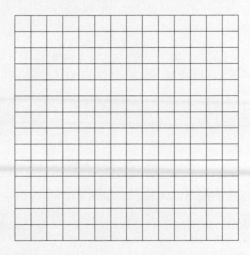

    c. Use your grapher to check your result in parts a and b.

    d. Determine the x-intercept.

    e. What is the domain of the function?

    f. What is the range?

g. Does the graph have a vertical or horizontal asymptote?

h. Use your grapher to determine $f(23)$.

i. Use your grapher to determine x when $f(x) = 2.46$.

13. Use the change of base formula and your calculator to approximate the following.

a. $\log_7 21$ **b.** $\log_{15} \frac{8}{9}$

14. Write each of the following as a sum, difference, or multiple of logarithms. Assume that x, y, and z are all greater than 0.

a. $\log_2 \dfrac{x^3 y}{z^{\frac{1}{2}}}$ **b.** $\log \sqrt[3]{\dfrac{x^4 y^3}{z}}$

15. Rewrite the following as the logarithm of a single quantity.

a. $\log x + \frac{1}{4} \log y - 3 \log z$ **b.** $\frac{1}{3}\left(\log x - 2 \log y - \log z\right)$

16. Solve the following algebraically.

a. $3^{3+x} = 7$ **b.** $\log_2 (4x + 9) = 4$

c. $50 + 6\ln x = 85$

17. a. Sketch the graph of the function using the data from the given table.

x	0.1	0.5	1	2	4	16
$f(x)$	-1.66	-0.5	0	.5	1	2

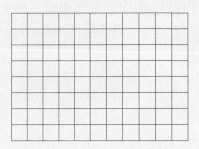

b. Use the table and the graphing feature of your calculator to verify that the equation that defines function f is $f(x) = 0.5 \log_2 x$

c. Use the function to determine the value of $f(54)$.

d. If $f(x) = 2.319$, determine the value of x.

e. Use your grapher to verify that the function $g(x) = 4^x$ is the inverse of f.

18. The population (in millions) of New York State and Florida can be modeled by the following:

New York State: $P_N = 18.98e^{0.0055t}$

Florida: $P_F = 15.98e^{0.0235t}$

where t represents the number of years since 2000.

a. Determine the population of New York and Florida in 2000 ($t = 0$).

b. Sketch a graph of each function on the same coordinate axes.

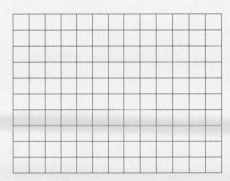

c. Determine graphically the year when the population of Florida will equal the population of New York State.

d. Determine algebraically the year when the population of Florida will first exceed 25 million.

Quadratic and Higher-Order Polynomial Functions

CLUSTER 1   Introduction to Quadratic Functions

ACTIVITY 4.1

Baseball and the
Sears Tower

OBJECTIVES

1. Identify functions of the
   form $f(x) = ax^2 + bx + c$
   as quadratic functions.

2. Explore the role of a as it
   relates to the graph of
   $f(x) = ax^2 + bx + c$.

3. Explore the role of b as it
   relates to the graph of
   $f(x) = ax^2 + bx + c$.

4. Explore the role of c as it
   relates to the graph of
   $f(x) = ax^2 + bx + c$.

Imagine yourself standing on the roof of the 1450-foot-high Sears Tower in Chicago. When you release and drop a baseball from the roof of the tower, the ball's height above the ground, H (in feet), can be described as a function of the time, t (in seconds), since it was dropped. This height function is defined by

$$H(t) = -16t^2 + 1450.$$

1. a. Complete the following table.

TIME, t (sec)	$H(t) = -16t^2 + 1450$
0	
1	
2	
3	
4	
5	
6	
7	
8	
9	
10	

b. How far does the baseball fall during the first second?

c. How far does it fall during the second second?

2. Using the height function, $H(t) = -16t^2 + 1450$, determine the average rate of change of H with respect to t over the given interval. Remember:

$$average\ rate\ of\ change = \frac{change\ in\ ouptut}{change\ in\ input}.$$

   a. $0 \le t \le 1$ **b.** $1 \le t \le 2$

   c. Based on the results of parts a and b, do you believe that $H(t) = -16t^2 + 1450$ is a linear function? Explain.

3. Graph the height function setting the window parameters at Xmin $= -10$ and Xmax $= 10$ for the input and Ymin $= -50$ and Ymax $= 1500$ for the output. Your graph should appear as follows:

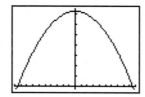

   a. Describe the important features of the graph of $H(t) = -16t^2 + 1450$. Discuss the shape, symmetry, and intercepts.

   b. What are the practical domain and range of the height function?

   c. Is the graph of the height function the actual path of the object when the ball is dropped? Explain.

The graph of the height function is a parabola. The graph of a parabola is a ∪-shaped figure that opens upward, ∪, or downward, ∩.

DEFINITION

Any function defined by an equation of the form $y = ax^2 + bx + c$ or $f(x) = ax^2 + bx + c$, where a, b, and c represent real numbers and $a \ne 0$, is called a **quadratic function**. The output variable y is defined by an expression having three terms: the **quadratic term**, ax^2, the **linear term**, bx, and the **constant term**, c. The numerical factors of the quadratic and linear terms, a and b, are called the **coefficients** of the terms.

Example 1 $H(t) = -16t^2 + 100$ *defines a quadratic function. The quadratic term is* $-16t^2$. *The linear term is 0t, although it is not written as part of the expression defining H(t). The constant term is 100. The numbers* -16 *and 0 are the coefficients of the quadratic and linear terms respectively. Therefore,* $a = -16, b = 0,$ *and* $c = 100$.

4. For each of the following quadratic functions, identify the value of a, b, and c.

QUADRATIC FUNCTION	a	b	c
$y = 3x^2$			
$y = -2x^2 + 3$			
$y = x^2 + 2x - 1$			
$y = -x^2 + 4x$			

The Effects of the Coefficient *a*

5. a. Graph the quadratic function defined by $g(t) = 16t^2 + 1450$ on the same screen as $H(t) = -16t^2 + 1450$. Use the window settings Xmin $= -10$, Xmax $= 10$, Ymin $= -50$, and Ymax $= 3000$.

b. What effect does the sign of the coefficient of t^2 appear to have on the graph of the parabola?

6. Graph the functions $h(t) = -16t^2 + 100$, $f(t) = -6t^2 + 100$, $g(t) = -40t^2 + 100$ in the same window. What effect does the magnitude of the coefficients of t^2 (namely,$|-16| = 16$, $|-6| = 6$, and $|-40| = 40$) appear to have on the graph of that particular parabola?

The results from Problems 5 and 6 regarding the effects of the coefficient a can be summarized as follows:

The graph of a quadratic function defined by $f(x) = ax^2 + bx + c$ is called a parabola.

- If $a > 0$, the parabola opens upward.
- If $a < 0$, the parabola opens downward.

- The magnitude of a affects the width of the parabola. The larger the absolute value of a, the narrower the parabola.

7. a. Is the graph of $h(x) = 0.3x^2$ wider or narrower than the graph of $f(x) = x^2$?

b. How do the output values of h and the output values of f compare for the same input value?

c. Is the graph of $g(x) = 3x^2$ wider or narrower than the graph of $f(x) = x^2$?

d. How do the output values of g and f compare for the same input value?

e. Describe the effect of the magnitude of the coefficient a on the width of the graph of the parabola.

f. Describe the effect of the magnitude of the coefficient of a on the output value.

The Effects of the Coefficient b

Assume for the time being that you are back on the roof of the 1450-foot Sears Tower. Instead of merely releasing the ball, suppose you *throw it down* with an initial velocity of 40 feet per second. Then the function describing its height above ground as a function of time is modeled by:

$$H_{down}(t) = -16t^2 - 40t + 1450.$$

If you tossed the ball up with an initial velocity of 40 feet per second, then the function describing its height above ground as a function of time is modeled by:

$$H_{up}(t) = -16t^2 + 40t + 1450.$$

8. Predict what features of the graphs of H_{down} and H_{up} have in common with $H(t) = -16t^2 + 1450$.

9. **a.** Graph the three functions—$H(t)$, $H_{\text{down}}(t)$, and $H_{\text{up}}(t)$—using the same window setting.

   b. What effect do the $-40t$ and $40t$ terms seem to have upon the graphs?

   If $b = 0$, the turning point of the parabola is located on the vertical axis. If $b \neq 0$, the turning point will not be on the vertical axis.

10. Set the window of your grapher to Xmin $= -8$, Xmax $= 8$, Ymin $= -20$, and Ymax $= 20$, and graph the parabolas defined by the following quadratic equations. Note the differences among the graphs, paying careful attention to the effects of the signs of the coefficients a, b, and c on the graph.

   a. $f(x) = x^2$ **b.** $g(x) = x^2 - 4x$

   c. $h(x) = x^2 - 8x$ **d.** $p(x) = x^2 + 6x$

   e. $q(x) = -x^2 + 6x$ **f.** $r(x) = -x^2 - 8x$

The Constant Term c

Consider once again the height function $H(t) = -16t^2 + 1450$ from the beginning of the activity.

11. a. What is the vertical intercept of the graph?

    b. What is the practical meaning of the vertical intercept in this situation?

    c. Predict what the graph of $h(t) = -16t^2 + 1450$ would look like if the constant term 1450 were changed to 800. Verify your prediction by graphing $y = -16t^2 + 800$. What does the constant term tell you about the graph of the parabola?

> The constant term c of a quadratic function $f(x) = ax^2 + bx + c$ *always* indicates the vertical intercept of the parabola. The vertical intercept of any quadratic function is $(0, c)$.

12. Graph the parabolas defined by the following quadratic equations. Note the similarities and differences among the graphs, especially the vertical intercepts. Be careful in your choice of a window.

    a. $f(x) = 1.5x^2$ **b.** $g(x) = 1.5x^2 + 7$

    c. $q(x) = 1.5x^2 + 4$ **d.** $s(x) = 1.5x^2 - 4$

13. Match each function with its corresponding graph below, and then verify using your graphing calculator.

    a. $f(x) = x^2 + 4x + 4$ **b.** $g(x) = 0.2x^2 + 4$ **c.** $h(x) = -x^2 + 3x$

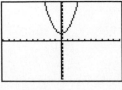 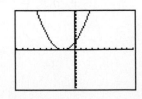 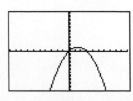

        i.   **ii.**   **iii.**

SUMMARY
Activity 4.1

1. The function that has a parabola as its graph is called a **quadratic function**. The equation of a quadratic function with x as the input variable and y as the output variable has the standard form

$$y = ax^2 + bx + c,$$

where a, b, and c represent real numbers and $a \neq 0$.

2. The graph of a quadratic function is called a **parabola**.

3. For the quadratic function defined by $f(x) = ax^2 + bx + c$:

   - If $a > 0$, the parabola opens upward.

   - If $a < 0$, the parabola opens downward.

   The magnitude of a affects the width of the parabola. The larger the absolute value of a, the narrower the parabola.

4. If $b = 0$, the turning point of the parabola is located on the vertical axis. If $b \neq 0$, the turning point will not be on the vertical axis.

5. The constant term, c, of a quadratic function $f(x) = ax^2 + bx + c$ always indicates the vertical intercept of the parabola. The vertical intercept of any quadratic function is $(0, c)$.

EXERCISES
Activity 4.1

1. **a.** Complete the following input/output table for $y = x^2$.

INPUT	OUTPUT
−3	
−2	
−1	
0	
1	
2	
3	

b. Use the results of part a to sketch a graph $y = x^2$.

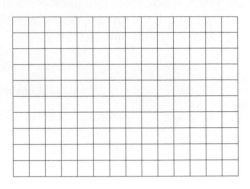

c. Use your graphing calculator to compare and check the information for parts a and b. Are the graphs the same?

d. What is the coefficient of the term x^2?

e. From the graph, determine the domain and range of the function.

2. a. Create a table similar to the one in Exercise 1a to show the output for $y = -x^2$.

INPUT	−3	−2	−1	0	1	2	3
OUTPUT							

b. Sketch the graph of $y = -x^2$.

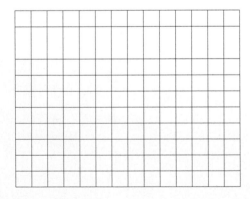

c. Use your graphing calculator to compare and check the information for parts a and b. Are the graphs the same?

d. What is the coefficient of the term $-x^2$?

3. In each of the following functions defined by an equation of the form $y = ax^2 + bx + c$, identify the value of a, b, and c.

    a. $y = -2x^2$

    b. $y = \frac{2}{5}x^2 + 3$

    c. $y = -x^2 + 5x$

    d. $y = 5x^2 + 2x - 1$

4. Predict what the graph of each of the following quadratic functions will look like. Use your graphing calculator to verify your prediction.

    a. $f(x) = 3x^2 + 5$ **b.** $g(x) = -2x^2 + 1$

    c. $h(x) = 0.5x^2 - 3$

5. Graph the following pairs of functions, and describe any similarities as well as any differences that you observe in the graphs.

    a. $f(x) = 3x^2, g(x) = -3x^2$ **b.** $h(x) = \frac{1}{2}x^2, f(x) = 2x^2$

    c. $g(x) = 5x^2, h(x) = 5x^2 + 2$ **d.** $f(x) = 4x^2 - 3, g(x) = 4x^2 + 3$

    e. $f(x) = 6x^2 + 1, h(x) = -6x^2 - 1$

6. Use your graphing calculator to graph the two functions $y_1 = 3x^2$ and $y_2 = 3x^2 + 2x - 2$.

   a. What is the vertical intercept of the graph of each function?

   b. Compare the two graphs to determine the effect of the linear term $2x$ and the constant term -2 on the graph of $y_1 = 3x^2$.

For Exercises 7–11, determine

   a. whether the parabola opens upward or downward and

   b. the vertical intercept.

7. $f(x) = -5x^2 + 2x - 4$

8. $g(t) = \frac{1}{2}t^2 + t$

9. $h(v) = 2v^2 + v + 3$

10. $r(t) = 3t^2 + 10$

11. $f(x) = -x^2 + 6x - 7$

12. Does the graph of $y = -2x^2 + 3x - 4$ have any horizontal intercepts? Explain.

13. a. Is the graph of $y = \frac{3}{5}x^2$ wider or narrower than the graph of $y = x^2$?

    b. For the same input value, which graph would have a larger output value?

14. Put the following in order from narrowest to widest.

    a. $y = 0.5x^2$ b. $y = 8x^2$ c. $y = -2.3x^2$

ACTIVITY 4.2
The Shot Put

OBJECTIVES

1. Determine the vertex or turning point of a parabola.
2. Determine the axis of symmetry of a parabola.
3. Identify the domain and range.
4. Determine the vertical intercept of a parabola.
5. Determine the horizontal intercept(s) of a parabola graphically.

Parabolas are good models for a variety of situations that you encounter in everyday life. Examples include the path of a golf ball after it is struck, the arch (cable system) of a bridge, the path of a baseball thrown from the outfield to home plate, the stream of water from a drinking fountain, and the path of a cliff diver.

Consider the 2000 men's Olympic shot put event, which was won by Finland's Arsi Harju with a throw of 69 feet $10\frac{1}{4}$ inches. The path of his winning throw can be approximately modeled by the quadratic function

$$H(x) = -0.015545x^2 + x + 6,$$

where x is the horizontal distance in feet from the point of the throw and $H(x)$ is the vertical height in feet of the shot above the ground.

1. **a.** After inspecting the equation for the path of the winning throw, which way do you expect the parabola to open? Explain.

   b. What is the vertical intercept of the graph of the parabola? What practical meaning does this intercept have in this situation?

2. Use your graphing calculator to produce a plot of the path of the winning throw. Be sure to adjust your window settings so that all of the important features of the parabola (including horizontal intercepts) appear on the screen. Your graph should resemble the following.

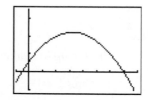

3. From the graph, what are the appropriate values for the input, x (practical domain), and the output, $H(x)$ (practical range)?

4. Use the table feature of your graphing calculator to complete the following table.

x	10	20	30	40	50
H(x)					

Vertex of a Parabola

An important feature of the graph of any quadratic function defined by $f(x) = ax^2 + bx + c$ is its **turning point**, also called the **vertex**. The turning point of a parabola that opens downward or upward is the point at which the parabola changes direction from increasing to decreasing or decreasing to increasing.

5. a. Use the results of Problem 4 to estimate the coordinates of the vertex of the shot put function H.

b. Use the TRACE feature of your graphing calculator to approximate the vertex of the shot put function H.

6. The vertex is often very important in a situation. What is the significance of the coordinates of the turning point in this problem?

The coordinates of the vertex of a parabola having equation $y = ax^2 + bx + c$ can be determined from the values of a and b in the equation, but first, you need an equivalent way of writing the equation $y = ax^2 + bx + c$.

An equivalent form of the equation $y = ax^2 + bx + c$ that make it easy to identify the vertex of the parabola is $y = a(x - h)^2 + k$ where $h = -\dfrac{b}{2a}$ and $k = -\dfrac{b^2 - 4ac}{4a}$.

 For a discussion of why this is true see Solving Quadratic Equations by Completing the Square in Appendix A.

Appendix

7. a. Identify the values of a, b, and c in the quadratic equation
$$y = 2x^2 - 12x + 11.$$

b. Using the values of a, b, and c from part a, determine the values of $h = -\dfrac{b}{2a}$ and $k = -\dfrac{b^2 - 4ac}{4a}$.

c. Write the equation $y = 2x^2 - 12x + 11$ in the form $y = a(x - h)^2 + k$.

d. Enter the equation $y = 2x^2 - 12x + 11$ into your graphing calculator as Y1 and the equation from part c into your graphing calculator as Y2. Graph both equations in the same screen. What do you notice about the graphs of the two equations?

e. Use the TRACE feature of your graphing calculator to determine the vertex of the parabola.

f. Describe the relationship between the coordinates of the vertex and the quadratic equation written in the form $y = a(x - h)^2 + k$.

DEFINITION

The **vertex** or turning point of a parabola having equation $y = ax^2 + bx + c$ has coordinates

$$\left(-\frac{b}{2a}, -\frac{b^2 - 4ac}{4a}\right),$$

where a is the coefficient of the x^2 term and b is the coefficient of the x term.

Note that the y-coordinate (output) of the vertex is determined by substituting the x-coordinate of the vertex into the equation of the parabola and evaluating the resulting expression.

Example 1 *Determine the vertex of the parabola defined by the equation*
$y = -3x^2 + 12x + 5$.

SOLUTION

Step 1. Determine the x-coordinate of the vertex by substituting the values of a and b into the formula $x = \frac{-b}{2a}$.

Because $a = -3$ and $b = 12$, you have

$$x = \frac{-(12)}{2(-3)} = \frac{-12}{-6} = 2.$$

Step 2. The y-coordinate of the vertex can be determined two ways.

i. Substitute $a = -3$, $b = 12$, and $c = 5$ into the expression $c - \frac{b^2 - 4ac}{4a}$ and evaluate as follows.

$$-\frac{12^2 - 4(-3)(5)}{4(-3)} = -\frac{144 + 60}{-12} = \frac{-204}{-12} = 17$$

ii. The y-value of the vertex is the corresponding output value for $x = 2$. Substituting 2 for x in the equation, you have

$$y = -3(2)^2 + 12(2) + 5 = 17.$$

Therefore, the vertex is (2, 17).

Because the parabola in Example 1 opens downward $(a = -3 < 0)$, the vertex is the high point (maximum) of the parabola as demonstrated by the following graph of the parabola.

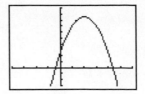

8. Determine the vertex of the parabola defined by
$H(x) = -0.015545x^2 + x + 6$ (the shot put function).

Rather than using the TRACE feature to approximate the vertex of a parabola, you can determine the vertex of a parabola by selecting the maximum option in the CALC menu of your graphing calculator. Follow the prompts to obtain the coordinates of the maximum point (vertex).

 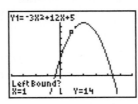 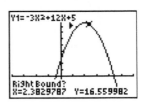

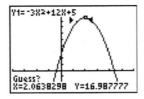 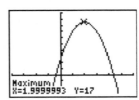

 For further help with the TI-83 Plus, see Appendix C.

9. **a.** Use your graphing calculator to determine the vertex of the parabola having equation $H(x) = -0.015545x^2 + x + 6$ (the shot put function H).

b. How do the coordinates you determined using your graphing calculator (see Problem 8) compare with your results in part a?

c. What is the practical meaning of the coordinates of the vertex in this situation?

Axes of Symmetry of a Parabola

DEFINITION

The **axis of symmetry** is a vertical line that divides the parabola into two symmetrical parts that are mirror images in the line.

Example 2   *Consider the parabola from Example 1. The axis of symmetry of the parabola is x = 2. Note that the line of symmetry passes through the vertex of the parabola.*

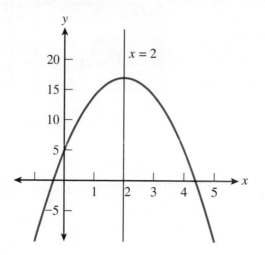

Because the vertex (turning point) of a parabola lies on the axis of symmetry, the equation of the axis of symmetry is

$$x = \frac{-b}{2a},$$

10. What is the axis of symmetry of the shot put function, H?

Intercepts of the Graph of a Parabola

The y-intercept (vertical intercept) of the graph of the parabola defined by $y = -3x^2 + 12x + 5$ (see Example 1) can be determined directly from the equation. If $x = 0$, then

$$y = -3(0)^2 + 12(0) + 5 = 5$$

and the y-intercept is $(0, 5)$.

In general, the y-intercept of the parabola defined by $y = ax^2 + bx + c$ is $(0, c)$.

Because the vertex $(2, 17)$ of the parabola having equation $y = -3x^2 + 12x + 5$ is a point above the x-axis (the y-coordinate is positive) and the parabola opens downward, the parabola must intersect the horizontal axis in two places. This is verified by the following graph.

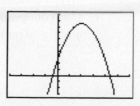

The x-intercepts can be determined using the zero option in the CALC menu of your graphing calculator. Follow the prompts to obtain one x-intercept at a time. The screens should appear as follows.

 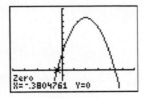 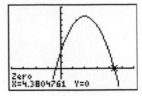

11. a. Use your graphing calculator to determine the x-intercept(s) for the shot put function having equation $H(x) = -0.015545x^2 + x + 6$. The right-most intercept appears in the following screen.

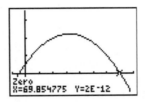

b. Is either x-intercept determine in part a significant to the problem situation? Explain.

The graph of $H(x) = -0.0155x^2 + x + 6$ has two x-intercepts. Does the graph of every parabola have x-intercepts? Problems 11 and 12 will help answer this question.

12. a. Use the values of a, b, and c to determine the coordinates of the vertex of the graph of $y = x^2 + 6x + 12$.

b. Use the y-coordinate of the vertex to determine if the vertex is above or below the x-axis.

c. Use the value of a in $y = x^2 + 6x + 12$ to determine if the parabola opens upward or downward.

d. Use the results from parts b and c to determine if the parabola has x-intercepts.

e. Use your graphing calculator to verify your answer to part d.

13. a. Use the values of a, b, and c to determine the coordinates of the vertex of the graph of $y = -x^2 + 8x - 21$.

b. Use the y-coordinate of the vertex to determine if the vertex is above or below the x-axis.

c. Use the value of a in $y = -x^2 + 8x - 21$ to determine if the parabola opens upward or downward.

d. Use the results from parts b and c to determine if the parabola has x-intercepts.

e. Use your graphing calculator to verify your answer to part d.

> If a parabola opens upward and the vertex is above the x-axis there are no x-intercepts. If a parabola opens downward and the vertex is below the x-axis, there are no x-intercepts.

14. a. Use your result from Problem 10 to determine the practical domain of the shot put function. How does this compare with your answer in Problem 3?

b. Sketch the path of the winning throw of the shot put. Be sure to label all key points, including the vertex and intercepts.

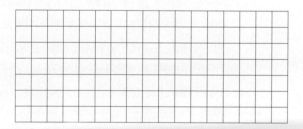

c. From the graph of the winning throw, over what horizontal distance (x-interval) is the height of the shot put increasing?

d. Determine the x-interval over which the height of the shot put is decreasing.

e. What is the practical range?

15. Now consider the function $H(x) = -0.015545x^2 + x + 6$ as a general function that is not restricted by the physical situation in the activity.

a. What is the domain of the general function?

b. Over what x-interval does the general function increase?

c. Over what x-interval does the general function decrease?

d. What is the range?

SUMMARY
Activity 4.2

The following characteristics are commonly used in analyzing the quadratic function defined by $f(x) = ax^2 + bx + c$, $a \neq 0$, and its graph.

1. The **axis of symmetry** is a vertical line that separates the parabola into two mirror images. The equation of the vertical axis of symmetry is given by $x = \frac{-b}{2a}$.

2. The **vertex** (turning point) always falls on the axis of symmetry. The x-coordinate of the vertex is given by $\frac{-b}{2a}$. Its y-coordinate is determined by evaluating the function at this value. In other words, the y-coordinate of the vertex is given by $f\left(-\frac{b}{2a}\right)$.

3. The **y-intercept**, the point where the parabola crosses the y-axis (that is, where its x-coordinate is zero), is always given by $(0, c)$.

4. The **x-intercept** is the point or points (if any) where the parabola crosses the x-axis (that is, where its y-coordinate is zero).

5. If a parabola opens upward and the vertex is above the x-axis there are no x-intercepts. If a parabola opens downward and the vertex is below the x-axis, there are no x-intercepts.

6. The **domain** of the general quadratic function is the set of all real numbers.

7. If the parabola opens upward, the **range** is all real numbers greater than or equal to the output value of the vertex. If the parabola opens downward, the *range* is all real numbers less than or equal to the output value of the vertex.

EXERCISES
Activity 4.2

For Exercises 1–8, determine the following characteristics of each quadratic function.

    a. the direction in which the graph opens

    b. the axis of symmetry

    c. the turning point (vertex)

    d. the y-intercept

1. $f(x) = x^2 - 3$

2. $g(x) = x^2 + 2x - 8$

3. $y = x^2 + 4x - 3$

4. $f(x) = 3x^2 - 2x$

5. $h(x) = x^2 + 3x + 4$

6. $g(x) = -x^2 + 7x - 6$

7. $y = 2x^2 - x - 3$

8. $f(x) = x^2 + x + 3$

Exercise numbers appearing in color are answered in the Selected Answers appendix.

For Exercises 9–16, use your graphing calculator to sketch the graphs of the functions, and then determine each of the following.

    a. the coordinates of the x-intercepts for each function, if they exist

    b. the domain and range for each function

    c. the horizontal interval over which each function is increasing

    d. the horizontal interval over which each function is decreasing

9. $g(x) = -x^2 + 7x - 6$ **10.** $h(x) = 3x^2 + 6x + 4$

11. $y = x^2 - 12$ **12.** $f(x) = x^2 + 4x - 5$

13. $g(x) = -x^2 + 2x + 3$ **14.** $h(x) = x^2 + 2x - 8$

15. $y = -5x^2 + 6x - 1$ **16.** $f(x) = 3x^2 - 2x + 1$

17. You shoot an arrow vertically into the air from a height of 5 feet with an initial velocity of 96 feet per second. The height, h, in feet above the ground, at any time, t (in seconds), is modeled by

$$h(t) = 5 + 96t - 16t^2.$$

    a. Determine the maximum height the arrow will attain.

b. Approximately when will the arrow reach the ground?

c. What is the significance of the vertical intercept?

d. What are the practical domain and practical range in this situation?

e. Use your graphing calculator to determine the horizontal intercepts. Determine the practical meaning of these intercepts in this situation.

18. As part of a recreational waterfront grant, the city council plans to enclose a rectangular area along the waterfront of Lake Erie and create a park and swimming area. The budget calls for the purchase of 3000 feet of fencing. *Note:* There is no fencing along the lake.

a. Draw a picture of the planned recreational area. Let x represent the length of one of the two equal sides that are perpendicular to the water.

b. Write an expression that represents the width (side opposite the water) in terms of x. *Note:* You have 3000 feet of fencing.

c. Write an equation that expresses the area $A(x)$ of this rectangular site as a quadratic function of x.

d. Determine the value of x for which $A(x)$ is a maximum.

e. What is the maximum area that can be enclosed?

f. What are the dimensions of the enclosed area?

g. Use your graphing calculator to graph the area function. What point on the graph represents the maximum area?

h. What is the vertical intercept? Does this point have any practical meaning in this situation?

i. From the graph, determine the horizontal intercepts. Do they have any practical meaning in this situation? Explain.

19. The cost to produce metal statues for local parks is given by

$$C(x) = 2x^2 - 120x + 2000,$$

where x represents the number of statues produced and $C(x)$ is the cost of producing them.

a. Use your graphing calculator to graph the cost function and determine the coordinates of the turning point.

b. Determine the vertex algebraically.

c. How do your answers in parts a and b compare?

d. Is the vertex a minimum or maximum point?

e. What is the practical meaning of the vertex in this situation?

f. What is the vertical intercept? What is the practical meaning of this intercept?

20. You are manufacturing ceramic lawn ornaments. After several months, your accountant tells you that your profit $P(n)$ can be modeled by

$$P(n) = -0.002n^2 + 5.5n - 1200,$$

where n is the number of ornaments sold each month.

a. Use your graphing calculator to produce a graph of this function. Use the table feature set at TblStart $= 0$ and ΔTbl $= 500$ to help you set your window. Include the x-intercepts and the vertex.

b. Determine the x-intercepts of the graph of the profit function.

c. Determine the practical domain of the profit function.

d. Determine the practical range of the profit function.

e. How many ornaments must be sold to maximize the profit?

f. Write the equation that must be solved to determine the number of ornaments that must be sold to produce a profit of $2300.

g. Solve the equation in part f graphically.

ACTIVITY 4.3

Per Capita
Personal Income

OBJECTIVES

1. Solve quadratic equations numerically.

2. Solve quadratic equations graphically.

3. Solve quadratic inequalities graphically.

According to statistics from the U.S. Department of Commerce, the per capita personal income (or the average annual income) of each resident of the United States from 1960 to 2000 can be modeled by the equation

$$P(t) = 15.1442t^2 + 98.7687t + 1831.6909,$$

where $P(t)$ represents the per capita income and t represents the number of years since 1960.

1. What is the practical domain for the model represented by the function P?

2. Let $t = 0$ correspond to the year 1960. Use your graphing calculator to complete the following table of values for t, the number of years since 1960, and $P(t)$, the per capita income. Round your output to the nearest dollar.

YEAR	1960	1965	1970	1975	1980	1985	1990	1995	2000
t									
$P(t)$ ($)									

3. Sketch a graph of the function using your graphing calculator using the window Xmin $= -5$, Xmax $= 45$, Ymin $= -2000$, and Ymax $= 35,000$. The graph should appear as follows.

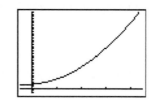

4. Estimate the per capita personal income in the year 1989 ($t = 29$).

5. You want to determine in which year the per capita personal income reached $20,500. Write an equation to determine the value of t when $P(t) = 20,500$.

The equation in Problem 5 is called a **quadratic equation**. Such equations involve polynomial expressions of degree 2. The standard form of a quadratic equation is $ax^2 + bx + c = 0, a \neq 0$. Examples of quadratic equations include $x^2 + 3x - 1 = 9, 2x^2 - 4x + 1 = 0$, and $6x^2 = 18$.

6. One method of approximating the solution to the equation in Problem 5 is numerical, using a table of appropriate data points (see Problem 2). What is your approximation using this approach?

7. a. Describe how to solve the quadratic equation $x^2 + 3x - 1 = 9$ numerically (using tables of data).

b. Solve the quadratic equation $x^2 + 3x - 1 = 9$ using the numerical approach described in part a.

Solving Quadratic Equations Graphically

A second method of solving the quadratic equation in Problem 5 is graphical, using your graphing calculator. Recall from Chapter 1 that you can solve the equation $15.1442x^2 + 98.7687x + 1831.6909 = 20,500$ by solving the following system of equations graphically:

$$y_1 = 15.1442x^2 + 98.7687x + 1831.6909$$
$$y_2 = 20,500$$

The expression for y_1 gives the per capita personal income in any given year. The value y_2 is the specific per capita personal income in which you are interested. The solution to the equation is the x-value for which $y_1 = y_2$. To do this, determine the point of intersection of these two graphs. If you use the intersect option under the CALC menu, the graph should appear as follows.

 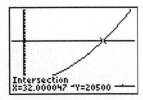

Another graphical method for solving the problem is to rearrange the quadratic equation

$$20,500 = 15.1442t^2 + 98.7687t + 1831.6909$$

so that the left-hand side is equal to zero. Subtracting 20,500 from each side, you have

$$0 = 15.1442t^2 + 98.7687t - 18,668.3091 \quad (1)$$

If you let $y = 15.1442t^2 + 98.7687t - 18,668.3091$, then the solution to the equation 1 is the x-value for which $y = 0$, if it exists. This is the x-value of the x-intercept (zero) of the graph.

8. a. Use your graphing calculator to sketch a graph of

$$y = 15.1442x^2 + 98.7687x - 18{,}668.3091.$$

The screen should appear as follows.

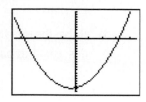

b. What are the x-intercepts of the new function defined by
$y = 15.1442t^2 + 98.7687t - 18{,}668.3091$?

c. Using the results from part b, determine the solutions to the equation
$20{,}500 = 15.1442t^2 + 98.7687t + 1831.6989$. Are both of the values
relevant to our problem? Explain.

9. Describe two different ways to solve the equation $2x^2 - 4x + 3 = 2$ using a
graphing approach. Solve the equation using each graphing method. How do
your answers compare?

Solving Quadratic Inequalities Graphically

You are interested in determining in which years the per capita personal income
was more than $15,000. To answer this problem, you need to solve the inequality

$$15.1442t^2 + 98.7687t + 1831.6909 > 15{,}000,$$

where t equals the number of years since 1960.

The following example demonstrates a procedure for solving an inequality similar
to the preceding one.

Example 1 *Solve the inequality $2x^2 - 4x + 3 > 7$ using a graphing approach.*

SOLUTION

Form the following system of equations

$$Y_1 = 2x^2 - 4x + 3$$
$$Y_2 = 7$$

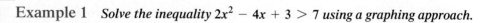

and graph each equation. The screen should resemble the following.

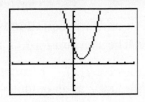

Use the intersection option on the CALC menu to determine where $Y_1 = Y_2$.

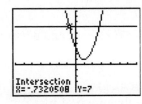 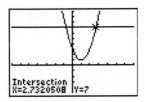

The solutions to the inequality are the values of x where $Y_1 > Y_2$, that is, the x-values of points where the graph of Y_1 is above the graph of Y_2.

Therefore, the solutions are $x < -0.732$ and $x > 2.732$.

10. **a.** Solve the inequality $15.1442t^2 + 98.7687t + 1831.6909 > 15,000$ using a graphing approach. Be careful. The practical domain is $t \geq 0$.

   b. In which years was the per capita personal income more than $15,000?

   c. There are negative values of x for which $Y1 > 15,000$. Determine them graphically.

   d. Explain why the values determined in part c are not relevant to the original problem situation.

11. Solve the following quadratic inequalities using a graphing approach.
   a. $x^2 - x - 6 < 0$

   b. $x^2 - x - 6 > 0$

SUMMARY
Activity 4.3

1. A **quadratic equation** is an equation involving polynomial expressions of degree 2. The standard form of a quadratic equation is $ax^2 + bx + c = 0$ $a \neq 0$.

2. To solve $f(x) = c$ **numerically**, construct a table, and determine the x-values that produce c as an output.

3. To solve $f(x) = c$ **graphically**:

   a. Graph $y = f(x)$, graph $y = c$, and determine the x-values of the points of intersection.

   b. Or graph $y = f(x) - c$, and determine the x-intercepts.

4. To solve $f(x) > c$ graphically:

   a. Graph $y = f(x)$, graph $y = c$, and determine all x-values for which the graph of f is above the graph of $y = c$.

   b. Or graph $y = f(x) - c$, and determine all x-values for which the graph of $f(x) - c$ is above the x-axis.

5. To solve $f(x) < c$ graphically:

   a. Graph $y = f(x)$, graph $y = c$, and determine all x-values for which the graph of f is below the graph of $y = c$.

   b. Or graph $y = f(x) - c$, and determine all x-values for which the graph of $f(x) - c$ is below the x-axis.

EXERCISES
Activity 4.3

In Exercises 1–4, solve the quadratic equation numerically (using tables of x- and y-values). Verify your solutions graphically.

1. $-4x = -x^2 + 12$

2. $x^2 + 9x + 18 = 0$

3. $2x^2 = 8x + 90$

4. $x^2 - x - 3 = 0$

In Exercises 5–8, solve the quadratic equation graphically using at least two different approaches. When necessary, give your solutions to the nearest hundredth.

5. $x^2 + 12x + 11 = 0$

6. $2x^2 - 3 = 2x$

7. $16x^2 - 400 = 0$

8. $4x^2 + 12x = -4$

Exercise numbers appearing in color are answered in the Selected Answers appendix.

In Exercises 9–12, solve the equation by using either a numeric or a graphic approach.

9. $x^2 + 2x - 3 = 0$

10. $x^2 + 11x + 24 = 0$

11. $x^2 - 2x - 8 = x + 20$

12. $x^2 - 10x + 6 = 5x - 50$

In Exercises 13–14, solve the given inequality using a graphing approach.

13. a. $x^2 - 4x - 1 < 11$

b. $x^2 - 4x - 1 > 11$

14. a. $2x^2 + 5x - 3 < 0$

b. $2x^2 + 5x - 3 \geq 0$

15. The stopping distance, d (in feet), for a car moving at a velocity (speed) v miles per hour is modeled by the equation

$$d(v) = 0.04v^2 + 1.1v.$$

a. What is the stopping distance for a velocity of 55 miles per hour?

b. What is the speed of the car if it takes 200 feet to stop?

16. An international rule for determining the number, n, of board feet (usable finished lumber) in a 16-foot log is modeled by the equation

$$n(d) = 0.22d^2 - 0.71d,$$

where d is the diameter of the log in inches.

a. How many board feet can be obtained from a 16-foot log with a 14-inch diameter?

b. Sketch a graph of this function. What is the practical domain of this function?

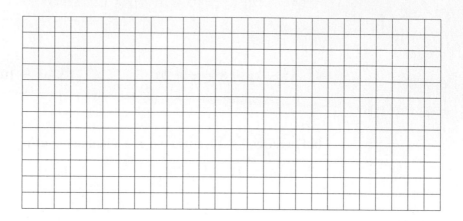

c. Use the graph to approximate the horizontal intercept(s). What is the practical meaning in this situation?

d. What is the diameter of a 16-foot log that has 200 board feet?

e. What inequality would you solve to determine the diameter when the board feet is at most 200?

f. Solve the inequality by using the graph of the function.

ACTIVITY 4.4

Sir Isaac Newton

OBJECTIVES

1. Factor expressions by removing the greatest common factor.

2. Factor trinomials using trial and error.

3. Use the zero-product principle to solve equations.

4. Solve quadratic equations by factoring.

Sir Isaac Newton XIV, a descendant of the famous physicist and mathematician, takes you to the top of a building to demonstrate a physics property discovered by his famous ancestor. He throws your math book into the air. The book's distance, s, above the ground as a function of time, t, is modeled by

$$s = -16t^2 + 16t + 32.$$

1. When the book strikes the ground, what is the value of s?

2. Write the equation that you must solve to determine when the book strikes the ground.

The quadratic equation in Problem 2 can be solved by using a numerical or a graphical approach. However, an algebraic technique is efficient in this case and will give an exact answer. The algorithm is based on the algebraic principle known as the **zero-product principle**.

If a and b are any numbers and $a \cdot b = 0$, then either a or b, or both, must be equal to zero.

Example 1 *Solve the equation $x(x + 5) = 0$.*

SOLUTION

The two factors in this equation are x and $x + 5$. The zero-product principle says one of these factors must equal zero. That is,

$$x = 0 \text{ or } x + 5 = 0.$$

The first equation tells you that $x = 0$ is a solution. To determine a second solution, solve $x + 5 = 0$.

$$x + 5 = 0$$
$$\underline{-5 \quad -5}$$
$$x = -5$$

There are two solutions, $x = 0$ and $x = -5$.

Your graphing calculator verifies the solutions as follows.

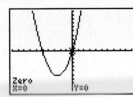
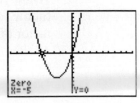

3. Solve each of the following equations using the zero-product principle.

a. $3x(x - 2) = 0$ **b.** $(2x - 3)(x + 2) = 0$

c. $(x + 2)(x + 3) = 0$

For the zero-product principle to be applied, one side of the equation must be zero. Therefore, at first glance, the zero-product principle can be used to solve the quadratic equation $3x^2 - 6x = 0$. However, a second condition must be satisfied. The nonzero side of the equation must be written as a product.

The process of writing an expression such as $3x^2 - 6x$ as a product is called factoring.

> **DEFINITION**
> Rewriting an expression as a product is called **factoring**.

Factoring Common Factors

A **common factor** is a number or an expression that is a factor of each term of the entire expression. Whenever you wish to factor a polynomial, look first for a common factor.

> **PROCEDURE: Removing a Common Factor from a Polynomial**
> First, identify the common factor, and then apply the distributive property in reverse.

Example 2   *Given the binomial* $3x + 6$, *you can identify 3 as a common factor because 3 is a factor of both* $3x$, *and 6. Applying the distributive property in reverse, you write*

$$3x + 6 \text{ as } 3(x + 2).$$

You may always check the factored binomial by multiplying:

$$3(x + 2) = 3(x) + 3(2) = 3x + 6.$$

When you look for a common factor, determine the largest or **greatest common factor** (or GCF). You can see that 3 is a common factor of $6x + 24$ because 3 is a factor of both 6 and 24. However, there is a larger common factor, 6. Therefore,

$$6x + 24 = 6(x + 4).$$

Example 3   *Given* $6x^2 + 14x - 30$, *you can see that 2 is a common factor. Is 2 the greatest common factor? Yes, because no larger number is a factor of every term.*

If you divide each term by 2, you obtain $3x^2 + 7x - 15$. The expression $6x^2 + 14x - 30$ can now be written in factored form as: $2(3x^2 + 7x - 15)$. Check the factored trinomial by multiplying.

Example 4   *Factor $4x^3 - 8x^2 + 28x$.*

SOLUTION

The greatest common factor is $4x$. You remove the GCF by dividing each term by $4x$. This leads to the factored form $4x(x^2 - 2x + 7)$.

You can check your factoring by applying the distributive property.

4. Factor the following polynomials by removing the greatest common factor.

a. $9a^6 + 18a^2$

b. $21xy^3 + 7xy$

c. $3x^2 - 21x + 33$

d. $4x^3 - 16x^2 - 24x$

Factoring Trinomials

With patience, you can factor trinomials by trial and error, using the FOIL method in reverse.

> **PROCEDURE: Factoring Trinomials by Trial and Error**
>
> 1. Remove the greatest common factor, GCF.
> 2. To factor the resulting trinomial into the product of two binomials, try combinations of factors for the first and last terms in two binomials.
> 3. Check the outer and inner products to match the middle term of the original trinomial.
>   a. If the constant term, c, is positive, both factors are positive or both are negative.
>   b. If the constant term is negative, one factor is positive and one is negative.
> 4. If the check fails, repeat steps 2 and 3.

Example 5   *Factor $6x^2 - 7x - 3$.*

SOLUTION

Step 1. There is no common factor, so go to step 2.

Step 2. You could factor the first term $6x^2$ as $6x(x)$ or as $2x(3x)$. The last term -3 has factors $3(-1)$ or $-3(1)$. Try $(2x + 1)(3x - 3)$.

Step 3. The outer product is $-6x$. The inner product is $3x$. The sum is $-3x$, not $7x$. The check fails.

Step 4. Try $(2x - 3)(3x + 1)$. The outer product is $2x$. The inner product is $-9x$. The sum is $-7x$. It checks.

5. Factor the following trinomials.

   a. $x^2 - 7x + 12$ **b.** $x^2 - 8x - 9$

   c. $x^2 + 14x + 49$ **d.** $25 + 10w + w^2$

Solving Quadratic Equations by Factoring

The following example demonstrates the procedure for solving quadratic equations written in standard form $ax^2 + bx + c = 0$ by factoring.

Example 6 *Solve the equation $3x^2 - 2 = -x$ by factoring.*

Step 1. Rewrite the equation in the form $ax^2 + bx + c = 0$ (called standard form).

$$3x^2 - 2 = -x$$
$$\underline{+\ x \qquad\qquad +\ x}$$
$$3x^2 + x - 2 = 0$$

Step 2. Factor the expression on the nonzero side of the equation.

$$(x + 1)(3x - 2) = 0$$

Step 3. Use the zero-product principle to set each factor equal to zero, and then solve each equation.

$$(x + 1)(3x - 2) = 0$$

$x + 1 = 0$	$3x - 2 = 0$
$x = -1$	$3x = 2$
	$x = \frac{2}{3}$

Therefore, the solutions are $x = -1$ and $x = \frac{2}{3}$.

These solutions can be verified graphically as follows.

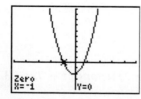 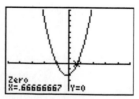

6. a. Returning to the math book problem from the beginning of this activity, solve the equation from Problem 2 by factoring.

b. Are both solutions to the equation $(t = 2$ and $t = -1)$ also solutions to the question, "At what time does the book strike the ground"? Explain.

7. a. You want to know at what time is the book 32 feet above the ground. Write a quadratic equation that represents this situation.

b. Solve the quadratic equation in part a by factoring.

8. Solve each of the following quadratic equations by factoring.

a. $2x^2 - x - 6 = 0$ **b.** $3x^2 - 6x = 0$ **c.** $x^2 + 4x = -x - 6$

SUMMARY
Activity 4.4

1. To remove a **common factor** from a polynomial, first
   a. identify the common factor, and then
   b. apply the distributive property in reverse.

2. The **zero-product principle** says that if $ab = 0$ is a true statement, then either $a = 0$ or $b = 0$.

3. To factor trinomials by **trial and error**:
   a. Remove the greatest common factor.
   b. Try combinations of factors for the first and last terms in two binomials.
   c. Check the outer and inner products to match the middle term of the original trinomial.
      • If the constant term, c, is positive, both factors are positive or both are negative.
      • If the constant term is negative, one factor is positive and one is negative.
   d. If the check fails, repeat steps 2 and 3.

4. To solve equations by **factoring**:
   a. Use the addition principle to remove all terms from one side of the equation. This results in the equation being set equal to zero.
   b. Combine like terms, and then factor the nonzero side of the equation.
   c. Use the zero-product principle to set each factor containing a variable equal to zero, and then solve the equations.
   d. Check your solutions in the original equation.

EXERCISES
Activity 4.4

In Exercises 1–4, factor the polynomials by removing the GCF (greatest common factor).

1. $12x^5 - 18x^8$

2. $14x^6y^3 - 6x^2y^4$

3. $2x^3 - 14x^2 + 26x$

4. $5x^3 - 20x^2 - 35x$

In Exercises 5–10, completely factor the polynomials. Remember to look first for the GCF.

5. $x^2 + x - 6$

6. $12 + 8x + x^2$

7. $2x^2 + 7x - 15$

8. $3x^2 + 19x - 14$

9. $8x^4 - 47x^3 - 6x^2$

10. $20b^4 - 65b^3 - 60b^2$

In Exercises 11–14, solve each quadratic equation by factoring.

11. $3x^2 + 11x - 4 = 0$

12. $3x^2 - 12x = 0$

13. $x^2 - 7x = 18$

14. $3x(x - 6) - 5(x - 6) = 0$

15. Your neighbors have just finished installing a new swimming pool at their home. The pool measures 15 feet by 20 feet. They would like to plant a strip of grass of uniform width around three sides of the pool, the two short sides and one of the longer sides.

a. Sketch a diagram of the pool and the strip of lawn, using x to represent the width of the uniform strip.

b. Write an equation for the area A in terms of x that represents the lawn area around the pool.

c. They have enough seed for 168 square feet of lawn. Write an equation that relates the quantity of seed to the area of the uniform strip of lawn.

d. Solve the equation in part c to determine the width of the uniform strip that can be seeded.

ACTIVITY 4.5
Motorcycle
Deaths

OBJECTIVE

1. Solve quadratic equations by the quadratic formula.

The following data was contained in a news release from the Insurance Institute for Highway Safety. The following table gives the number of motorcycle deaths of cyclists aged 30–39 from 1991 to 2000.

					ANNUAL NUMBER OF DEATHS OF MOTORCYCLISTS AGED 30–39, 1991–2000					
YEAR	1991	1992	1993	1994	1995	1996	1997	1998	1999	2000
NUMBER OF DEATHS	711	638	647	584	562	541	547	599	601	687

1. a. Use your graphing calculator to create a scatterplot of the data in the preceding table. Let x represent the number of years since 1990. Let n represent the annual number of motorcyclist deaths.

b. This data can be modeled by the function

$$n = 6.875x^2 - 80.76x + 791,$$

where n represents the number of deaths and x represents the number of years since 1990. Verify that this is a good model by graphing the function on the same screen with your scatterplot. Use the data from the preceding table to determine an appropriate window. Your graph should appear as follows.

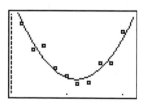

c. You want to know in what years the model predicts the number of deaths of motorcyclists aged 30–39 to be approximately 1000. Write the equation that must be solved to determine these years.

The Quadratic Formula

The equation in Problem 1c can be solved by using a numerical or a graphical approach. However, the algebraic technique of factoring cannot be applied.

The technique of solving quadratic equations by factoring is very limited. In most real-world applications involving quadratic equations, the quadratic is not factorable. In those cases, you can use a formula to solve the quadratic equation.

Beginning with the standard quadratic function defined by $y = ax^2 + bx + c$, $a \neq 0$, set $y = 0$ to obtain the equation

$$0 = ax^2 + bx + c.$$

The quadratic equation $ax^2 + bx + c = 0$ has two solutions,

$$x_1 = \frac{-b + \sqrt{b^2 - 4ac}}{2a} \text{ and } x_2 = \frac{-b - \sqrt{b^2 - 4ac}}{2a}.$$

These solutions are often written as a single expression,

$$x = \frac{-b \pm \sqrt{b^2 - 4ac}}{2a}.$$

This formula is known as the **quadratic formula**.

 For the details of solving the equation $0 = ax^2 + bx + c$ using the quadratic formula, see Appendix A. The section is called Derivation of the Quadratic Formula.

The following example demonstrates the procedure for using the quadratic formula to solve an equation of the form $ax^2 + bx + c = 0$.

Example 1   *Solve $x(3x + 4) = 5$ using the quadratic formula.*

SOLUTION

Step 1. Write the equation in standard form, $ax^2 + bx + c = 0$.

$$\begin{aligned}
x(3x + 4) &= 5 \qquad &\text{Apply the distributive property on the left side.} \\
3x^2 + 4x &= 5 \\
\underline{ -5 \quad -5} \qquad &\text{Subtract 5 from both sides.} \\
3x^2 + 4x - 5 &= 0
\end{aligned}$$

Step 2. Identify the coefficients a, b, and the constant term, c.

$$a = 3, b = 4, c = -5$$

Step 3. Substitute the values a, b, and c into the quadratic formula, and simplify.

$$x = \frac{-b \pm \sqrt{b^2 - 4ac}}{2a}$$

$$x = \frac{-4 \pm \sqrt{4^2 - 4(3)(-5)}}{2(3)}$$

$$x = \frac{-4 \pm \sqrt{16 - (-60)}}{6} = \frac{-4 \pm \sqrt{76}}{6}$$

$$x \approx \frac{-4 \pm 8.7178}{6} = -2.1196 \text{ or } 0.7863$$

Step 4. Check your solutions. The following graphs verify the solutions.

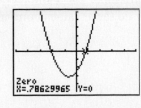   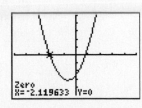

2. a. Solve the equation from Problem 1c, $6.875x^2 - 80.76x + 791 = 1000$, using the quadratic formula. Check your solution graphically.

b. Use the results from part a to predict the years in which the number of motorcycle deaths will be (was) approximately 1000.

3. a. Sketch a graph of $y = 2x^2 + 9x - 5$.

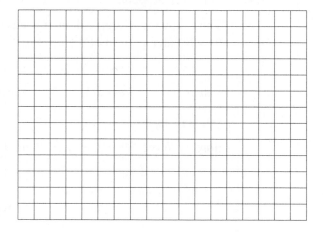

b. Write the equation that you would need to solve to determine the x-intercepts of the graph.

c. Solve the equation $0 = 2x^2 + 9x - 5$ using the quadratic formula.

d. Approximate the x-intercepts of the graph using your graphing calculator. How do these values compare to the results in part c?

4. The following data from the National Health and Nutrition Examination Survey indicates that the number of American adults who are overweight or obese is increasing.

YEARS SINCE 1960, t	1	12	18	31	39
PERCENT OF OVERWEIGHT OR OBESE AMERICANS, P(t)	45	47	47	56	64.5

This data can be modeled by the equation $P(t) = 0.017t^2 - 0.174t + 45.493$.

a. Use your graphing calculator to create a scatterplot of the data and a graph of the model P.

b. Does the model appear to be a good fit for the data? Explain.

c. Using the quadratic formula, determine the year when the model predicts that the percent of overweight or obese Americans will first exceed 75%.

Axis of Symmetry Revisited

If you write the quadratic formula in a slightly different form, you obtain

$$x = -\frac{b}{2a} \pm \frac{\sqrt{b^2 - 4ac}}{2a} \text{ or } x_1 = \frac{-b}{2a} + \frac{\sqrt{b^2 - 4ac}}{2a}, x_2 = \frac{-b}{2a} - \frac{\sqrt{b^2 - 4ac}}{2a}.$$

The next problem uses the rewritten form of the quadratic formula to help identify a relationship between the x-intercepts and the axis of symmetry of the graph of f.

5. Consider the function $f(x) = 2x^2 + 9x - 5$.

a. Determine the equation of the axis of symmetry of the graph of f.

b. What is the value of $\dfrac{\sqrt{b^2 - 4ac}}{2a}$?

c. Sketch a graph of the function f, and label the axis of symmetry. Show where the value computed in part b is located graphically. What are the x-intercepts of the graph?

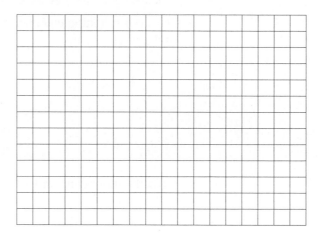

d. What is the relationship between the axis of symmetry and the x-intercepts of a parabola?

6. For each of the following quadratic functions, determine the x-intercepts of the graph. Solve the appropriate equation using the quadratic formula. Round your answers to the nearest hundredth.

a. $f(x) = 2x^2 - 6x - 3$ **b.** $h(x) = x^2 - 8x + 16$

SUMMARY
Activity 4.5

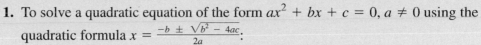

1. To solve a quadratic equation of the form $ax^2 + bx + c = 0, a \neq 0$ using the quadratic formula $x = \frac{-b \pm \sqrt{b^2 - 4ac}}{2a}$:

    a. Set the quadratic equation equal to zero.

    b. Identify the coefficients a and b, and the constant term, c.

    c. Substitute these values into the formula, and simplify.

    d. Check your solutions.

2. For a parabola with x-intercepts, the axis of symmetry is always midway between the x-intercepts of a parabola.

3. The distance from the axis of symmetry to either x-intercept is $\frac{\sqrt{b^2 - 4ac}}{2a}$.

EXERCISES
Activity 4.5

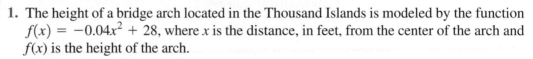

1. The height of a bridge arch located in the Thousand Islands is modeled by the function $f(x) = -0.04x^2 + 28$, where x is the distance, in feet, from the center of the arch and $f(x)$ is the height of the arch.

    a. Sketch a picture of this arch on a grid using the vertical axis as the center of the arch.

    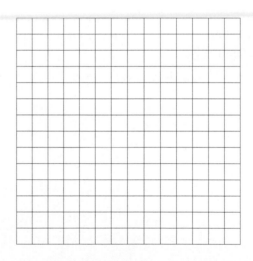

    b. Determine the vertical intercept. What is the practical meaning of this intercept in this situation?

    c. Determine the x-intercepts algebraically using the quadratic formula.

d. Graph the function on your graphing calculator and check the accuracy of the intercepts you found in part c.

e. If the arch straddles the river exactly, how wide is the river?

f. A sailboat is approaching the bridge. The top of the mast measures 30 feet. Will the boat clear the bridge? Explain.

g. You want to install a flagpole on the bridge at an arch height of 20 feet. Write the equation that you must solve to determine how far to the right of center the arch height is 20 feet.

h. Solve the equation in part g using the quadratic formula. Use your graphing calculator to check your result.

In Exercises 2–8, identify the values of a, b, and c, and then solve the equations using the quadratic formula. Round your answers to the nearest hundredth. Verify your solutions graphically.

2. $x^2 + 6x - 3 = 0$

3. $4x^2 + 4x + 1 = 0$

4. $x^2 + 5x = 13$

5. $2x^2 - 6x + 3 = 0$

6. $2x^2 - 3x = 5$

7. $(2x - 1)(x + 2) = 1$

8. $(x + 2)^2 + x^2 = 44$

In Exercises 9–11, determine the x-intercept of the graph algebraically. Then check your results graphically.

9. $y = 3x^2 + 6x$

10. $y = x^2 - x - 6$

11. $f(x) = 2x^2 - x + 5$

12. The number, n (in millions), of cellular phone subscribers in the United States from 1990 to 1999 is given in the following table.

YEAR	1990	1991	1992	1993	1994	1995	1996	1997	1998	1999
NUMBER OF SUBSCRIBERS (millions)	5.28	7.56	11.03	16.01	24.13	33.79	44.04	55.31	69.20	86.05

This data can be approximated by the quadratic model

$$n(t) = 0.846t^2 + 1.32t + 5.20,$$

where $t = 0$ corresponds to the year 1990.

a. Use your graphing calculator to sketch a graph of the function.

b. Use the graph in part a to estimate the year in which there will be 120 million cellular phone subscribers.

c. Use the quadratic formula to answer part b. How does your answer compare to the estimate you obtained using a graphical approach?

d. How confident are you in your prediction? Explain.

13. The quadratic function defined by the equation

$$d = 2r^2 - 16r + 34$$

gives the density of smoke, d, in millions of particles per cubic foot for a certain type of diesel engine. The input variable, r, represents the speed of the engine in hundreds of revolutions per minute.

a. Determine the density of smoke when $r = 3.5$ (350 revolutions per minute).

b. Determine the number of revolutions per minute for minimum smoke. What is the minimum output?

c. If the density of smoke is determined to be 100 million particles per cubic foot, determine the speed of the engine.

ACTIVITY 4.6
Air Quality in
Atlanta

OBJECTIVES

1. Determine quadratic regression models using the graphing calculator.

2. Solve problems using quadratic regression models.

The Air Quality Index, or AQI, measures how polluted the air is by measuring five major pollutants: ground-level ozone, particulate matter, carbon monoxide, sulfur dioxide, and nitrogen oxide. Based on the amount of each pollutant in the air, the AQI assigns a numerical value to air quality, as follows.

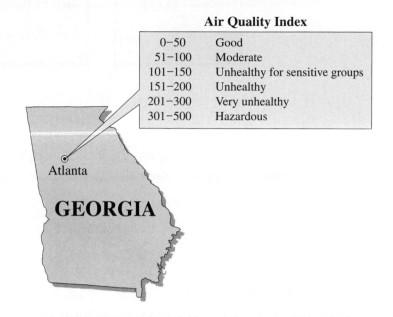

Air Quality Index

0–50	Good
51–100	Moderate
101–150	Unhealthy for sensitive groups
151–200	Unhealthy
201–300	Very unhealthy
301–500	Hazardous

Atlanta

GEORGIA

The following table indicates the number of days in which the AQI was greater than 100 in the city of Atlanta, Georgia.

YEAR	1990	1992	1994	1996	1998	1999
NUMBER OF DAYS AQI $>100, n$	42	20	15	25	50	61

1. Sketch a scatterplot of the data. Let t represent the number of years since 1990. Therefore, $t = 0$ corresponds to the year 1990. Does the data appear to be quadratic? Explain.

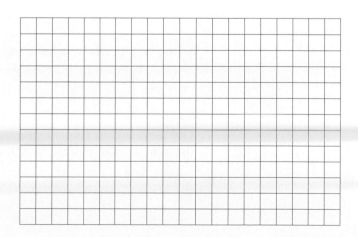

2. Use the regression feature of your graphing calculator to determine and plot a quadratic function that best fits these data. Your graph should appear as follows.

 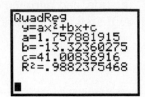 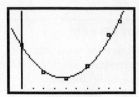

3. Do you believe that the quadratic regression model is a good model for the number of days the AQI exceeded 100 in Atlanta from 1990 to 1999?

4. What is the practical domain of this function?

5. a. Use the quadratic regression equation to estimate the number of days the AQI exceeded 100 in Atlanta in each of the following years.

    i. 1995   **ii.** 1988   **iii.** 2002

    b. Which, if any, of these estimates do you think is most reliable? Explain.

6. Estimate the years in which the number of days that the AQI exceeded 100 in Atlanta was less than or equal to 30 using:

    a. the given table (numerical method).

    b. the graph of the quadratic regression equation (graphical method).

7. Use the quadratic formula (the algebraic method) to estimate the year in which the number of days that the AQI exceeded 100 in Atlanta was equal to 17.

8. Estimate graphically the years between 1990 and 1999 when the number of days that the AQI exceeded 100 in Atlanta was greater than 25.

9. a. The number of days that the AQI exceeded 100 in Atlanta in 1997 was 31. Does this agree with the prediction from the quadratic regression model?

b. Include the data for 1997 from part a in the original data set, and then re-calculate the quadratic regression equation.

c. Predict the number of days that the AQI exceeded 100 in Atlanta in 1997 from this new quadratic regression equation from part b. Are the results any better?

10. The following data from the National Health and Nutrition Examination Survey appear in Problem 4 of Activity 4.5. The data indicates that the number of American adults who are overweight or obese is increasing.

YEARS SINCE 1960, t	1	12	18	31	39
PERCENT OF OVERWEIGHT OR OBESE AMERICANS, $P(t)$	45	47	47	56	64.5

Use the regression feature of your graphing calculator to verify that this data can be modeled by the equation $P(t) = 0.017t^2 - 0.174t + 45.493$.

SUMMARY
Activity 4.6

Parabolic data can be modeled by a **quadratic regression equation**.

EXERCISES
Activity 4.6

1. During one game, the Buffalo Bills punter was called upon to punt the ball eight times. On one of these punts, the punter struck the ball at his own 30-yard line. The height, h, of the ball above the field in feet as a function of time, t, in seconds can be partially modeled by the following table.

t	0	0.6	1.2	1.8	2.4	3.0
$h(t)$	2.50	28.56	43.10	46.12	37.12	17.60

   a. Sketch a scatterplot of the data using your graphing calculator.

   b. Use your grapher to obtain a quadratic regression function for these data. Round the values of a, b, and c to four decimal places.

   c. Graph the equation from part b on the same coordinate axes as the data points. Does the curve appear to be a good fit for the data? Explain.

   d. In this model, what is the practical domain of the quadratic regression function?

   e. Estimate the practical range of this model.

   f. How long after the ball was struck did the ball reach 35 feet above the field? Explain.

   g. How many results did you obtain for part f? Do you think you have all of the solutions? Explain.

2. Use the following data set to perform the tasks in parts a–e.

x	0	3	6	9	12
y	5	28	86	180	310

Exercise numbers appearing in color are answered in the Selected Answers appendix.

a. Determine an appropriate scale, and plot these points.

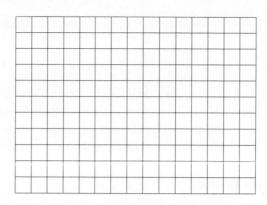

b. Use your graphing calculator to determine the quadratic regression equation for this data set.

c. Graph the regression equation on the same coordinate axes as the data points in part a.

d. Compare the predicted outputs with the outputs given in the table.

e. Predict the output for $x = 7$ and for $x = 15$.

3. The following table shows the stopping distance for a car at various speeds on dry pavement.

SPEED (mph)	25	35	45	55	65	75
DISTANCE (ft)	65	108	167	245	340	450

a. Use your graphing calculator to determine a quadratic regression equation that represents this data.

b. Use the regression equation to predict the stopping distance at 90 mph.

c. What speed would produce a stopping distance of 280 feet? (Round to the nearest tenth.) Explain how you arrived at your conclusion.

4. A downturn in the high-tech industry during the early 2000s has caused a similar down-turn in computer science and engineering enrollments for new undergraduates. The following data from the Computing Research Association represents new enrollments in computer science and engineering from 1995 to 2001.

YEARS SINCE 1995, t	0	1	2	3	4	5	6
NEW IN COMPUTER SCIENCE AND ENGINEERING UNDERGRADS (in thousands), N	8.2	11.9	16.2	17.1	16.6	18.9	18.0

Source: Computer Research Association

a. Sketch a scatterplot of the data using your graphing calculator.

b. Use your graphing calculator to obtain a quadratic regression function for this data. Round the values of a, b, and c to four decimal places.

c. Graph the equation from part b on the same coordinate axes as the data points. Does the curve appear to be a good fit for the data? Explain.

d. What does the regression equation predict for the new undergraduate enrollments in computer science and engineering in 2003? Does this seem reasonable?

e. Use the model to determine the year that new undergraduate computer science and engineering enrollments will drop below 15,000.

What Have I Learned?

1. In order for the graph of the equation $y = ax^2 + bx + c$ to be a parabola, the value of the coefficient of x^2 cannot be zero. Explain.

   b. What is the vertex of the parabola having an equation of the form $y = ax^2$?

   c. Describe the relationship between the vertex and the vertical intercept of the graph of $y = ax^2 + c$.

2. Determine if the vertex is a minimum point or a maximum point of $y = ax^2 + bx + c$ in each of the following situations.

   a. $a < 0$ b. $a > 0$

3. a. What are the possibilities for the number of vertical intercepts of a quadratic function?

   b. What are the possibilities for the number of horizontal intercepts of a parabola?

4. What is the relationship between the vertex and the x-intercept of the graph of $y = x^2 - 4x + 4$?

5. a. The vertex of a parabola is (3, 1). Using this information, complete the following table.

x	1	2	3	4	5
y	5	2			

   b. If the vertex of a parabola is (2, 4), complete the following table

x	−2	0	2	4	6
y	0	3			

6. a. Given the following graph, explain why choices i, ii, and iii do not fit the curve.

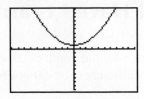

    i. $f(x) = ax^2 + bx$ with $a > 0, b < 0$

    ii. $g(x) = ax^2 + c$ with $a < 0, c > 0$

    iii. $h(x) = ax^2 + bx + c$ with $a < 0, b > 0, c < 0$

  b. What restrictions on a, b, and c are necessary to fit $y = ax^2 + bx + c$ to this graph?

7. Review the steps in the following solution. Is the solution correct? Explain why or why not.

$$x^2 - 3x - 4 = 6$$
$$(x - 4)(x + 1) = 6$$
$$x - 4 = 6 \qquad x + 1 = 6$$
$$x = 10 \qquad x = 5$$

8. Describe how you would determine the solutions to $ax^2 + bx + c > 5$ graphically?

CLUSTER 1

How Can I Practice?

1. Complete the following table.

EQUATION OF THE FORM $y = ax^2 + bx + c$	VALUE OF a	VALUE OF b	VALUE OF c
$y = 5x^2$			
$y = \frac{1}{3}x^2 + 3x - 1$			
$y = -2x^2 + x$			

For Exercises 2–7, determine the following characteristics for each graph.

    a. the direction in which the parabola opens

    b. the equation of the axis of symmetry

    c. the vertex

    d. the y-intercept

2. $y = -2x^2 + 4$ **3.** $y = \frac{2}{3}x^2$

4. $f(x) = -3x^2 + 6x + 7$ **5.** $f(x) = 4x^2 - 4x$

6. $y = x^2 + 6x + 9$ **7.** $y = x^2 - x + 1$

For Exercises 8–11, use your graphing calculator to sketch the graph of each quadratic function, and then determine the following for each function.

    a. the coordinates of the x-intercepts (if they exist)

    b. the domain and range

    c. the horizontal interval over which the function is increasing

    d. the horizontal interval over which the function is decreasing

8. $y = -x^2 + 4$ **9.** $y = x^2 - 5x + 6$

10. $y = -3x^2 - 6x + 8$ **11.** $y = 0.22x^2 - 0.71x + 2$

12. Use your graphing calculator to approximate the vertex of the graph of the parabola defined by the equation

$$y = -2x^2 + 3x + 25.$$

13. Completely factor the following polynomials.

    a. $9a^5 - 27a^2$ **b.** $24x^3 - 6x^2$

    c. $4x^3 - 16x^2 - 20x$ **d.** $5x^2 - 16x + 6$

    e. $x^2 - 5x - 24$ **f.** $y^2 + 10y + 25$

14. Determine one solution of the following quadratic equations numerically. That is, construct a table of (x, y) ordered pairs, and estimate the value of x (input) that results in the required y-value (output).

    a. $5x^2 = 7$ **b.** $x^2 - 7x + 10 = 5$ **c.** $3x^2 - 5x = 2$

a.

x	1	1.1	1.2	1.3	1.4	1.5
y						

b.

x	0.5	0.6	0.7	0.8	0.9	1
y						

c.

x	0	1	2	3	4	5
y						

15. Solve each of the equations from Exercise 14 using the quadratic formula. When necessary, round your solutions to the nearest tenth. Check your solutions by graphing.

16. Solve the following inequalities using a graphing approach.

    a. $x^2 + 6x - 16 < 0$ **b.** $x^2 + 6x - 16 > 0$

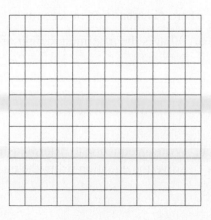

17. Solve each of the following equations by factoring.

    a. $4x^2 - 8x = 0$ **b.** $x^2 - 6 = 7x + 12$

    c. $2x(x - 4) = 6$ **d.** $x^2 - 8x + 16 = 0$

    e. $x^2 - 2x - 24 = 0$ **f.** $y^2 - 2y - 35 = -20$

    g. $a^2 + 2a + 1 = 3a + 7$ **h.** $4x^2 + 4x - 3 = -3x - 1$

18. A fastball is hit straight up over home plate. The ball's height, h (in feet), from the ground is modeled by

$$h(t) = -16t^2 + 80t + 5,$$

where t is measured in seconds.

    a. What is the maximum height of the ball above the ground?

    b. Write an equation to determine how long will it take for the ball to reach the ground. Solve the equation using the quadratic formula. Check your solution by graphing.

c. Write the equation you would need to determine when the ball is 101 feet above the ground.

d. Solve the equation you determined in part c algebraically to determine the time it will take for the ball to reach a height of 101 feet. Verify your results graphically.

19. A suspension bridge (shown in the accompanying figure) is 100 meters long. The bridge is supported by cables attached to the tops of towers 35 meters high at each end of the bridge. The cables hang from the towers approximately in the shape of a parabola. The height, $h(t)$ (in meters), of the cables above the surface of the roadway is modeled by

$$h(x) = 0.01x^2 - x + 35,$$

where x is the horizontal distance measured from the point where the tower and roadway meet.

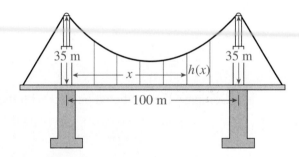

a. Use your graphing calculator to examine the height function. What is the practical domain of this function?

b. What is the minimum distance of the cables from the roadway?

20. Use the following data set to perform the tasks in parts a–f.

x	0	1	3	5	7	8
y	10	4	−18	−54	−107	−145

a. Determine an appropriate scale, and plot these points.

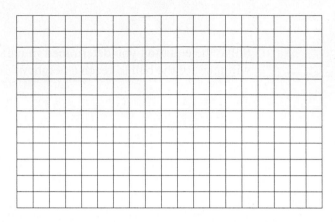

b. Use your graphing calculator to determine the quadratic regression equation for this data set.

c. Graph the regression equation on the same coordinate axes as the data points.

d. Compare the predicted outputs with the given outputs in the table.

e. What is the predicted output for $x = 4$ and for $x = 9$?

f. For what value of x is $y = -40$? Use the quadratic formula.

CLUSTER 2 **Complex Numbers and Problem Solving Using Quadratic Functions**

ACTIVITY 4.7

Complex
Numbers

OBJECTIVES

1. Identify the imaginary unit $i = \sqrt{-1}$.

2. Identify a complex number.

3. Determine the value of the discriminant $b^2 - 4ac$.

4. Determine the types of solutions to a quadratic equation.

5. Solve a quadratic equation in the complex number system.

Recall that the solutions to $ax^2 + bx + c = 0$ correspond to the x-intercepts of the parabola having equation $y = ax^2 + bx + c$.

Do all parabolas possess x-intercepts? Consider the graph of $y = 2x^2 + x + 5$. If you graph the function in the window Xmin $= -5$, Xmax $= 5$, Ymin $= -3$, and Ymax $= 15$, the graph will resemble the following.

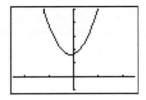

1. **a.** Based on what you know about parabolas, will the graph of $y = 2x^2 + x + 5$ have any x-intercepts? Explain.

   b. What can you say about the solutions to $2x^2 + x + 5 = 0$?

If you use the TI-92 graphing calculator to solve this equation, the following screen will appear.

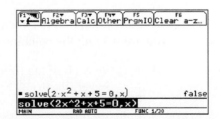

The result "false" indicates that there are no solutions using the solve command. This is consistent with the graph. Because there are no x-intercepts for the graph of $y = 2x^2 + x + 5$, there are no real-valued solutions to the equation $2x^2 + x + 5 = 0$. Would you have discovered this if you had tried to solve the equation $2x^2 + x + 5 = 0$ algebraically using the quadratic formula? Problem 2 addresses this question.

2. Use the quadratic formula to solve $2x^2 + x + 5 = 0$. Where does the solution process break down? Explain.

Complex Numbers

The solve command on the TI-92 tries to determine real-valued solutions to an equation. If you use the command cSolve $(2x^2 + x + 5 = 0, x)$, the following screen will appear.

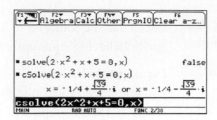

The TI-92 gives you two solutions:

$$x = -\frac{1}{4} + \frac{\sqrt{39}}{4}i \text{ and } x = -\frac{1}{4} - \frac{\sqrt{39}}{4}i$$

These are not **real numbers**. They are known as the **complex numbers** (an extension of the real numbers). The distinguishing characteristic of the complex numbers is the **imaginary unit**, $i = \sqrt{-1}$.

The quadratic formula solution to $2x^2 + x + 5 = 0$ uses the values $a = 2$, $b = 1$, and $c = 5$. The solution is

$$x = \frac{-1 \pm \sqrt{1^2 - 4(2)(5)}}{2(2)} = \frac{-1 \pm \sqrt{1 - 40}}{4} = -\frac{1}{4} \pm \frac{\sqrt{-39}}{4}.$$

The problem is that you cannot evaluate $\sqrt{-39}$ in the real-number system because any real number multiplied by itself is nonnegative. Therefore, you need to introduce the imaginary unit, $i = \sqrt{-1}$ and interpret $\sqrt{-39}$ as

$$\sqrt{39(-1)} = \sqrt{39}\sqrt{-1} = \sqrt{39} \cdot i = i\sqrt{39}.$$

With this interpretation, the cSolve solution of the TI-92 now makes sense because you are solving the equation in the complex-number system.

3. Rewrite each of the following in the form bi, where $i = \sqrt{-1}$.

   a. $\sqrt{-26}$ **b.** $\sqrt{-5}$

   c. $\sqrt{-64}$ **d.** $\sqrt{3 \cdot (-7)}$

   e. $\sqrt{-18}$ **f.** $\sqrt{-27}$

   g. $\sqrt{-\frac{3}{4}}$ **h.** $\sqrt{-\frac{15}{27}}$

> **DEFINITION**
>
> Numbers of the form bi, where b is a real number and $i = \sqrt{-1}$, are called **pure imaginary numbers**. Numbers of the form $a \pm bi$, where a and b are real and $i = \sqrt{-1}$, are called **complex numbers**. Imaginary numbers are complex numbers of the form $0 + bi$. Real numbers are complex numbers of the form $a + 0i$.

Example 1

a. *The numbers $-3i$, $\frac{2}{3}i$, and 7.4i are pure imaginary numbers.*

b. *The numbers $-4 + 3i$, $\frac{1}{2} - \frac{2}{3}i$, and 5 $-$ 6i are complex numbers.*

Note that the set of real numbers is contained within the set of complex numbers. A real number a may be thought of as the complex number $a + 0i$.

In the sixteenth century, complex numbers were used as solutions to polynomial equations. The notation $\sqrt{-1}$ was used during this time. Such numbers were called imaginary because their existence was not clearly understood. In 1777, Leonhard Euler introduced the notation i and wrote complex numbers in the form $a + bi$. Caspar Wessel in 1797 and Carl Friedrich Gauss in 1799 used the geometric interpretation of complex numbers as points in a plane. This made such numbers more concrete and less mysterious. Finally, in 1833, Sir William Hamilton showed that if the number i is defined to have the property

$$i^2 = -1,$$

then the set of real numbers can be extended to include numbers like $\sqrt{-1}$.

Today, complex numbers are used in a variety of applications, including chaos theory (fractals) and engineering.

Operations with Complex Numbers

The operations of addition, subtraction, and multiplication of complex numbers are demonstrated in the following example.

Example 2

a. *To add complex numbers, add the real parts and the imaginary parts.*
$(3 + 2i) + (5 - 7i) = 8 - 5i$

b. *To subtract complex numbers, add the opposite.*
$(2 - 2i) - (-6 + i) = 2 - 2i + 6 - i = 8 - 3i$

c. *To multiply complex numbers, multiply each term of the first by each term of the second, and simplify.*
$$(3 - 5i)(-1 + 8i) = -3 + 24i + 5i - 40i^2$$
$$= -3 + 29i + 40$$
$$= 37 + 29i$$
Remember that $i \cdot i = i^2 = -1$

The TI-83 Plus is capable of operations with complex numbers. You first need to change the mode of the calculator from Real to $a + bi$. Note that the i key is 2nd ⌷ period.

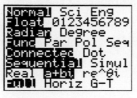 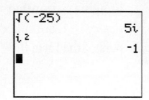 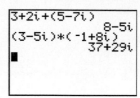

4. Perform the following operations with complex numbers. Use your graphing calculator to check your results.

   a. $(3 + 4i) + (-5 + 6i)$ **b.** $(5 - 7i) - (-2 + 5i)$

   c. $5i(2 - 4i)$ **d.** $(3 - 2i)(4 + 5i)$

Discriminant

In the complex-number system, a quadratic equation has exactly two solutions. In the quadratic formula

$$x = \frac{-b \pm \sqrt{b^2 - 4ac}}{2a},$$

the expression $b^2 - 4ac$ is called the **discriminant** because its value determines the number and type of solutions of a quadratic equation $ax^2 + bx + c = 0$. There are three possible cases, depending on whether the value of the discriminant is positive, zero, or negative. Problems 5, 6, and 7 investigate this relationship.

5. For each of the following quadratic functions, determine the sign of the discriminant. Then sketch a graph using your graphing calculator, and determine the number of x-intercepts.

   a. $y = 2x^2 - 7x - 4$

   $a =$

   $b =$

   $c =$

   $b^2 - 4ac$

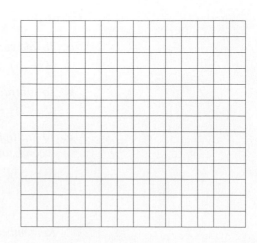

b. $y = 3x^2 + x + 1$

$a =$

$b =$

$c =$

$b^2 - 4ac$

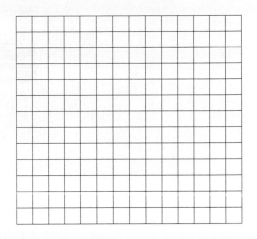

c. $y = x^2 + 2x + 1$

$a =$

$b =$

$c =$

$b^2 - 4ac$

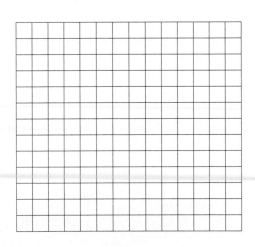

Recall that the solutions to the equation $ax^2 + bx + c = 0$ and the x-intercepts of the graph of $y = ax^2 + bx + c$ are the same. For example, the graph of $y = 2x^2 + x + 5$ has no x-intercept.

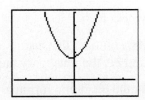

Because the quadratic equation $2x^2 + x + 5 = 0$ must have exactly two solutions in the complex number system, the two solutions must be complex and not real. Similarly, if the graph of a quadratic function has two x-intercepts, the solutions to the equation $ax^2 + bx + c = 0$ must be two real numbers.

6. If the graph of $y = ax^2 + bx + c$ has one x-intercept, what are the number and type (real or complex) of solutions to the equation $ax^2 + bx + c = 0$?

7. Return to Problem 5. Use the value of the discriminant and the number of x-intercepts in parts a, b, and c to complete the following table.

SOLUTIONS TO $ax^2 + bx + c = 0$ IN THE COMPLEX-NUMBER SYSTEM	
$b^2 - 4ac$	NUMBER AND TYPE OF SOLUTIONS
Positive	
Zero	
Negative	

8. a. Evaluate the discriminant for each of the equations below, and indicate the number and type of solutions to the equations.

    i. $2x^2 - 7x - 4 = 0$ **ii.** $3x^2 + x + 1 = 0$

    iii. $x^2 - 2x + 1 = 0$ **iv.** $3x^2 + 2x = -1$

b. Determine the solutions to each of the equations in part a in order to verify your results from part a.

    i.   **ii.**

    iii.   **iv.**

SUMMARY
Activity 4.7

1. The **imaginary unit** is the number $\sqrt{-1}$. The notation for the imaginary unit is i, where $i^2 = -1$.

2. Any number that can be written in the form $a + bi$, where a and b are real numbers and i is the imaginary unit, is called a **complex number**.

3. In the quadratic formula

$$x = \frac{-b \pm \sqrt{b^2 - 4ac}}{2a},$$

the expression $b^2 - 4ac$ is called the **discriminant**. Its value determines the number and type of solutions of a quadratic equation $ax^2 + bx + c = 0$.

4. Solutions of the quadratic equation $ax^2 + bx + c = 0$ in the complex-number system are summarized in the following table.

$b^2 - 4ac$	NUMBER AND TYPE OF SOLUTIONS
Positive	2 real solutions
Zero	1 real solution
Negative	2 complex solutions

EXERCISES
Activity 4.7

In Exercises 1–8, write each of the following in the form bi, where $i = \sqrt{-1}$

1. $\sqrt{-25}$ **2.** $\sqrt{-20}$

3. $\sqrt{-36}$ **4.** $\sqrt{-10}$

5. $\sqrt{-48}$ **6.** $\sqrt{-80}$

7. $\sqrt{-\frac{9}{16}}$ **8.** $\sqrt{\frac{-20}{75}}$

In Exercises 9–13, perform the operations, and express your answer in the form $a + bi$. Use your graphing calculator to verify the results.

9. $(2 + 8i) + (-7 + 2i)$ **10.** $(5 - 3i) - (2 - 6i)$

11. $5i + (3 - 7i)$ **12.** $3i(-2 + 4i)$

13. $(4 - 3i)(1 + 2i)$

Exercise numbers appearing in color are answered in the Selected Answers appendix.

14. Complex numbers are used in electronics to describe the current in an electric circuit. In an alternating current, the resistance, R, in ohms, is the measure of how much the circuit resists (or impedes) the flow of current through it. The resistance, R, is related to the voltage, V, and current, I, by Ohm's law:

$$V = IR$$

    a. If $I = (0.3 + 2i)$ amperes and $R = (0.5 - 3i)$ ohms, determine the voltage, V.

    b. If $I = (2 - 3i)$ amperes and $R = (3 + 5i)$ ohms, determine the voltage, V.

In Exercises 15–18, solve the quadratic equations in the complex-number system using the quadratic formula. Verify your solutions graphically.

15. $3x^2 - 2x + 7 = 0$ **16.** $x^2 + x = 3$

17. $2x^2 + 5x = 7$ **18.** $0.5x^2 - x + 3 = 0$

In Exercises 19–24, determine the number and type of solutions of each equation by examining the discriminant.

19. $2x^2 + 3x - 5 = 0$ **20.** $6x^2 + 7x - 5 = 0$

21. $4x^2 - 4x + 1 = 0$ **22.** $9x^2 + 6x + 1 = 0$

23. $12x^2 = 4x - 3$ **24.** $3x^2 = 5x + 7$

ACTIVITY 4.8
Airfare

OBJECTIVES

1. Build a quadratic model as a product of linear models.

2. Analyze a model contextually.

You are an assistant to the president of a small commuter airline. You have been asked to develop a strategy for increasing the revenue from your primary route. The current fare for this route is $160 per person, and each flight averages 40 passengers.

1. What is the average revenue from each flight?

A recent marketing analysis suggests that each $2 increase in fare will result in one less passenger per flight, and that each $2 reduction in fare will produce one additional passenger per flight. Your job is to use this information to set an airfare that maximizes the revenue from these flights.

You might first decide to adjust the fare up or down by $2 increments and then determine the projected revenues. Do this by completing the accompanying table, where each positive value in the first column represents the number of upward fare adjustments and each negative value represents the number of downward fare adjustments (In other words, $n = -3$ means you have decreased the fare by $6; $n = 2$ means you have increased the fare by $4.)

Fare Play

NUMBER AND DIRECTION OF FARE ADJUSTMENTS	FARE ($)	NUMBER OF PASSENGERS	ANTICIPATED REVENUE ($)
−3			
−2			
−1			
0			
1			
2			
3			

2. a. Determine an equation for the airfare, $F(x)$, as a function of x, where x represents the number and direction of the $2 fare adjustments.

b. Determine the value of $F(4)$. What is its practical meaning?

c. Determine the value of $F(-5)$. What is its practical meaning?

3. a. Determine an equation for the average number of passengers, $P(x)$, as a function of x.

b. Determine the value of $P(4)$. What is its practical meaning?

c. Determine the value of $P(-5)$. What is its practical meaning?

4. a. The revenue for the flight is the product of the airfare and the number of passengers. Determine an equation that represents the revenue, $R(x)$, as a function of x.

b. Determine the value of $R(2)$. What is its practical meaning?

c. What is the largest value of x for which the revenue function has practical meaning?

d. What is the smallest value of x for which the revenue function has practical meaning?

5. Use your algebraic model for the revenue function in Problem 4a to answer the following.

a. What type of function describes the revenue?

b. What is its vertical intercept? What does the intercept signify in terms of the flight revenue?

c. Where is the vertex of this function located? What is the significance of each coordinate of the vertex in terms of the flight revenue?

d. What are the x-intercepts, and how are they significant in terms of the revenue?

6. Set the window of your graphing calculator so that you can see all the features of the revenue function that you examined algebraically in the previous question. Graphically confirm your results from Problem 5.

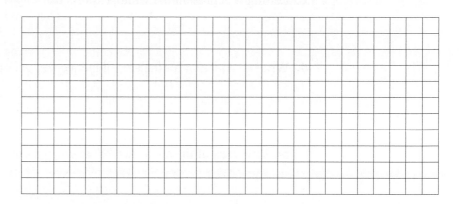

7. From your graph, determine what airfares will result in revenue greater than $700.

8. Summarize the strategy that you will recommend to your boss for maximizing the revenue from this flight.

EXERCISES
Activity 4.8

On your job as assistant to the foreman at a construction site, you have been asked to build an enclosure to store valuable materials and equipment. You have a single 500-foot roll of 10-foot high heavy chain-link fencing. Your boss would like as much storage area as possible. Your job is to figure out the location of the corner posts for this enclosure.

1. a. Assuming that the enclosure will be rectangular and that you will use all of the fencing, make a table showing the dimensions (width and length) of some possible enclosures, with the resulting areas.

WIDTH	25	50	75	100	125	150	175	200	225
LENGTH									
AREA									

    b. From your table, choose the dimensions that would maximize the area.

    c. Letting w represent the width, express the length as a function of w.

    d. Write an equation for the area of the rectangular enclosure as a function of w.

    e. With your graphing calculator, graph the area function and determine the maximum possible area of the enclosure. What dimensions yield this largest enclosure?

    f. Determine the w-intercepts of the graph. Why must you reject these values in this situation?

    g. What are the practical domain and range of this area function?

2. The space for an editorial in a college newspaper is in the shape of a rectangle. The height of the rectangle is 2 inches more than three times the base. If the area of the rectangle is 56 square inches, determine its dimensions.

3. A boat leaves Virginia Beach and sails due east for one hour and then due north for two hours. The boat is then 10 miles from its departure point. If the average rate sailing east was 2 miles per hour faster than the average rate sailing north, determine the rate of speed in each direction.

**PROJECT
ACTIVITY 4.9**
Chemical-Waste
Holding Region

OBJECTIVE

1. Solving problems using
   quadratic functions.

Your architecture firm is designing a rectangular chemical-waste holding region for a local chemical manufacturing company. The holding area is to be located on a rectangular lot that is 200 meters wide and 80 meters long. Federal regulations require that the holding region be 10,000 square meters in area. A safety zone of uniform width around the perimeter of the holding area is also required.

The questions raised here are, can these federal regulations be met if the chemical-waste holding region is constructed on the available rectangular lot? And if so, what would be the width of the safety zone?

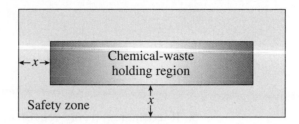

1. Let x represent the width of the safety zone.

   a. Because the width of the lot is 200 meters, what is the width of the holding region in terms of x?

   b. Write an expression for the length of the holding region in terms of x.

2. What is the practical domain of the variable x in this situation? That is, what are the smallest and the largest values of x for which the problem makes sense?

3. The width and length of the waste holding region are both expressed in terms of x. Therefore, the area, A, of the holding region is a function of x. Using the results from Problem 1, write an equation that expresses A as a function of x.

4. In this situation, the area of the holding region is required to be 10,000 square meters.

   a. Write an equation to determine the value of x that gives an output of 10,000. Solve the resulting equation using the quadratic formula. Approximate your solutions to the nearest hundredth.

   b. Verify your results in part a graphically.

5. a. Are both solutions to the equation in Problem 4 practical in this situation? Explain.

   b. What are the dimensions of the chemical-waste holding region?

   c. What is the width of the safety zone?

6. If federal regulations require a safety zone 15 meters wide around the perimeter of the holding region, can the holding region be built on the given rectangular lot? Explain.

7. Consider the holding region area function developed in Problem 3.

   a. What are the horizontal intercepts of the graph of this function? What practical meaning do these intercepts have in this situation?

   b. What is the vertex of the graph (a parabola) of the area function? Does the vertex have any significance in this situation?

CLUSTER 2

What Have I Learned?

1. Which of the following statements are true? In each case justify your decision.

   a. $3 + 2i$ is a pure imaginary number.

   b. $\sqrt{-7}$ is a complex number.

   c. 0 is a complex number.

2. **a.** Describe the relationship between the x-intercepts (if they exist) of the graph of $y = ax^2 + bx + c$ and the solutions to the equation $ax^2 + bx + c = 0$.

   b. Describe the relationship between the x-intercepts (if they exist) of the graph of $y = ax^2 + bx + c$ and the discriminant $b^2 - 4ac$.

3. Consider the quadratic equation $ax^2 + bx + c = 0$. If the quadratic expression $ax^2 + bx + c$ is factorable, what can you say about the sign of the discriminant, $b^2 - 4ac$? Is it positive, negative, or zero? Explain.

4. For what values of c are the solutions to $2x^2 - 5x + c = 0$ imaginary?

5. For what values of k does $x^2 - kx + k = 0$ have only one solution? (*Hint:* Examine the discriminant.)

CLUSTER 2

How Can I Practice?

Write each of the following in the form bi, where $i = \sqrt{-1}$.

1. $\sqrt{-49}$

2. $\sqrt{-45}$

3. $\sqrt{-121}$

4. $\sqrt{-15}$

5. $\sqrt{-112}$

6. $\sqrt{-125}$

7. $\sqrt{-\frac{16}{25}}$

8. $\sqrt{\frac{-24}{42}}$

Perform the following operations, and express your answer in the form $a + bi$. Use your graphing calculator to verify the results.

9. $(7 + 5i) + (-3 - 2i)$

10. $(3 - 3i) - (8 - 9i)$

11. $3i + (6 - 7i)$

12. $-3i(8 - 4i)$

13. $(3 - 4i)(-1 + 2i)$

14. Perform the following tasks.

    a. Identify the values of a, b, and c in $ax^2 + bx + c = 0$.

    b. Determine the type of solution by examining the sign of the discriminant.

    c. Solve the given equation using the quadratic formula. If necessary, round your solutions to the nearest hundredth.

    d. Check your solutions by graphing as well as by substitution.

    a. $3x^2 - x = 7$ **b.** $x^2 - 4x + 10 = 0$

c. $2x^2 - 3x = 2x + 3$ **d.** $3x(3x - 2) + 1 = 0$

15. Each of the following graphs represents a quadratic function. For each graph, determine if the discriminant is positive, negative, or zero. Explain your decision.

a. 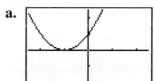   **b.** 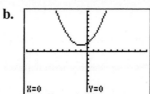   **c.**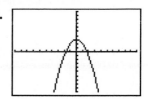

i. graph a

ii. graph b

iii. graph c

16. You are planning to rebuild your garage, which currently measures 20 feet by 30 feet. You want the new garage to be larger than the existing one, but in the same proportions as the original. If you increase the width by x feet, you will have to increase the length by $\frac{3}{2}x$ feet.

   a. Write the floor area of the new garage in terms of the variable x.

   b. Write a quadratic function for the new floor area using your expression from part b.

   c. Which part of the function you determined in part c represents the increase in floor area from the original garage?

d. Use the result of part c to determine the total increase in the floor area if you extend the width by 2 feet and retain the original proportions.

e. You have determined that you want to expand the floor by 264 square feet. Write the equation you need to solve to determine the new dimensions of the garage that are proportional to the original.

f. Solve the equation in part f algebraically. Verify your solution graphically.

g. Are all of your solutions in part g relevant to the garage situation? Explain.

CLUSTER 3 Curve Fitting and Higher Order Polynomial Functions

ACTIVITY 4.10
The Power of
Power Functions

OBJECTIVES

1. Identify a direct variation function.

2. Determine the constant of variation.

3. Identify the properties of graphs of power functions defined by $y = kx^n$, where n is a positive integer, $k \neq 0$.

You are traveling in a hot air balloon when suddenly your binoculars drop from the edge of the balloon's basket. At that moment, the balloon is maintaining a constant height of 500 feet. The distance of the binoculars *from the edge of the basket* is modeled by

$$s = 16t^2.$$

The following table gives the distance, s (in feet), from the drop point at various

t, TIME	0	1	2	3	4
s, DISTANCE	0	16	64	144	256

times, t (in seconds).

As the input values (units of time) increase, the corresponding output values (units of distance) increase. Let us look more closely at how this increase takes place.

Because $s = 16t^2$, you can say that the output, s, varies directly as the square of the input, t. Therefore, as t doubles in value from 1 to 2 or from 2 to 4, the corresponding output values become four times as large: 16 to 64 or 64 to 256.

1. **a.** As t triples from 1 to 3, the corresponding s-values become _____ times as large.

   b. In general, if y varies directly as the square of x, then when x becomes n-times as much, the corresponding y-values become _____ times as large.

The volume, V, of a sphere is given by $V = \frac{4}{3}\pi r^3$. In this situation, you can say that the output, V, varies directly as the cube of the radius, r.

2. **a.** Complete the following table. Leave your answers for V in terms of π.

r	1	2	3	4	8
V					

   b. As r doubles from 2 to 4 or from 4 to 8, the corresponding V-values become _____ times as large.

   c. In general, if y varies directly as the cube of x, then when x becomes n times as large, the corresponding y-values become _____ times as large.

d. Sketch a graph of the volume function. What is the practical domain of this function?

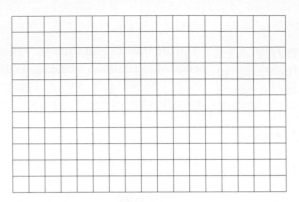

DEFINITION

The equation

$$y = kx^n,$$

where $k \neq 0$ and n is a positive integer, defines a **direct variation** function in which y varies directly as x^n. The constant, k, is called the **constant of variation**.

 Example 1 *The constant of variation, k, in the free-falling object situation defined by $s = 16t^2$ is 16.*

3. What is the constant of variation for the direct variation function defined by $V = \frac{4}{3}\pi r^3$?

In the falling binocular situation, you are given the direct variation equation. Suppose you only know that the distance, s, varies directly as the square of t and one data pair. Would you be able to determine the direct variation equation? Example 2 demonstrates the process.

 Example 2 *Let s vary directly as the square of t. If $s = 68$ when $t = 2$, determine the direct variation equation.*

SOLUTION

Because s varies directly as the square of t, you have

$$s = kt^2,$$

where k is the constant of variation. Substituting 68 for s and 2 for t, you have

$$68 = k(2)^2 \quad \text{or} \quad 68 = 4k \text{ or } k = 17.$$

Therefore, the direct variation equation is

$$s = 17t^2.$$

4. For each table, determine the pattern and complete the table. Then write a direct variation equation for each table.

   a. y varies directly as x.

x	1	2	4	8	12
y		12			

   b. y varies directly as x^3.

x	1	2	3
y		32	

   c. y varies directly as x.

x	1	2	3	4	5
y			3		

5. The length, l, of skid marks left by a car varies directly as the square of the initial velocity, v (in miles per hour), of the car.

   a. Write a general equation for l as a function of v. Let k represent the constant of variation.

   b. Suppose a car traveling at 40 miles per hour leaves skid marks of 60 feet. Use this information to determine the value of k.

   c. Use the function to determine the length of the skid marks left by the car traveling at 60 miles per hour.

Power Functions

The direct variation functions that have equations of the form $y = kx^n$, where n is a positive integer and $k \neq 0$, are also called **power functions**. The graphs of this family of functions are very interesting and are useful in problem solving.

6. Sketch a graph of each of the following power functions. Use your graphing calculator to verify the graph.

a. $y = x$

b. $y = x^2$

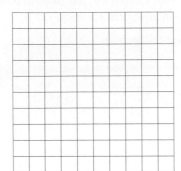
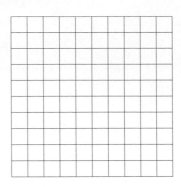

c. $y = x^3$

d. $y = x^4$

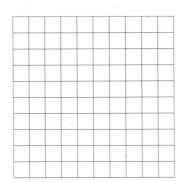
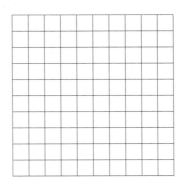

e. $y = x^5$

f. $y = x^6$

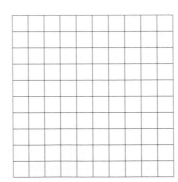
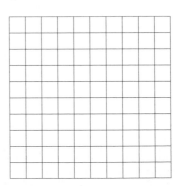

7. Each graph in Problem 6 has an equation of the form $y = x^n$, where n is a positive integer.

a. What is the basic shape of the graph if

i. n is even?

ii. n is odd?

b. If n is even, what happens to the graph as n gets larger in value?

c. If n is odd, is the function increasing or decreasing?

8. Use the patterns from Problem 7 in combination with graphing techniques you have learned previously to sketch a graph of each of the following without using a graphing calculator.

a. $y = x^2 + 1$

b. $y = -2x^4$

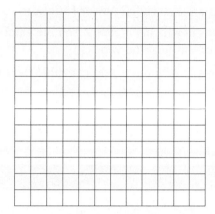
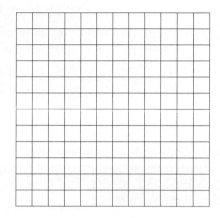

c. $y = 3x^8 + 1$

d. $y = -2x^5$

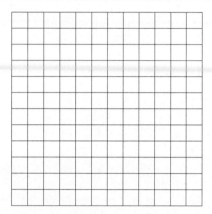
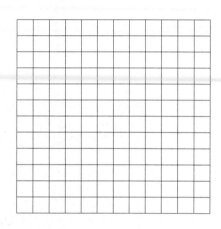

e. $y = x^{10}$

f. $y = 5x^3 + 2$

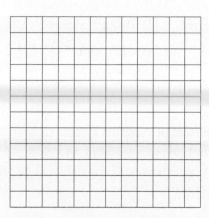
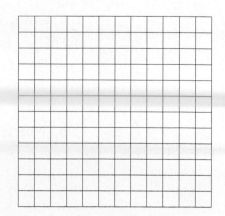

SUMMARY
Activity 4.10

1. The equation $y = kx^n$, where $k \neq 0$ and n is a positive integer, defines a **direct variation function**. The constant, k, is called the **constant of variation**.

2. The direct variation functions that have equations of the form $y = kx^n$, where n is a positive integer, are also called **power functions**.

   a. Power functions in which n is even resemble parabolas. As n increases in value, the parabola flattens near the vertex.

   b. Power functions in which n is odd resemble the graph of $y = kx^3$. If k is positive, the graph is increasing. If k is negative, the graph is decreasing.

EXERCISES
Activity 4.10

1. For each table, determine the pattern and complete the table. Then write a direct variation equation for each table.

   a. y varies directly as x.

x	$\frac{1}{4}$	1	4	8
y		8		

   b. y varies directly as x^3.

x	$\frac{1}{2}$	1	3	6
y		1		

2. The area, A, of a circle is given by the function $A = \pi r^2$, where r is the radius of the circle.

   a. Does the area vary directly as the radius? Explain.

   b. What is the constant of variation k?

3. Assume that y varies directly as the square of x, and that when $x = 2$, $y = 12$. Determine y when $x = 8$.

4. The distance, d, that you drive at a constant speed varies directly as the time, t, that you drive. If you can drive 150 miles in three hours, how far can you drive in six hours?

5. The number of meters, d, that a skydiver falls before her parachute opens varies directly as the square of the time, t, that she is in the air. A skydiver falls 20 meters in two seconds. How far will she fall in 2.5 seconds?

In Exercises 6–10, sketch a graph of the given power function. Verify your graphs using your graphing calculator.

6. $y = -3x^2$

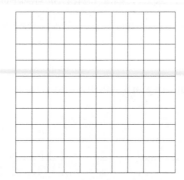

7. $y = x^4 + 1$

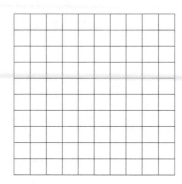

8. $y = -2x^5$

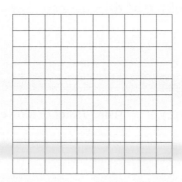

9. $f(x) = x^6$

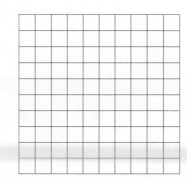

10. $g(x) = 3x^3 - 3$

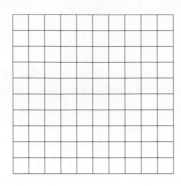

11. Determine the x-interval over which the function $f(x) = \frac{1}{2}x^4$ is increasing.

12. Does the function $g(x) = -\frac{1}{2}x^6$ have a maximum or a minimum point? Explain.

13. For $x > 1$, is the graph of $y = x^2$ rising faster or slower than the graph of $y = x^3$? Explain.

14. Is the graph of $y = \frac{3}{2}x^4$ wider or narrower than the graph of $y = x^4$?

15. How are the graphs of $y = -2x^3$ and $y = 2x^3 + 1$ different? How are the graphs similar?

16. a. For $x > 0$, is the graph of $y = x^2$ rising faster or slower than the graph of $y = 2^x$? Explain.

b. For $x > 0$, is the graph of $y = x^5$ rising faster or slower than the graph of $y = 2^x$? Explain.

ACTIVITY 4.11

Hot Air Balloon

OBJECTIVES

1. Identify equations that define polynomial functions.

2. Determine the degree of a polynomial function.

3. Determine the intercepts of the graph of a polynomial function.

4. Identify the properties of the graphs of polynomial functions.

Returning to the hot air balloon situation, you are relieved because the binoculars you dropped did not strike anyone on the ground. As you continue your ride in the balloon, you enter into a conversation with the pilot concerning why the balloon rises. He says that because the hot air is lighter than the surrounding air, the balloon will rise if the upward lift (force) provided by the hot air is great enough to overcome the downward force on the balloon—its weight.

The upward force varies directly as the volume of the spherical balloon. Because $V = \frac{4}{3}\pi r^3$, the upward force varies directly as the cube of the radius.

1. Write a general equation that represents the upward force, denoted by U, as a function of the radius, r, of the sphere. Let k_1 represent the constant of variation.

The total weight of the balloon itself, the basket, and the heat generator represent the downward force. The weight of the basket and heat generator can be considered a constant, represented by C.

The weight of the balloon material is directly proportional to the surface area of the balloon when inflated. Because $S = 4\pi r^2$, you can say that the weight of the balloon varies directly as the square of the radius, r.

2. **a.** Write a general equation that represents the weight of the balloon as a function of the radius, r. Let k_2 represent the constant of variation.

   b. The downward force, denoted by D, is the sum of the weight of the balloon material, denoted by W, and the weight of the basket and heat generator, denoted by C. Write an equation that expresses D as a function of r and C.

The total force F acting on the balloon is given by

$$F = U - D.$$

Substituting your results from Problem 1 and Problem 2b into this equation gives you

$$F = k_1 r^3 - (k_2 r^2 + C)$$

or

$$F(r) = k_1 r^3 - k_2 r^2 - C.$$

This equation expresses the total force acting on the hot air balloon as a function of r. Because the largest exponent on the input variable r is 3, this function is a **third-degree polynomial function** or a **cubic function**.

Polynomial Functions

Polynomial functions are defined by equations of the form

output = polynomial expression involving the input.

The largest exponent on the input variable n is called the **degree** of the function.

If x and y represent the input and output, respectively, then y must equal sums and differences of terms of the form ax^n, where n is a positive integer. The following example gives several types of polynomial functions.

Example 1   *Examples of polynomial functions are listed in the following table.*

POLYNOMIAL FUNCTION	DEGREE OF THE POLYNOMIAL	NAME
$y = 3x - 2$	1	linear
$y = 2x^2 + 3x - 4$	2	quadratic
$y = 3x^3 - x - 4$	3	cubic
$y = 0.2x^4 - 2x^2 + 7x - 1$	4	quartic
$y = -2x^5 + 3x^4 + 2x - 6$	5	quintic

Note that the cubic function defined by $y = 3x^3 - x - 4$ can be written as $y = 3x^3 + 0x^2 - x - 4$.

Polynomial Functions of Degree 3 or Greater

You have already studied polynomial functions of degree 1 (linear) and degree 2 (quadratic). What are some of the properties and shapes of the graphs of polynomial functions having degree 3 or greater?

3. **a.** What is the domain of the cubic function defined by $y = 2x^3 - 8x^2 - 10x$?

b. Determine the y-intercept of the graph of the cubic function in part a.

c. Use your graphing calculator to sketch a graph. Use the window Xmin $= -5$, Xmax $= 8$, Ymin $= -50$, Ymax $= 10$, Yscl $= 5$. The graph should appear as follows.

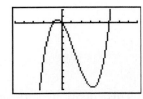

d. Write an equation to determine the x-intercepts of the graph.

e. Solve the equation in part d using a graphing approach. Use the zero option in the CALC menu of your grapher.

Can the equation $2x^3 - 8x^2 - 10x = 0$ be solved using an algebraic approach? Yes! The solution process is demonstrated in the following example.

Example 2   *Solve the equation* $2x^3 - 8x^2 - 10x = 0$ *using an algebraic approach.*

SOLUTION

Step 1. The equation is already written in the form *polynomial expression* = 0.

Step 2. Completely factor the cubic expression on the left side of the equation.

$2x^3 - 8x^2 - 10x = 0$ Factor out the GCF.

$2x(x^2 - 4x - 5) = 0$ Factor the trinomial.

$2x(x + 1)(x - 5) = 0$

Step 3. Apply the zero-product principle, and set each factor equal to zero. Solve each resulting equation.

$2x(x + 1)(x - 5) = 0$

| $2x = 0$ | $x + 1 = 0$ | $x - 5 = 0$ |
| $x = 0$ | $x = -1$ | $x = 5$ |

4. Using an algebraic approach (factoring), determine the horizontal intercepts of each of the following polynomial functions. Verify your results using a graphing approach.

   a. $y = 2x^3 + 5x^2 - 12x$

   b. $f(x) = x^2(x^2 - 5) + 4$

   c. $g(x) = 2x^5 - 18x^3$

5. **a.** Returning to the cubic function defined by $y = 2x^3 - 8x^2 - 10x$, the graph shows a high point (maximum point) in quadrant II and a low point (minimum point) in quadrant IV. Note that the maximum and minimum points occur at turning points of the graph. Use the maximum and minimum options in the CALC menu of your graphing calculator to approximate the coordinates of each of these points. Your screens should appear as follows.

 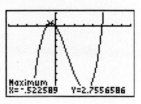

   b. Using the results from part a, determine the interval along the *x*-axis where the function is

   i. increasing.

   ii. decreasing.

6. Using your graphing calculator, plot the following third-degree polynomials. Be careful of your choice of windows.

   a. $f(x) = x^3$

   b. $i(x) = 3x^3 - x - 4$

   c. $g(x) = 0.2x^3 - 2x + 7$

   d. $j(x) = -5x^3 + 1$

   e. $h(x) = -0.6x^3 + 2x^2 - 1$

   f. Use the graphs gathered in parts a–e to write a few sentences comparing and contrasting the graph of the general quadratic equation, $y = ax^2 + bx + c$, $a \neq 0$, and the general cubic equation, $y = ax^3 + bx^2 + cx + d$, $a \neq 0$. Include comments on turning points and general trends, such as increasing and decreasing intervals.

7. Using your graphing calculator, plot the following fourth-degree polynomials. Be careful of your choice of windows.

   a. $f(x) = x^4$

   b. $i(x) = 3x^4 - x - 4$

c. $g(x) = 0.2x^4 - 2x^2 + 7x - 1$ **d.** $j(x) = -5x^4 + 1$

e. $h(x) = -0.6x^4 + 2x^3 - x + 1$

f. Use the graphs gathered in parts a–e to write a few sentences comparing and contrasting the graph of the general quadratic equation, $y = ax^2 + bx + c$, $a \neq 0$, and the general quartic equation, $y = ax^4 + bx^3 + cx^2 + dx + e$, $a \neq 0$. Include comments on turning points and general trends, such as increasing and decreasing intervals.

SUMMARY
Activity 4.11

1. **Polynomial functions** are defined by equations of the form

   output = polynomial expression involving the input.

2. The largest exponent on the input variable n is called the **degree** of the function.

3. Polynomial functions are continuous, with the domain of all real numbers.

4. **Polynomial equations** of the form

   polynomial expression = 0

   can be solved graphically by locating the x-intercepts of the graph of the function defined by

   y = polynomial expression

   or algebraically using factoring (if possible) and the zero-product principle.

EXERCISES
Activity 4.11

In Exercises 1–3, determine the x-intercept(s) of the graphs of each polynomial function using an algebraic approach (factoring). Verify your answer using your graphing calculator.

1. $f(x) = x^3 + 3x^2 + 2x$

2. $g(x) = 2x^2(x^2 - 4)$

3. $h(x) = x^4 - 13x^2 + 36$

4. Determine the vertical intercept of each of the functions in Exercises 1–3.

5. Sketch a graph of the function $f(x) = x^4 - 6x^3 + 8x^2 + 1$. Does the function have a maximum and/or a minimum point(s)? If yes, determine these points.

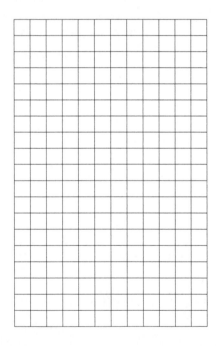

6. Describe any symmetry of the graph of $y = x^4 - 4x^2 - 2$.

7. As the value of the input variable x increases without bound (say 10 to 100 to 1000 and so on), do the output values decrease without bound for the function $y = x^3 + 3x^2 - x - 4$? Use a graph of the function to help answer the question.

8. Is the graph of $y = -1x^3 - x + 3$ increasing or decreasing?

9. Consider the following graph of $y = f(x)$.

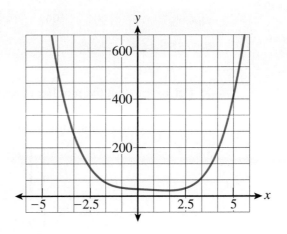

a. As x decreases without bound, the corresponding y-values _____.

b. Is the function f increasing or decreasing for $-2 < x < 2$?

c. How many turning points does the curve have?

10. Sketch a graph of $y = (x - 2)^4$. What is the relationship between the minimum point of the graph and its horizontal intercept?

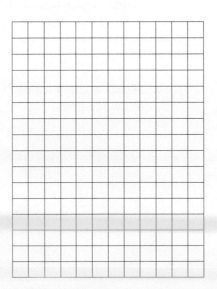

ACTIVITY 4.12

Stolen Bases

OBJECTIVES

1. Determine the regression equation of a polynomial function that best fits the data.

2. Distinguish between a discrete function and a continuous function.

There is probably no sport in which more statistics are gathered and analyzed by non-participants than baseball. Numbers are generated in all kinds of offensive and defensive categories. Some of these categories provide numbers that can be related quite nicely to polynomial functions. Consider the following data on stolen base leaders in the National League in recent years.

YEAR	1991	1992	1993	1995	1997	1999	2000	2001
BASES STOLEN	76	78	58	56	60	72	62	46

1. Let x represent the number of years since 1990. Let y represent the number of stolen bases by the National League individual champion. Plot these points on your graphing calculator. Your scatterplot should resemble the following.

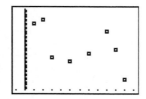

2. **a.** Using your graphing calculator determine the regression equations of the first-, second-, third-, and fourth-degree curves of best fit. Which of these curves seems to best represent the data? Explain.

First degree (linear):

Second degree (quadratic):

Third degree (cubic):

Fourth degree (quartic):

b. Sketch a graph for each of your best-fit curves and the scatterplot on the following grids.

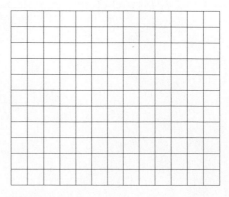

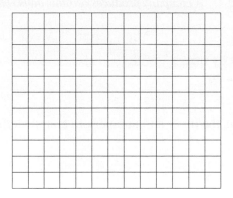

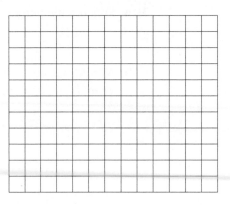

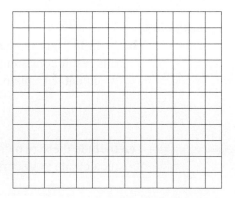

Discrete versus Continuous Functions

Functions are **discrete** if they are defined only at isolated input values and do not make sense for input values between those values. Functions are **continuous** if they are defined for all input values, with the possible exception of a few isolated values.

Example 1 *A company produces ceramic dolls. The monthly production levels are given in the following table.*

MONTH	Mar	Apr	May	June	July	Aug	Sept	Oct	Nov	Dec
NUMBER OF DOLLS (in thousands)	8	15	20	25	28	30	28	25	20	15

If month is the input and number of ceramic dolls is the output, this is a discrete function.

The data in Example 1 can be modeled by $y = -0.86x^2 + 8.58x + 7.38$, where x represents the number of months since March. This quadratic function is continuous because any real number can be used as an input. However, when it is used to model this situation, the only input values that make sense are the integers 3, 4, 5, 6, ... , 12 because the production function is discrete.

3. a. What is the practical domain of the stolen base problem?

b. Would you consider this a discrete situation (consisting of separate, isolated points) or a continuous situation? Explain.

c. What is the practical range of this problem?

4. a. In 1996 Eric Young led the National League in stolen bases with 53. Is this result consistent with the curve you chose as the best to describe the given data? Explain.

b. In 1998 Tony Womack led the National League in stolen bases with 58. Is this result consistent with the curve you chose as the best to describe the given data? Explain.

c. Include these two data points on the lists in your graphing calculator, and recalculate your curve of best fit. Describe the changes.

d. Using the function generated in Problem 4c, predict the number of stolen bases the National League will have in 2005. Is this a realistic prediction? Explain.

SUMMARY
Activity 4.12

1. The graphing calculator can be used to model data with cubic and quartic polynomial functions as well as linear and quadratic polynomial functions.

2. Functions are **discrete** if they are defined only at isolated input values and do not make sense for input values between those values.

3. Functions are **continuous** if they are defined for all input values, with the possible exception of a few isolated values.

EXERCISES
Activity 4.12

1. According to the U.S. Department of Transportation, the following table summarizes the average number of gallons of fuel consumed per vehicle in the U.S. from 1960 to 1998.

YEAR	1960	1970	1980	1990	1995	1996	1997	1998
AVERAGE GALLONS OF FUEL CONSUMED PER VEHICLE, g	784	830	712	677	700	700	711	719

a. Sketch a scatterplot of the data. Let t represent the number of years since 1960. Does the data appear to be linear? Explain.

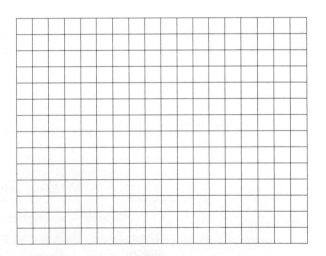

b. Use your graphing calculator to determine and plot the quadratic, cubic, and quartic regression equations for this data.

2. a. Using the results from Exercise 1b, select the equation that best expresses the average number of gallons of fuel consumed per vehicle, g, as a function of the number of years since 1960, t.

b. What is the practical domain of this function?

c. What is the practical range of this function? Explain.

d. Use your equation in part a to estimate the average number of gallons of fuel consumed per vehicle in 1955, 1985, and 2002. In which estimates do you have the most confidence? Explain.

3. The following graphic gives the annual consumption of cigarettes (in billions) in the United States for specific years.

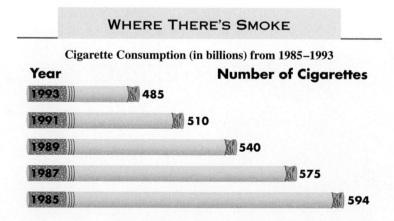

WHERE THERE'S SMOKE

Cigarette Consumption (in billions) from 1985–1993

Year	Number of Cigarettes
1993	485
1991	510
1989	540
1987	575
1985	594

a. Let $t = 0$ correspond to the year 1985. Sketch a scatterplot of the data.

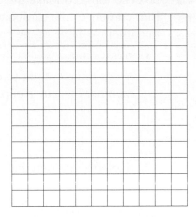

b. Determine a linear model (equation) for the data.

c. Determine a quadratic model (equation) for the data.

d. Which model best represents the data? Explain.

e. Use each model to predict the consumption of cigarettes in the year 2005.

f. How confident are you in your predictions? Explain.

**PROJECT
ACTIVITY 4.13**
Finding the
Maximum Volume

OBJECTIVE

1. Problem solving using polynomial functions.

You have an 8.5-by-11-inch piece of cardboard that you want to make into an open box (no top). To make the open box, you must cut squares of equal sizes from all four corners of the cardboard and then fold up the sides. Your goal is to obtain the maximum volume of the box.

1. Before doing any cutting or calculating, estimate what size square you think needs to be cut out to make a box with the largest (maximum) volume. Each member of your group should come up with a different estimate.

2. Sketch a diagram that represents the problem.

3. Cut a square from each corner of your cardboard. Measure as carefully as you can to the size you chose in Problem 1. Fold and tape the sides to form your box. Measure the dimensions, and calculate the volume of your box.

4. Make a table showing the size of the cut square, the other two dimensions of the box and the volume of the box. Enter the data for your box in the first column, followed by the data from other persons in the class.

SQUARE (Height, in.)						
BOX LENGTH (in.)						
BOX WIDTH (in.)						
VOLUME (in.)						

5. Use several other sizes for the cut square, calculate the resulting volume, and enter it in the table above.

6. From the data in your table, what do you think is the maximum volume? What size of cut square results in this largest box?

7. Let x represent the length of each side of the cut squares. Therefore, x is the height of the box. Write two expressions: one for the length of the box in terms of x, and one for the width of the box in terms of x. Then use these expressions to write an equation for the volume, $V(x)$, as a function of x.

8. Graph the volume function on your graphing calculator. Use the maximum command under the CALC menu to determine the maximum volume. How closely does it agree with your answer in Problem 6? How close was your original estimate in Problem 1?

9. As accurately as you can, give the dimensions for the box with maximum volume. Could you reasonably cut out the size square needed to make the maximum volume? Explain.

10. What are the practical domain and range of this volume function?

11. If you changed the length of your cardboard from 11 to 14 inches, would the maximum volume change? If so, what would be the value of the cut size needed to obtain the maximum volume? Explain the method you used to get your answer.

CLUSTER 3 **What Have I Learned?**

1. In a hurricane, the wind pressure varies directly as the square of the wind velocity (speed). If the wind speed doubles in value, what change in the wind pressure do you experience?

2. Is the graph of $y = 3x^4$ narrower or wider than the graph of $y = x^2$? Explain.

3. The graph of any cubic (third-degree polynomial) function must have one of the four following general shapes:

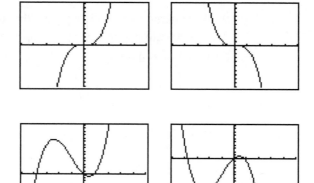

   a. Complete the following table, which gives the maximum number of turning points for a given family of polynomial functions.

DEGREE OF POLYNOMIAL FUNCTION	MAXIMUM NUMBER OF TURNING POINTS
1 (linear)	
2 (quadratic)	
3 (cubic)	
4 (quartic)	

   b. If n represents the degree, then write an expression that represents the maximum number of turning points.

4. a. Sketch a graph of $y = x^4 - 4x^2$. Describe any symmetry that you observe.

   b. Do all graphs of quartic (fourth-degree) functions have symmetry? Explain.

5. a. Does the graph of any cubic function have a horizontal intercept? Can the graph have more than one horizontal intercept? Explain.

   b. Does the graph of any cubic function have at least one vertical intercept? Explain.

6. Given the following graph, determine whether any of the three functions in parts a–c fit the curve. Explain.

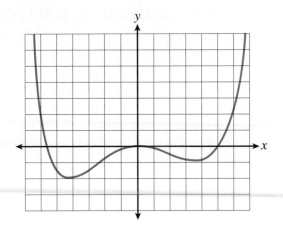

   a. $f(x) = ax^4 + bx^3 + cx^2 + dx + e, \quad a > 0, e = 0$

   b. $g(x) = ax^3 + bx^2 + cx + d, \quad a > 0, e = 0$

   c. $h(x) = ax^4 + bx^3 + cx^2 + dx + e, \quad a < 0, e = 0$

How Can I Practice?

1. y varies directly as x^2. When $x = 3$, $y = 45$. Determine y when $x = 6$.

2. Have you ever noticed that, during a thunderstorm, you see lightning before you hear the thunder? This is true because light travels faster than sound. If d represents the distance (in feet) of the lightning from the observer, then d varies directly as the time, t (in seconds), it takes to hear the thunder. The relationship is modeled by

$$d = 1080t.$$

   a. As the time t doubles (say from 3 to 6), the corresponding d-values _____.

   b. What is the value of k, the constant of variation, in this situation? What significance does k have in this problem?

3. The velocity, v, of a falling object varies directly to the time, t, of the fall. After three seconds, the velocity of the object is 60 feet per second. What will be its velocity after four seconds?

4. Sketch a graph of each of the following pairs of functions. Describe the differences and the similarities in the graphs.

   a. $y = 3x^2$, $y = 3x^2 + 5$

b. $y = 5x^4, y = -5x^4$

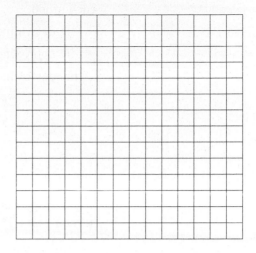

c. $y = 2x^3 + 1, y = 2x^3 - 4$

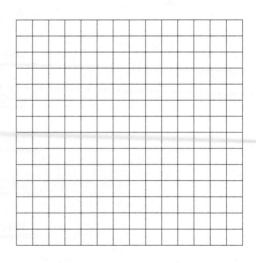

d. $y = 4x^2, y = 4(x - 1)^2$

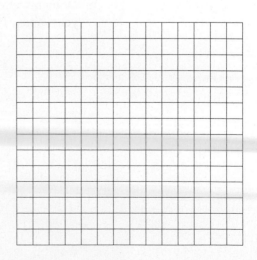

5. Using your graphing calculator, graph each of the following polynomial functions. For each graph,

    i. determine the vertical intercepts.

    ii. approximate the horizontal intercepts (if they exist).

    iii. determine the coordinates of any turning points.

    a. $y = x^3 + 2x^2 - 8x$ **b.** $y = -1x^4 + 2x + 3$

6. The average miles per gallon (mpg) for U.S. cars has steadily increased over the past several years. The following table gives the average miles per gallon for selected years.

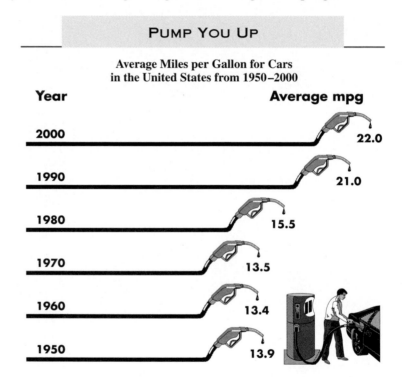

PUMP YOU UP

Average Miles per Gallon for Cars in the United States from 1950–2000

Year	Average mpg
2000	22.0
1990	21.0
1980	15.5
1970	13.5
1960	13.4
1950	13.9

a. Draw a scatterplot of the data points, where the input, x, represents the number of years since 1950.

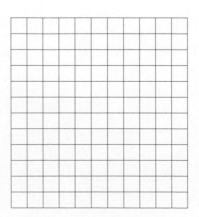

b. Determine the equation of a quadratic function that best fits the data.

c. Use the regression equation to predict the average miles per gallon in the year 2010.

d. Use your model to determine the year when the average miles per gallon is 30.

7. Your bathtub is partially filled. You finish filling the tub and settle in for a nice hot bath. The drain plug is broken, and water is slowly leaking out. The amount of water (in gallons) in the bathtub is given by

$$W(t) = 10 + 7t^2 - t^3 \quad (t \geq 0),$$

where t is time in minutes, and $W(t)$ represents the amount of water in gallons.

a. With the aid of your graphing calculator, sketch the graph of $W(t)$. Don't go beyond $t = 10$. Why?

b. How much water was in the bathtub at the time you began to fill it?

c. Determine the maximum amount of water in the tub from the graph. Explain your result.

d. Use the zero features of your graphing calculator to determine when the tub will be completely empty, to the nearest 0.01 minute.

CHAPTER 4

Summary

The bracketed numbers following each concept indicate the activity in which the concept is discussed.

CONCEPT/SKILL	DESCRIPTION	EXAMPLE
Quadratic function [4.1]	The quadratic function with x as the input variable has the standard form $y = ax^2 + bx + c$, where a, b, and c represent real numbers and $a \neq 0$.	$y = 2x^2 - 3x - 2$
Graph a quadratic function (a parabola) [4.1]	For the quadratic function defined by $f(x) = ax^2 + bx + c$; if $a > 0$, the parabola opens upward; if $a < 0$, the parabola opens downward.	The graph of $y = 2x^2 - 3x - 2$ is a parabola that opens upward.
Vertical intercept of the graph of a quadratic function [4.1]	The constant term c of a quadratic function $f(x) = ax^2 + bx + c$ always indicates the vertical intercept of the parabola. The vertical intercept of any quadratic function is $(0, c)$.	The vertical intercept of the graph of $y = 2x^2 - 3x - 2$ is $(0, -2)$.
Axis of symmetry [4.2]	The axis of symmetry of a parabola is a vertical line that separates the parabola into two mirror images. The equation of the vertical axis of symmetry is given by $x = \frac{-b}{2a}$.	The axis of symmetry of the parabola defined by $y = 2x^2 - 3x - 2$ is $x = \frac{3}{4}$.
Vertex (turning point) [4.2]	The vertex of a parabola defined by $f(x) = ax^2 + bx + c$ is the point where the graph changes direction. It is given by $\left(\frac{-b}{2a}, f\left(-\frac{b}{2a} \right) \right)$.	The vertex of the parabola defined by $y = 2x^2 - 3x - 2$ is $\left(\frac{3}{4}, -\frac{25}{8} \right)$.
x-intercept(s) [4.2]	An x-intercept is the point or points (if any) where the parabola crosses the x-axis (that is, where its y-coordinate is zero).	The x-intercepts of the parabola defined by $y = 2x^2 - 3x - 2$ are $(2, 0)$ and $(-0.5, 0)$.
Domain of the quadratic function [4.2]	The domain of all quadratic functions is all real numbers.	The domain of $y = 2x^2 - 3x - 2$ is all real numbers.
Range of the quadratic function [4.2]	If the parabola opens upward, the range is {output value of turning point, ∞}. If the parabola opens downward, the range is {$-\infty$, output value of turning point}.	The range of the parabola defined by $y = 2x^2 - 3x - 2$ is $\left(-\frac{25}{8}, \infty \right)$.
Solving $f(x) = c$ graphically [4.3]	Graph $y = f(x)$, graph $y = c$, and determine the x-values of the points of intersection. Or graph $y = f(x) - c$ and determine the x-intercepts.	

Solving $f(x) > c$ graphically [4.3]	Graph $y = f(x)$, graph $y = c$, and determine all x-values for which the graph of f is above the graph of $y = c$. Or graph $y = f(x) - c$ and determine all x-values for which the graph of $f(x) - c$ is above the x-axis.	Example 1, Activity 4.3
Solving $f(x) < c$ graphically [4.3]	Graph $y = f(x)$, graph $y = c$, and determine all x-values for which the graph of f is below the graph of $y = c$. Or graph $y = f(x) - c$ and determine all x-values for which the graph of $f(x) - c$ is below the x-axis.	
Greatest common factor (or GCF) [4.4]	The GCF is the largest quantity common to all terms in an expression.	The GCF of $3x^4 - 6x^3 + 18x^2$ is $3x^2$.
Zero-product principle [4.4]	If a and b are any numbers and $a \cdot b = 0$, then either a or b, or both, must be equal to zero.	Example 1, Activity 4.4
Factoring trinomials by trial and error [4.4]	To factor trinomials by trial and error: 1. Remove the GCF. 2. Try combinations of factors for the first and last terms in two binomials. 3. Check the outer and inner products to match middle term of the original trinomial. 4. If the check fails, repeat steps 2 and 3.	Example 5, Activity 4.4
Solving quadratic equations by factoring [4.4]	To solve a quadratic equation by factoring: 1. Use the addition principle to remove all terms from one side of the equation. This results in the equation being set equal to zero. 2. Combine like terms, and then factor the nonzero side of the equation. 3. Use the zero-product principle to set each factor containing a variable equal to zero, and then solve the equations. 4. Check your solutions in the original equation.	Example 6, Activity 4.4
Quadratic formula [4.5]	$$x = \frac{-b \pm \sqrt{b^2 - 4ac}}{2a}$$	Example 1, Activity 4.5

Solving a quadratic equation of the form $ax^2 + bx + c = 0$, $a \neq 0$, **using the quadratic formula** [4.5]	To solve a quadratic equation of the form $ax^2 + bx + c = 0$, $a \neq 0$ using the quadratic formula $$x = \frac{-b \pm \sqrt{b^2 - 4ac}}{2a},$$ 1. Set the quadratic equation equal to zero. 2. Identify the coefficients a and b and the constant term c. 3. Substitute these values into the formula, and simplify. 4. Check your solutions.	Example 1, Activity 4.5
Imaginary unit [4.7]	The imaginary unit is the number $\sqrt{-1}$. The notation for the imaginary unit is i.	
Complex number [4.7]	Any number that can be written in the form $a + bi$, where a and b are real numbers and i is the imaginary unit, is called a complex number.	$2 + 6i$
Discriminant [4.7]	In the quadratic formula $$x = \frac{-b \pm \sqrt{b^2 - 4ac}}{2a},$$ the expression $b^2 - 4ac$ is called the discriminant. Its value determines the number and type of solutions of a quadratic equation $ax^2 + bx + c = 0$.	For the quadratic equation $y = 2x^2 - 7x - 4$, the discriminant is $49 - 4(2)(-4) = 81$. The equation has two real solutions.
Direct variation function [4.10]	The equation $y = kx^n$, where $k \neq 0$ and n is a positive integer, defines a direct variation function in which y varies directly as x^n.	$y = 4x^3$
Constant of variation [4.10]	In the direct variation equation $y = kx^n$, the constant, k, is called the constant of variation.	In $y = 4x^3$, the constant of variation is 4.
Power functions [4.10]	The direct variation function having an equation of the form $y = kx^n$, where n is a positive integer, is also called a power function.	$y = 4x^3$ is a third-power function.
Polynomial functions [4.11]	Polynomial functions are defined by equations of the form $$\text{Output} = \frac{\text{polynomial expression}}{\text{involving the input.}}$$	$y = 5x^4 + 7x^2 - 3x + 1$

Degree of a polynomial [4.11]	The largest exponent on the input variable n is called the degree of the function.	$y = 5x^4 + 7x^2 - 3x + 1$ is a fourth-degree polynomial function.
Discrete functions [4.12]	Functions are discrete if they are defined only at isolated input values and do not make sense for input values between those values.	Example 1, Activity 4.12
Continuous functions [4.12]	Functions are continuous if they are defined for all input values, with the possible exception of a few isolated values.	The quadratic function $y = 2x^4 + 3x - 1$ is continuous for all real numbers.

Gateway Review

In Exercises 1–8, determine the following characteristics of each quadratic function by inspecting its equation.

    a. the direction in which the graph opens

    b. the equation of the axis of symmetry

    c. the vertex

    d. the *y*-intercept

1. $f(x) = x^2 + 2$ **2.** $F(x) = -3x^2$

3. $g(x) = -3x^2 + 4$ **4.** $f(x) = 2x^2 - x$

5. $h(x) = x^2 + 5x + 6$ **6.** $F(x) = x^2 - 3x + 4$

7. $f(x) = x^2 - 2x + 1$ **8.** $g(x) = -x^2 + 5x - 6$

In Exercises 9–15, sketch the graph of each quadratic function using your graphing calculator. Then determine each of the following using the graph.

    a. the coordinates of the *x*-intercepts; if they exist

    b. the domain and the range of the function

    c. the horizontal interval in which the function is increasing

    d. the horizontal interval in which the function is decreasing

9. $g(x) = x^2 + 4x + 3$ **10.** $f(x) = x^2 + 2x - 3$

Answers to all Gateway exercises are included in the Selected Answers appendix.

11. $F(x) = x^2 - 3x + 1$ **12.** $h(x) = 2x^2 + 8x + 5$

13. $F(x) = -2x^2 + 8$ **14.** $f(x) = -3x^2 + 4x - 1$

15. $g(x) = 4x^2 + 5$

In Exercise 16–19, solve the quadratic equation numerically (using tables). Verify your solutions graphically.

16. $x^2 + 4x + 4 = 0$ **17.** $x^2 - 5x + 6 = 0$

18. $3x^2 = 18x + 10$ **19.** $-x^2 = 3x - 10$

In Exercises 20–21, solve the equation using two different approaches. Round your answer to the nearest tenth when necessary.

20. $8x^2 = 10$ **21.** $5x^2 + 25x = -5$

22. Completely factor the following polynomials.

    a. $9a^5 - 27a^2$ **b.** $24x^3 - 6x^2$ **c.** $4x^3 - 16x^2 - 20x$

    d. $5x^2 - 16x + 6$ **e.** $x^2 - 5x - 24$ **f.** $t^2 + 10t + 25$

In Exercises 23–27, solve each equation by factoring. Verify your answer graphically or by using the substitution method.

23. $x^2 - 9 = 0$ **24.** $-x^2 + 36 = 0$ **25.** $x^2 - 7x + 12 = 0$

26. $x^2 - 6x = 27$ **27.** $x^2 = -x$

In Exercises 28–32, write each of the equations in the form $ax^2 + bx + c = 0$. Then identify a, b, and c, and solve the equation using the quadratic formula. Verify your solutions by substitution.

28. $x^2 + 5x + 3 = 0$ **29.** $2x^2 - x = -3$ **30.** $x^2 = 81$

31. $3x^2 + 5x = 12$ **32.** $2x^2 = 3x + 5$

33. For the quadratic function $f(x) = 2x^2 - 8x + 3$, determine the x-intercepts of the graph, if they exist. First, approximate the intercepts using your graphing calculator. Second, solve the equation using the quadratic formula. Approximate your answers to the nearest hundredth.

34. Write each of the following using the imaginary unit, i.

     a. $\sqrt{-49}$ **b.** $\sqrt{-48}$ **c.** $\sqrt{-9}$

     d. $\sqrt{-23}$ **e.** $\sqrt{-\frac{5}{9}}$ **f.** $\sqrt{-\frac{17}{16}}$

35. Perform the following operations with complex numbers. Use your graphing calculator to check your results.

     a. $(2 + 7i) + (-7 + 10i)$ **b.** $(4 - 9i) - (-1 + 7i)$

     c. $4i(3 - 8i)$ **d.** $(4 - i)(6 + 3i)$

In Exercises 36–39, determine the type of solution to each of the equations considering only its discriminant.

36. $2x^2 - 3x + 1 = 0$ **37.** $4x^2 + 16x = 0$

38. $x^2 - 9 = 0$ **39.** $3x^2 + 2x + 2 = 0$

40. Solve the equation in Problem 39 in the complex-number system using the quadratic formula. Verify your solution graphically.

41. Solve the following inequalities using a graphing approach.

    a. $x^2 - x - 6 < 0$ **b.** $x^2 - x - 6 > 0$

42. a. Suppose y varies directly as x. When $x = 3$, $y = 12$. Determine y when $x = 5$.

    b. Suppose y varies directly as x^2. When $x = 4$, $y = 8$. Determine y when $x = 8$.

    c. Suppose y varies directly as x^3. When $x = 1$, $y = 5$. Determine y when $x = 2$.

In Exercises 43–47, graph the function using your graphing calculator. Then answer the following questions referring to the graphing calculator.

    a. Determine the x-intercept, of the function, if it has any.

    b. Determine the domain and range of the function.

    c. Determine the values of x for which the function is increasing and the values of x for which the function is decreasing.

43. $y = x^3 - 8$ **44.** $y = -2x^3 - 2$

45. $y = x^4 - 8$ **46.** $y = x^4 + 2x$

47. $y = x^4 + 5$

48. The height, h (in feet), of a golf ball is a function of the time, t (in seconds), it has been in flight. A golfer strikes a golf ball with an initial velocity of 80 feet per second. The flight path of the ball is a parabola. The approximate height of the ball above the ground is modeled by

$$h(t) = -16t^2 + 80t.$$

a. Sketch a graph of the function. What is the practical domain in this situation?

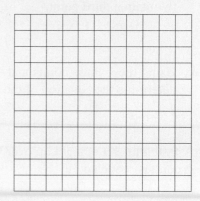

b. Determine the vertex of the parabola. What is the practical meaning of this point?

c. What is the vertical intercept, and what is its practical meaning in this situation?

d. Determine the horizontal intercepts. What is the significance of these intercepts?

e. What assumption are you making in this situation about the elevation of the spot where the ball is struck and the point where the ball lands?

49. To use the regression feature of your calculator to determine the equation of a parabola, you need three distinct points. The stream of water flowing out of a water fountain is in the shape of a parabola. Suppose you let the origin of a coordinate system correspond to the point where the water begins to flow out of the nozzle (see figure).

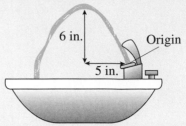

The maximum height of water stream occurs approximately 5 inches measured horizontally from the nozzle. The maximum height of the stream of water is measured to be approximately 6 inches.

a. What is the vertex of the parabola?

b. You already have two points that lie on the parabola. What are they? Use symmetry to obtain a third point.

c. Using these three points and the regression feature of your graphing calculator, determine the equation of the stream of water.

50. A fastball is hit straight up over home plate. The ball's height, h (in feet), from the ground is modeled by

$$h(t) = -16t^2 + 80t + 5,$$

where t is measured in seconds.

a. What is the maximum height of the ball above the ground?

b. How long will it take for the ball to reach the ground?

51. Safe automobile spacing, S (in feet), is modeled by

$$S(v) = 0.03125v^2 + v + 18,$$

where v is average velocity in feet per second.

a. Suppose a car is traveling at 44 feet per second. To be safe, how far should it be from the car in front of it?

b. If the car is following 50 feet behind a van, what is a safe speed for the car to be traveling? How fast is this in miles per hour (60 miles per hour $\approx$ 88 feet per second)?

Rational and Radical Functions

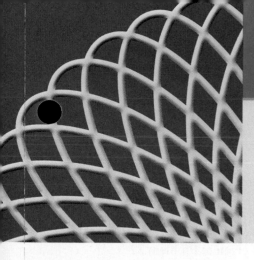

Rational Functions

ACTIVITY 5.1

Speed Limits

OBJECTIVES

1. Determine the domain and range of a function defined by $y = \frac{k}{x}$, k is a nonzero real number.

2. Determine the vertical and horizontal asymptotes of the graph of $y = \frac{k}{x}$.

3. Sketch a graph of functions of the form $y = \frac{k}{x}$.

4. Determine the properties of graphs having equation $y = \frac{k}{x}$.

The speed limit on the New York State Thruway is 65 miles per hour.

1. If you maintain an average speed of 65 miles per hour, how long will it take you to make a 200-mile trip on the thruway? Recall that *distance = rate · time*,
$$time = \frac{distance}{rate}.$$

2. Complete the following table, in which the input variable r represents the average speed in miles per hour and the output variable t represents the time in hours to complete a 200-mile trip.

r (mph)	20	30	40	50	60	70	80
t (hr)							

3. Write an equation that defines travel time t as a function of the average speed r.

4. As the average speed, r, increases, what happens to the travel time, t? What does this mean in practical terms?

5. During a winter storm, a combination of drifting snow and icy conditions reduces your average speed to almost a standstill. Complete the following table for a 200-mile trip on the New York State Thruway.

r (mph)	10	7	5	3	2	1
t (hr)						

6. As the average speed r get closer to zero, what happens to the travel time t? Explain what this means in practical terms.

7. Can zero be used as an input value? Explain.

8. a. What is the practical domain of the function given in Problem 3?

b. Sketch a graph of this function using the table values in Problem 2 and 5.

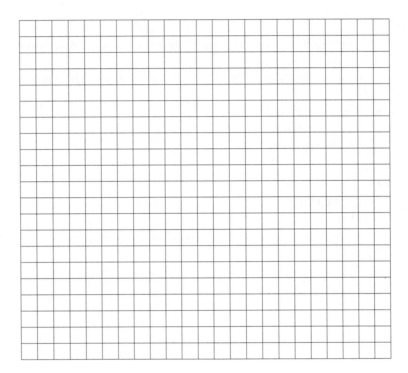

9. a. What are the horizontal and vertical intercepts of the graph?

b. Describe the relationship between the horizontal axis $(t = 0)$ and the graph of the function as the values of r get very large.

In this situation, the horizontal line $t = 0$ is called a **horizontal asymptote**. Recall that a horizontal asymptote is a horizontal line that a graph approaches as the input values of r get very large in a positive direction (or very small in a negative direction).

c. Describe the relationship between the vertical axis $(r = 0)$ and the graph of the function as the values of r get close to zero.

> In this situation, the vertical line $r = 0$ is called a **vertical asymptote**. A vertical asymptote is a vertical line that the graph approaches as the input values get close to the line.

10. Using the graph in Problem 8, approximate your average speed if the 200-mile trip takes 3.5 hours.

Functions Defined by $y = \frac{k}{x}$, Where k is a Nonzero Constant

The function rule $t = \frac{200}{r}$ gives the relationship between the average speed, r, the time, t, and the given value for distance (200). This function belongs to a family of functions having a general rule of the form $f(x) = \frac{k}{x}$, where k represents some nonzero constant.

Example 1   *Examples of this type of function are* $f(x) = \frac{1}{x}, g(x) = \frac{5}{x}, and\ h(x) = \frac{10}{x}.$

11. a. What is the domain of functions f, g, and h defined in Example 1?

b. Complete the following table.

x	−20	−10	−5	−1	−0.5	−0.1	0	0.1	0.5	1	5	10	20
f(x)													
g(x)													
h(x)													

c. Sketch graphs of f, g, and h on the same coordinate system. Verify using your graphing calculator having window Xmin = -5, Xmax = 5, Ymin = -25, and Ymax = 25.

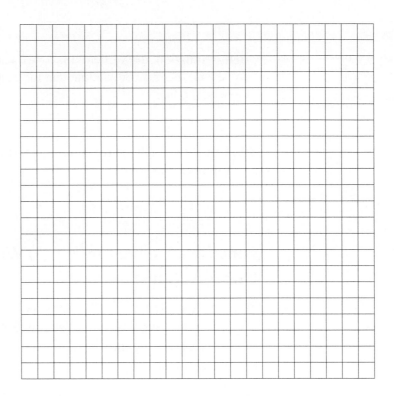

12. Using the table and graphs in Problem 11, answer each of the following questions.

   a. What happens to the output values as the input increases infinitely (without bound) in both the positive and negative directions?

   b. What is the horizontal asymptote for each graph?

   c. What happens to the output as the positive input values get closer to zero?

   d. What happens to the output as the negative input values get closer to zero?

   e. What is the vertical asymptote for each graph?

13. a. Do the graphs of f, g, or h in Problem 11 have x- or y-intercepts?

b. Do the functions f, g, and h have a maximum function value or a minimum function value? Explain.

14. a. Complete the following table, where $Q(x) = \frac{-1}{x}$.

x	−10	−5	−1	−0.5	−0.1	0	0.1	0.5	1	5	10
Q(x)											

b. Sketch a graph of Q. Verify using your graphing calculator having window Xmin = −4, Xmax = 4, Ymin = −4, and Ymax = 4.

c. Describe the effect of the negative sign on the graphs of $f(x) = \frac{1}{x}$ and $Q(x) = \frac{-1}{x}$.

SUMMARY
Activity 5.1

Functions defined by $f(x) = \dfrac{k}{x}$, where k represents some nonzero constant, have the following properties:

1. The domain and the range consist of all real numbers except zero.

2. If $k > 0$, the graph of $f(x)$ has the following general shape.

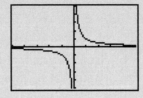

3. If $k < 0$, the graph of $f(x)$ has the following general shape.

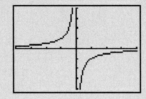

4. The vertical line $x = 0$ is the vertical asymptote.

5. The horizontal line $y = 0$ is the horizontal asymptote.

6. The graph does not intersect either axis (there are no intercepts).

7. There is no maximum or minimum output value.

EXERCISES
Activity 5.1

1. You are a member of a group of distance runners who compete in races ranging in length from 5 to 25 kilometers. In these races, each runner who finishes is told his or her time. Given the time and the length of the race, you can calculate your average running speed.

   a. If you finish a 20-kilometer race in 1 hour 15 minutes, what is your average speed?

   b. Complete the following table for a 20-kilometer race.

t (hr)	1.00	1.25	1.50	1.75	2.00	2.25	2.50
s (km/hr)							

c. Write an equation that expresses the average speed *s* as a function of time *t* in a 20-kilometer race.

d. i. What is the domain of the function?

   ii. What is the practical domain?

   iii. Sketch a graph of the function using the practical domain.

e. As the running time *t* gets longer, what happens to the average speed, *s*?

f. As the running time *t* is reduced (gets closer to zero), what happens to the average speed, *s*?

2. a. Sketch the graphs of the following pair of functions on the same coordinate system. Use labels or colors to differentiate the graphs. Verify your sketches using your graphing calculator.

$$f(x) = \frac{5}{x}, \; g(x) = \frac{-5}{x}$$

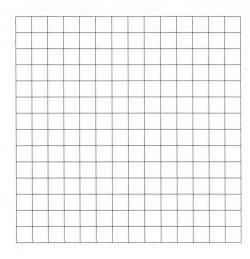

b. Describe how the graph of g can be obtained from the graph of function f.

3. A commercial refrigerator has an initial cost, C, and a scrap value, V. If the life of the refrigerator is N years, then the amount, D, that can be depreciated each year is given by the formula

$$D = \frac{C - V}{N}.$$

a. If the initial cost is $1400 and the scrap value is $200, write a rule for D as a function of N.

b. Complete the following table.

N	1	2	3	6	12	24
D						

c. Sketch a graph of the function.

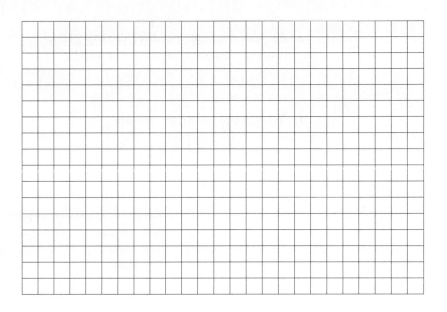

d. If the refrigerator is well constructed, it should have a long, useful life. Will an increase in the useful life of the refrigerator increase or decrease the amount, D, that can be depreciated each year? Explain.

4. The speed limit on Route 66 in Arizona is 75 miles per hour.

a. If you maintain an average speed of 75 miles per hour, how long will it take you to make a 350-mile trip on Route 66?

b. Write an equation that defines t as a function of r, in which r represents the average speed in miles per hour and t represents the time in hours to complete the 350-mile trip.

c. Complete the following table for the equation you determined in part b.

INPUT, r, mph	25	35	45	55	65	75	85
OUTPUT, t, hr							

d. As your average speed increases, what happens to the time it takes to complete the trip?

e. As your average speed for the trip gets closer to zero, what happens to the time it takes to complete the 350-mile trip?

f. What is the practical domain?

g. Sketch a graph.

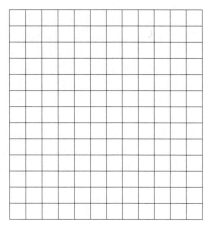

h. What are the vertical and horizontal asymptotes of the graph of the function?

ACTIVITY 5.2
Loudness of a Sound

OBJECTIVES

1. Graph an inverse variation function defined by an equation of the form $y = \frac{k}{x^n}$, where n is any positive integer.

2. Describe the properties of graphs having equation $y = \frac{k}{x^n}$.

3. Determine the constant of proportionality (also called the constant of variation).

The loudness (or intensity) of any sound is a function of the listener's distance from the source of the sound. In general, the relationship between the intensity I and the distance d can be modeled by an equation of the form

$$I = \frac{k}{d^2},$$

where I is measured in decibels, d is measured in feet, and k is a constant determined by the source of the sound and the surroundings.

1. The intensity, I, of a human voice can be given by the formula $I = \frac{1500}{d^2}$. Complete the following table.

d (ft)	0.1	0.5	1	2	5	10	20	30
I (dB)								

2. **a.** What is the practical domain of the function?

   b. Sketch a graph that shows the relationship between intensity of sound and distance from the source of the sound. Use the table in Problem 1 to help determine an appropriate scale.

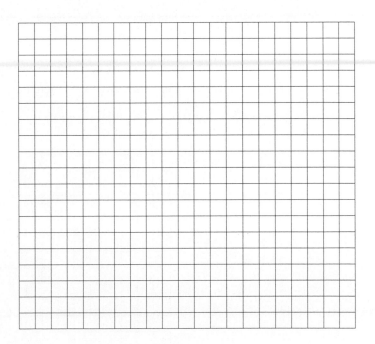

3. As you move closer to the person speaking, what happens to the intensity of the sound?

4. As you move away from the person speaking, what happens to the intensity of the sound?

Functions Defined by $y = \frac{k}{x^2}$, Where k is a Nonzero Constant

The function defined by $I = \dfrac{1500}{d^2}$ belongs to a family of functions having an equation of the form $y = \dfrac{k}{x^2}$, where k represents some nonzero constant.

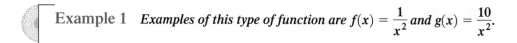

Example 1 *Examples of this type of function are* $f(x) = \dfrac{1}{x^2}$ *and* $g(x) = \dfrac{10}{x^2}$.

5. a. What is the domain of functions f and g defined in Example 1?

b. Complete the following table.

x	−20	−10	−5	−1	−0.5	−0.1	0	0.1	0.5	1	5	10	20
f(x)													
g(x)													

c. Sketch the graphs of f and g on the same coordinate system. Verify your sketch using your graphing calculator.

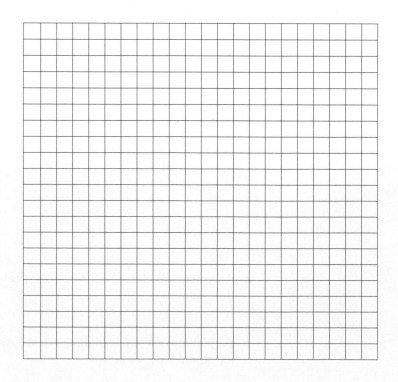

d. Explain why no part of each graph appears below the x-axis.

e. What happens to the output values as the input increases infinitely in the positive direction or decreases infinitely in the negative direction?

f. What is the horizontal asymptote for each graph?

g. What happens to the output values as the positive input values get closer to zero?

h. What happens to the output values as the negative input values get closer to zero?

i. What is the vertical asymptote for each graph?

j. Do the functions have a maximum function value or a minimum function value?

k. For a given input, how is the output of g related to the output of f? Describe this relationship graphically.

6. Describe how the graphs of $y = \dfrac{1}{x}$ and $y = \dfrac{1}{x^2}$ are similar and how they are different.

7. a. Complete the following table.

x	-20	-10	-5	-1	-0.5	-0.1	0	0.1	0.5	1	5	10	20
$g(x) = \dfrac{10}{x^2}$													
$h(x) = \dfrac{-10}{x^2}$													

b. Sketch graphs of functions g and h on the same coordinate system. Use labels or different colors to differentiate the graphs. Verify your sketch using your graphing calculator.

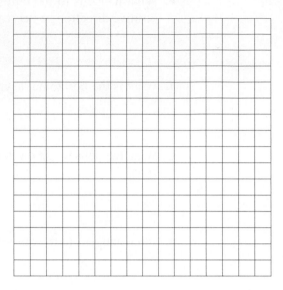

c. Describe how to obtain the graph of function h using the graph of function g.

8. Sketch a graph of $h(x) = \dfrac{1}{x^3}$. Is the graph similar to the graph of $f(x) = \dfrac{1}{x}$ or $g(x) = \dfrac{1}{x^2}$?

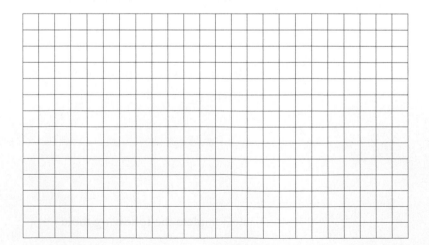

9. Sketch a graph $R(x) = \dfrac{1}{x^4}$. Is the graph similar to the graph of $f(x) = \dfrac{1}{x}$ or $g(x) = \dfrac{1}{x^2}$?

Inverse Variation Functions

> **DEFINITION**
>
> Functions defined by equations of the form $y = \dfrac{k}{x^n}$, where k is a nonzero constant and n is a positive integer, belong to the family of functions called **rational functions**. The rational functions of the form $y = \dfrac{k}{x^n}$ are also called **inverse variation functions**.

Example 2 *For the inverse variation function given by $y = \dfrac{4}{x}$, y varies inversely as x, or y is inversely proportional to x. The number 4 is called the constant of variation or the constant of proportionality. The following table demonstrates that as x doubles in value, the corresponding y-values are reduced by half.*

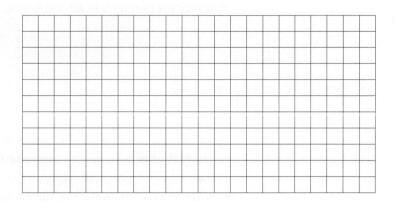

x	2	4	8	16
$y = \frac{4}{x}$	2	1	$\frac{1}{2}$	$\frac{1}{4}$

10. For the function defined by $I = \dfrac{1500}{d^2}$ (see Problem 1), answer the following questions.

a. *I* varies inversely as what quantity?

b. What is the constant of proportionality?

    c. If d is doubled, what is the effect on I?

11. Using the patterns of inverse variation, complete the following table if y is inversely proportional to the cube of x.

x	0.5	1	2	6
y		8		

12. In this activity, the relationship between the intensity, I, of a human voice and the distance, d, from the individual was given by $I = \frac{1500}{d^2}$, where I is measured in decibels, d is measured in feet, and 1500 is the constant of proportionality. The constant of proportionality depends on the source of the sound and the surroundings. If the source of the sound changes, the value of the constant of proportionality will also change.

    a. The intensity of the sound made by a heavy truck 60 feet away is 90 decibels. Determine the constant of proportionality.

    b. Write a formula for the intensity, I, of the sound made by a truck when it is d feet away.

    c. Use the formula from part b to determine the intensity of the sound made by the truck when it is 100 feet away.

SUMMARY
Activity 5.2

- Functions defined by equations of the form $f(x) = \frac{k}{x^n}$, where k is a positive integer, have the following properties.

  1. The domain and range consist of all real numbers except zero.

  2. The graph of f has the following general shape:

  a. Where $k > 0$, and n is an even integer.

  b. Where $k > 0$, and n is an odd integer.

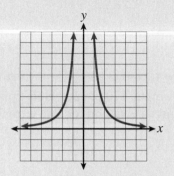
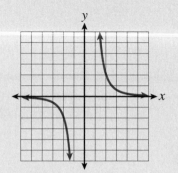

  c. Where $k < 0$, and n is an even integer.

  d. Where $k < 0$, and n is an odd integer.

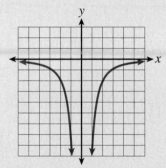
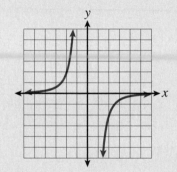

  3. The vertical asymptote is the vertical line $x = 0$.

  4. The horizontal asymptote is the horizontal line $y = 0$.

  5. There are no vertical or horizontal intercepts.

  6. There is no maximum or minimum function value.

- Functions defined by $y = \frac{k}{x^n}$ are called **inverse variation functions** in which

  1. y is said to vary inversely as the nth power of x.

  2. k is called the constant of variation or constant of proportionality.

1. Doctors sometimes use a patient's body-mass index to determine whether or not the patient should lose weight. The formula for the body-mass index, B, is

$$B = \frac{705w}{h^2},$$

where w is the weight in pounds and h is height in inches.

a. What is your body-mass index?

b. Suppose your friend weighs 170 pounds. Substitute this value into the body-mass index formula to obtain an equation for B in terms of height.

c. What is the practical domain of the body-mass index function in part b?

d. Complete the following table using the formula for the body-mass index of a 170-pound person.

h, HEIGHT IN INCHES	60	64	68	72	76	80
B						

e. Sketch a graph of the function defined by $B = \dfrac{119{,}850}{h^2}$. Use the data values in part d to help determine an appropriately scaled axis.

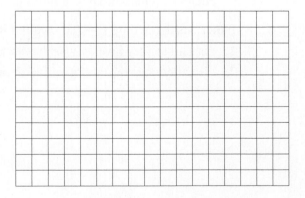

f. What happens to the body-mass index as height increases? Does this make sense in the context of the situation? Explain why or why not.

g. It is recommended that a person's body-mass index be between 19 and 25. Use the graph and the trace key on your calculator to approximate the values of h for which $19 < B < 25$.

2. Sketch a graph of the functions $f(x) = \dfrac{3}{x^2}$ and $g(x) = \dfrac{-3}{x^2}$ on the same coordinate system. Describe how the graph of g is related to the graph of f.

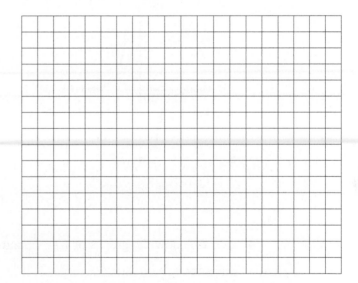

3. Match the following functions with the accompanying graphs.

    i. $f(x) = \dfrac{10}{x^4}$ **ii.** $g(x) = \dfrac{100}{x^5}$

    iii. $h(x) = \dfrac{-10}{x^3}$ **iv.** $F(x) = \dfrac{-1}{x^2}$

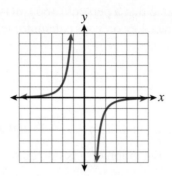

a.

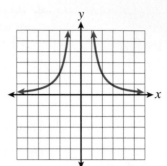

b.

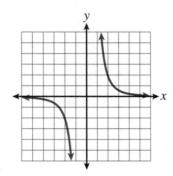

c.

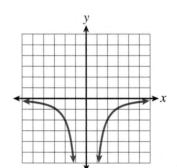

d.

4. Describe how the graphs of $y = \dfrac{1}{x^2}$ and $y = \dfrac{1}{x^3}$ are similar and how they are different.

5. Consider the family of functions of the form $f(x) = \dfrac{k}{x^n}$, where k is a nonzero constant and n is a positive integer.

    a. What is the domain of f?

b. Use several different values of k and n, where $k > 0$ and n is an odd positive integer, to determine the general shape of the graph of f.

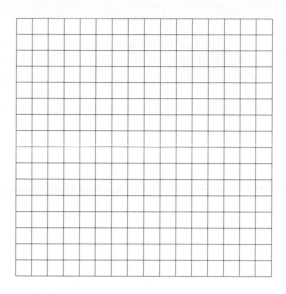

c. Use several different values of k and n, where $k > 0$ and n is an even positive integer, to determine the general shape of the graph of f.

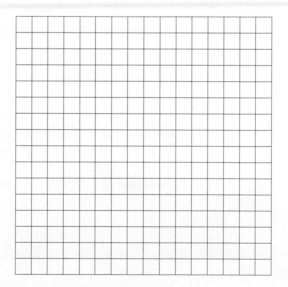

6. How will the general shapes of the graphs in Exercise 5 change if $k < 0$?

7. Complete the following tables of ordered pairs for the given inverse variations.

   a. y varies inversely as x.

x	y
$\frac{1}{2}$	
1	2
2	
6	

   b. y varies inversely as x^3.

x	y
$\frac{1}{2}$	
1	8
2	1
6	

8. If y varies inversely as the cube of x, determine the constant of proportionality if $y = 16$ when $x = 2$.

9. The amount of current, I, in a circuit varies inversely as the resistance R. A circuit containing a resistance of 10 ohms has a current of 12 amperes. Determine the current in a circuit containing a resistance of 15 ohms.

10. The intensity, I, of light varies inversely as the square of the distance, d, between the source of light and the object being illuminated. A light meter reads 0.25 unit at a distance of 2 meters from a light source. What will the meter read at a distance of 3 meters from the source?

11. You are investigating the relationship between the volume, V, and pressure, P, of a gas. In a laboratory, you conduct the following experiment: While holding the temperature of a gas constant, you vary the pressure and measure the corresponding volume. The data that you collect appears in the following table.

P (psi)	20	30	40	50	60	70	80
V (ft³)	82	54	41	32	27	23	20

a. Sketch a graph of the data.

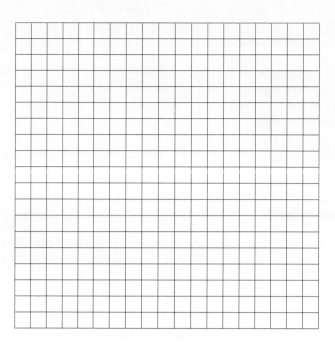

b. One possible model for the data is that V varies inversely as the square of P. Does the data fit the model $V = \dfrac{k}{P^2}$? Explain.

c. Another possible model for the data is that V varies inversely as P. Does $V = \dfrac{k}{P}$ model the data? Explain.

d. Predict the volume of the gas if the pressure is 65 pounds per square inch.

ACTIVITY 5.3

Percent Markup

OBJECTIVES

1. Determine the domain of a rational function defined by an equation of the form $y = \frac{k}{g(x)}$, where k is a nonzero constant and $g(x)$ is a polynomial expression.

2. Identify the vertical and horizontal asymptotes of $y = \frac{k}{g(x)}$.

3. Sketch a graph of rational functions defined by $y = \frac{k}{g(x)}$.

You are a buyer for a national chain of retail stores. You purchase merchandise at a wholesale cost. The merchandise is then sold at a retail price (called the selling price). The retailer's markup is the difference between the selling price (what the consumer pays) and the wholesale cost.

1. a. You acquire a line of sports jackets at a wholesale cost of $80 per jacket. If the jackets sell for $120 each at the retail level, what is the amount of the markup?

b. The markup is what percent of the selling price? (This percent is called the percent markup of the selling price.)

2. The relationship between the selling price, S, the wholesale cost, C, and the percent markup, P, of the selling price (expressed as a decimal) is given by

$$S = \frac{C}{1 - P}.$$

If the wholesale cost of a sports jacket is $80, write an equation for S in terms of P.

3. a. Complete the following table for $S = \frac{80}{1 - P}$.

P (% markup)	0	0.01	0.05	0.10	0.25	0.50	0.75	0.95
S (selling price)								

b. As the values of P approach 1, what happens to the values of S? What does this mean in practical terms?

4. a. Can the percent markup of the selling price be 100% (i.e., can $P = 1$)? Explain.

b. What is the practical domain of this function?

5. Sketch a graph of the function. Use the table of data pairs in Problem 4 to help determine an appropriate scale. Verify your sketch using your graphing calculator.

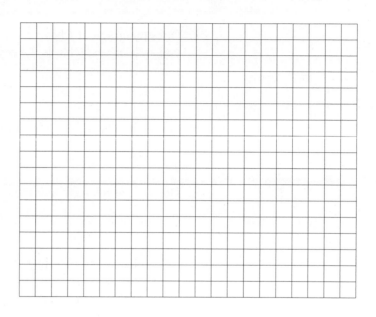

Graphs of $f(x) = \frac{k}{g(x)}$

Because S and P represent real-world quantities, the practical domain limits our investigation of the function defined by $S = \frac{80}{1 - P}$.

6. Consider the general function f defined by $f(x) = \frac{80}{1 - x}$.

   a. What is the domain of function f?

   b. Complete the following table.

x	−10	−5	0	0.50	0.75	0.90	1	1.10	1.25	1.50	2	5	10
f(x)													

   c. Sketch a graph of the function f using your graphing calculator. Use the table in part b to determine an appropriate scale.

d. Does the graph of *f* have a horizontal asymptote? Explain why or why not. If you answer yes, what is the equation of the horizontal asymptote?

e. Does the graph of *f* have a vertical asymptote? Explain why or why not. If you answer yes, what is the equation of the vertical asymptote?

f. For what value of *x* is $f(x)$ maximum?

g. Does the graph have any intercepts?

7. Consider the function defined by $g(x) = \dfrac{80}{x + 1}$.

a. What is the domain of *g*?

b. Construct a table of data points for *g*.

x	−4	−3	−2	−1	0	1	2
g(x)							

c. Sketch a graph of *g* using an appropriate scale.

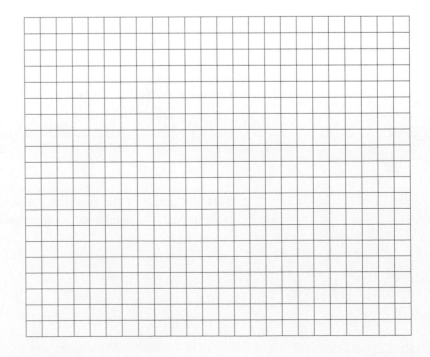

d. Determine the equation of the vertical asymptote. As a graphing aid, if the vertical asymptote is not the y-axis ($x = 0$), the asymptote is drawn as a dotted vertical line. If you have not done so, draw the vertical asymptote in the graph in part c.

e. Determine the equation of the horizontal asymptote.

f. Does the graph of g have any intercepts?

8. a. Sketch a graph of $g(x) = \dfrac{80}{1 + x}$ using your graphing calculator. Your screens should appear as follows.

 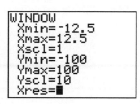 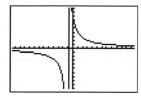

b. The vertical asymptote ($x = -1$) appears to be part of the graph. What do you think has happened?

The calculator was in connected mode. In this mode, it will connect all the points in plots. If the window is such that the calculator does not try to plot a point at $x = -1$, then it will connect a point for an x-value slightly less than -1 to one slightly greater than -1. This creates the appearance of an asymptote. To avoid this, change the mode to dot mode, as follows:

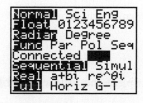 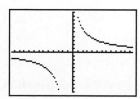

9. How are the graphs of $f(x) = \dfrac{80}{1 - x}$ and $g(x) = \dfrac{80}{x + 1}$ similar? How are they different?

Rational Functions

A function Q, defined by an equation of the form

$$Q(x) = \frac{k}{g(x)},$$

where k is a nonzero constant and $g(x)$ is a polynomial, and $g(x) \neq 0$, belongs to the family of functions known as **rational functions**. The inverse variation function in Activity 5.2 is a special case of rational function, where $g(x) = x^n$. The only value at which the function is not defined is any value for which the denominator is zero. If $g(a) = 0$, then $x = a$ is a vertical asymptote of the graph of Q. The horizontal asymptote is the x-axis ($y = 0$).

Example 1 *Determine the domain and the vertical and horizontal asymptotes for each of the following rational functions.*

a. $f(x) = \frac{3}{x}$ **b.** $g(x) = \frac{10}{x - 4}$ **c.** $h(x) = \frac{-5}{2x + 6}$

SOLUTION

FUNCTION	DOMAIN	VERTICAL ASYMPTOTE	HORIZONTAL ASYMPTOTE
a. $f(x) = \frac{3}{x}$	all real numbers except $x = 0$	$x = 0$	$y = 0$
b. $g(x) = \frac{10}{x - 4}$	all real numbers except $x = 4$	$x = 4$	$y = 0$
c. $h(x) = \frac{-5}{2x + 6}$	all real numbers except $x = -3$	$x = -3$	$y = 0$

10. Consider the function defined by $f(x) = \dfrac{5}{3x - 6}$.

     a. Determine the domain of f.

     b. Complete the following table.

x	-10	-5	0	1.5	1.9	2	2.1	2.5	3	8	13
$f(x)$											

     c. Determine the vertical and horizontal asymptotes of the graph of f.

d. Sketch a graph of f.

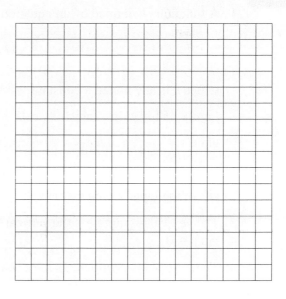

e. Verify the graph using your graphing calculator.

11. Without using your graphing calculator, match the following function rules with the accompanying graphs. Use your graphing calculator to verify your matches.

i. $f(x) = \dfrac{5}{x}$ **ii.** $g(x) = \dfrac{5}{x^2}$ **iii.** $h(x) = \dfrac{-5}{x}$

iv. $F(x) = \dfrac{5}{x + 2}$ **v.** $G(x) = \dfrac{5}{2x - 4}$

a.
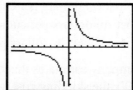

b.
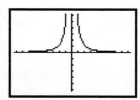

c.
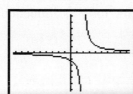

d.
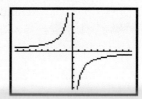

e.
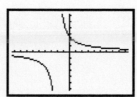

SUMMARY
Activity 5.3

1. A function Q, defined by an equation of the form

$$Q(x) = \frac{k}{g(x)},$$

where k is any nonzero constant, $g(x)$ is a polynomial, and $g(x) \neq 0$, belongs to a family of functions known as **rational functions**. Examples include $f(x) = \frac{10}{x^2}$, $g(x) = \frac{10}{x-4}$, and $h(x) = \frac{-5}{2x+6}$.

2. The domain of the rational function is the set of all real numbers except those values of the input x such that $g(x) = 0$.

3. The vertical asymptote is the vertical line that passes through the x-value for which $g(x) = 0$.

4. The horizontal asymptote is the x-axis $(y = 0)$.

EXERCISES
Activity 5.3

1. To obtain an estimate of the required volume, V, of timber that must be harvested for a logging company to break even, use the following model:

$$V = \frac{Y + L}{P - S - F - T},$$

where: V is the required annual logging volume (in cubic meters),
  Y is the yard cost (in dollars),
  L is the loading cost (in dollars),
  P is the selling price (in dollars per cubic meter),
  S is the skidding cost (in dollars per cubic meter),
  F is the falling cost (in dollars per cubic meter), and
  T is the transportation cost (in dollars per cubic meter).

a. Suppose a logging company estimates that the yard cost will be $25,000, the loading cost will be $55,000, the skidding cost will be $1.50 per cubic meter, the falling cost will be $0.40 per cubic meter, and the transportation cost will be $0.60 per cubic meter. Write a rule for V as a function of P.

b. Complete the following table.

P ($)	2.50	3.00	5.00	10.00	25.00
V (m³)					

c. As the selling price per cubic meter increases, what happens to the corresponding required logging volume, V?

d Determine the value of V when $P = 2$. What is the practical meaning of the negative value of V?

e. What is the practical domain of this function?

f. Sketch a graph of the function. Use the table in part b to determine an appropriate scale.

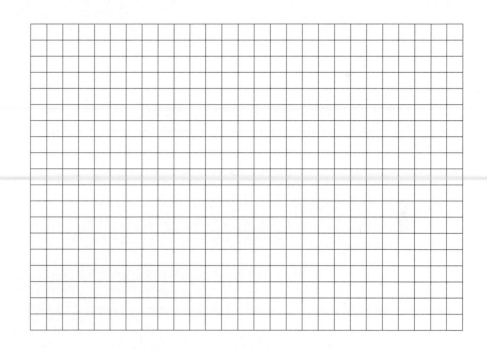

2. Two functions are defined by $f(x) = \dfrac{10}{x - 5}$ and $g(x) = \dfrac{10}{5 - x}$.

a. Describe how you can determine the vertical asymptote without graphing.

b. Determine the vertical and horizontal asymptote for the graph of each function.

c. Verify your answers by graphing each function on your graphing calculator.

3. Without graphing, determine the domain of each of the following functions. Then determine the equation of the vertical asymptote for each function. Verify your answers using your graphing calculator.

a. $f(x) = \dfrac{6}{x - 7}$

b. $g(x) = \dfrac{20}{25 - x}$

c. $h(x) = \dfrac{3}{2x - 10}$

d. $F(x) = \dfrac{13}{0.5x - 7}$

e. $G(x) = \dfrac{-4}{2x + 5}$

4. Give examples of two different rational functions that have a vertical asymptote at $x = 10$.

5. As the input value of a rational function gets closer to a vertical asymptote, the output becomes larger in magnitude, approaching either positive or negative infinity. Consider the functions $f(x) = \dfrac{10}{x - 5}$ and $g(x) = \dfrac{10}{5 - x}$.

a. Determine the equations of the vertical asymptotes for functions f and g.

b. Describe what happens to the output value when x is near the vertical asymptote but to the right of it.

c. Describe what happens to the output value when x is near the vertical asymptote but to the left of it.

ACTIVITY 5.4

Blood-Alcohol Levels

OBJECTIVES

1. Solve an equation involving a rational expression using an algebraic approach.

2. Solve an equation involving a rational expression using a graphing approach.

3. Determine horizontal asymptotes of the graph of $y = \frac{f(x)}{g(x)}$, where $f(x)$ and $g(x)$ are first-degree polynomials.

In 1992, the U.S. Department of Transportation recommended that states adopt 0.08% blood-alcohol concentration as the legal measure of drunk driving. If you assume that a regular 12-ounce beer is 5% alcohol by volume and that the normal bloodstream contains 5 liters (or 169 ounces) of fluid, your maximum blood-alcohol concentration, B, can be approximately modeled by the function having the equation

$$B = \frac{600n}{w(169 + 0.6n)},$$

where n is the number of beers consumed in one hour and w is your body weight in pounds.

1. a. Replace w with your body weight. Write an equation for B in terms of n.

b. Complete the following table using your equation from part a.

NUMBER OF BEERS, n	1	2	3	4	5	6	7	8	9	10
BLOOD-ALCOHOL CONCENTRATION, B										

2. According to this model, how many beers can you consume in one hour without exceeding the recommended legal measure of drunk driving?

3. a. A football player friend of yours weighs 232 pounds. Rewrite the equation for B in terms of n. What is his maximum blood-alcohol level if he drinks four beers in one hour?

b. Complete the following table using your equation from part a.

NUMBER OF BEERS, n	1	2	3	4	5	6	7	8	9	10
BLOOD-ALCOHOL CONCENTRATION, B										

c. What is the practical domain of the blood-alcohol function of part a?

d. Does the weight of a person have any impact on the practical domain? Explain.

e. What is the vertical intercept? Does this seem reasonable within the context of the problem?

f. Sketch the graph of your blood-alcohol function over the practical domain identified in part c. Use the indicated scale.

Note that the graph of the blood-alcohol function in Problem 3f looks like a line when drawn in the practical domain. The graph of the general function defined by

$$f(x) = \frac{600x}{232(169 + 0.6x)}$$

appears as follows.

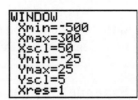

Solving Equations Involving a Rational Expression

4. a. Your 232-pound football player friend is given a breathalyzer test. The result is a blood-alcohol concentration of 0.05%. Using the blood-alcohol concentration function, write an equation that can be solved to determine the number of beers your friend consumed in the previous hour.

b. Solve the equation in part a using the graph in Problem 3f.

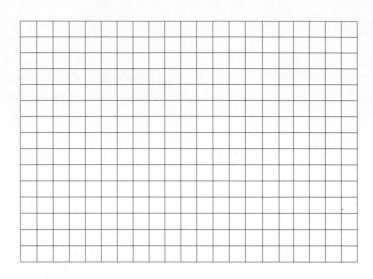

c. Use your graphing calculator to check the answer in part b. Your screen(s) should appear as follows.

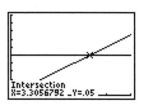

The equation in Problem 4 can be solved using an algebraic approach. If a variable appears in the denominator of a fraction, a general approach is to multiply both sides of the equation by the denominator and solve the resulting equation. Example 1 demonstrates this approach as well as two other methods.

Example 1 Method 1. General Case

To solve the equation $\dfrac{16}{x + 3} = 2$, first multiply both sides of the equation by the denominator $x + 3$, as follows:

$$(x + 3) \cdot \frac{16}{x + 3} = 2(x + 3)$$

$$16 = 2x + 6.$$

Solving for x, you have

$$10 = 2x$$

$$5 = x.$$

Solutions to an equation involving rational functions should always be checked:

$$\frac{16}{5 + 3} = \frac{16}{8} = 2$$

Therefore, 5 is a solution.

Method 2. Cross Multiplication

You can also solve the equation $\dfrac{16}{x+3} = 2$ by applying the following property.

> If two ratios, $\frac{a}{b}$ and $\frac{c}{d}$, represent the same value, then $\frac{a}{b} = \frac{c}{d}$ is equivalent to $ad = bc$.

This process is called cross multiplication. Therefore,

$$\dfrac{16}{x+3} = \dfrac{2}{1} \qquad \text{Cross multiply}$$

$$2(x+3) = 16 \cdot 1$$

$$2x + 6 = 16$$

$$\underline{-6 \quad -6}$$

$$2x = 10$$

$$x = 5$$

Method 3. Graphical Approach

You can verify that 5 is a solution to the given equation by graphing $y_1 = \dfrac{16}{x+3}$ and $y_2 = 2$ and determining the x-value of the point of intersection of the two graphs.

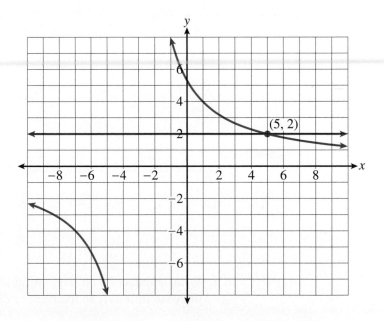

5. Solve each of the following equations using an algebraic approach. Verify your answer graphically.

   a. $\dfrac{45}{x} = 9$

   b. $\dfrac{23}{x+2} = 15$

   c. $\dfrac{13}{x} = \dfrac{2}{5}$

   d. $\dfrac{16}{x^2} = \dfrac{1}{4}$

6. a. Solve the equation in Problem 4a using an algebraic approach.

b. How does your solution compare with the result in Problem 4c using a graphical approach?

Horizontal Asymptotes

The graphs you have studied so far in this chapter have at least one feature in common. The horizontal asymptote is the horizontal axis. As the input values increase infinitely in the positive direction or decreases infinitely in the negative direction, the output values have always approached zero.

7. Consider the function defined by $y = \frac{2x}{x+5}$. This equation is in the form $y = \frac{f(x)}{g(x)}$, where $f(x) = 2x$ and $g(x) = x + 5$.

a. What is the domain of this function?

b. What is the vertical asymptote?

c. Complete the following table.

x	-15	-10	-6	-5.5	-5.1	-5	-4.9	-4.5	-4	0	5
y											

d. What appears to be happening to the output values as the input values increase infinitely to the right or decrease infinitely to the left?

e. What is the horizontal asymptote?

f. Sketch a graph of the function.

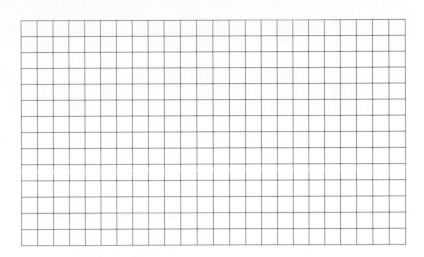

g. Verify the graph in part f using your graphing calculator. As a graphing aid, also sketch the graph of $y = 2$.

8. a. In the blood-alcohol function, as the positive values of n increase in value, what happens to the corresponding values of B?

b. Extend the window of your graphing calculator until you can see the graph leveling off (becoming horizontal) for large values of n. Estimate the equation of the horizontal asymptote of the graph of the blood-alcohol function.

c. Does this asymptote have any practical significance for this application?

9. When comparing smokers and nonsmokers between the ages of 55 and 64, a recent study determined that smokers in the age group had an incidence ratio of 10 for death due to lung cancer. An incidence ratio of 10 means that smokers in this age group are 10 times more likely than nonsmokers to die of lung cancer.

For a given ratio, x, the percent, P, in decimal form of deaths due to a certain illness caused by smoking can be modeled by

$$P = \frac{x - 1}{x}.$$

a. Determine what percent, P, of the deaths due to lung cancer is caused by smoking x in the age group (between 55 and 64.)

b. Complete the following table.

x	1	2	5	10	30
P					

c. As the incidence ratio, x, increases in value, what happens to the corresponding values of the percent P?

d. Determine the horizontal asymptote. Does this make sense in this situation? Explain.

SUMMARY
Activity 5.4

1. To solve an equation of the form $\frac{f(x)}{g(x)} = \frac{a}{b}$, where $g(x) \neq 0$ and $b \neq 0$, using an algebraic approach:

    i. Method 1. Multiply both sides of the equation by the product $b \cdot g(x)$, and solve the resulting equation for x.

    ii. Method 2. Cross multiply to obtain $b \cdot f(x) = a \cdot g(x)$ and solve the resulting equation for x.

2. If the output values of a function R get closer to a number a as the input values increase infinitely in the positive direction or decrease infinitely in the negative direction, then the graph of the function R has a horizontal asymptote. The equation of the horizontal asymptote is $y = a$. For example, the horizontal asymptote for $R(x) = \frac{3x + 1}{x - 2}$ is $y = 3$. Note that the vertical asymptote is $x = 2$.

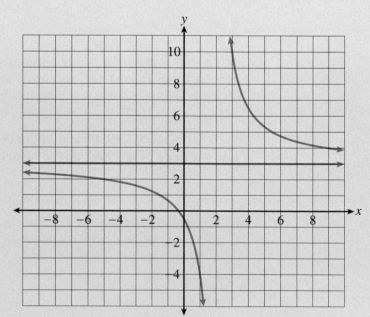

EXERCISES
Activity 5.4

1. You are on the five-year reunion committee for your high school. The committee selects a restaurant that charges $500 to rent a large room to accommodate your group and $50 per person for dinner and a one-hour open bar. Other expenses include $600 for a DJ, $500 for printing invitations, a program, and name tags; and $400 for decorations.

   a. What are the total fixed costs?

   b. The total cost of the event is a function of the number, n, of people who will attend. Write an expression that represents the total cost in terms of n.

   c. Your committee decides to divide the total cost evenly among the people attending. Let m represent the cost per person if n people attend. Write an equation for m in terms of n.

   d. What is the cost per person if 100 people attend?

   e. What is the practical domain of the cost function?

   f. Complete the following table.

n, NUMBER OF PEOPLE ATTENDING	50	100	150	200	250
m, MEAN COST PER PERSON ($)					

   g. Does the graph of the cost function have a horizontal asymptote? Does it make sense in this situation? Explain.

h. Sketch a graph of the cost function over its practical domain.

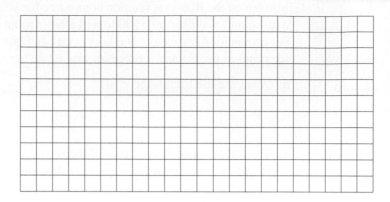

2. For each rational function:
   i. Determine the domain.
   ii. Determine the vertical asymptotes.
   iii. Graph the function using your graphing calculator.
   iv. Determine the horizontal asymptote by inspecting the graph of your function.

   a. $y = \dfrac{4x}{x + 2}$ **b.** $y = \dfrac{1 - x}{x + 1}$

   c. $y = \dfrac{3x}{x - 4}$ **d.** $y = 12 - \dfrac{6x}{1 - 2x}$

3. Solve the following equations algebraically and check your results by graphing.

   a. $\dfrac{3x}{2x-1} = 3$

   b. $\dfrac{x+1}{5x-3} = 2$

   c. $\dfrac{-7x}{2.8+x} = 3.1$

4. In a 20-kilometer race, a runner's average rate (in kilometers per hour) can be expressed as a function of time (in hours) by the equation $r = \dfrac{20}{t}$.

   a. Determine your time to complete the race if you average 16 kilometers per hour.

   b. What is your time if you average 18 kilometers per hour?

5. The intensity of the human voice is inversely proportional to the square of the distance from the source. This is given by the formula from Activity 5.2, Loudness of a Sound: $I = \dfrac{1500}{d^2}$, where I is decibels and d is distance in feet.

   a. Determine the distance from the source when the intensity of the sound is 15 decibels.

   b. What is the distance from the source when the intensity is 8000 decibels?

6. As a fund-raising project, the international club at your college decides to publish and sell a calendar. The cost of photographs and typesetting is $450. It costs $3 to print and assemble each calendar.

a. What is the total cost of printing 200 calendars?

b. What is the average cost per calendar of printing the 200 calendars?

c. Write an expression for the total cost of printing n calendars.

d. Let A represent the average cost per calendar. Write an equation that gives A as a function of n.

e. Complete the following table.

n (number)	50	75	100	500	750	1000
A (average cost)						

f. As the input n increases, what happens to the output A?

g. What is the horizontal asymptote of this function?

h. Verify your answer in part g graphically.

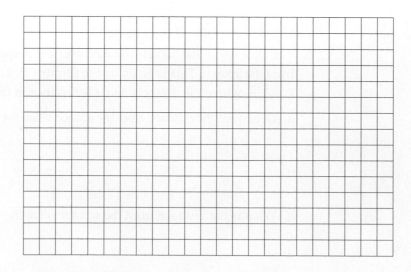

i. Interpret what the horizontal asymptote means in the context of the problem.

j. Suppose you want the average cost to be less than \$3.20. Model this problem with an inequality, solve it algebraically for n, and verify it graphically.

7. The average waiting time in a line before being served is modeled by the formula

$$W = \frac{S(S - A)}{A},$$

where W represents the average waiting time in line before being served
  A represents the average arrival time to the restaurant
  S represents the average service time

a. At Jumpin Jack's fast-food restaurant, the average service time is three minutes. Write an equation for the average waiting time in line W at Jumpin Jack's as a function of A, the average arrival time at the restaurant.

b. Use the equation to determine W for an average arrival time of two minutes.

c. What is the practical domain?

d. Complete the following table.

A	3	2.5	2	1.5	1	0.5	0.2	0.1
W								

e. Determine the horizontal intercept. What is the practical meaning of this intercept in the context of the problem?

f. Determine the vertical asymptote.

g. Sketch a graph of the average waiting time function over its practical domain.

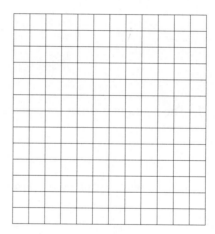

8. In a predator-prey model from wildlife biology, the rate, R, at which prey are consumed by one predator is approximated by the function

$$R = \frac{0.623n}{1 + 0.046n},$$

measured in prey per week, where n is the number of prey available per square mile.

a. If the number of prey available per square mile is 30, what is R?

b. Approximately how many prey must be available per square mile for the predator to consume 10 prey per week?

c. Suppose you want to maintain the prey population to ensure that a predator may obtain between 6 and 10 prey per week. One of the necessary equations for this situation was solved in part b above. Write the other equation and solve it.

d. Use the graph of this function to illustrate your solutions to parts b and c.

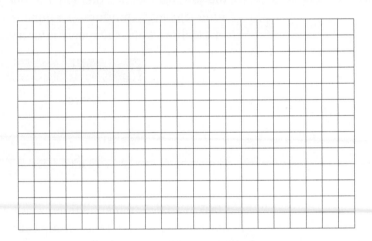

ACTIVITY 5.5

Traffic Flow

OBJECTIVES

1. Determine the least common denominator (LCD) of two or more rational expressions.

2. Solve an equation involving rational expressions using an algebraic approach.

3. Solve a formula for a specific variable.

You are an intern at an architecture firm. The company is designing an auditorium that will be annexed to the local high school building. The rate of traffic flow through the exits is an important consideration. The auditorium will have three exit doors. Two exits are single doors of slightly different sizes. The first exit, by itself, can be used to empty the auditorium in ten minutes. The second exit can be used to empty the room in eight minutes. The third exit is a double-wide door that, by itself, can be used to empty the auditorium in five minutes.

1. If only the first door is open, how much of the auditorium can be emptied in five minutes? Two minutes? One minute?

2. The rate at which a door can be used to empty the auditorium is the fraction of the job that can be completed in one minute. In this case, the units of measurement are auditoriums per minute. Determine the rate of emptying for each of the three exits, and record your answers in the following table.

EXIT	RATE OF EMPTYING
First	
Second	
Third	

Your task is to determine the time, T, it takes to empty the auditorium if all three exit doors are open. The relationship between this time, T, and the individual emptying times is given by the formula

$$\frac{1}{t_1} + \frac{1}{t_2} + \frac{1}{t_3} = \frac{1}{T},$$

where t_1, t_2, and t_3 represent the times that each exit door can be used by itself to empty the auditorium.

Note that it is the rates that are added to determine the rate at which the three doors working together can be used to empty the auditorium.

3. **a.** Write the equation that can be used to determine the time, T, that it takes for the auditorium to be emptied if all three exits are open.

b. Solve this equation graphically. Use the window Xmin = 0, Xmax = 10, Ymin = 0, Ymax = 2, and Yscl = 0.5.

Solution Using an Algebraic Approach

An algebraic approach to solving the equation $\frac{1}{10} + \frac{1}{8} + \frac{1}{5} = \frac{1}{T}$ is to eliminate the fractions from the equation. This can be accomplished by first determining the least common denominator (LCD) of the fractions involved in the equation. The following example demonstrates the procedure for determining the LCD.

Example 1 **a.** Determine the LCD for $\frac{5}{12}$ and $\frac{7}{45}$.

SOLUTION

Step 1. Write the prime factorization of each denominator. Express repeated factors as powers.

$$12 = 2 \cdot 2 \cdot 3 = 2^2 \cdot 3$$

$$45 = 3 \cdot 3 \cdot 5 = 3^2 \cdot 5$$

Step 2. Identify the different bases (factors) in step 1.

2, 3, 5

Step 3. Write LCD as the product of the highest power of each of the different factors from step 2.

$$\text{LCD} = 2^2 \cdot 3^2 \cdot 5 = 4 \cdot 9 \cdot 5 = 180$$

The smallest number that both 12 and 45 will divide evenly is 180.

b. Determine the LCD for $\frac{11}{6xy^3}$ and $\frac{5a}{9x^2y}$.

SOLUTION

Step 1. $6xy^3 = 2 \cdot 3 \cdot x^1 \cdot y^3$

$9x^2y = 3^2 \cdot x^2 \cdot y$

Step 2. $2, 3, x, y$

Step 3. $\text{LCD} = 2 \cdot 3^2 \cdot x^2 \cdot y^3 = 18x^2y^3$

$18x^2y^3$ is evenly divided by both $6xy^3$ and $9x^2y$.

You are now ready to solve the equation $\frac{1}{10} + \frac{1}{8} + \frac{1}{5} = \frac{1}{T}$ using an algebraic approach.

4. a. Determine the LCD of the rational expressions in the equation $\frac{1}{10} + \frac{1}{8} + \frac{1}{5} = \frac{1}{T}$.

b. Multiply each side of the equation by the LCD, and solve the resulting equation.

c. How does this solution compare to the solution you determined graphically in Problem 3?

5. The auditorium is to be equipped with two ventilation fans. The first fan can exchange the air in the room in four hours. The building code requires a complete exchange of air in the room every three hours. To model this situation, use the equation

$$\frac{1}{t_1} + \frac{1}{t_2} = \frac{1}{T},$$

where t_1 and t_2 are the exchange times for the fans working alone, and T is the exchange time for the two fans working together.

a. Let x represent the exchange time for the second ventilation fan. Write an equation that can be used to determine x so that the fans working together will satisfy the building code.

b. Solve this equation graphically.

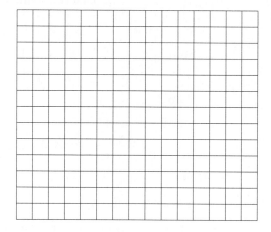

c. Solve this equation algebraically.

6. Suppose the first fan can exchange the air in the auditorium twice as fast as the second fan.

a. If x represents the exchange time for the second fan, write an expression that represents the exchange time for the first fan.

b. Working together, the two fans can exchange the air in the room in four hours. Using the formula $\frac{1}{t_1} + \frac{1}{t_2} = \frac{1}{T}$, write an equation that can be solved to determine x.

c. Solve this equation algebraically. Verify your solution graphically.

d. Determine the rate for each fan.

Solving a Formula for a Specified Letter

An alternate approach to solving problems involving formulas of the form $\frac{1}{t_1} + \frac{1}{t_2} = \frac{1}{T}$ is to use an equivalent formula that has been solved for the variable T.

Example 2 *Solve $\frac{1}{t_1} + \frac{1}{t_2} = \frac{1}{T}$ for T.*

SOLUTION

Step 1. Determine the LCD.

$$\text{LCD} = t_1 t_2 T$$

Step 2. Multiply each side of the equation by the LCD.

$$\frac{t_1 t_2 T}{1} \cdot \left(\frac{1}{t_1} + \frac{1}{t_2} \right) = \frac{1}{T} \cdot \frac{t_1 t_2 T}{1}$$

$$\frac{t_1 t_2 T}{t_1} + \frac{t_1 t_2 T}{t_2} = \frac{t_1 t_2 T}{T}$$

Simplifying, you have $t_2 T + t_1 T = t_1 t_2$.

Step 3. Solve the resulting equation for T.

$$T(t_2 + t_1) = t_1 t_2 \qquad \text{\small T is the common factor on the left side.}$$

$$T = \frac{t_1 t_2}{t_2 + t_1} \qquad \text{\small Divide both sides by } t_2 + t_1.$$

7. a. If t_1 and t_2 are the exchange times for the fans working alone, and T is the exchange time for the two fans working together, determine T if $t_1 = 4$ hours and $t_2 = 12$ hours. Use the formula $\frac{1}{t_1} + \frac{1}{t_2} = \frac{1}{T}$.

b. Repeat part a using the formula $T = \dfrac{t_1 t_2}{t_2 + t_1}$.

   c. Compare the results in parts a and b.

8. The following formula is used in work with lenses and mirrors: $\frac{1}{p} + \frac{1}{q} = \frac{1}{f}$. Solve the formula for f.

SUMMARY
Activity 5.5

1. To determine an LCD of two or more expressions:

   Step 1. Write the prime factorization of each denominator. Express repeated factors as powers.

   Step 2. Identify the different bases (factors) in step 1.

   Step 3. Write the LCD as the product of the highest power of each of the different factors from step 2.

2. To solve an equation involving rational expressions:

   Step 1. Determine the LCD of all denominators in the equation.

   Step 2. Multiply each side of the equation by the LCD, and simplify the resulting equation.

   Step 3. Solve the resulting equation for the desired variable.

EXERCISES
Activity 5.5

Many real-life applications involve equations of the form

$$\frac{1}{a} + \frac{1}{b} = \frac{1}{c}, \text{ where } a, b, \text{ and } c \neq 0.$$

For example,

$\frac{1}{R_1} + \frac{1}{R_2} = \frac{1}{R}$ is used in the area of electrical circuits,

$\frac{1}{p} + \frac{1}{q} = \frac{1}{f}$ is used in work with lenses and mirrors, and

$\frac{1}{t_1} + \frac{1}{t_2} = \frac{1}{T}$ is used to calculate the time it takes to complete a task when two machines or people are working together.

These formulas can also be extended to three or more resistors, lenses, or machines by simply adding additional fractions in each case.

1. Two pumps are working together to empty a gasoline tank.

   a. The emptying times for pump 1 and pump 2 are 30 minutes and 45 minutes, respectively. Determine the time required to empty the tank if both pumps are working. Use the formula $\frac{1}{t_1} + \frac{1}{t_2} = \frac{1}{T}$, where t_1 and t_2 are the emptying times for pump 1 and pump 2 respectively and T is the total time required to empty the tank.

   b. Solve the equation $\frac{1}{t_1} + \frac{1}{t_2} = \frac{1}{T}$ for T.

   c. Using the equation developed in part b, determine T if $t_1 = 20$ minutes and $t_2 = 15$ minutes.

   d. If one pump can empty the tank in 40 minutes, how fast must a second pump work for the pumps working together to empty the tank in 10 minutes? Use $\frac{1}{t_1} + \frac{1}{t_2} = \frac{1}{T}$.

   e. Suppose three pumps are working together to empty the tank, with emptying times of 25 minutes, 30 minutes, and 50 minutes. How long will it take to empty the tank if all three pumps are working simultaneously? Use the formula $\frac{1}{t_1} + \frac{1}{t_2} + \frac{1}{t_3} = \frac{1}{T}$.

2. Solve each of the following equations using an algebraic approach. Verify your answers using a graphing approach.

   a. $\frac{10}{x+1} = 4$

   b. $\frac{2}{x} + \frac{3}{x} = 1$

   c. $\frac{1}{x} + \frac{1}{3x} = \frac{1}{5}$

   d. $\frac{3}{x} + \frac{2}{x} = \frac{4}{x}$

In Exercises 3–5, use the formula $\frac{1}{t_1} + \frac{1}{t_2} = \frac{1}{T}$.

3. The custodian in the mathematics building can buff the main floor 2 minutes faster than his supervisor. If working together they can buff the floor in 35 minutes, how long does it take the supervisor working alone to buff the main floor of the mathematics building? *Hint:* Let t represent the supervisor's time, and $t - 2$ represent the custodian's time.

4. It takes you 4 hours working alone to clean your apartment, and your spouse takes 5 hours and 15 minutes. If you begin at noon and work together, will you complete the cleaning in time to leave for the game at 2:30? How late or early will you be?

5. One of your jobs as a work-study student at your college is sending out mailings (stuffing, sealing, and stamping envelopes). You can work twice as fast as your supervisor. If working together you complete a job in 7 hours, how long would it have taken you to complete the job by yourself?

Hint: Let t represent your total time working alone and $2t$ the time of your supervisor working alone.

6. Solve each of the following formulas for the indicated variable. Express your answer as a single fraction.

a. $\frac{1}{a} + \frac{2}{b} = \frac{3}{c}$, solve for a

b. $\frac{1}{x + y} = \frac{1}{z}$, solve for x

c. $\frac{1}{a} + \frac{2}{b} = \frac{3}{c}$, solve for c

ACTIVITY 5.6
Electrical Circuits

OBJECTIVES

1. Add and subtract rational expressions.

2. Simplify a complex fraction.

In performing some technical work in your new job at a local electronics firm, you need to be familiar with resistors combined in a circuit. The total resistance, R, of two resistors in a parallel circuit can be calculated from the formula

$$R = \frac{1}{\frac{1}{R_1} + \frac{1}{R_2}},$$

where R_1 and R_2 are the two resistors in the circuit, measured in ohms.

1. Calculate the total resistance for each pair of resistors.

R_1 (OHMS)	R_2 (OHMS)	R
10	10	
10	5	
15	5	
20	10	

If one resistor has to be 10 ohms, then the formula

$$R = \frac{1}{\frac{1}{10} + \frac{1}{R_2}}$$

expresses the total resistance as a function of the second resistor's value.

Now the total resistance, R, is a function of R_2. The right side of the equation is a fraction in which fractions also appear in the denominator.

> **DEFINITION**
>
> A fraction that contains fractions in either its numerator or denominator, or both, is called a **complex fraction**.

Example 1 *The following are examples of complex fractions.*

$$\frac{1}{\frac{1}{10} + \frac{1}{R_2}}, \quad \frac{\frac{50}{x}}{\frac{100}{x^2 + 5x}}, \quad \frac{4 + \frac{1}{x}}{\frac{10}{x^2} - \frac{2}{x}}$$

2. To make the equation $R = \dfrac{1}{\frac{1}{10} + \frac{1}{R_2}}$ less cumbersome to work with, simplify the right side of the equation so that it is written as a single fraction (with only one dividing line). This can be accomplished as follows.

   a. Determine the LCD of the fractions $\frac{1}{10}$ and $\frac{1}{R_2}$.

   b. Add the fractions $\frac{1}{10}$ and $\frac{1}{R_2}$ by writing each fraction as an equivalent fraction that has the LCD as the denominator.

c. Divide the numerator by the denominator, and simplify.

3. a. Using your grapher, sketch a graph of the function defined by

$$R = \frac{10R_2}{R_2 + 10},$$

where R is the total resistance of two resistors in a parallel circuit, one having resistance 10 ohms and the second having a resistance represented by R_2. Your screens should resemble the following.

 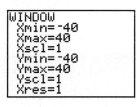 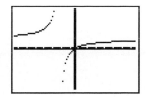

b. What is the domain of this function?

c. What is the practical domain?

d. What is the horizontal asymptote of this function?

e. Interpret its practical meaning.

4. a. Write an equation that you can use to determine the size of the second resistor you would need to add to the circuit to make a total resistance of 7 ohms.

b. Solve this equation graphically.

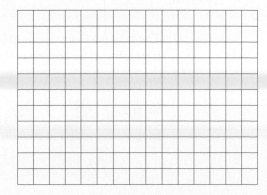

c. Solve the equation in part a algebraically.

5. If resistors are available only in increments of 0.1 ohms, what size would you use to get as close as possible, but still have a total resistance of at least 7 ohms?

Operations with Rational Expressions

As you discovered in Problem 2, simplifying complex fractions involves a lot of work with rational expressions. The following examples illustrate how to perform operations with algebraic fractions. Appendix A contains several additional examples and practice exercises involving operations with rational expressions.

Example 2 *Simplify the following complex fractions.*

a. $\dfrac{\frac{50}{x}}{\frac{100}{x^2 + 5x}}$

Both the numerator and the denominator of the complex fraction contain single rational expressions. Therefore, write the complex fraction as a division problem, and divide.

$$\dfrac{\frac{50}{x}}{\frac{100}{x^2 + 5x}} = \frac{50}{x} \div \frac{100}{x^2 + 5x} \qquad \text{Divide using the division rule } \frac{a}{b} \div \frac{c}{d} = \frac{a}{b} \cdot \frac{d}{c}.$$

$$= \frac{50}{x} \cdot \frac{x^2 + 5x}{100} \qquad \text{Simplify, if possible.}$$

$$= \frac{\cancel{50}}{\cancel{x}} \cdot \frac{\cancel{x}(x + 5)}{2 \cdot \cancel{50}}$$

$$= \frac{x + 5}{2}$$

b. $\dfrac{4 + \frac{1}{x}}{\frac{10}{x^2} - \frac{2}{x}}$

The rational expressions in both the numerator and denominator of the complex fraction can be combined into a single rational expression.

Step 1. $\dfrac{4}{1} + \dfrac{1}{x} = \dfrac{4}{1} \cdot \dfrac{x}{x} + \dfrac{1}{x} = \dfrac{4x}{x} + \dfrac{1}{x} = \dfrac{4x + 1}{x}$

Step 2. $\dfrac{10}{x^2} - \dfrac{2}{x} = \dfrac{10}{x^2} - \dfrac{2x}{xx} = \dfrac{10}{x^2} - \dfrac{2x}{x^2} = \dfrac{10 - 2x}{x^2}$

Step 3. Now divide the numerator of the complex fraction by the denominator.

$$\dfrac{4 + \frac{1}{x}}{\frac{10}{x^2} - \frac{2}{x}} = \frac{4x + 1}{x} \div \frac{10 - 2x}{x^2} = \frac{(4x + 1)}{\cancel{x}} \cdot \frac{\cancel{x} \cdot x}{(10 - 2x)} = \frac{4x^2 + x}{10 - 2x}$$

7. Simplify the following complex fraction

$$\frac{\dfrac{4}{x+3}}{\dfrac{1}{x+2}+\dfrac{3}{x}}$$

SUMMARY
Activity 5.6

1. To **multiply or divide** rational expressions:

  a. Factor the numerator and denominator of each fraction completely.

  b. Divide out the common factors (cancel).

  c. Multiply remaining factors.

  d. In division, proceed as above after inverting the divisor (the fraction after the division sign).

2. To **add or subtract** rational expressions:

  a. Determine the LCD (least common denominator).

  b. Build each fraction to have the LCD.

  c. Add or subtract numerators.

  d. Place the numerator over the LCD, and simplify if necessary.

3. To simplify a complex fraction by simplifying the numerator and denominator:

  a. Express the numerator as a single fraction.

  b. Express the denominator as a single fraction.

  c. Divide the numerator by the denominator.

  d. Simplify, if possible.

1. An EMS vehicle has a single-tone siren with a pitch of approximately 330 hertz (Hz). You are standing on the street corner as the vehicle approaches you at 40 miles per hour. Due to the Doppler effect, the actual pitch you hear is not 330 hertz. The pitch, h, that you hear is given by

$$h = \frac{a}{1 - \dfrac{s}{770}},$$

where a is the actual pitch and s is the speed of the source of the sound in miles per hour.

a. Is the pitch you hear due to the Doppler effect lower or higher than the actual pitch?

b. Simplify the original equation by rewriting the complex fraction on the right side as a single fraction (using only one dividing line).

c. Redo part a using the new equation obtained from part b. How do the results compare?

d. What pitch sound would you hear if the EMS vehicle were traveling at 60 miles per hour?

2. If three resistors having resistances R_1, R_2, and R_3 are connected in parallel, their combined resistance, R, is given by the formula

$$R = \frac{1}{\dfrac{1}{R_1} + \dfrac{1}{R_2} + \dfrac{1}{R_3}}.$$

a. Determine R if R_1 is 4 ohms, R_2 is 8 ohms, and R_3 is 12 ohms.

b. Simplify the complex fraction on the right side of the original formula.

Exercise numbers appearing in color are answered in the Selected Answers appendix.

c. Redo part a using the new formula from part b. How do the answers compare?

3. Simplify the following complex fractions.

a. $\dfrac{\frac{1}{2} - \frac{2}{x}}{x - 4}$

b. $\dfrac{\frac{1}{x} - \frac{1}{2}}{\frac{1}{x^2} - \frac{1}{4}}$

c. $\dfrac{x + \frac{2x - 6}{x - 1}}{\frac{x}{3} - \frac{3}{x}}$

d. $\dfrac{\frac{x}{2} - 1}{x - \frac{4}{x}}$

4. You decide to buy a new car, but you are concerned about the size of the monthly payments. The amount, A, of each monthly payment is given by the formula

$$A = \dfrac{Pi}{1 - \dfrac{1}{(1 + i)^n}},$$

where P represents the principle or the amount borrowed,
   i is the monthly interest rate, and
   n is the number of monthly payments.

a. The car you are looking at costs \$16,000 at 1% monthly interest. If you want to pay for the car over 60 months, how much will your monthly payment be?

b. Simplify the complex fraction on the right side of the original formula.

c. Use the simplified formula to determine the monthly payments. How does your answer compare to the answer in part a?

CLUSTER 1   # What Have I Learned?

1. Make a list of some of the special features of the graphs of rational functions.

2. Describe the connection, if any, between the domain of a rational function and the equation of its vertical asymptote.

3. Describe the algebraic steps required to determine the vertical asymptote(s) of the graph of a rational function.

4. For what values of k will the graph of $y = \dfrac{k}{x}$ be in the second and fourth quadrants?

5. **a.** Describe the algebraic steps required to solve $10 = \dfrac{35}{1 + 5x}$.

   b. Explain the technique for solving the equation in part a graphically.

6. Explain how you would determine the horizontal asymptote of the rational function $f(x) = \dfrac{6x + 1}{2 - 3x}$.

How Can I Practice?

1. Describe the relationship between the graphs of $f(x) = \dfrac{1}{x}$ and $g(x) = \dfrac{-1}{x}$.

2. Describe the relationship between the graphs of $f(x) = \dfrac{1}{x}$ and $g(x) = \dfrac{1}{x-5}$.

3. Describe the relationship between the graphs of $f(x) = \dfrac{1}{x}$ and $g(x) = \dfrac{1}{x^3}$.

4. **a.** Suppose you are taking a trip of 145 miles. Assume that you drive the entire distance at a constant speed. Express your time to take this trip as a function of your speed.

   b. What is the practical domain of this function?

   c. Using the equation for the function, determine the domain.

5. Determine the domain of each of the following functions. Then give the equation of the vertical asymptote of each function.

   a. $g(x) = \dfrac{10}{x+5}$ 　　　　 **b.** $f(x) = \dfrac{5}{13-2x}$

   c. $g(x) = \dfrac{-3}{5x-8}$ 　　　　 **d.** $h(x) = \dfrac{0.02}{5.7x-3.2}$

6. The weight of a body above the surface of the Earth varies inversely with the square of the distance from the center of the Earth. If an object weighs 100 pounds when it is 4000 miles from the center of the Earth, how much will it weigh when it is 4500 miles from the center?

7. A manufacturer of lawn mowers uses the function $C(x) = \dfrac{132x + 75{,}250}{x}$ to model the average cost per lawn mower, in dollars, where x is the number of lawn mowers produced.

a. What is the practical domain of this function?

b. What is the minimum number of lawn mowers that must be manufactured to bring the average cost per lawn mower down to $199? Solve algebraically and graphically.

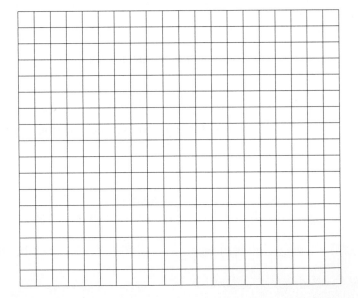

8. The concentration of a drug in the bloodstream, measured in milligrams per liter, can be modeled by the function

$$C = \frac{14t}{3t^2 + 2.5},$$

where t is the number of minutes after injection of the drug.

a. How long after injection will it take for the concentration to equal 0.05 milligrams per liter? Solve algebraically and check graphically.

b. Use the graph to determine when the drug be at its highest concentration?

9. Solve each equation algebraically. Verify your answer graphically.

a. $\dfrac{3}{x+1} = 4$

b. $\dfrac{3x}{2x-5} = 10$

c. $\dfrac{4}{x+3} + 12 = 52$

d. $\dfrac{2.4x}{1 + 0.3x} = 5.8$

10. As an object rises, the effect of Earth's gravitational pull on the object is reduced. The weightlessness that astronauts experience in the space shuttle as it orbits the Earth is due to the distance of the shuttle above Earth's surface.

If an object weighs E kilograms at sea level, then the weight, W (also in kilograms), of the object at a distance of h kilometers above sea level is given by the function rule

$$W = \frac{E}{\left(1 + \frac{h}{6400}\right)^2}.$$

a. Suppose you are flying in a commercial jetliner 15 kilometers above sea level. Replace E with your body weight, measured in kilograms (1 kilogram = 2.2 pounds), and calculate your weight at 15 kilometers above sea level.

b. If an astronaut weighs 70 kilograms at sea level, write a function rule that expresses the astronaut's weight as a function of his or her distance above sea level.

c. Complete the following table using the function rule from part b.

h	0	10	100	1000	1500	2000	10,000	20,000
W								

d. As the height, h, of the space shuttle increases, what happens to the corresponding weight of the astronaut?

e. As the space shuttle reaches its orbiting altitude of 40,000 kilometers above sea level, what is the weight of the astronaut?

f. What is the practical domain of the weight function?

g. Use your graphing calculator to sketch a graph of the weight function.

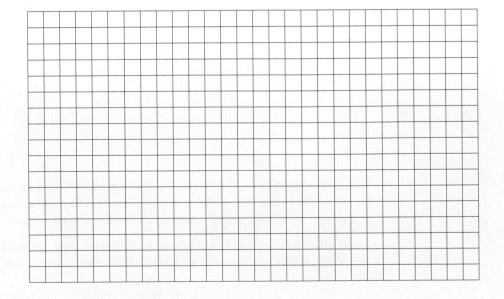

h. At what altitude does the astronaut's weight equal one-half of what it is at sea level?

11. An electrical circuit has three resistors. The total resistance of the circuit R is related to the individual resistances R_1, R_2, and R_3 by the equation

$$\frac{1}{R_1} + \frac{1}{R_2} + \frac{1}{R_3} = \frac{1}{R}.$$

a. You know that $R_1 = 4$ ohms, $R_2 = 6$ ohms, and the total resistance of the circuit is 2 ohms. Determine R_3.

b. Solve the equation $\frac{1}{R_1} + \frac{1}{R_2} + \frac{1}{R_3} = \frac{1}{R}$ for R.

c. Using the equation developed in part b, determine R if $R_1 = 4$ ohms, $R_2 = 6$ ohms, and $R_3 = 12$ ohms.

12. Solve each of the following equations using an algebraic approach. Verify your answers using a graphing approach.

a. $\frac{3}{x-1} = 10$

b. $\frac{2}{x} - \frac{4}{x} = 2$

c. $\frac{1}{x} + \frac{1}{4x} = \frac{1}{4}$

d. $\frac{3}{x} - 4 = \frac{2}{x}$

13. The average speed, s, of your roundtrip commute from home to campus is given by

$$s = \frac{2d}{\dfrac{d}{r_1} + \dfrac{d}{r_2}},$$

where d is the one-way distance from home, r_1 is your average morning commute speed, and r_2 is your average afternoon commute speed.

a. Your one-way commute to campus is 15.3 miles. Your average morning commute speed is 45 miles per hour. Your average afternoon commute speed is 40 miles per hour. What is your average speed, s?

b. Simplify the original equation by rewriting the complex fraction on the right side as a single fraction.

c. Redo part a using the new equation obtained in part b. How do the results compare?

14. Simplify the following complex fractions.

a. $\dfrac{4 + \dfrac{2}{x}}{1 - \dfrac{3}{x}}$

b. $\dfrac{\dfrac{x}{5} - \dfrac{5}{x}}{\dfrac{1}{5} + \dfrac{1}{x}}$

c. $\dfrac{\dfrac{1}{x + 2}}{1 + \dfrac{1}{x + 2}}$

CLUSTER 2   **Radical Functions**

ACTIVITY 5.7

Hang Time

OBJECTIVES

1. Determine the domain of a radical function defined by $y = \sqrt{g(x)}$, where $g(x)$ is a polynomial.

2. Graph functions having equation $y = \sqrt{g(x)}$ and $y = -\sqrt{g(x)}$.

3. Identify the properties of the graph of $y = \sqrt{g(x)}$ and $y = -\sqrt{g(x)}$.

During the seventeenth century, many mathematicians were interested in projectiles, primarily for their military applications. More recently, studies have been done in the area of sports, where the shot put, the javelin, and a punted football are the projectiles of interest. Dr. Peter Brancazio, a physics professor at Brooklyn College, has spent considerable time studying one of the most famous human projectiles, Michael Jordan. In particular, Brancazio has been interested in Jordan's hang time, the length of time elapsed from the instant that Jordan leaves the floor to the instant that he touches it again. Using basic physics, Brancazio knows that the hang time, T, of any jump (in seconds) is related to the height, H, of that jump (in feet) by the formula

$$H = 4T^2, T \geq 0.$$

1. Using the formula with the table feature of your calculator, complete the following table.

T (sec)	0	0.2	0.4	0.6	0.8	1.0	1.2	1.4	1.6
H (ft)									

2. In the formula $H = 4T^2$, the height, H, is a function of the hang time, T. Determine the practical domain and range of this function.

3. Sketch a graph of the function $H = 4T^2$.

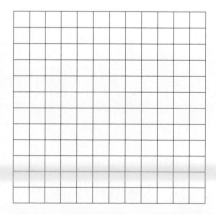

Although the formula $H = 4T^2$ is useful, the form of the equation is not entirely satisfactory. The formula implies that the height of the jump, H, depends on hang time, T. Therefore, T is the input and H is the output. The equation may be true in a relational sense, but it may be more logical to say that the hang time depends on the height of the jump. Therefore, it makes more sense to input values for H and find corresponding output values of T.

4. Suppose you measure the height of a jump and obtain a value of 1.44 feet. Determine the corresponding hang time, T.

5. The process of determining hang times for given jump heights would be greatly simplified if you had a formula that expressed T as a function of H.

   a. Solve the formula $H = 4T^2$ for T, where $T \geq 0$.

   b. Complete the following table using the formula you obtained in part a.

H (ft)	0	0.16	0.64	1.44	2.56	4	5.76	7.84	10.24
T (sec)									

   c. What are the practical domain and range of this new function?

6. Sketch the graph of $T = \frac{1}{2}\sqrt{H}$, where H is the input, T is the output, and $0 \leq\ = H \leq 4$.

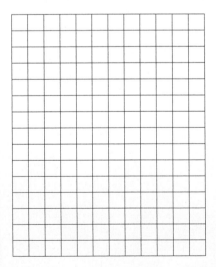

7. a. How are the ordered pairs in the tables in Problems 1 and 5 related?

   b. What is the relationship between the domains and ranges of the two functions defined by $H = 4T^2$ and $T = \frac{1}{2}\sqrt{H}$, where $T \geq 0$?

Recall that a fundamental characteristic of *inverse* functions is that their domains and ranges are interchanged. Therefore, these functions are inverses.

Radical Functions

The investigation of the properties of the function defined by $T = \frac{1}{2}\sqrt{H}$ is somewhat limited because of the restrictions placed on the variables T and H, which represent real-world quantities.

8. a. Consider the general function defined by $F(x) = \frac{1}{2}\sqrt{x}$.
   What is the domain of F?

   b. Sketch a graph of F.

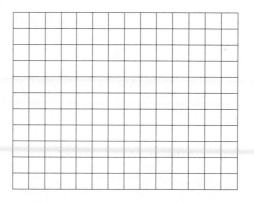

   c. What is the range of F?

> Recall that the square root of a negative number is not a real number. Therefore, the domain of a function defined by an equation of the form $y = \sqrt{g(x)}$, where $g(x)$ is a polynomial, is the set of all real numbers for x such that $g(x) \geq 0$.

Example 1 *Determine the domain of the function defined by $f(x) = \sqrt{2x - 10}$.*

SOLUTION

You need to determine all values of x such that $2x - 10 \geq 0$. Therefore,

$$2x - 10 \geq 0$$
$$2x \geq 10$$
$$x \geq 5.$$

The domain is all real numbers greater than or equal to 5.

9. Consider the functions defined by the following equations.

$$f(x) = \sqrt{x} \quad g(x) = \sqrt{x + 2} \quad h(x) = \sqrt{x - 3}$$

a. Determine the domain of each function.

b. Sketch graphs of all three functions on the same coordinate system.

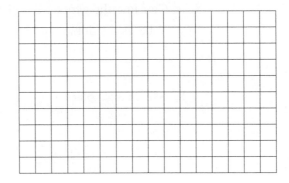

c. Are the functions f, g, and h increasing or decreasing?

10. a. Determine the domain and range of the functions defined by $f(x) = \sqrt{x}$ and $g(x) = -\sqrt{x}$.

b. Sketch graphs of f and g on the same coordinate system.

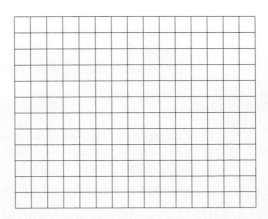

c. How would you obtain the graph of g from the graph of f?

11. a. Determine the domain of the function defined by $h(x) = \sqrt{x - 3}$ and $F(x) = \sqrt{3 - x}$.

b. Complete the following tables.

x	3	4	7	12	19	28
h(x)						

x	3	2	−1	−6	−13	−22
F(x)						

c. Sketch the graphs of h and F on the same coordinate system.

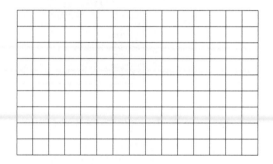

d. Determine whether the functions h and F are increasing or decreasing.

12. Consider the function defined by $f(x) = -\sqrt{3x - 6}$.

a. Determine the domain of f.

b. Determine the x-intercept of f.

c. Complete the following table. If necessary, approximate $f(x)$ to the nearest tenth.

x	2	3	4	5	8	10
f(x)						

d. Sketch a graph of f.

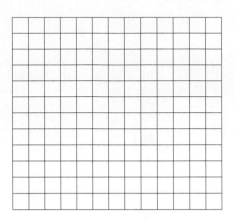

e. Use your graphing calculator to sketch a graph of f. Your screens should appear as follows.

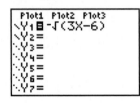 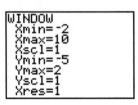 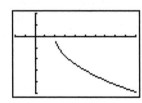

f. Determine the range of f.

g. Does the graph of f have a maximum value? If so, what is it?

13. Consider the function defined by $g(x) = \sqrt{x^2 + 4}$.

a. Determine the domain of g.

b. Complete the following table.

x	-4	-2	0	2	4	6
$g(x)$						

c. Sketch a graph of g.

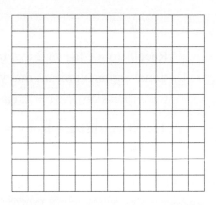

d. Verify your graph in part c using your graphing calculator.

e. Determine the range of g.

f. Does the graph of g have a maximum or minimum value? If so, what is it?

14. a. Complete the following table for $f(x) = \sqrt{x^2} + \sqrt{4}$.

x	−4	−2	0	2	4	6
f(x)						

b. How do the outputs in the table in part a compare with the outputs for $g(x) = \sqrt{x^2 + 4}$ in Problem 13b?

c. Use your graphing calculator to sketch a graph of $f(x) = \sqrt{x^2} + \sqrt{4}$.

d. How does the graph of $f(x) = \sqrt{x^2} + \sqrt{4}$ compare to the graph of $g(x) = \sqrt{x^2 + 4}$ in Problem 13c?

e. Is the expression $\sqrt{x^2 + 4}$ equivalent to $\sqrt{x^2} + \sqrt{4}$? Explain.

Problem 14 demonstrates the following important fact about radicals:

$$\sqrt{a + b} \neq \sqrt{a} + \sqrt{b}$$

Recall from Chapter 2 that the expression $\sqrt{x}$ can also be written as $x^{\frac{1}{2}}$. The fractional exponent means that you are taking the positive square root of x. Therefore, the expression $\sqrt{3x - 2}$ can also be written as $(3x - 2)^{\frac{1}{2}}$.

15. Sketch the graph of $y = (3x - 2)^{\frac{1}{2}}$. Use your graphing calculator to verify that this is the same as $y = \sqrt{3x - 2}$. What is the domain?

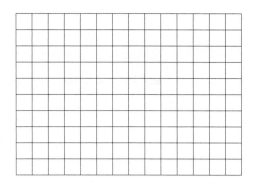

Space and Radicals

16. It is not unreasonable to imagine that some day travel in space will be a common occurrence. According to Einstein's theory of relativity, time would pass more quickly on Earth than it would for someone who is traveling in a space-craft at a velocity close to the speed of light. As a result, a person on Earth would age more rapidly than a space traveler. The formula

$$A = F\sqrt{1 - \frac{v^2}{c^2}}$$

gives the relationship between the aging rate, A, of an astronaut and the aging rate, F, of a person on Earth. The variable v represents the astronauts' velocity in miles per second, and c represents the speed of light (approximately 186,000 miles per second).

a. Suppose you are on a spaceship that is traveling 80% of the speed of light. What is your aging rate compared to a person on Earth?

b. If you are gone for one year, approximately how much time has passed for a person on Earth?

c. Suppose you are traveling at a velocity very close to the speed of light. Substitute c (speed of light) for v in the formula, and simplify. Interpret your results.

17. Escape velocity is the minimum speed that an object must attain to escape a planet's pull of gravity. Escape velocity, V, is given by the formula

$$V = \sqrt{\frac{2Gm}{r}},$$

where G is the universal gravitational constant,

        m is the mass of the planet, and

        r is the radius of the planet.

If the Earth has mass 5.97×10^{24} kilograms and radius 6.37×10^6 meters, then determine the escape velocity for Earth. Round your answer to the nearest whole number. Use $G = 6.67 \times 10^{-11}$ m^3/kg · s^2.

SUMMARY
Activity 5.7

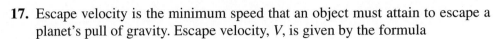

Square Root Notation and Terminology

1. $\sqrt[2]{n}$, or simply $\sqrt{n}$, represents the square root of a nonnegative number n. The 2 is called the *index*. In general, when you are working with square roots, the 2 is omitted.

2. The symbol $\sqrt{}$ is called the *radical sign*. The expression under the radical is called the *radicand*.

3. $\sqrt{n} \geq 0$

4. $\sqrt{a \cdot b} = \sqrt{a} \cdot \sqrt{b}$

5. $\sqrt{\dfrac{a}{b}} = \dfrac{\sqrt{a}}{\sqrt{b}}$

6. $\sqrt{a + b} \neq \sqrt{a} + \sqrt{b}$

Properties of Radical Functions

1. The function defined by $y = \sqrt{g(x)}$ has domain $g(x) \geq 0$. This function is increasing over its entire domain.

2. The function defined by $y = -\sqrt{g(x)}$ has domain $g(x) \geq 0$. This function is decreasing over its domain. The graph of $y = -\sqrt{g(x)}$ is the reflection of the graph of $y = \sqrt{g(x)}$ about the x-axis.

EXERCISES
Activity 5.7

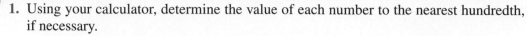

1. Using your calculator, determine the value of each number to the nearest hundredth, if necessary.

   a. $\sqrt{30}$ **b.** $6^{\frac{1}{2}}$ **c.** $(\sqrt{13})^4$ **d.** $(9^{\frac{1}{2}})^3$

2. Determine the domain of each function.

   a. $f(x) = \sqrt{x - 5}$ **b.** $g(x) = \sqrt{3x + 2}$

   c. $h(x) = \sqrt{6 - 2x}$ **d.** $R(x) = -\sqrt{2x}$

3. The following table gives the amount of new money (in billions of dollars) loaned to students in each of the given years.

YEAR	1993	1994	1995	1996	1997	1998	1999	2000
x, NUMBER OF YEARS SINCE 1993	0	1	2	3	4	5	6	7
A(x), AMOUNT OF NEW MONEY FOR STUDENT LOANS (in $ billions)	12.0	18.0	22.0	24.0	25.5	27.2	28.7	30.0

   This data can be modeled by the function

   $$A(x) = 6.8\sqrt{x} + 12.$$

   a. Determine the A-intercept. What does this intercept represent in this situation?

   b. Complete the following table using the given equation. How well does the equation represent the actual data?

x	0	1	2	3	4	5	6	7
A(x)								

c. Sketch a graph of the student loan function represented by $A(x) = 6.8\sqrt{x} + 12$.

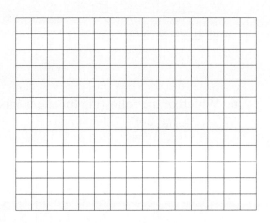

d. Use the model to predict the amount of new money to be loaned to students in 2005.

4. Describe how to obtain the graph of the second function from the graph of the first.

   a. $g(x) = \sqrt{x},\ h(x) = \sqrt{x} + 1$

   b. $f(x) = \sqrt{x},\ g(x) = -\sqrt{x}$

   c. $h(x) = 2\sqrt{x},\ H(x) = 2\sqrt{x} + 1$

5. a. Sketch a graph of $f(x) = x^2$ and $g(x) = \sqrt{x}$ on the same coordinate system. Use the graphs to answer parts b and c.

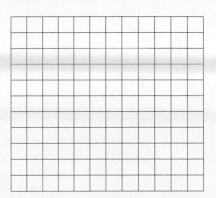

b. Is $x^2 > \sqrt{x}$ for $0 < x < 1$? Explain.

c. Is $x^2 > \sqrt{x}$ for $x > 1$? Explain.

6. Which of the following functions increases more rapidly for $x > 1$: $f(x) = \sqrt{x}$ or $g(x) = \ln(x)$?

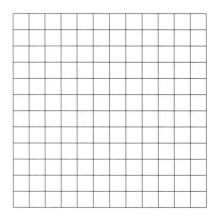

7. For each of the given functions,
    i. determine the domain.
    ii. determine the x- and y- intercepts.
    iii. sketch a graph.

a. $f(x) = -\sqrt{4x + 8}$

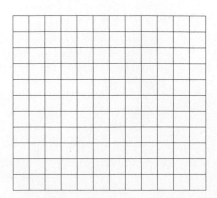

b. $g(x) = \sqrt{5 - x}$

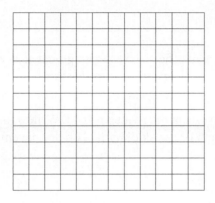

c. $h(x) = \sqrt{x^2 + 9}$

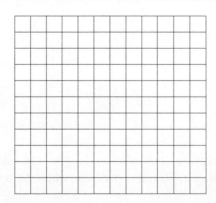

8. The surface area, S, of a cone is given by the formula

$$S = \pi r \sqrt{r^2 + h^2},$$

where r is the radius of the base and h is the height.

An umbrella is in the shape of a cone having radius 4 feet and height 2 feet. Determine the amount of material needed to make the umbrella.

9. A shipping carton in the shape of a rectangular box and has dimensions 12 inches × 24 inches × 17 inches. The diagonal, d, of a rectangular box is given by

$$d = \sqrt{w^2 + l^2 + h^2},$$

where w is the width, l is the length, and h is the height.

Will an umbrella measuring 34 inches long fit in the carton?

10. Law enforcement officers investigating a car accident often use the formula $s = \sqrt{30fl}$ to estimate a car's speed, s, in miles per hour based upon the length, l, of the skid marks. The f in the formula represents the road condition at the time of the accident.

   a. On dry pavement the f-value is 0.85. Write a function rule for speed, s, as a function of the length of the skid marks l on dry pavement.

   b. Estimate the speed of a car if the skid marks on dry pavement are 90 feet long.

   c. What is the practical domain for the function in part a?

   d. Sketch a graph of this function using your graphing calculator.

   e. Use the graph to determine the length of the skid marks on dry pavement if the car was traveling at 70 miles per hour when the brakes were applied.

ACTIVITY 5.8
Falling Objects

OBJECTIVE

1. Solve an equation involving a radical expression using a graphical and algebraic approach.

If an object is dropped from a tall building, the time, t, in seconds, it takes for the object to strike the ground is given by

$$t = \frac{\sqrt{d}}{4} = \frac{1}{4}\sqrt{d},$$

where the input, d, is the distance traveled in feet. The time it takes for the object to hit the ground is directly proportional to the square root of the distance traveled. The number $\frac{1}{4}$ or 0.25 is the constant of proportionality or constant of variation.

1. **a.** How long will it take an object to fall from the top of the Sears Tower in Chicago, a distance of 1450 feet? Round to the nearest hundredth of a second.

b. Complete the following table.

d (ft)	0	100	200	300	500	750	1000
t (sec)							

c. Sketch a graph of the given function. Use the table in part b to determine an appropriate scale.

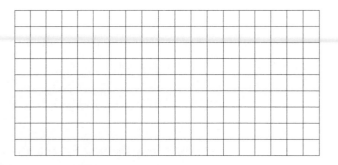

d. How tall must a building be for an object to take 8 seconds to fall to the ground? Use the graph from part c to approximate your answer.

Solving Equations Involving Radical Expressions

Suppose you are interested in determining the value of d for many different values of t. In such a situation, the process could be simplified if you had a rule for d as a function of t. The equation $t = \frac{\sqrt{d}}{4}$ gives t as a function of d. You need to solve this equation for d.

DEFINITION

An equation in which at least one side contains a radical with a variable in the radicand is called a radical equation.

Example 1 *Radical equations include* $t = \frac{\sqrt{d}}{4}$, $\sqrt{2x + 1} = 5$, *and* $2\sqrt{3x} = \sqrt{5x - 7}$.

Solving an equation when the variable appears under a radical involves using the following property of equations.

If a and b are two quantities such that $a = b$, then $a^n = b^n$, where n is a positive integer.

Example 2 *If* $t = \sqrt{s}$, *then apply the preceding property by squaring both sides of the equation.*

$t^2 = (\sqrt{s})^2$ Rewrite $\sqrt{s}$ as $s^{\frac{1}{2}}$.

$t^2 = (s^{\frac{1}{2}})^2$ Apply the property of exponents $(a^m)^n = a^{mn}$.

$t^2 = s^{\frac{1}{2} \cdot 2}$ Simplify.

$t^2 = s^1$

Therefore, if $t = \sqrt{s}$, then $t^2 = s$.

2. a. Now solve the equation $t = \frac{\sqrt{d}}{4}$ for d by first squaring both sides of the equation.

b. Using the new formula from part a, determine how tall a building must be for an object to take 8 seconds to fall to the ground.

c. How does your answer compare to the result in Problem 1d?

d. You could also answer part b by solving the equation $8 = \frac{\sqrt{d}}{4}$. Solve the equation. How does your answer compare to the result in part b.

The following example demonstrates a general algebraic procedure for solving radical equations.

Example 3 *Solve for x:* $2\sqrt{3x} - \sqrt{5x + 7} = 0$

SOLUTION

Step 1. If the equation involves more than one radical term, isolate one radical term on one side of the equation.

$$\begin{aligned} 2\sqrt{3x} - \sqrt{5x + 7} &= 0 \\ + \sqrt{5x + 7} &+ \sqrt{5x + 7} \\ \hline 2\sqrt{3x} &= \sqrt{5x + 7} \end{aligned}$$

Step 2. Square both sides of the equation.

$$(2\sqrt{3x})^2 = (\sqrt{5x + 7})^2$$
$$4 \cdot 3x = 5x + 7$$

Step 3. If a radical remains, repeat steps 1 and 2. Solve the resulting equation.

$$4 \cdot 3x = 5x + 7$$
$$12x = 5x + 7$$
$$7x = +7$$
$$x = 1$$

Step 4. Check all solutions in the original equation.

$$2\sqrt{3(1)} - \sqrt{5(1) + 7} = 2\sqrt{3} - \sqrt{12}$$
$$= 2\sqrt{3} - \sqrt{4 \cdot 3} = 2\sqrt{3} - 2\sqrt{3} = 0$$

You can also check your answer by solving the equation graphically.

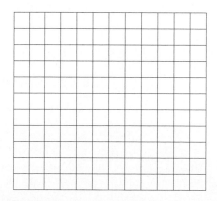

3. Suppose two different objects are dropped: a marble and a large beach ball. Because of air resistance, the beach ball will take longer than the marble to fall the same distance. Assume that the marble falls according to $t = 0.25\sqrt{d}$, as in Problem 1. The time for the beach ball to hit the ground is modeled by $t = k\sqrt{d}$, where the positive constant, k, is determined by experiment.

   The beach ball is dropped from a height of 250 feet, and it takes 4.11 seconds to hit the ground. Determine the constant, k, accurate to the hundredths place. Remember that $(ab)^2 = a^2b^2$.

4. Now suppose the beach ball in Problem 3 is dropped from a height 50 feet lower than the marble. Then $t = 0.26\sqrt{d - 50}$ is the time for the beach ball to drop $d - 50$ feet, where d is the height the marble falls.

   a. Write an equation that can be used to determine from what height the marble must be dropped so the beach ball and marble will hit the ground at the same time.

   b. Solve this equation using an algebraic approach.

   c. Verify your solution in part b using your graphing calculator.

5. Consider the following algebraic solution of the equation $\sqrt{x + 3} + 5 = 0$.

$$\sqrt{x + 3} + 5 = 0$$
$$\sqrt{x + 3} = -5$$
$$(\sqrt{x + 3})^2 = (-5)^2$$
$$x + 3 = 25$$
$$x = 22$$

a. It appears that $x = 22$ is a solution to the given equation. Check the solution by substituting 22 for x in the original equation. Does it check?

b. What happened in the solution process to cause an **extraneous solution** (an apparent solution that does not check) to appear?

c. Does the equation $\sqrt{x + 3} + 5 = 0$ have a solution? Include a graph to help support your answer.

6. The following table gives the amount of new money (in billions of dollars) loaned to students in each of the given years.

	YEAR							
	1993	**1994**	**1995**	**1996**	**1997**	**1998**	**1999**	**2000**
x, NUMBER OF YEARS SINCE 1993	0	1	2	3	4	5	6	7
A(x), AMOUNT OF NEW MONEY FOR STUDENT LOANS (in $ billions)	12.0	18.0	22.0	24.0	25.5	27.2	28.7	30.0

This data can be modeled by the function

$$A(x) = 6.8\sqrt{x} + 12.$$

Determine the year in which the amount of new money for student loans will first exceed $40 billion.

SUMMARY
Activity 5.8

1. If a and b are two quantities such that $a = b$, then $a^n = b^n$ where n is a positive integer.

2. To solve an equation involving one radical expression:

   a. Isolate the radical term on one side of the equation.

   b. Square both sides of the equation.

   c. Solve the resulting equation.

   d. Check all solutions in the original equation.

2. To solve an equation involving more than one radical expression:

   a. If the equation involves more than one radical term, isolate one radical term on one side of the equation.

   b. Square both sides of the equation.

   c. If a radical remains, repeat steps a and b. Solve the resulting equation.

   d. Check all solutions in the original equation.

EXERCISES
Activity 5.8

1. Without actually solving any of the following equations determine which does **not** have a solution. Explain your reasoning, and then verify your answer by solving all three equations.

   a. $\sqrt{2x + 1} = 3$ 　　　 b. $\sqrt{x + 1} + 5 = 1$

   c. $\sqrt{2 - x} = -x$

2. Solve each of the following equations algebraically. Then verify your answers graphically.

a. $\sqrt{x} = 2.5$ **b.** $\sqrt{x} - 3 = 0$ **c.** $\sqrt{2x} = 14$

d. $3\sqrt{x} = 243$ **e.** $4 - 5\sqrt{3x} = 1$ **f.** $\sqrt{x + 1} = 9$

3. Solve each of the following equations algebraically and by graphing. Be aware of any extraneous roots.

a. $\sqrt{x + 5} = 1$ **b.** $10\sqrt{x + 2} = 20$

c. $\sqrt{5 - x} = x + 1$

4. Solve algebraically and graphically: $\sqrt{1.4x + 3.2} = \sqrt{3.8x - 1}$

5. The time, t, in seconds, that it takes for a pendulum to complete one complete period (to swing back and forth one time) is modeled by

$$t = 2\pi\sqrt{\frac{L}{32}},$$

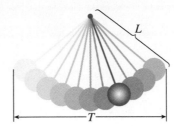

where L is the length of the pendulum, in feet. How long is the pendulum of a clock with a period of 1.95 seconds?

6. In a certain population, there are 28,520 births on a particular day. The number, N, of these people surviving to age x can be modeled by the function $N = 2850\sqrt{100 - x}$.

   a. According to this model, how many of the 28,520 babies will survive to age 5?

   b. What is the practical domain of this function?

   c. When only 5000 of this group are still alive, how old do you expect them to be?

   d. After how many years will half of the original population of 28,520 people remain alive?

7. a. A pressure gauge on a bridge indicates a wind pressure, P, of 10 pounds per square foot. What is the velocity, V, of the wind if

$$V = \sqrt{\frac{1000P}{3}},$$

where velocity is measured in miles per hour.

b. What is the wind pressure if the wind is blowing at 70 miles per hour?

8. Artificial gravity can be created in a space station by revolving the station. The number of revolutions required can be determined by

$$N = \frac{1}{2\pi}\sqrt{\frac{a}{r}},$$

where N is measured in revolutions per second, a is the artificial gravity produced (measured in meters per second squared), and r is the radius of the space station in meters.

a. To produce an artificial acceleration simulating gravity on Earth, a must equal 9.8 meters per second squared. If the space station must revolve at the rate of one revolution every five minutes, what must its radius be? Solve both algebraically and graphically. Be careful, N is measured in revolutions per second.

b. Solve the original formula for r.

c. Use the formula in part b to answer part a again. How do the answers compare?

9. The Masteller formula for calculating the adult body surface area, A, is

$$A = \sqrt{\frac{hw}{3131}},$$

where h is the person's height in inches and w is the adult's weight in pounds. A is the surface area in square meters.

a. Determine the body surface area, A, of an adult who is 70 inches tall and weighs 200 pounds.

b. Solve the formula for w.

ACTIVITY 5.9
Propane Tank

OBJECTIVES

1. Determine the domain of a function defined by an equation of the form $y = \sqrt[n]{g(x)}$, where n is a positive integer and $g(x)$ is a polynomial.

2. Graph $y = \sqrt[n]{g(x)}$.

3. Identify the properties of graphs of $y = \sqrt[n]{g(x)}$.

4. Solve radical equations that contain radical expressions with an index other than 2.

A propane tank is in the shape of a sphere. The radius, r, of the spherical tank is given by the formula

$$r = \sqrt[3]{\frac{3V}{4\pi}},$$

where V is the volume of the tank.

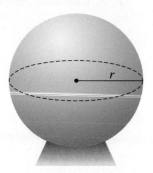

1. **a.** What is the radius of a propane tank having volume 50 cubic feet (round the answer to the nearest tenth)?

   b. What is the practical domain of the radius function?

   c. Complete the following table.

V (ft³)	0	5	10	15	20	25
r (ft)						

   d. Sketch a graph of the radius function over its practical domain.

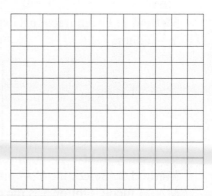

Graphs of $y = \sqrt[n]{g(x)}$, $n = 1, 2, 3, 4, 5$

The investigation of the properties of the radius function defined by $r = \sqrt[3]{\dfrac{3v}{4\pi}}$ is somewhat limited by the restrictions placed on the variables V and r.

2. a. Consider the cube root function defined by $f(x) = \sqrt[3]{x}$. What is the domain of f?

b. Complete the following table.

x	-10	-7	-4	-1	0	1	4	7	10
$f(x)$									

c. Sketch a graph of f. Verify your sketch using your graphing calculator.

d. Is the function f increasing or decreasing?

There are two different ways to enter the cube root function on your calculator.

Method 1: Using fractional exponents, enter the cube root of x as $x^\wedge(1/3)$ in the Y = editor.

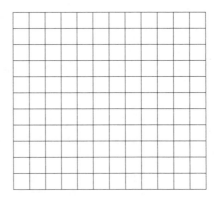

Method 2: Select the Y = editor, highlight the Y_n you want, and then select the MATH menu. Option 4 is the cube root.

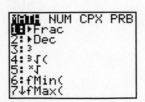

Select option 4 and press (ENTER). Insert the argument x and the right parenthesis to complete the function.

3. Use your graphing calculator to verify that the graphs of $y = x^{\frac{1}{3}}$ and $y = \sqrt[3]{x}$ are identical. Your screens should appear as follows:

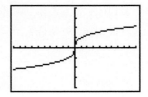

4. Consider functions defined by $f(x) = x^3$ and $g(x) = \sqrt[3]{x}$.

   a. Determine the domain of each function.

   b. Complete the following table.

x	−10	−7	−4	−1	0	1	4	7	10
f(x)									
g(x)									

   c. Sketch a graph of f and g on the same coordinate system. Verify your sketch using your graphing calculator.

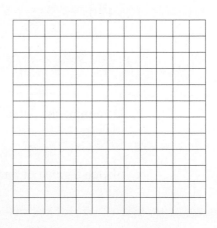

   d. Are f and g inverse functions? Explain.

   e. Determine the composition of f and g. That is, determine $f(g(x))$ and $g(f(x))$.

5. a. Sketch a graph of each of the following.

$$f(x) = \sqrt[3]{x}, \; g(x) = \sqrt[3]{x + 1} \text{ and } h(x) = \sqrt[3]{x - 2}$$

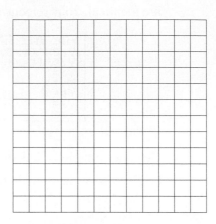

b. How are the graphs of f, g, and h similar, and how are they different?

6. Given the functions f and g defined by

$$f(x) = \sqrt[4]{x - 1} \text{ and } g(x) = \sqrt[5]{x}.$$

a. Complete the following table.

x	−10	−5	−1	0	1	5	10
f(x)							
g(x)							

b. Determine the domains of f and g.

c. Sketch a graph of f and g on the same coordinate axes.

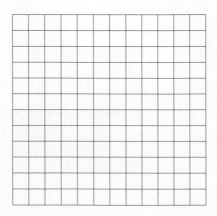

Solving Equations Involving Radical Expressions

Solving an equation such as $\sqrt[3]{x + 1} + 5 = 9$ is similar to solving equations involving square roots.

Example 1   *Solve* $\sqrt[3]{x + 1} + 5 = 9$.

SOLUTION

Step 1. Isolate the radical term on one side of the equation.

$$\sqrt[3]{x + 1} + 5 = 9$$
$$\sqrt[3]{x + 1} = 4$$

Step 2. Raise each side of the equation to the power that matches the index of the radical. In this situation, cube each side and simplify.

$$(\sqrt[3]{x + 1})^3 = 4^3$$

Step 3. If a radical remains, repeat steps 1 and 2. Solve the resulting equation.

$$x + 1 = 64$$
$$x = 63$$

Step 4. Check all solutions in the original equation.

$$\sqrt[3]{63 + 1} + 5 = 9$$
$$\sqrt[3]{64} + 5 = 9$$
$$4 + 5 = 9$$
$$9 = 9$$

You can also verify your solutions graphically.

 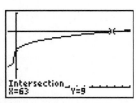

7. **a.** Returning to the propane tank situation, suppose the amount of space available for a propane tank is 10 feet. Write an equation to determine the volume of a spherical tank that can fit into the given space.

   b. Solve this equation using an algebraic approach.

c. Verify the solution using your graphing calculator. The screen should appear as follows.

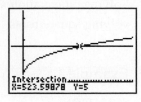

8. Solve the formula $V = I^3$ for I.

9. The basal metabolic rate (BMR) is the number of calories per day a person needs to maintain life. A person's basal metabolic rate is a function of his or her weight and is modeled by

$$B(w) = 70\sqrt[4]{w^3},$$

where $B(w)$ represents the basal metabolic rate measured in calories per day and w is the person's weight in kilograms.

a. Write the expression $70\sqrt[4]{w^3}$ using fractional exponents.

b. If your friend weighs 50 kilograms (approximately 110 pounds), determine her basal metabolic rate. Round your answer to the nearest calorie.

c. Determine your basal metabolic rate. Be sure to convert your weight to kilograms.

d. Suppose a person is on a 2000-calorie-per-day diet. If the number of calories represents the person's basal metabolic rate, write an equation to determine the weight that is associated with this number of calories per day.

e. Solve the equation in part d. To help determine to what power you need to raise each side, first consider how you simplify the expression $\left(x^{\frac{3}{4}}\right)^{\frac{4}{3}}$.

f. If the person weighs 210 pounds, is the 2000-calorie diet healthy?

SUMMARY
Activity 5.9

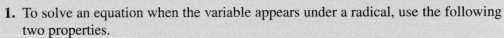

1. To solve an equation when the variable appears under a radical, use the following two properties.

   i. If a and b are two quantities such that $a = b$, then $a^n = b^n$, where n is a positive integer.

   ii. $\left(b^{\frac{1}{n}}\right)^n = b^1$ and $\left(b^{\frac{m}{n}}\right)^{\frac{n}{m}} = b^1$

2. **i.** The domain of $f(x) = \sqrt[x]{x}$, where x is a positive odd integer that is all real numbers.

   ii. The domain of $f(x) = \sqrt[n]{x}$, where n is a positive even integer and $x \geq 0$.

3. To solve equations involving radicals:

   Step 1. Isolate the radical term on one side of the equation.

   Step 2. Raise each side of the equation to the power that matches the index of the radical.

   Step 3. If a radical remains, repeat steps 1 and 2. Solve the resulting equation.

   Step 4. Check all solutions in the original equation.

EXERCISES
Activity 5.9

1. If possible, determine the exact value of each of the following.

   a. $\sqrt[3]{64}$ **b.** $\sqrt[4]{16}$ **c.** $(-27)^{\frac{1}{3}}$ **d.** $(625)^{\frac{1}{4}}$

   e. $\sqrt{\frac{1}{36}}$ **f.** $(-81)^{\frac{1}{4}}$ **g.** $(100{,}000)^{\frac{1}{5}}$ **h.** $(-1)^{\frac{1}{6}}$

2. If the volume of a cube is 728 cubic centimeters, to the nearest tenth of a centimeter, what is the length of one edge?

3. If the volume of a cube is decreased from 1450 cubic inches to 1280 cubic inches (and still remains a cube), by how much has the length of one edge decreased?

4. The volume of a sphere is 520 cubic meters. What is the diameter of the sphere?

5. What is the domain of each function?

  a. $y = \sqrt[3]{x + 6}$

  b. $f(x) = \sqrt[4]{x - 3}$

  c. $g(x) = \sqrt[5]{2 - x}$

  d. $f(x) = (2 - x)^{\frac{1}{6}}$

6. Solve each of the following algebraically and graphically.

  a. $\sqrt[3]{x + 4} = 3$

  b. $\sqrt[4]{x + 5} = 2$

  c. $\sqrt[3]{2x - 3} + 4 = 3$

  d. $\sqrt[4]{2 - x} = 5$

7. Solve each of the following algebraically, and verify your results graphically.

    a. $x^{\frac{2}{3}} = 16$ **b.** $2x^{\frac{3}{4}} = 54$

8. **a.** The diameter d of a sphere is given by the formula

$$d = \sqrt[3]{\frac{6v}{\pi}},$$

    where v represents the volume of a sphere. Approximate the diameter of a sphere having a volume of 10 cubic inches.

    b. Determine the volume of a sphere having diameter 5 feet.

9. The radius, r, of a sphere is given by

$$r = \sqrt[3]{\frac{3V}{4\pi}},$$

where V is the volume of the sphere.

    a. Determine the radius of a sphere that has a volume equal to 40 cubic centimeters.

    b. Determine the volume of a sphere that has a radius equal to 3.5 feet.

    c. Solve the formula for V, expressing volume as a function of the radius.

CLUSTER 2 **What Have I Learned?**

1. Explain the steps involved in solving the equation $\sqrt{2x + 3} = 5$ using an algebraic approach. Why must you be sure to check your solution?

2. What is the domain for the variable b in each of the following?

   a. $\sqrt[n]{b}$, where n is an even positive integer

   b. $\sqrt[n]{b}$, where n is an odd positive integer

3. Is it possible for an extraneous solution to appear when the equation $\sqrt[3]{2x + 1} = -3$ is being solved? Explain.

4. Describe two ways to check for extraneous solutions when you are solving an equation by squaring both sides.

5. Determine whether the following statements are true or false. Explain your answer.

   a. If two numbers are equal, then their squares are equal.

   b. If the squares of two numbers are equal, then the two numbers are equal.

6. a. For a given value of x, which is greater, $\sqrt[5]{x}$ or $x^{\frac{1}{3}}$? Explain how you determine your answer.

   b. In part a, did you assume that $x > 0$? Does your answer change if $x < 0$?

CLUSTER 2   How Can I Practice?

1. Solve each of the following equations algebraically and check graphically.

   a. $\sqrt{x + 2} = 10$ **b.** $(x - 5)^{\frac{1}{2}} = 6$ **c.** $\sqrt{2x + 1} - 5 = 0$

   d. $\sqrt[3]{x^2 + 3} = 4$ **e.** $\sqrt{x} = \sqrt{x + 2}$ **f.** $(2 - x)^{\frac{1}{3}} = -2$

   g. $\sqrt[4]{2x - 5} = 2$ **h.** $(2.3x + 1.9)^{\frac{1}{3}} = 1.6$

2. Identify the domain of each of the following functions.

   a. $f(x) = \sqrt{6 - x}$ **b.** $g(x) = (2x - 9)^{\frac{1}{3}}$ **c.** $h(x) = (x^2 - 4)^{\frac{1}{4}}$

3. If the volume of a cube is 458 cubic inches, what is the length of one edge? Determine the value to the nearest hundredth of an inch.

4. If the volume of a sphere is 620 cubic centimeters, what is its radius?

5. When a stone is dropped to the ground, its velocity is modeled by the function $v = \sqrt{64d}$, where d is the distance the stone has fallen, in feet, and v is its velocity in feet per second. If the stone hits the ground at 100 feet per second, from what height was it dropped?

6. A cardboard box with a square bottom has a height of 10 inches and a volume of 422.5 cubic inches. What are the dimensions of the bottom of the box?

7. Describe the similarities and differences between the graphs of $y = \sqrt{2 - x}$ and $y = \sqrt{x - 2}$.

Answers to all How Can I Practice exercises are included in the Selected Answers appendix.

Summary

The bracketed numbers following each concept indicate the activity in which the concept is discussed.

CONCEPT/SKILL	DESCRIPTION	EXAMPLE
Domain and range of $y = \frac{k}{x}$ [5.1]	The domain and range consist of all real numbers except zero.	$y = \frac{3}{x}$
Graph of $y = \frac{k}{x}$ [5.1]	The graph is in the first and third quadrants if $k > 0$ and in the second and fourth if $k < 0$.	$y = \frac{3}{x}$ 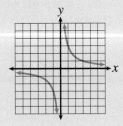
Asymptotes of the graph of $y = \frac{k}{x}$ [5.1]	The y-axis, $x = 0$, is the vertical asymptote. The x-axis, $y = 0$, is the horizontal asymptote.	$y = \frac{3}{x}$ 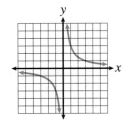
Domain of $y = \frac{k}{x^n}$ [5.2]	The domain consists of all real numbers except zero.	$y = \frac{4}{x^3}$
Graph of $y = \frac{k}{x^n}$ [5.2]	The graphs will vary depending on the values of k and n.	See the summary at the end of Activity 5.2.
Asymptotes of the graph of $y = \frac{k}{x^n}$ [5.2]	The y-axis, $x = 0$, is the vertical asymptote. The x-axis, $y = 0$, is the horizontal asymptote.	$y = \frac{4}{x^3}$ 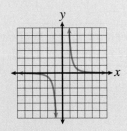
Inverse variation functions [5.2]	Functions defined by $y = \frac{k}{x^n}$ are called inverse variation functions in which y is said to vary inversely as the nth power of x; k is called the constant of variation.	For the function $y = \frac{4}{x^3}$, y varies inversely as the cube of x, and 4 is the constant of variation.
Rational function [5.3]	A function Q, defined by an equation of the form $Q(x) = \frac{k}{g(x)}$, where k is a nonzero constant and $g(x)$ is a polynomial, belongs to the family of functions known as rational functions.	$f(x) = \frac{10}{x - 3}$

Domain of a rational function [5.3]	The domain of the rational function is the set of all real numbers except those values of the input x such that $g(x) = 0$.	The domain of $f(x) = \frac{10}{x-3}$ is all real numbers except 3.
Vertical asymptote of a rational function [5.3]	The vertical asymptote is the vertical line that passes through the x-value for which $g(x) = 0$.	The vertical asymptote of $f(x) = \frac{10}{x-3}$ is $x = 3$.
Horizontal asymptote of a rational function [5.3]	The horizontal asymptote is the x-axis $(y = 0)$.	$f(x) = \frac{10}{x-3}$
Rational equations [5.4]	Method 1. To solve an equation of the form $\frac{f(x)}{g(x)} = \frac{a}{b}$, where $g(x) \neq 0$, and $b \neq 0$, multiply both sides of the equation by the product $b \cdot g(x)$, and solve the resulting equation for x.	See Example 1, Activity 5.4, page 564
Rational equations [5.4]	Method 2. To solve an equation of the form $\frac{f(x)}{g(x)} = \frac{a}{b}$, where $g(x) \neq 0$, and $b \neq 0$, cross-multiply to obtain $b \cdot f(x) = a \cdot g(x)$, and solve the resulting equation for x.	See Example 1, Activity 5.4, page 564
Horizontal asymptotes of rational functions [5.4]	Suppose the output values of a rational function R get closer and closer to a number a as the input values increase infinitely in both the positive and negative directions. Then, the graph of the function R has a horizontal asymptote. The equation of the horizontal asymptote is $y = a$.	The horizontal asymptote of $R(x) = \frac{3x+1}{x-2}$ is $y = 3$.
Determine an LCD of two or more expressions [5.5]	1. Write the prime factorization of each denominator. Express repeated factors as powers. 2. Identify the different bases (factors) in step 1. 3. Write LCD as the product of the highest power of each of the different factors from step 2.	Example 1, Activity 5.5, page 577
Solving an equation involving rational expressions [5.5]	1. Determine the LCD of all denominators in the equation. 2. Multiply each side of the equation by the LCD, and simplify the resulting equation. 3. Solve the resulting equation for the desired variable.	See Appendix A.

Simplifying rational expressions [5.6]	1. Factor the numerator and the denominator. 2. Divide the numerator and the denominator by the common factors.	See Appendix A.
Multiplying or dividing rational expressions [5.6]	1. Factor the numerator and denominator of each fraction completely. 2. Divide out the common factors (cancel). 3. Multiply remaining factors. 4. In division, proceed as above after inverting the divisor (the fraction after the division sign).	See Appendix A.
Adding or subtracting rational expressions [5.6]	1. Determine the LCD. 2. Build each fraction to have the LCD. 3. Add or subtract numerators. 4. Place the numerator over the LCD, and simplify if necessary.	See Appendix A.
Simplifying a complex fraction [5.6]	1. Express the numerator as a single fraction. 2. Express the denominator as a single fraction. 3. Divide the numerator by the denominator. 4. Simplify, if possible.	See Example 2, Activity 5.6, page 586
Radical functions [5.7]	The function defined by $y = \sqrt{g(x)}$ has domain $g(x) \geq 0$. This function increases over its entire domain.	$y = \sqrt{2x + 1}$ 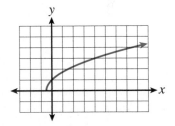
Radical functions [5.7]	The function defined by $y = -\sqrt{g(x)}$ has domain $g(x) \geq 0$. This function decreases over its domain.	$y = -\sqrt{2x + 1}$ 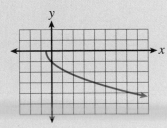
Solving an equation involving one radical expression [5.8]	1. Isolate the radical term on one side of the equation. 2. Square both sides of the equation. 3. Solve the resulting equation. 4. Check all solutions in the original equation.	See Appendix A.

Solving an equation involving more than one radical expression [5.8]	1. If the equation involves more than one radical term, isolate one radical term on one side of the equation.	See Example 3, Activity 5.8, page 613
	2. Square both sides of the equation.	
	3. Solve the resulting equation. If a radical remains, repeat steps 1 and 2.	
	4. Check all solutions in the original equation.	

| **Domain of a function defined by an equation of the form** $y = \sqrt[n]{g(x)}$, **where** n **is a positive integer and** $g(x)$ **is a polynomial** [5.9] | The domain of $y = \sqrt[n]{g(x)}$ is the domain of g if n is an odd positive integer. The domain of $y = \sqrt[n]{g(x)}$ is all real numbers for which $g(x) \geq 0$ if n is an even positive integer. | $y = \sqrt[3]{x + 3}$ 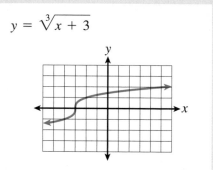 |

Solving radical equations that contain radical expressions with an index other than 2 [5.9]	1. Isolate the radical term on one side of the equation.	See Example 1, Activity 5.9, page 625
	2. Raise each side of the equation to the power that matches the index of the radical.	
	3. If a radical remains, repeat steps 1 and 2. Solve the resulting equation.	
	4. Check all solutions in the original equation.	

Gateway Review

1. According to the blueprint, the floor area of the stage in the new auditorium at your college must be rectangular and equal to 1200 square feet. The width of the stage is key to all of the theater productions. Therefore, in this situation, the stage's depth is a function of its width.

   a. Let d represent the depth and w represent the width. Write an equation that expresses d as a function of w.

   b. Complete the following table using the equation from part a.

w	30	35	40	50	60
d					

   c. What happens to the depth as the width increases?

   d. What happens if the width is 100 feet? Is this realistic? Explain.

   e. Can the width be zero? Explain.

   f. What do you think is the practical domain for this function?

   g. What type of a function do you have in this situation?

   h. What is the domain of the general function?

   i. What is the vertical asymptote?

   j. What is the horizontal asymptote? Explain in words how you determined it.

Answers to all Gateway exercises are included in the Selected Answers appendix.

2. Sketch the following graphs without using your graphing calculator.

    a. $f(x) = \dfrac{1}{x^2}$ **b.** $g(x) = \dfrac{-1}{x^3}$

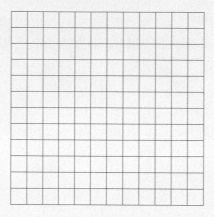
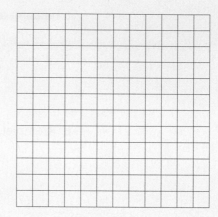

    c. Describe how the graphs are similar and how they are different.

3. a. If y varies inversely as x, and $x = 10$ when $y = 12$, then determine the value of y when $x = 30$.

    b. The loudness, in decibels, of a stereo is inversely proportional to the square of the distance from the speaker to the person listening. If the loudness is 32 decibels at a distance of 4 feet, then what is the loudness when the listener is 10 feet from the speaker?

    c. When the volume of a cylinder is constant, the height varies inversely as the square of the radius. If the radius is 2 inches when the height is 8 inches, determine the height when the radius is 5 inches.

4. Determine the horizontal and vertical asymptotes and the intercepts of each of the following. Then sketch a graph of each function.

a. $f(x) = \dfrac{10}{x^3}$ **b.** $g(x) = \dfrac{4}{x-3}$ **c.** $f(x) = \dfrac{2x}{x+2}$

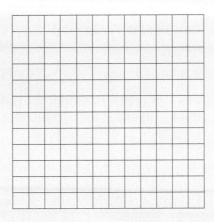

5. Students from the local community college plan to celebrate their ten-year reunion. They rent a restaurant for an evening of entertainment. The fee for the band is $600, and food will cost each person $45.

a. If $f(n)$ represents the total cost for n people to participate in the reunion, write an equation for the total cost.

b. If 100 people attend, what will be the total cost of the event?

c. Determine a function, $A(n)$, that will represent the average cost per person to attend the event.

d. If 100 people attend the reunion, how much will each person pay?

639

e. Use your graphing calculator to complete the following table.

n (no. of people attending)	50	100	150	200	250
A(n) (cost per person)					

f. If the committee thinks that each person would pay at most $50, how many would have to attend for the cost to be $50? Show your answer algebraically, and check it using your graphing calculator.

g. Determine the practical domain of your function.

h. From the graph, determine the vertical asymptote. Is there a practical meaning of this asymptote in this situation?

i. From the graph, determine the horizontal asymptote. Is there a practical meaning of this asymptote in this situation?

6. a. Solve the equation $\frac{4}{x-2} + 3 = 9$ using an algebraic approach. Verify your answer graphically.

b. When you graph the function $f(x) = \frac{4}{x-2} - 6$, what do you discover about the solution to the equation in part a and the x-intercept of the graph of the function f? Explain.

7. Solve each of the following equations using an algebraic approach. Verify your solutions graphically.

a. $\dfrac{3}{x+2} = 5$

b. $\dfrac{-2x}{3x-4} = 2$

8. The local grocery store has just hired you and your friend. You can stock a shelf in 15 minutes. Your friend will take 20 minutes to do the shelf. How long will it take to stock the shelf if you both work together?

9. You work in the admissions office at your community college. You must assemble all of the packets for the placement test sessions. You work with a friend who takes twice as long as you do to assemble the packets. If you work together, the packets can be completed in 45 minutes. On the day you must assemble the packets, you have a big exam. How many hours does it take your friend to do the job?

10. Solve each of the following equations algebraically. Verify your answers graphically.

   a. $\frac{1}{6} - \frac{3}{2x} = \frac{1}{5x}$ **b.** $\frac{2}{x} + \frac{3}{4x} = \frac{1}{12}$

11. Solve each of the following equations for the indicated variable. Express your answer as a single fraction.

   a. Solve $S = \frac{C}{1-r}$ for r.

   b. Solve $\frac{1}{a} + \frac{3}{b} = \frac{4}{c}$ for b.

12. Simplify each of the following complex expressions.

   a. $\dfrac{\frac{1}{a} + \frac{2}{b}}{\frac{2}{a} + \frac{1}{b}}$ **b.** $\dfrac{1 + \frac{1}{x-2}}{1 - \frac{3}{x+2}}$

13. A camera lens possesses a measurement called the focal length, f. When an object is in focus, the focal length is related to the distance of the object from the lens, p, and the image distance from the lens, q, by the formula

$$f = \frac{1}{\frac{1}{p} + \frac{1}{q}}.$$

a. Determine f if p is 4 meters and q is 3 meters.

b. Simplify the complex fraction on the right side of the original formula.

c. Redo part a using the new formula from part b. How do the answers compare?

14. a. What is the domain of the function defined by $f(x) = \sqrt{x + 4}$.

b. Sketch the graph of the function f.

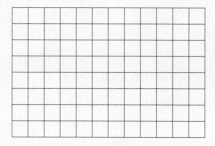

c. As the input increases, what happens to the output values?

d. What is the range of this function?

e. Are there any intercepts? If so, what are they?

f. How is the function f similar to the function $g(x) = \sqrt{x - 4}$?

g. How is the graph of the function f similar to $h(x) = -\sqrt{x + 4}$?

15. a. Draw the graph of $f(x) = \sqrt{x}$.

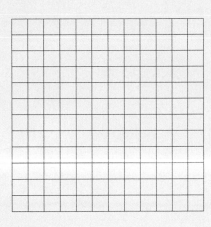

b. Determine the equation of the inverse of the function f.

c. Sketch the graph of the inverse on the same axes as the graph of f.

d. From the graphs, describe how you know that they are inverses.

e. Show that f and f^{-1} are inverses algebraically.

16. For each of the given functions,
  i. determine the domain and range.
  ii. determine the x- and y-intercepts.
  iii. sketch a graph.

a. $f(x) = \sqrt{x} + 4$ **b.** $f(x) = \sqrt{x + 4}$

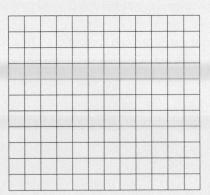 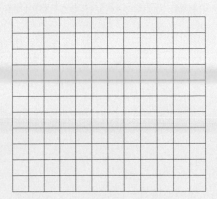

17. Solve each of the following equations using an algebraic approach. Verify your solutions graphically.

a. $\sqrt{3x - 2} - 6 = -2$

b. $\sqrt{2x + 1} - \sqrt{x + 7} = 0$

c. $\sqrt[3]{5x + 4} = 3$

d. $\sqrt{4x + 8} - 3 = -5$

e. $x^{\frac{4}{3}} = 81$

18. What is the domain of each function?

a. $y = \sqrt[3]{x + 8}$

b. $y = \sqrt[4]{x - 6}$

c. $y = (x + 1)^{\frac{1}{6}}$

19. A submarine periscope must be a certain distance above the water for it to be used to locate a ship a certain number of miles away. The equation for the distance (in miles) that the submarine periscope can see is $d = \sqrt{1.5h}$, where h represents the height (in feet) above the surface of the water.

How far above the surface of the water would the periscope have to be to see a ship that is 6 miles away?

20. A ring is dropped from the American span of the Thousand Island Bridge and hits the water 3.1 seconds later. What is the height of the bridge? Use the formula $T = \sqrt{\frac{d}{16}}$, where d represents the distance in feet and T represents the time in seconds.

Introduction to the Trigonometric Functions

Introducing the Sine, Cosine, and Tangent Functions

ACTIVITY 6.1

The Leaning Tower of Pisa

OBJECTIVES

1. Identify the sides and corresponding angles of a right triangle.

2. Determine the length of the sides of similar right triangles using proportions.

3. Determine the sine, cosine, and tangent of an angle using a right triangle.

4. Determine the sine, cosine, and tangent of an acute angle by using the graphing calculator.

During a trip to Italy, you visit a wonder of the world, the Leaning Tower of Pisa. Your guidebook explains that the tower now makes an 85-degree angle with the ground and measures 179 feet in high. If you drop a stone straight down from the top of the tower, how far from the base will it land?

To answer this question more accurately, you need a branch of mathematics called **trigonometry**. Although the development of trigonometry is generally credited to the ancient Greeks, there is evidence that the ancient Egyptian cultures used trigonometry in constructing the pyramids.

Right Triangles

As you will see, the accurate answer to the Leaning Tower question requires some knowledge of **right triangles**. Consider the following right triangle with angles A, B, and C and sides a, b, and c.

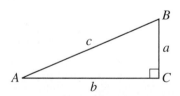

DEFINITION

Angle C is the right angle, the angle measuring 90°. The side opposite the right angle, c, is called the **hypotenuse**. Side a is said to be **opposite** angle A because it is not part of angle A. Side b is said to be **adjacent** to angle A because it and the hypotenuse form angle A. Similarly, side b is the side opposite angle B, and side a is the side adjacent to angle B.

Note that in any right triangle, the lengths of the sides are related by the Pythagorean theorem:

$$c^2 = a^2 + b^2.$$

In words, the square of the hypotenuse is equal to the sum of the squares of the other two sides.

Example 1 *Consider the following right triangle.*

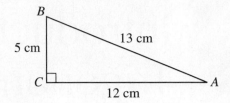

The side opposite angle B is 12 centimeters long. The side opposite angle A is 5 centimeters long. The side adjacent to angle B is 5 centimeters long. The side adjacent to angle A is 12 centimeters long. The hypotenuse is 13 centimeters long.

1. Consider the following right triangle.

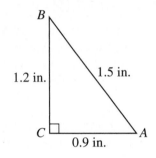

Determine the length of each of the following.

a. the side opposite angle B **b.** the side adjacent to angle B

c. the side opposite angle A **d.** the side adjacent to angle A

e. the hypotenuse

Similar Triangles

Consider the following right triangles whose angles are the same but whose sides are different lengths. These three triangles are called **similar triangles**.

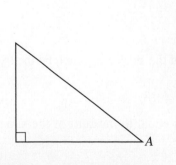
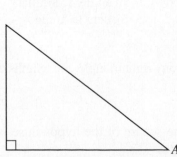
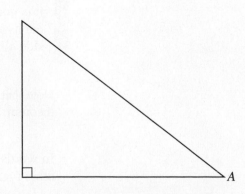

Appendix

2. a. Using a protractor, estimate the measure of angle A to the nearest degree. (For a review of angles measured in degrees, see Appendix B.) Because $0° \leq A < 90°$, angle A is called an **acute angle**.

b. Use a metric or English ruler to complete the following table.

	LENGTH OF THE HYPOTENUSE	LENGTH OF THE SIDE OPPOSITE A	LENGTH OF THE SIDE ADJACENT TO A
SMALL TRIANGLE			
MIDSIZE TRIANGLE			
LARGE TRIANGLE			

3. a. Use the information in the preceding table in Problem 2b to complete the ratios in the following table with respect to angle A. Write each ratio as a decimal rounded to the nearest tenth.

	LENGTH OF THE SIDE OPPOSITE ANGLE A / LENGTH OF THE HYPOTENUSE	LENGTH OF THE SIDE ADJACENT TO ANGLE A / LENGTH OF THE HYPOTENUSE	LENGTH OF THE SIDE OPPOSITE ANGLE A / LENGTH OF THE SIDE ADJACENT TO ANGLE A
SMALL TRIANGLE	0.6	0.8	0.8
MIDSIZE TRIANGLE			
LARGE TRIANGLE			

b. What do you observe about the ratio $\dfrac{length\ of\ the\ side\ opposite\ angle\ A}{length\ of\ the\ hypotenuse}$ for each of the three right triangles?

The table in Problem 3a illustrates the geometric principle that **corresponding sides of similar triangles are proportional**. Recall that the ratios $\frac{a}{b}$ and $\frac{c}{d}$ are proportional if $\frac{a}{b} = \frac{c}{d}$.

4. Consider another right triangle in which the measure of angle A is not the same as the measure of angle A in the three similar triangles from Problems 2 and 3.

a. Using a protractor, estimate the measure of angle A to the nearest degree.

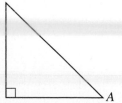

b. For this new triangle, use your ruler to complete the following table with respect to angle A.

	LENGTH OF THE SIDE OPPOSITE ANGLE A / LENGTH OF THE HYPOTENUSE	LENGTH OF THE SIDE ADJACENT TO ANGLE A / LENGTH OF THE HYPOTENUSE	LENGTH OF THE SIDE OPPOSITE ANGLE A / LENGTH OF THE SIDE ADJACENT TO ANGLE A
NEW TRIANGLE			

c. Are the ratios for the new triangle the same as the ratios for the three similar triangles?

d. What changed from the similar triangles to the new triangle to make the ratios change?

Sine, Cosine, and Tangent Functions

As demonstrated in Problem 4, the ratios of the sides of a right triangle are dependent on the size of the angle A. If the angle changes, the ratios change. This fact is fundamental to trigonometry.

The ratios of the sides of a right triangle are a function of the size of an acute angle.

The ratios are given special names and are defined as follows.

DEFINITION

Let A be an acute angle (less than $90°$) of a right triangle. The **sine**, **cosine**, and **tangent** of angle A are defined by

$$\text{Sine of } A = \sin A = \frac{length\ of\ the\ side\ opposite\ A}{length\ of\ the\ hypotenuse},$$

$$\text{Cosine of } A = \cos A = \frac{length\ of\ the\ side\ adjacent\ to\ A}{length\ of\ the\ hypotenuse}, \text{ and}$$

$$\text{Tangent of } A = \tan A = \frac{length\ of\ the\ side\ opposite\ A}{length\ of\ the\ side\ adjacent\ to\ A},$$

where sin, cos, and tan are the standard abbreviations for sine, cosine, and tangent, respectively.

• Sine, cosine and tangent are called **trigonometric functions**.
• Note that the input values for the trigonometric functions are angles and the output values are ratios.

Example 2   *Consider the following large triangle with the given dimensions*

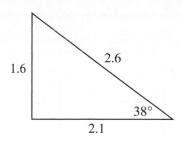

$$\sin 38° = \frac{1.6}{2.6} \approx 0.615 \text{ or } 0.6 \text{ (nearest tenth)}$$

$$\cos 38° = \frac{2.1}{2.6} \approx 0.807 \text{ or } 0.8 \text{ (nearest tenth)}$$

$$\tan 38° = \frac{1.6}{2.1} \approx 0.761 \text{ or } 0.8 \text{ (nearest tenth)}$$

Therefore, for any size right triangle with a 38° angle as one of its acute angles, the sine of 38° is always approximately 0.615, the cosine of 38° is always approximately 0.807 and the tangent of 38° is always approximately 0.761.

You can use your graphing calculator to evaluate sin 38°. Make sure the calculator is in degree mode. Press the ⌈SIN⌉ key, followed by ③, ⑧, ⌐⌐ and ⌈ENTER⌉.

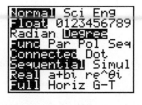 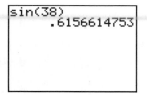

5. a. The sum of the three angles in a right triangle is 180°. If one of the acute angles measures 38°, the other acute angle is 52°. Use the accompanying figure to determine each of the following.

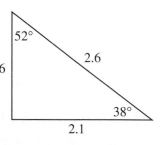

   i. sin 52°

   ii. cos 52°

   iii. tan 52°

b. Verify your answers in part a using your graphing calculator.

Example 3 *Consider the following right triangle where A and B are acute angles.*

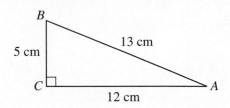

$$\sin A = \frac{5}{13}, \cos A = \frac{12}{13}, \tan A = \frac{5}{12}, \sin B = \frac{12}{13}, \cos B = \frac{5}{13}, \tan B = \frac{12}{5}$$

6. Consider the following right triangle.

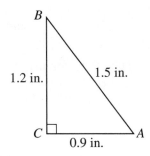

Calculate the following.

a. $\sin A$ **b.** $\cos A$ **c.** $\tan A$

d. $\sin B$ **e.** $\cos B$ **f.** $\tan B$

Tower of Pisa Problem Revisited

You are now ready to answer the original Tower of Pisa problem.

7. a. Recall that the tower makes an 85° angle with the ground and the tower is 179 feet tall. Construct a right triangle that satisfies these conditions.

b. With respect to the 85° angle, is the height of the tower represented by the length of an opposite side, the length of an adjacent side, or the length of the hypotenuse of your triangle?

c. You want to determine how far from the base of the tower the stone hits the ground. Therefore, you want to determine the length of which side of the triangle with respect to the 85° angle?

d. Which trigonometric function relates the side with the length you know and the side with the length you want to know?

e. Write an equation using the information in parts a–d.

f. Using your calculator to evaluate $\cos 85°$, solve the equation in part e.

8. Solve the following equations. Round your answers to the nearest tenth.

    a. $\sin 24° = \frac{x}{10}$ **b.** $\cos 63° = \frac{x}{23.5}$ **c.** $\tan 48° = \frac{16}{x}$

SUMMARY
Activity 6.1

1. The **trigonometric functions** are functions whose inputs are measures of the acute angles of a right triangle and whose outputs are ratios of the lengths of the sides of the right triangle.

2. The three sides of a right triangle are the **adjacent** side, the **opposite** side, and the **hypotenuse**. The hypotenuse is always the side opposite the right (90°) angle. The other two sides vary, depending on which angle is used as the input.

3. The **sine**, **cosine**, and **tangent** of the acute angle A of a right triangle are defined by

$$\sin A = \frac{\text{length of the side opposite A}}{\text{length of the hypotenuse}}$$

$$\cos A = \frac{\text{length of the side adjacent to A}}{\text{length of the hypotenuse}}, \text{ and}$$

$$\tan A = \frac{\text{length of the side opposite A}}{\text{length of the side adjacent to A}}.$$

For additional practice working with the trigonometric functions in right triangles and special triangles, see Appendix B.

1. Triangle ABC is a right triangle.

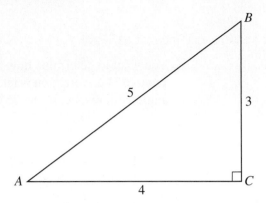

Determine each of the following.

a. $\sin A$ **b.** $\sin B$ **c.** $\cos A$

d. $\cos B$ **e.** $\tan A$ **f.** $\tan B$

2. In a certain right triangle, $\sin A = \frac{24}{25}$.

a. Determine possible lengths of the three sides of a right triangle. *Hint:* Use the Pythagorean theorem, $c^2 = a^2 + b^2$, to determine the length of any unknown side.

b. Determine $\cos A$.

c. Determine $\tan A$.

3. In a certain right triangle, $\tan B = \frac{7}{4}$.

a. Determine possible lengths of the three sides of a right triangle.

b. Determine $\sin B$.

c. Determine $\cos B$.

4. Consider the accompanying right triangle.

   a. Which of the trigonometric functions relates angle A and sides a and x?

   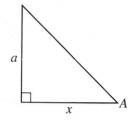

   b. What equation involving angle A and side a would you solve to determine the value of x?

5. Given angle B and side c in the accompanying diagram, answer the questions in parts a and b.

   a. Which of the trigonometric functions relates angle B and sides c and y?

   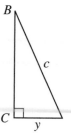

   b. What equation involving angle B and side c would you solve to determine the value of y?

6. Consider the accompanying right triangle.

   a. Which of the trigonometric functions relates angle A and sides b and z?

   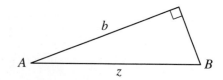

   b. What equation involving angle A and side b would you solve to determine the value of z?

7. Solve the following equations.

   a. $\sin 49° = \frac{x}{12}$ **b.** $\tan 84° = \frac{x}{9}$ **c.** $\sin 22° = \frac{23}{x}$

8. A friend asks you to help build a ramp at his mother-in-law's house. There are three 7-inch-high steps leading to the front door. Another friend donates a 15-foot ramp. The building inspector informs you that any handicapped ramp can have an inclination no greater than 5°.

   a. Sketch a diagram that assumes the land in front of the steps is level and that the ramp makes a 5° angle with the top of the steps.

   b. What is the increase in height from one end of the ramp to the other?

   c. Would the donated ramp be long enough to meet the code? Explain.

ACTIVITY 6.2
A Gasoline
Problem

OBJECTIVES

1. Identify complementary angles.

2. Demonstrate that the sine and cosine of complementary angles are equal.

You and a friend take a camping trip to the American Southwest. On the way home, you realize that your Jeep is running low on gas, so you stop at a filling station. Unfortunately, the attendant informs you that his station has been without gas for several days. However, he is certain that there is a station in the county, on a side road up ahead, that should have gas to sell.

A quick phone call confirms this information, and you begin to ask about the exact location of the station. The attendant's map of the area indicates that the station is about 10 miles away at an angle of 35° with the road you are currently traveling.

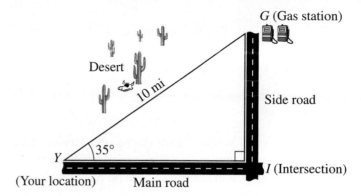

Your Jeep is equipped with a compass, so driving through the desert is not a great problem, although you would prefer to stay on the road. You estimate that you have enough gas for 15 miles. Do you have enough gas to make it to the other station without going through the desert?

This problem can be solved using the triangle above and some trigonometry to determine the distance from you to the intersection, side *YI*, and from the intersection to the gas station, side *IG*.

  1. With respect to angle *Y*,

    a. the side *YI* is the _____.

    b. the side *YG* is the _____.

    c. the function that relates *YI*, *YG*, and angle *Y* is the _____.

  2. a. Set up an equation indicated by Problem 1c.

    b. Solve the equation for the length of *YI*.

  3. With respect to angle *Y*,

    a. the side *IG* is the _____.

    b. the side *YG* is the _____.

    c. the function that relates *IG*, *YG*, and angle *Y* is the _____.

4. a. Set up an equation indicated by Problem 3c.

b. Solve the equation for the length of *IG*.

5. Do you have enough gas to make it without journeying through the desert? Explain.

Meanwhile at the second service station, the attendant is also making some calculations. He is aware of your situation and is trying to anticipate which option you will choose in case he has to go look for you.

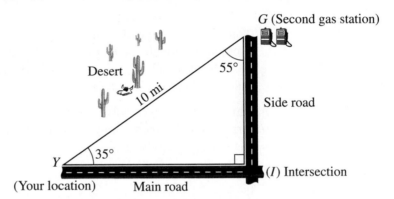

He knows that the line to the first station makes a 55° angle with his road.

6. With respect to angle *G*,

   a. the side *YI* is called the _____.

   b. the side *YG* is called the _____.

   c. the function that relates *YI*, *YG*, and angle *G* is the _____.

7. a. Set up an equation indicated by Problem 6c.

   b. Solve the equation for the length of *YI*.

8. With respect to angle G,

    a. the side IG is called the _____.

    b. the side YG is called the _____.

    c. the function that relates IG, YG, and angle G is the _____.

9. a. Set up an equation indicated by Problem 8c.

    b. Solve the equation for the length of IG.

Complementary Angles

Although you used different angles of the triangle YGI in Problems 1–9, your results for the lengths of YI and IG should have been the same. The angles involved in these calculations were 35° and 55°. Because their sum is 90°, they are called **complementary angles**.

> **DEFINITION**
>
> Two angles A and B, whose measures sum to 90°, are called **complementary angles**. If x represents the measure of an acute angle, then $90 - x$ represents the measure of the complementary angle.

Example 1

x, ACUTE ANGLE	$90 - x$, COMPLEMENTARY ANGLE
35	55
55	35
72	18
18	72

10. a. The solution to Problem 2b involved $\cos 35°$ and the solution to Problem 7b involved $\sin 55°$. Compare the values of $\cos 35°$ and $\sin 55°$.

    b. The solution to Problem 4b involved $\sin 35°$ and the solution to Problem 9b involved $\cos 55°$. Compare the values of $\sin 35°$ and $\cos 55°$.

This situation demonstrates another fundamental principle of trigonometry.

Cofunctions of complementary angles are equal. Symbolically, if x is an acute angle, then

$$\sin x = \cos(90 - x)$$

and

$$\cos x = \sin(90 - x).$$

In fact, the name *cosine* is derived from the words *sine* and *complement*.

Example 2   **a.** $\sin 35° = \cos 55°$ **b.** $\sin 55° = \cos 35°$

c. $\sin 72° = \cos 18°$ **d.** $\sin 18° = \cos 72°$

11. Complete the following table, where x represents the measure of an angle in degrees. Use your graphing calculator to determine values of $\sin x$ and $\cos(90 - x)$.

x	$90 - x$	$\sin x$	$\cos(90 - x)$
0			
15			
30			
45			
60			
75			
90			

SUMMARY
Activity 6.2

1. **Complementary** angles are two acute angles, whose measures sum to 90°.

2. The *co* in cosine is from the word *complement*.

3. **Cofunctions** of complementary angles are equal.

EXERCISES
Activity 6.2

1. You are in a rowboat on Devil Lake in Ontario, Canada. Your lakeside cabin has no running water, so you sometimes go to a fresh spring at a different point on the lakeshore. You row in a direction 60° north of east, for half a mile. You are not yet tired, and it is a lovely day.

   a. How far would you now have to row if you were to return to your cabin by rowing directly south and then directly west?

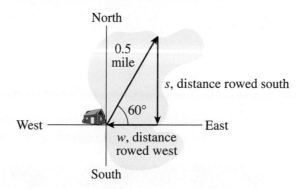

   b. Check your solution to part a, using the complementary angle and the cofunction of those used in part a.

2. One afternoon you don't pay enough attention while rowing. You row 20° off course, too far north, instead of directly west as you had intended. You row off course for 300 meters.

   a. Draw a diagram of this situation.

   b. How far west have you gone?

c. How far north are you of where you had originally planned to be?

d. Check your solutions to parts b and c, using the complementary angle and the cofunctions of those used in parts b and c.

3. a. Complete the following table, where x represents the measure of an angle in degrees. Use your graphing calculator to determine values of $\sin x$ and $\cos(90 - x)$.

x	$90 - x$	$\sin x$	$\cos(90 - x)$
7			
17			
24			
33			
48			
67			
77			

b. What trigonometric property does this table illustrate?

ACTIVITY 6.3
The Sidewalks of
New York

OBJECTIVES

1. Determine the inverse tangent of a number.

2. Determine the inverse sine and cosine of a number using the graphing calculator.

3. Identify the domain and range of the inverse sine, cosine, and tangent functions.

A friend of yours is having a party in her Manhattan apartment, which borders Central Park. She gives you the following directions from your place, which also borders the park:

i. If you are coming after dark, head east for three blocks, going around the park, and then go north for two blocks.

ii. If you can come early, you can cut through the park, a shorter route.

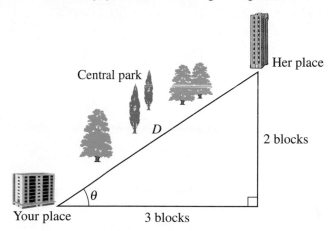

You leave your place early and decide to cut through the park. You want to compute the shortest distance, D (in blocks), between your apartments.

1. Using the Pythagorean theorem, $c^2 = a^2 + b^2$, determine the distance, D, in blocks, across the park.

Now you want to determine the direction (angle) you need to go to get to your friend's place. You can represent the angle by θ, the Greek letter *theta*. Because you now have the lengths of all three sides of the triangle, you can use any of the trigonometric functions to help determine θ. Begin with the tangent function.

Inverse Tangent Function

Recall that the input for the tangent function is an angle and the output is the ratio of the length of the side opposite to the length of the side adjacent in a right triangle:

$$tangent\ of\ an\ angle = \frac{opposite\ side}{adjacent\ side}.$$

2. Determine the value of the tangent of θ, written tan θ, for the Central Park triangle.

If you want to determine the tangent of a known angle, you can use the TAN key on your calculator. From Problem 2, you know the value of the tangent of the angle but do not know the value of the angle θ. That is, you know the output for the tangent function, but you do not know the input.

3. Use the table feature on your calculator to approximate θ from Problem 2. You should compute the tangent of several possible values of θ to get as close as possible to the desired answer. Complete the following table. Round your answers to five decimal places.

θ	$\tan \theta$
30°	0.57735
	0.83910
35°	
	0.67451
33.7°	
	0.66666

You don't have to experiment every time to determine θ when you know the tangent of θ. There is a more direct method, using the inverse tangent function. Recall that with inverse functions, the inputs and outputs are interchanged.

DEFINITION

The input of the **inverse tangent** function is the ratio of the length of the side opposite to the length of the side adjacent to the angle. The output is the acute angle. The inverse tangent function is denoted by $\tan^{-1}$ or **arctan** and defined by

$$\tan^{-1} x = \theta \text{ if and only if } \tan \theta = x,$$

where x represents the ratio of the length of the opposite side to the length of the adjacent side.

Example 1 *Consider the following triangle.*

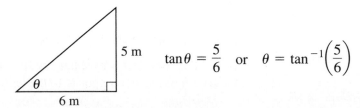

$$\tan \theta = \frac{5}{6} \quad \text{or} \quad \theta = \tan^{-1}\left(\frac{5}{6}\right)$$

Using the calculator, you can determine that $\theta \approx 39.8°$. See the following TI-83 Plus screen. Be careful, your calculator must be in degree mode.

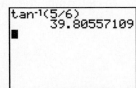

The inverse tangent function is located on your calculator as second function to tangent. That is, you will need to press the [2nd] key before you press the tangent key. Note that your calculator uses the notation $\tan^{-1}$ rather than arctan.

4. Use your calculator to determine the inverse tangent of the answer to Problem 2. (Make sure your calculator is in degree mode.) Compare this answer with your approximation from Problem 3.

Remember: $\tan^{-1}(\text{ratio}) = \text{angle}$

$\tan(\text{angle}) = \text{ratio}$

Inverse Sine and Cosine Function

There are similar definitions for the inverse sine and inverse cosine functions.

DEFINITIONS

The input of the **inverse sine** function is the ratio of the length of the side opposite the angle to the length of the hypotenuse of the triangle. The output is the acute angle. The inverse sine function is denoted by $\sin^{-1}$ or arcsin and defined by

$$\sin^{-1} x = \theta, \text{ if and only if } \sin\theta = x.$$

The input of the **inverse cosine** function is the ratio of the length of the side adjacent to the angle to the hypotenuse of the triangle. The output is the acute angle. The inverse cosine function is denoted by $\cos^{-1}$ or arccos and defined by

$$\cos^{-1} x = \theta, \text{ if and only if } \cos\theta = x.$$

Example 2   *Consider the following triangle.*

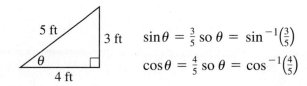

$\sin\theta = \frac{3}{5}$ so $\theta = \sin^{-1}\left(\frac{3}{5}\right)$

$\cos\theta = \frac{4}{5}$ so $\theta = \cos^{-1}\left(\frac{4}{5}\right)$

The calculator tells us $\theta \approx 36.9°$. See the following TI-83 Plus screen.

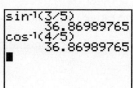

5. a. In Problems 2 and 4 you determined the values of θ in the Central Park situation using the inverse tangent function. Now use the inverse sine function to determine the value of θ in the Central Park situation.

    b. Use the inverse cosine function to determine the value of θ in the Central Park situation.

6. The term used by highway departments when describing the steepness of a hill is **percent grade**. For example, a hill with a 5% grade possesses a slope of $\frac{5}{100}$ or $\frac{1}{20}$.

    a. You are driving along Route 17 in the Catskill Mountains of New York State. Just before coming to the top of a hill, you spot a sign that reads "7% Grade Next 3 Miles Trucks Use Lower Gear." Draw a triangle, and label the appropriate parts to model this situation.

    b. Use an inverse trigonometric function to determine the angle that the base of the hill makes with the horizontal.

    c. Use trigonometry to determine how many feet of elevation you will lose from the top of the hill to the bottom.

SUMMARY
Activity 6.3

1. The domain (inputs) of the **inverse trigonometric functions** is the set of ratios of the lengths of the sides of a right triangle.

2. The range (outputs) of the inverse trigonometric functions is the set of angles.

3. The inverse trigonometric functions are defined by

$$\sin^{-1} x = \theta \text{ if and only if } \sin\theta = x,$$
$$\cos^{-1} x = \theta \text{ if and only if } \cos\theta = x, \text{ and}$$
$$\tan^{-1} x = \theta \text{ if and only if } \tan\theta = x.$$

EXERCISES
Activity 6.3

1. For each of the following, use your calculator to determine θ to the nearest $0.01°$.

   a. $\theta = \arcsin\left(\frac{1}{2}\right)$ **b.** $\theta = \cos^{-1}\left(\frac{3}{7}\right)$

   c. $\theta = \arctan(2.36)$ **d.** $\theta = \sin^{-1}(0.8974)$

   e. $\tan\theta = \frac{7}{3}$ **f.** $\cos\theta = \frac{3}{7}$

   g. $\sin\theta = 0.3791$ **h.** $\tan\theta = 0.3791$

2. Complete the accompanying table, which refers to the following right triangle.

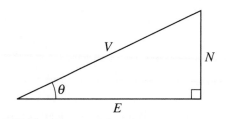

V	θ	E	N
32	65°	13.5	29.0
		12	20
4.1		2.4	
26			18
4.5	45°		

3. You are in a rowboat again, as in Exercise 2 in Activity 6.2. You have gone 300 meters in a direction 20° north of west, instead of going directly west. You turn counterclockwise to go directly south. At what angle must you turn your rowboat?

4. While hiking, you see an interesting rock formation on the side of a vertical cliff. You want to describe for a friend how he might see it when he walks down the path. If you stand on the path at a certain place, 50 feet from the base of the cliff, the rock formation is visible about 30 feet up the cliff. At what angle should you tell your friend to look?

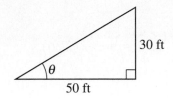

5. A warehouse access ramp claims to have a 10% grade. The ramp is 15 feet long.

   a. Draw a diagram of this situation.

   b. What angle does the ramp make with the horizontal?

   c. How much does the elevation change from one end of the ramp to the other?

ACTIVITY 6.4

Solving a Murder

OBJECTIVE

1. Determine the measure of all sides and all angles of a right triangle.

There has been a fatal shooting 110 feet from the base of a 25-story building. Each story measures approximately 12 feet. Two likely suspects live in the building, one on the seventh floor, the other on the twentieth. Both suspects were in their apartments at the time of the murder. Forensic specialists report that the bullet was fired from somewhere in the building and entered the body at an angle of approximately 58° with the ground.

1. Draw a diagram of this situation.

2. Use the appropriate trigonometric function to determine which of the two suspects could not have committed the murder.

3. On what other floors of the building should the police question additional possible suspects? Explain.

> **DEFINITION**
>
> To **solve a triangle** means to determine the measure of all sides and all angles. This process can be especially useful for architects, surveyors, and navigators.

Example 1 *Solve the right triangle ABC, with $A = 33.0°$, $C = 90.0°$, and $c = 12.2$ inches.*

SOLUTION

Consider the following diagram.

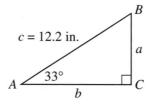

You need to determine the measurements of the remaining sides and angles; angle B, side a, and side b. Because $A + B = 90°$ and $A = 33.0°$,

$$33 + B = 90 \text{ or}$$
$$B = 90° - 33° = 57°.$$

To determine the length of side a, you can use angle A, side c, and the sine function.

$$\sin A = \frac{a}{c}$$
$$\sin 33° = \frac{a}{12.2}$$

So $12.2 \sin 33° = a$, $a = 6.6$ inches to the nearest tenth. See the following calculator screen.

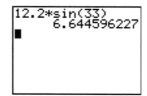

To determine the length of side b, you can use angle A, side c, and the cosine function.

$$\cos A = \frac{b}{c}$$
$$\cos 33° = \frac{b}{12.2}$$
$$b = 12.2 \cos 33° \approx 12.2(.8387) \approx 10.2$$

4. Use the given information to solve each of the following right triangles.

a.

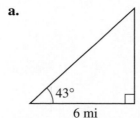

b.

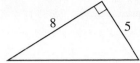

c.

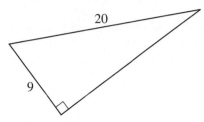

SUMMARY
Activity 6.4

1. Many trigonometric problems involve solving right triangles, that is, determining the measures of all sides and angles.

2. Following is a trigonometric problem-solving strategy:
   a. Draw a diagram of the situation using right triangles.
   b. Identify all known sides and angles.
   c. Identify sides and/or angles you want to know.
   d. Identify functions that relate the known and unknown.
   e. Write and solve the appropriate trigonometric equation(s).

EXERCISES
Activity 6.4

1. Use the given information to solve each of the following right triangles.

   a.

   6.5 ft

   57°

   b.

   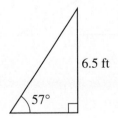

   18

   6

c.

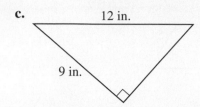

12 in.

9 in.

2. You need to construct new steps for your deck and read that stringers are on sale at the local lumber company. Stringers are precut side supports to which you nail the steps; they are made in three-, four-, five-, six-, or seven-step sizes. Each step on the stringers is 7 inches high and 12 inches deep.

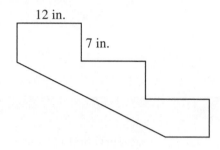

12 in.

7 in.

a. If the vertical rise of your deck measures 3.5 feet, which size stringer should you buy? Explain.

b. How far out will your steps extend from the porch? Explain.

c. What angle will your steps make with the ground? Explain.

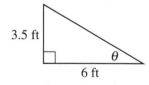

3.5 ft

6 ft

θ

d. What angle will your steps make with the house? Explain.

3. Some application problems involve a horizontal line of sight, which is used as a reference line. An angle measured *above* the sight line is called an **angle of elevation**. An angle measured *below* the sight line is called an **angle of depression**.

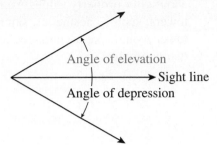

You and some friends take a trip to Colton Point State Park in the Grand Canyon of Pennsylvania. Some of your group goes white-water rafting, while some of your friends join you for a hike. You reach the observation deck in time to see the rest of your party battling the white water. Someone in the group asks you how close you think the rafts actually get to you as they float by. You have no idea, but you ask a nearby park ranger.

She doesn't know either, but she does tell you that the canyon is approximately 800 feet deep at Colton Point and that the angle of depression to the creek is about 22°.

a. Draw a diagram of this situation.

b. Use trigonometry to estimate how close the rafters get to you on the observation deck.

ACTIVITY 6.5
How Stable Is
That Tower?

OBJECTIVE

1. **Solve problems using right-triangle trigonometry.**

Situation 1: Stabilizing a Tower

You are considering buying property near a cell-phone tower and are concerned about your property values. You decide to do some reading about issues involving towers, such as aesthetics, safety, and stability. Of course, anyone living near the tower would like a guarantee that it could not blow down. Guy wires are part of that guarantee.

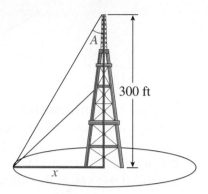

The tower rises 300 feet and is supported by several pairs of guy wires all attached on the ground at the same distance from the base of the tower. In each pair, one guy wire extends from the ground to the top of the tower, and the other attaches half way up. The accompanying diagram illustrates one pair of wires.

1. New guidelines for stability recommend that the angle (*A*) the guy wires make with the line through the center of the tower (not the angle of elevation) must be at least 40°. Therefore, the existing guy wires may need to be replaced. Because you are concerned about how close the wires will come to your property, you need to compute the shortest distance from the tower at which the guy wires may be attached. Use trigonometry to compute this distance.

2. To improve stability, the authors of the guidelines propose increasing the minimum angle the guy wire makes with the tower from 40° to 50°. What is the effect on the shortest distance from the tower at which the guy wires may be attached?

Situation 2: Climbing a Mountain

You are camping in the Adirondacks and decide to climb a mountain that dominates the local area. You are curious about the vertical rise of the mountain and recall from your mathematics course that surveyors can measure the angle of elevation of a mountain summit with a theodolite. You are able to borrow a theodolite from the local community college to gather some pertinent data. You find a level field and take a first reading of an angle of elevation of 23°. Then you walk 100 feet towards the mountain summit and take a second reading of 24° angle of elevation (see accompanying illustration).

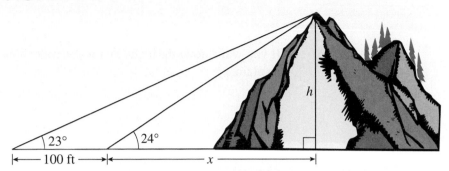

3. Use *h* to represent the vertical rise of the mountain, and write expressions for tan 23° and tan 24° in terms of *h* and *x*.

a.   **b.**

4. a. Solve the system of equations in Problem 3 to determine *x*.

b. Use the value of *x* from part a to determine the vertical rise, *h*, of the mountain.

EXERCISES
Activity 6.5

1. You are driving on a straight highway at sea level and begin to climb a hill with a 5% grade. (Recall that a grade is given in a percent but may be expressed as a fraction. In that way you can view grade as a slope.)

Note: Not to scale

a. What is the slope of the highway with a grade of 5%?

   b. What is the **angle of elevation**, *A*? That is, what angle does the highway make with the horizontal? Explain.

   c. If you walk along the highway for one mile, how many feet above sea level are you? Explain.

2. You are standing 92 meters from the base of the CN Tower in Toronto, Canada. You are able to measure the angle of elevation to the top of the tower as 80.6°.

   a. Draw a diagram and indicate the angle of elevation.

   b. What is the height of the tower?

3. You are in a spy satellite, equipped with a measuring device like a theodolite, orbiting five miles above Earth. Your mission is to discover the length of a secret airport runway. You measure the angles of depression to each end of the runway as 30° and 25° respectively. What is the length of the runway?

4. The Empire State Building rises 1414 feet above the ground, and you are standing across 34th Street, approximately 80 feet from the base of the building.

    a. If you look up to the top of the building, what angle of elevation do you make with the ground?

    b. How far from the building must you be to make your angle of elevation with the ground 85°?

**PROJECT
ACTIVITY 6.6**

Seeing Abraham
Lincoln

OBJECTIVE

1. Solve optimization
problems using right-
triangle trigonometry and
by analyzing graphs.

You are traveling to South Dakota and plan to see Mount Rushmore. In preparation for your trip, you do some research and discover that from the observation center, the vertical rise of the mountain is approximately 500 feet and the height of Abraham Lincoln's face is 60 feet (see accompanying diagram). To get the best view, you want to position yourself so that your viewing angle of Lincoln's face is as large as possible.

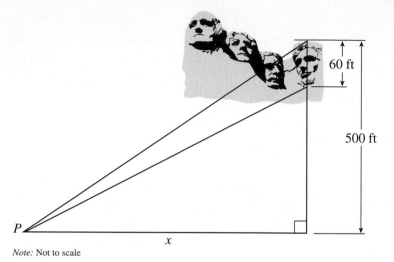

60 ft

500 ft

P

x

Note: Not to scale

1. **a.** Label your viewing angle if you are standing at point *P* on the diagram. Use the appropriate trigonometric function and its inverse to write an expression that defines the viewing angle as a function of the distance, *x*, you are standing from the base.

   b. What is the practical domain of the function that describes the viewing angle in part a?

2. Use your graphing calculator to graph this function over its practical domain.

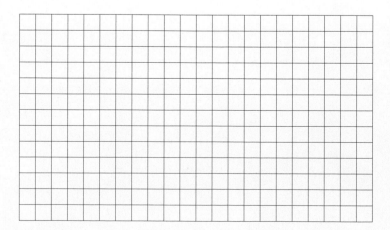

3. What is the largest value of the viewing angle? Justify your conclusion.

4. How far should you stand from the mountain to obtain this maximum value of the viewing angle? Explain.

5. Consider the function defined by $f(x) = \arctan \frac{20}{x} - \arctan \frac{10}{x}$ defined for $0.01 \le x \le 100$.

    a. Use your graphing calculator to sketch a graph of this function.

    b. Over the given domain, what is the maximum value of the function, and where does it occur?

    c. Over the given domain, what is the minimum value, and where does it occur?

EXERCISES
Activity 6.6

1. You are interested in constructing a feeding trough for your cattle that can hold the largest amount of feed. You buy a 15 by 50-foot piece of aluminum to construct a 50-foot-long trapezoidal trough with a base of 5 feet. You bend up the two 5-foot sides through an angle of $t°$ with the horizontal. Each cross section is a trapezoid (see accompanying diagram). You need to determine the angle t that produces the largest volume for the trough.

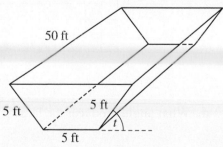

50 ft

5 ft   5 ft

t

5 ft

Note: Not to scale

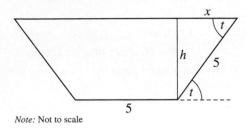

a. Recall the formula for the area of a trapezoid (see Appendix A, if necessary), and write the area of the trapezoidal cross section in terms of h and x (see accompanying diagram).

Note: Not to scale

b. Using right-triangle trigonometry, write h in terms of t. In a similar way, write x in terms of t. Using this information, write an equation for the area of the trapezoid as a function of t.

c. Use your graphing calculator to graph the area function you constructed in part b with $0° < t < 90°$. Label your axes, and remember to indicate units.

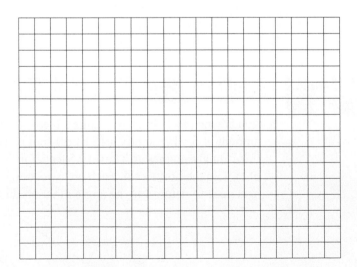

d. What angle produces the largest area for the trapezoid? Explain.

e. What is that area? Explain.

f. Write the volume of the trough as a function of the angle *t*.

g. Use your graphing calculator to graph the volume function you constructed in part f with $0° < t < 90°$. Label axes and show your units.

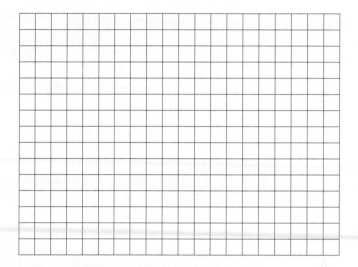

h. What angle produces the largest volume, and what is that volume? Justify your conclusion.

i. What do you conclude about the angle that produces the largest cross-sectional area and the angle that produces the largest volume for the trough?

What Have I Learned?

1. Classical right-triangle trigonometry was developed by the ancient Greeks to solve problems in surveying, astronomy, and navigation. For purposes of computation, the side opposite the angle θ, side O, is called the opposite, the side opposite the right angle, side H, is called the hypotenuse, and the third side, side A, is called the adjacent.

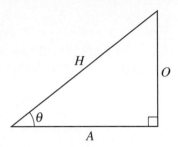

Define the three major trigonometric functions—$\sin\theta$, $\cos\theta$, and $\tan\theta$—in terms of H, A, and O.

2. **a.** On any right triangle, which trigonometric function would you use to determine the opposite side if you knew the angle measure and the length of the hypotenuse?

   b. Which trigonometric function would you use to determine the adjacent side if you knew the angle measure and the length of the hypotenuse?

   c. Which trigonometric function would you use to determine the adjacent side if you knew the angle measure and the length of the opposite side?

   d. Which trigonometric function would you use to determine the opposite side if you knew the angle measure and the length of the adjacent side?

3. **a.** Suppose for a right triangle you know the length of the side opposite an angle and you know the length of the hypotenuse. How can you determine the angle?

   b. There is another way to solve part a. Describe this alternative technique.

c. If you know the lengths of any two sides of a right triangle, how can you determine all the angles in the triangle? Make up an example, and determine all the angles. Remember that all the interior angles of a triangle add up to 180°.

4. Consider the following right triangle.

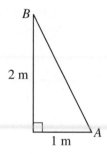

Calculate each of the following.

a. sin *A*   **b.** cos *A*   **c.** tan *A*

d. sin *B*   **e.** cos *B*   **f.** tan *B*

g. What trigonometric property is illustrated by parts a–f? Explain.

5. Consider the following two calculator screens.

 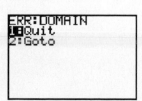

a. The second screen indicates that 1.257 is not in the domain of the inverse sine function. Do you agree? Explain.

b. The following screen indicates that you do not have the same problem for the inverse tangent function. Why not? Explain.

6. Using diagrams, explain the difference between angle of depression and angle of elevation.

CLUSTER 1 ## How Can I Practice?

1. Triangle ABC is a right triangle.

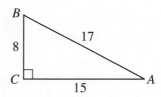

   Calculate each of the following. Write your answer as a ratio.

   a. $\tan A$ **b.** $\tan B$ **c.** $\cos A$

   d. $\cos B$ **e.** $\sin A$ **f.** $\sin B$

2. Use your graphing calculator to determine the values of each of the following. Round your answers to the nearest thousandths.

   a. $\sin 47° =$ **b.** $\cos 55° =$

   c. $\tan 31° =$ **d.** $\tan 80° =$

3. Given $\sin A = \frac{5}{13}$, determine $\cos A$ and $\tan A$ exactly.

4. Given $\tan B = \frac{7}{4}$, determine $\sin B$ and $\cos B$ exactly.

5. Using your calculator, evaluate the following for θ to the nearest $0.1°$, where $0° \le \theta \le 90°$.

   a. $\sin \theta = \frac{3}{4}$ **b.** $\cos \theta = 0.9172$ **c.** $\theta = \arctan \frac{7}{2}$

   d. $\theta = \sin^{-1} \frac{2}{7}$ **e.** $\theta = \tan^{-1} 0.9714$ **f.** $\theta = \arccos 0.9714$

Answers to all How Can I Practice exercises are included in the Selected Answers appendix.

6. Solve the following right triangle. That is, determine all the missing sides and angles.

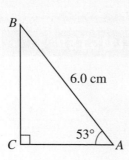

7. You are building a new garage to be attached to your home and investigate several 30-foot-wide trusses to support the roof. You narrow down your choices to three: One has an angle of 45° with the horizontal, and the others have angles of 35° and 25° with the horizontal. You determine that the walls of the garage must be 10 feet high. To match the height of the rest of the house, the peak of the garage should be approximately 20 feet high. Which truss should you buy?

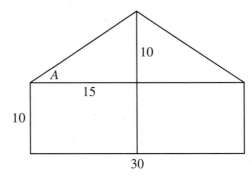

8. The side view of your swimming pool is shown here.

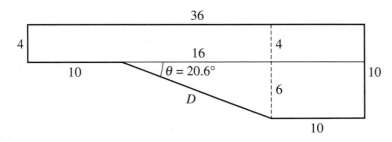

a. What is the angle of depression, θ?

b. What is the length of the inclined side, D?

9. a. As part of your summer vacation, you rent a cottage on a large lake. One day, you decide to visit a small island that is six miles east and two and a half miles north of your cottage. Draw a diagram for this situation. How far from your cottage is the island?

b. At what angle with respect to due east should you direct your boat to make the trip from your cottage to the island as short as possible?

CLUSTER 2   # Why Are the Trigonometric Functions Called Circular Functions?

ACTIVITY 6.7

Learn Trig or Crash!

OBJECTIVES

1. Determine the coordinates of points on a unit circle using sine and cosine functions.

2. Sketch a graph of $y = \sin x$ and $y = \cos x$.

3. Identify the properties of the graphs of the sine and cosine functions.

You are piloting a small plane and want to land at the local airport. Due to an emergency on the ground, the air traffic controller places you in a circular holding pattern at a constant altitude with a radius of one mile. Because your fuel is low, you are concerned with the distance traveled in the holding pattern. Of course, you communicate with air traffic control about your coordinates so that you do not collide with another airplane.

The given diagram shows your path in the air. The airport is located at the center $(0, 0)$ of the circle on the ground. The radius of the circle is one mile. A circle centered at $(0, 0)$ and having radius 1 is called a **unit circle**. You begin your holding pattern at $(1, 0)$.

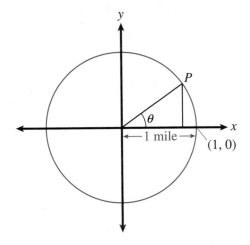

Let's examine the beginning of your first circular loop.

1. What distance (in miles) do you fly in one loop?

2. Let P represent your position after flying only one-tenth of a loop. (See preceding diagram.)

   a. Compute the distance traveled from $(1, 0)$ to P.

   b. Determine the number of degrees of the central angle, θ, when you fly one-tenth of a loop (see preceding diagram). Recall that there are $360°$ in a circle.

The angle in Problem 2b is called a central angle.

DEFINITION
A **central angle** is an angle with its vertex at the center of a circle.

3. Now refer back to the right triangle in the preceding diagram. The measure of the central angle is 36°.

    a. What is the length of the hypotenuse of the right triangle?

    b. If (x, y) represents the coordinates of point P, then which letter x or y, represents the length of the side opposite the 36° angle? Which letter represents the length of the side adjacent to the 36° angle?

    c. Use the appropriate trigonometric function to determine the value of x.

    d. Use the appropriate trigonometric function to determine the value of y.

    e. What are the coordinates of point P?

Problem 3 demonstrates that if an object is moving a distance d counterclockwise on the unit circle from the starting point $(1, 0)$, and if $P (x, y)$ represents the position of the object on the unit circle after it has moved distance d, and if θ represents the corresponding central angle, then the coordinates of P are given by $(\cos \theta, \sin \theta)$.

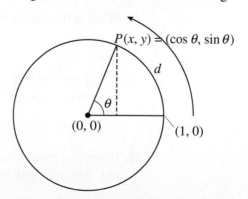

4. Repeat the procedure demonstrated in Problem 3 for the following fractions of a loop, and record your results in the table and on the diagram.

LOOP	CENTRAL ANGLE, θ	DISTANCE TRAVELED	$(\cos \theta, \sin \theta)$
$\frac{1}{5}$	72°	$\frac{2\pi}{5}$ mi, or 1.26 mi	$(0.31, 0.95)$
$\frac{1}{8}$			
$\frac{1}{20}$			

Trigonometric Functions of Angles Greater than 90°

5. a. Your odometer tells you that you have traveled two miles in the holding pattern. What is the central angle, θ? (*Hint:* What fraction of the loop have you traveled?)

For $\theta > 90°$, the coordinates of P are $(\cos\theta, \sin\theta)$ as before. Depending on where you are on the circle, these coordinates may be positive or negative.

6. a. From Problem 5, you know that the central angle is 114.6° (See accompanying diagram). What is the measure of the central angle contained within the right triangle?

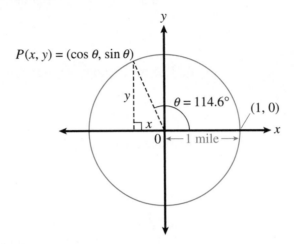

b. What is the length of the hypotenuse?

c. Using the sine and cosine functions, determine the lengths of the remaining two sides of the right triangle?

d. Using the results from part c, what are the coordinates of point P? Remember, the point is in quadrant II.

7. Your graphing calculator can be used to determine the coordinates of point in a direct manner. It involves calculating the sine and cosine of the central angle θ.

a. Determine the value of each of the following:

cos 114.6°

sin 114.6°

b. How do the results in part a compare to the coordinates of point P in Problem 6d?

In general, the position $P(x, y)$ of an object moving on a unit circle is given by $x = \cos\theta$ and $y = \sin\theta$, where θ is a central angle with initial side the positive x-axis and terminal side OP.

8. Give an example of where on the unit circle (circle of radius 1) both coordinates of P are negative. Place θ and the coordinates of P on the following diagram.

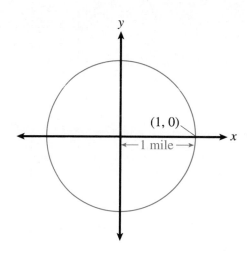

9. **a.** By the time you have traveled five miles in the holding pattern, what is the central angle, θ?

b. What are the coordinates of point P?

Graphs of Sine and Cosine Functions

10. Complete the following table. Be sure you are in degree mode.

DISTANCE TRAVELED	CENTRAL ANGLE θ	cos θ	sin θ
0	0°		
	30°		
	45°		
	60°		
1.57			
	120°		
3.14			
	220°		
6			
	360°		

11. Plot the data pairs $(\theta, \sin\theta)$ on the following grid. Draw a smooth curve through these points. Verify the results with your graphing calculator by sketching a graph of $y = \sin x$. Use the window Xmin $= 0$, Xmax $= 360$, Ymin $= -1$, and Ymax $= 1$ and degree mode.

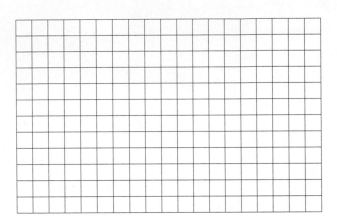

12. Repeat Problem 11 for data pairs $(\theta, \cos\theta)$.

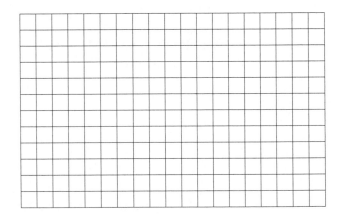

13. You notice the odometer reads 11.28 miles from the start of the holding pattern. What is the central angle, θ, and what are the coordinates of point P?

14. Suppose you are at some particular point in the holding pattern. How many miles will you travel in the loop to return to the same coordinates?

15. a. Complete the following table.

θ (degrees)	0	90	180	270	360	450	540	630	720
$\sin\theta$									
$\cos\theta$									

b. Sketch a graph of $y = \sin\theta$ for $0° \leq \theta \leq 720°$. Verify using your graphing calculator.

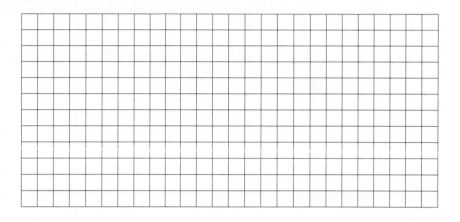

c. Sketch a graph of $y = \cos\theta$ for $0° \leq \theta \leq 720°$.

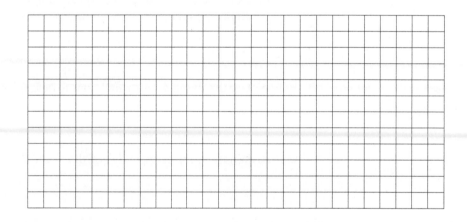

d. What pattern do you observe in each of the graphs?

Note that these repeating graphs in Problem 15 are the basis for the **periodic** or **cyclic** behavior of the trigonometric functions. Because many real-world phenomena involve this repeating behavior, the trigonometric functions are very useful in modeling these phenomena.

DEFINITION

The shortest time it takes for one cycle to be completed is called the **period**. The period is 2π for $y = \sin x$ and $y = \cos x$.

16. A negative angle means that the object is moving along the circumference of a unit circle in a clockwise direction.

a. Locate the point P that has a central angle of $-30°$ on the following diagram.

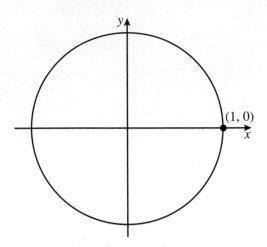

b. Construct the appropriate right triangle. What is the measure of the central angle within this right triangle? What is the length of the hypotenuse?

c. Use the sine and cosine functions to determine the lengths of the other two sides.

d. What are the coordinates of the point P?

e. Use your graphing calculator to determine $\cos(-30°)$ and $\sin(-30°)$. How do these results compare to the coordinates of point P in part d?

17. a. Complete the following table.

θ	$\cos\theta$	$\sin\theta$
0		
-30		
-90		
-180		
-270		
-360		

b. Graph the points of the form $(\theta, \sin\theta)$ using the values from the table in part a.

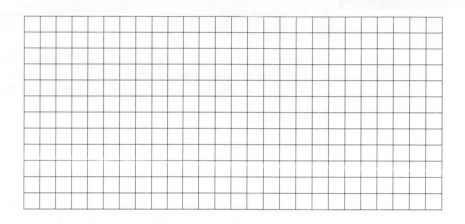

c. Graph the points of the form $(\theta, \cos\theta)$ using the values from the table in part a.

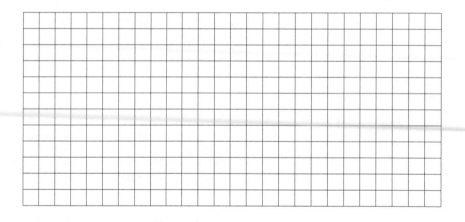

d. Use your graphing calculator to sketch a graph of $y = \sin x$ and $y = \cos x$ for $-720° \le \theta \le 720°$.

18. a. What is the domain of the sine and cosine functions?

b. What is the range of the sine and cosine functions?

SUMMARY
Activity 6.7

1. A **central angle** is an angle whose vertex is the center of a circle.

2. The position $P(x, y)$ of an object moving on the unit circle from the point $(0, 1)$ can be defined by $x = \cos\theta$, $y = \sin\theta$, where θ is a central angle formed by the positive x-axis and the line segment OP. Because of this connection to the unit circle, the sine and cosine functions are often called **circular functions**.

3. The **domain** of the sine and cosine functions is all angles, both positive and negative.

4. The **range** of the sine and cosine functions is all values of N such that $-1 \leq N \leq 1$.

5. The graphs (one cycle) of $y = \sin x$ and $y = \cos x$ look like the following.

sin x

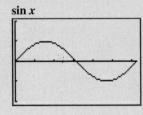

cos x

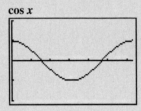

6. The **period** is 2π for $y = \sin x$ and $y = \cos x$.

EXERCISES
Activity 6.7

1. Determine the coordinates of the point on the unit circle corresponding to the following central angles.

   a. $72°$ **b.** $310°$ **c.** $270°$

   d. $111°$ **e.** $212°$ **f.** $435°$

   g. $-70°$

2. For each of the points on the unit circle determined in Exercise 1, determine the distance traveled from $(1, 0)$ to the point along the circle.

   a.   **b.**

   c.   **d.**

   e.   **f.**

g.

3. Consider the graph of the following function.

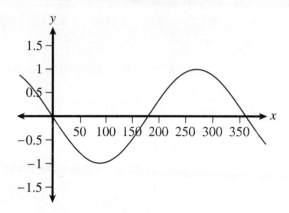

a. Compare this graph to graphs studied in this activity.

b. What is the motion along the unit circle described by the graph?

4. Consider the graph of the following function.

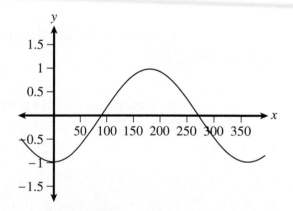

a. How does this graph compare to the graphs studied in this activity?

b. What is the motion around the unit circle described by the graph?

5. The following table represents the number of daylight hours for a certain city in the Western Hemisphere on the dates indicated.

MAR. 21	APR. 21	MAY 21	JUNE 21	JULY 21	AUG. 21	SEPT. 21	OCT. 21	NOV. 21	DEC. 21	JAN. 21	FEB. 21	MAR. 21
11.9	10.5	9.6	8.7	9.7	10.6	12.1	13.5	14.7	15.7	14.7	13.4	11.9

a. Plot the data on the following grid.

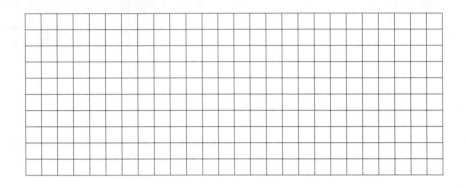

b. Do the data indicate a circular function (a function defined by points on the unit circle)? Explain.

c. How does this function compare with the others in this activity?

d. Is the city in question north or south of the equator? Explain.

ACTIVITY 6.8
It Won't Hertz

OBJECTIVES

1. Convert between degree and radian measure.

2. Identify the period and frequency of a function defined by $y = a \sin (bx)$ or $y = a \cos (bx)$ using the graph.

Household electric current is called alternating current, or AC, because it changes magnitude and direction with time. The household current through a 60-watt light bulb is given by the equation

$$A = 2\sin(120\pi t),$$

where A is the current in amperes and t is time in seconds.

Note that the input of the sine function in this activity is time measured by real numbers (seconds) and not angles measured in degrees. In order for this to make sense, an alternate real-number method for measuring angles must be introduced. This method is called **radian measure**.

> Degree and radian measure correspond in the following way:
>
> $$180° \text{ is the same as } 1\pi, \text{ or about } 3.14 \text{ radians.}$$
>
> Therefore, $1° = \frac{\pi}{180°}$ radians and 1 radian $= \frac{180°}{\pi}$.

Whenever you see an angle measure without a degree symbol, assume that the angle is measured in radians.

Example 1 *Ninety degrees is the same as $\frac{\pi}{2}$ radians. You can convert degrees to radians as follows:*

$$90° = \left(\frac{90°}{1}\right)\left(\frac{1\pi \text{ radians}}{180°}\right) = \left(\frac{90° \pi \text{ radians}}{180°}\right) = \frac{\pi}{2} \text{ radians}$$

1. In the following table, convert degree measures to radian measure.

DEGREE MEASURE	RADIAN MEASURE
10	
20	
30	
60	
120	
360	

2 **a.** Describe a procedure to convert radians to degrees.

   b. How many degrees are there in 1.5π radians?

c. How many degrees are there in 2π radians?

d. How many degrees are there in $\frac{\pi}{10}$ radians?

 For more practice converting degree measure to radian measure and vice versa, see Appendix B.

Periodic Behavior of Graphs of Sine and Cosine Functions

As stated, the function defined by $A = 2\sin(120\pi t)$, where A is the current in amperes and t is time in seconds, gives the household current through a 60-watt lightbulb.

3. Graph the equation $A = 2\sin(120\pi t)$ using your calculator. Your calculator must be in radian mode instead of degree mode. Using the indicated window, the graph should appear as follows:

Xmin = 0   Ymin = −3

Xmax = $\frac{1}{20}$ Ymin = 3

Xscl = $\frac{1}{240}$ Yscl = 1

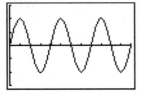

4. Using the graph, determine the maximum current. Explain.

5. What do you think is happening to the current when the graph drops below the horizontal axis? (Reread the description of alternating current.)

There is a pattern on the graph of $A = 2\sin(120\pi t)$ that repeats. This pattern is called a **cycle**.

Recall from Activity 6.7 that the shortest time it takes for one cycle to be completed is called the **period**.

Example 2 *Determine the period of each of the following using its graph.*

a. $y = \sin(2x)$ **b.** $y = \sin\left(\frac{1}{2}x\right)$

SOLUTION

a.
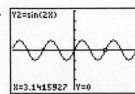

The period is π.

b.
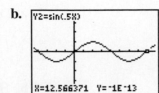

The period is 4π.

6. What is the *period* of the electric current function from Problem 3? (The tick marks on the horizontal axis are at $\frac{1}{240}$-second intervals)

7. Determine the period (shortest interval in which the graph repeats) of the following functions. Use the window Xmin = 0, Xmax = 6π, Xscl = $\frac{\pi}{4}$, Ymin = -2, Ymax = 2, and Yscl = 1.

   a. $y = \sin x$ **b.** $y = \sin\left(\frac{1}{2}x\right)$

   c. $y = \cos(3x)$ **d.** $y = \cos\left(\frac{2}{3}x\right)$

DEFINITION

The **frequency** of $y = a\sin(bx)$ or $y = a\cos(bx)$ is the number of cycles completed in 2π units.

Example 3   *Determine the frequency of each of the following using its graph.*

   a. $y = \sin(2x)$ **b.** $y = \sin\left(\frac{1}{2}x\right)$

SOLUTION

a. 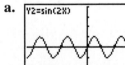   b.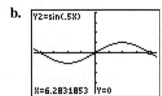

The frequency is 2 because the graph completes two complete cycles in 2π units.

The frequency is 0.5 because the graph completes half of one cycle in 2π units.

8. Determine the frequency, the number of cycles completed in 2π units, for each of the following functions.

   a. $y = \sin x$ **b.** $y = \cos\frac{1}{2}x$ **c.** $y = \cos 2x$

9. For normal household current described by $A = 2\sin(120\pi t)$, how many cycles occur in one second?

SUMMARY
Activity 6.8

1. **Radian** measure is used when the input of a repeating function is better defined by real numbers than angles measured in degrees.

2. To convert degree measure to radian measure, multiply the degree measure by $\dfrac{\pi \text{ radians}}{180°}$.

3. To convert radian measure to degree measure, multiply the radian measure by $\dfrac{180°}{\pi \text{ radians}}$.

4. The pattern of a graph that is repeated is called the **cycle**.

5. The smallest interval of input necessary for the graph of a function to repeat is called the **period**.
   Note: formula for period done in Activity 6.10

6. The **frequency** of a periodic or cyclic function is the number of cycles completed over a given interval of input.

EXERCISES
Activity 6.8

1. Convert the following degree measures to radian measures.

   a. 45°   **b.** 140°

   c. 330°   **d.** −36°

2. Convert the following radian measures to degree measures.

   a. $\frac{3\pi}{4}$ **b.** 2.5π

   c. 6π **d.** 1.8π

For Exercises 3–9, be sure your calculator is in radian mode.

3. Complete the following table.

DEGREE MEASURE	0°	30°		60°	90°		180°	210°		360°
RADIAN MEASURE	0		$\pi/4$			$3\pi/4$			$3\pi/2$	

Exercise numbers appearing in color are answered in the Selected Answers appendix.

4. a. Complete the following table.

RADIAN MEASURE, x	0	$\pi/4$	$\pi/2$	$\frac{3\pi}{4}$	$\pi/2$	$5\pi/4$	$\frac{3\pi}{2}$	$7\pi/4$	2π
sin x									

b. Sketch a graph of $f(x) = \sin x$ using the input/output table in part a.

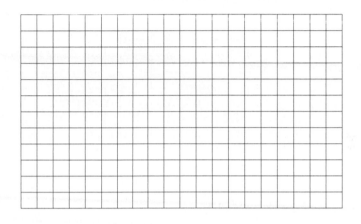

c. Use your graphing calculator to sketch a graph of $f(x) = \sin x$ for $0 \le x \le 2\pi$

5. How do the graphs of each pair of the following functions compare? Use your graphing calculator.

a. $y = 2\cos x$, $y = \cos 2x$

b. $y = \cos\frac{1}{3}x$, $y = \cos 3x$

For Exercises 6–9, graph each function, and determine the following for each function:
  a. the largest (maximum) value of the function.
  b. the smallest (minimum) value of the function.
  c. the period (the shortest interval for which the graph repeats).

6. $y = 0.5 \sin 2x$

7. $y = -3 \sin 3x$

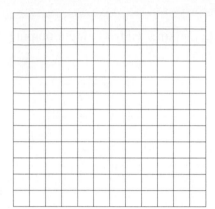

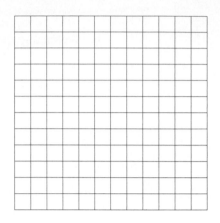

8. $f(x) = 2.3 \cos(0.5x)$

9. $y = 2 \sin x + 3 \cos x$

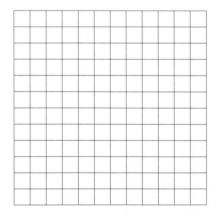

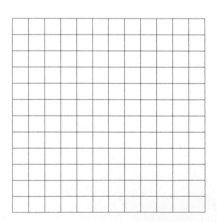

10. You are setting up a budget for the new year. Your utility bill for natural gas and electric usage is a large part of your budget. To help determine the amount you might need to spend on gas and electricity, you examine your previous three years' bills. Because you live in a rural area, you are billed every two months instead of every month. The data you obtain appears in the following graph.

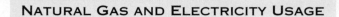

NATURAL GAS AND ELECTRICITY USAGE

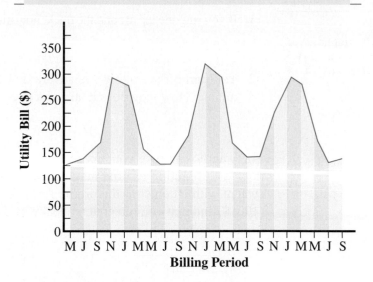

You notice that a pattern develops, which is repeated. This function, though not exactly periodic, models a periodic function for practical purposes. Label the horizontal axis with bimonthly periods, beginning with the June and July bill from three years ago.

a. For what months of the year is the utility bill the highest? How much is this bill?

b. What months of the year is the utility bill the lowest? How much is this bill?

c. What is the largest value for the function whose graph is given here?

d. What is the period of the graph?

e. Your power company announces that its rates will increase 5% beginning in April of the coming year. How will this change affect the graph of this function? Will it affect the periodic nature of the function?

f. You heat your house and your water with natural gas and use electricity for all other purposes. You do not currently have air conditioning in your house. If you were to install a central air conditioning unit next summer, what changes do you think might occur in the shape of the graph?

ACTIVITY 6.9

Get in Shape

OBJECTIVE

1. Determine the amplitude of the graph of $y = a\sin(bx)$ or $y = a\cos(bx)$.

You decide to try jogging to shape up. You are fortunate to have a large neighborhood park nearby that has a circular track with a radius of 100 meters.

1. If you run one lap around the track, how many meters have you traveled? Explain.

2. You start off averaging a relatively slow rate of approximately 100 meters per minute. How long does it take you to complete one lap?

You want to improve your speed. Gathering data describing your position on the track as a function of time may be useful. You sketch the track on a coordinate system with a center at the origin. Assume that the starting line has coordinates (100, 0) and that you run counterclockwise.

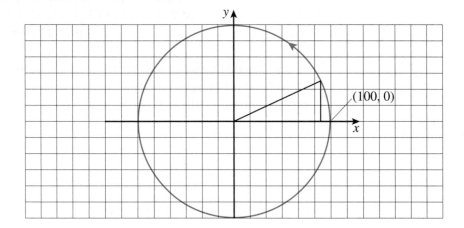

3. You begin to gather data about your position at various times. Using the preceding diagram and the results from Problem 2, complete the following table to locate your coordinates at selected times along your path.

t (min)	YOUR x-COORDINATE	YOUR y-COORDINATE
0		
$\frac{\pi}{2}$		
π		
$\frac{3\pi}{2}$		
2π		
$\frac{5\pi}{2}$		
3π		

4. What patterns do you notice about the numerical data in Problem 3? Predict your coordinates as your time increases.

5. To better analyze your position at times other than those listed in the previous table, you decide to make some educated guesses. Let's consider $t = \frac{\pi}{4}$ minutes. Use the graph to approximate the coordinates of your position when $t = \frac{\pi}{4}$, and label these coordinates on the graph.

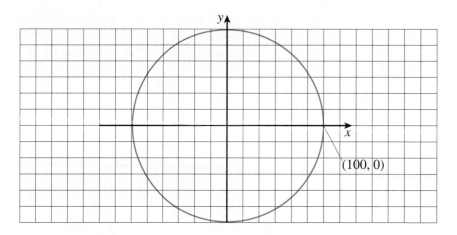

$(100, 0)$

Appendix

6. To check your guess, you recall that right-triangle trigonometry gives some useful information about special right triangles. (If you are not familiar with special right triangles, see Appendix B.) First, however, you need to compute θ, given in the graph of Problem 5. Calculate θ and explain how you arrive at your answer.

7. You are now able to gather more data about your position. Complete the following table to give the coordinates of your position at the additional special points along your path.

t (min)	YOUR x-COORDINATE	YOUR y-COORDINATE
$\frac{\pi}{4}$	70.71	70.71
$\frac{3\pi}{4}$		
$\frac{5\pi}{4}$		
$\frac{7\pi}{4}$		
$\frac{9\pi}{4}$		

8. On the following grid, use the data from Problems 3 and 7 to plot (t, y), your
y-coordinate as a function of t. Connect your data pairs to make a smooth graph,
and predict what will happen to the graph for values of t before and after the
values of t in the tables.

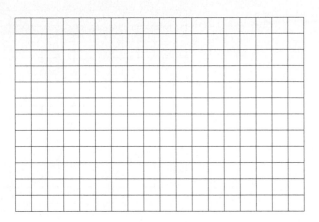

9. Use your graphing calculator to plot $y = 100\sin(x)$. Make sure your graphing
calculator is in radian mode. Note that the name of the input t has been changed
to x to conform with the calculator.

10. Compare your answers to Problems 8 and 9.

11. a. What is the maximum value of the sine function in Problem 9?

b. What is the minimum value of the sine function in Problem 9?

Another important feature of the graphs of the sine and cosine functions is called
the **amplitude**.

DEFINITION

The **amplitude** of a periodic function equals

$$\tfrac{1}{2}(M - m),$$

where M is the maximum output value of the function and m is the minimum output value of the function.

Example 1   *The amplitude of the function defined by $y = \sin x$ is 1. The maximum*
output value is 1. The minimum output value is -1. The amplitude is
$\frac{1}{2}(1 - (-1)) = \frac{1}{2}(1 + 1) = \frac{1}{2}(2) = 1.$

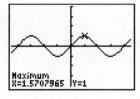
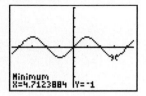

12. **a.** What is the amplitude of the sine function in the equation $y = 100 \sin x$?

b. Is there a relationship between the amplitude of $y = 100 \sin x$ and the coefficient 100? Explain.

13. Determine by inspection the amplitude of the following functions. Then verify your answers using your graphing calculator.

a. $y = 1.5 \sin x$ **b.** $f(x) = 15 \sin(2x)$ **c.** $y = 3 \cos\left(\frac{1}{3}x\right)$

14. **a.** Is -2 the amplitude of $y = -2 \sin x$? Explain.

b. What is the amplitude of $y = -2 \sin x$?

c. Use your graphing calculator to sketch a graph of $y = -2 \sin x$.

d. How does the graph of $y = -2 \sin x$ compare to the graph of $y = 2 \sin x$?

e. What is the general effect of the negative sign of the coefficient a in $y = a \sin x$?

SUMMARY
Activity 6.9

The **amplitude** of trigonometric functions defined by $y = a \sin(bx)$ or $y = a \cos(bx)$ is defined by:

$$\tfrac{1}{2}(M - m),$$

where M represents the maximum function value and m represents the minimum function value. The amplitude is equivalent to $|a|$, the absolute value of the coefficient of the function. Therefore,

$$|a| = \tfrac{1}{2}(M - m).$$

EXERCISES
Activity 6.9

1. On the following grid, repeat Problem 8 to plot the data pairs (t, x).

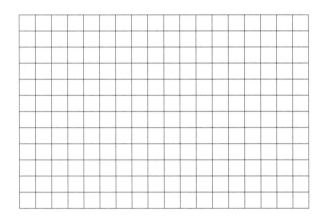

2. Use your graphing calculator to plot $y = 100 \cos(x)$. Make sure your graphing calculator is in radian mode. Note that the names of both the input t and output (your x-coordinate) have been changed to conform with your calculator.

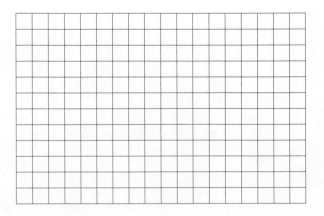

3. **a.** What is the maximum function value of the cosine function in Problem 2?

b. What is the minimum function value of the cosine function in Problem 2?

4. What is the connection between the function values in Problem 3, the 100-meter radius of the circle, and the coefficient 100 of the function?

5. If you are forced to run on a larger circular track of radius 150 meters, and you increase your speed to 150 meters per minute, predict the equations of the functions describing the x- and y-coordinates of your position. Explain.

6. Determine by inspection the amplitude of the following functions. Then verify the results with your graphing calculator. (*Remember:* Amplitude *cannot* be negative.)

a. $y = 3 \sin x$ **b.** $y = 0.4 \cos x$

c. $f(x) = -2 \cos x$ **d.** $g(x) = -2.3 \sin x$

e. $y = 2 \sin (3x)$ **f.** $h(x) = -4 \cos (x)$

ACTIVITY 6.10
Speeding Up

OBJECTIVE

1. Determine the period of the graph of $y = a \sin(bx)$ and $y = a \cos(bx)$ using a formula.

After a lot of practice, you begin to speed up on your circular track of radius 100 meters.

You finally achieve your personal goal of 200 meters per minute.

1. If you run 200 meters per minute, how long does it take you to complete one lap? Note that this amount of time to complete one lap will be important in defining the key concept of period for the trigonometric functions in the following problems.

2. Complete the following table to give your coordinates at selected special points along your path.

t (min)	YOUR x-COORDINATE	YOUR y-COORDINATE
0	100	0
$\frac{\pi}{2}$		
π		
2π		

3. On the following grid, use the data from Problem 2 to plot (t, y), your y-coordinate as a function of t. Connect your data pairs to make a smooth graph, as in the previous activity, and predict what will happen to the graph for values of t before and after the values of t in the table.

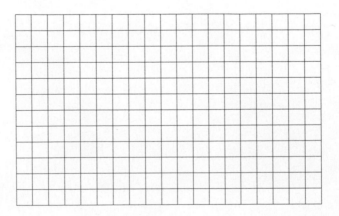

4. What is the amount of time it takes for the graph to complete one full cycle (that is, for you to complete one full lap)?

5. What effect does doubling your speed have on the amount of time to complete one cycle? Explain.

6. Use your graphing calculator to plot $y = 100 \sin (2x)$. Note that the name of the input has been changed to conform with your calculator.

7. Compare your results in Problems 3 and 6.

8. Recall that the period of a trigonometric function is the shortest time (distance) to complete one full cycle. Determine the period of each of the following using the graph.

    a. $y = 100 \sin x$ **b.** $y = 100 \sin (2x)$

The period of the graph of the sine function can be determined directly from the equation that defines the function. For example, the period of $y = 100 \sin (2x)$ is π units, which is half of the period of $y = 100 \sin (x)$. It appears that the coefficient of x in the function rule affects the period of the function. In general, if $y = a \sin (bx)$, $b > 0$, then the period is $2\pi/b$.

9. On the following grid, repeat Problem 3 to plot the data pairs (t, x).

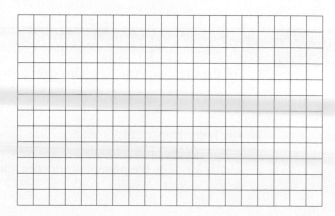

10. a. What equation should you enter into your calculator to produce the graph from Problem 9?

b. Enter the equation from part a into your calculator, and obtain a graph.

11. a. If you were to triple your speed on the track from the original 100 meters per minute, what effect would this have on the amount of time to complete one lap?

b. What equations would describe the *x*- and *y*-coordinates of your position?

12. Determine the period of each of the following functions.

a. $y = 100 \cos x$

b. $g(x) = 100 \cos(2x)$

c. $h(x) = 100 \cos(3x)$

SUMMARY
Activity 6.10

1. For the trigonometric functions defined by $y = a \sin(bx)$ or $y = a \cos(bx)$, the **period** is $\dfrac{2\pi}{b}$, where $b > 0$.

2. The **frequency** of trigonometric functions defined by $y = a \sin(bx)$ or $y = a \cos(bx)$ is the number of cycles completed by the graphs over intervals of length 2π. The **frequency** of trigonometric functions defined by $y = a \sin(bx)$ or $y = a \cos(bx)$ is b.

EXERCISES
Activity 6.10

1. For the following functions, identify by inspection both the amplitude and the period. Note that amplitude is always positive.

   a. $y = 3\cos 1.5x$

   b. $y = 0.5\sin 2x$

   c. $f(x) = -2.3\cos 0.4x$

   d. $g(x) = 36\sin 2\pi x$

2. Is there any relationship between the amplitude and the period of the sine function?

3. For each of the following tables, identify a function of the form $y = a\sin bx$ or $y = a\cos bx$ that approximately satisfies the table.

   a.
x	0	0.7854	1.5708	2.3562	3.1416
y	0	-15	0	15	0

   b.
x	0	2.244	4.488	6.732	8.976
y	1.3	0	-1.3	0	1.3

4. For each of the following graphs, identify a function of the form $y = a\sin bx$ or $y = a\cos bx$ that the graph approximates.

   a.

   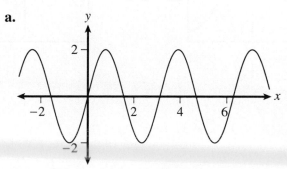

b.

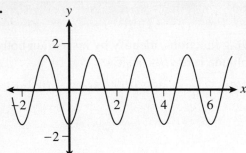

5. Match the given equation to one of the graphs that follow. Assume that Xscl = 1 and Yscl = 1.

a. $y = 2\cos 0.5x$ **b.** $y = -0.5\sin 2x$

c. $y = 0.5\cos 2x$ **d.** $y = 2\sin 0.5x$

i.

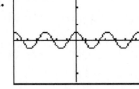

ii.

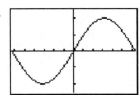

iii.

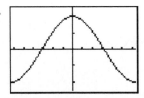

iv.

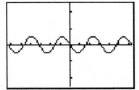

ACTIVITY 6.11

Running with a
Friend

OBJECTIVE

1. Determine the displacement
   of $y = a\sin(bx + c)$ and
   $y = a\cos(bx + c)$ using a
   formula.

While jogging, you become good friends with another runner who started out, like you, with a speed of 100 meters per minute on the 100-meter track. You enjoy running together but prefer to keep a healthy distance between each other during the run. You start at (100, 0), and when you arrive at (0, 100), your friend starts at (100, 0).

1. Assume that you both maintain the same speed of 100 meters per minute.

   a. How far ahead of your friend are you?

   b. How long after you start does your friend wait before starting?

2. Complete the following table to give coordinates for both you and your friend at selected special points along the track.

t (min)	YOUR x-COORDINATE	YOUR y-COORDINATE	YOUR FRIEND'S x-COORDINATE	YOUR FRIEND'S y-COORDINATE
0	100	0	—	—
$\frac{\pi}{2}$			100	0
π				
$\frac{3\pi}{2}$				
2π				
$\frac{5\pi}{2}$				

3. On the following grid, use the data from Problem 2 to plot (t, y) for both you and your friend. Connect the points to smooth out your graphs.

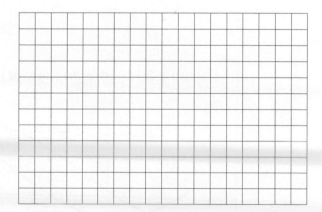

4. What is the relationship between the two graphs?

> **DEFINITION**
>
> The **displacement**, or **phase shift**, of the graph of $y = a\sin(bx + c)$ is the smallest movement (left or right) necessary for the graph of $y = a\sin(bx)$ to match the graph of $y = a\sin(bx + c)$ exactly.

Example 1 *Consider the graphs of $y = \sin x$ and $y = \sin(x + 1)$.*

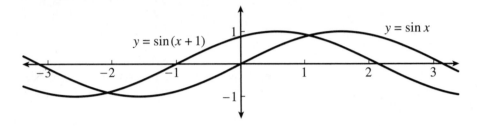

The graph of $y = \sin x$ must be moved 1 unit to the left to match the graph of $y = \sin(x + 1)$ exactly, so the displacement, or phase shift, is -1.

5. a. Which graph is displaced in Problem 3?

b. What is the displacement?

6. Would you expect the same type of relationship between the two graphs representing the x-coordinates? Explain.

Displacement, or phase shift, is defined for the cosine function in the same manner it is defined for sine.

Example 2 *Consider the graphs of $y = 3\cos 2x$ and $y = 3\cos(2x - 1)$.*

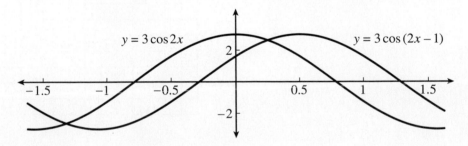

The graph of $y = 3\cos(2x - 1)$ appears to be about $\frac{1}{2}$ unit to the right of the graph of $y = 3\cos 2x$, so the displacement is approximately $\frac{1}{2}$.

7. a. If your *x*-coordinate is given by $100 \cos t$, predict the defining equation of your friend's *x*-coordinate.

b. Use your graphing calculator to test your prediction.

In general, in the functions defined by $y = a \sin(bx + c)$ and $y = a \cos(bx + c)$, the values for *b* and *c* affect the displacement, or phase shift, of the function. The phase shift is given by $-\frac{c}{b}$. Note that if $-\frac{c}{b}$ is negative, the shift is to the left. If $-\frac{c}{b}$ is positive, the shift is to the right.

8. a. Using the expression $-\frac{c}{b}$, what is the displacement of $y = 3 \cos(2x - 1)$?

b. Is your result consistent with the graph in Example 2?

9. In this activity, your friend's *y*-coordinate is given by $y = a \sin(bx + c)$, where $a = 100$ and $b = 1$. Calculate *c*.

SUMMARY
Activity 6.11

1. The **displacement**, or **phase shift**, of the graph of $y = a \sin(bx + c), b > 0$ is the smallest movement (left or right) necessary for the graph of $y = a \sin bx$ to match the graph of $y = a \sin(bx + c)$ exactly.

2. The **displacement**, or **phase shift**, of the graph of $y = a \cos(bx + c), b > 0$ is the smallest movement (left or right) necessary for the graph of $y = a \cos(bx)$ to match the graph of $y = \cos(bx + c)$ exactly.

3. For the functions $y = a \sin(bx + c)$, and $y = a \cos(bx + c)\, b > 0$, the phase shift is given by $-\frac{c}{b}$.

4. If $-\frac{c}{b}$ is negative, the shift is to the left. If $-\frac{c}{b}$ is positive, the shift is to the right.

EXERCISES
Activity 6.11

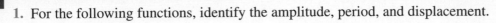

1. For the following functions, identify the amplitude, period, and displacement.

   a. $y = 0.7\cos\left(2x + \frac{\pi}{2}\right)$ **b.** $y = 3\sin(x - 1)$

   c. $f(x) = -2.5\sin\left(0.4x + \frac{\pi}{3}\right)$ **d.** $g(x) = 15\sin(2\pi x - 0.3)$

2. For each of the following tables, identify a function of the form $y = a\sin(bx + c)$ or $y = a\cos(bx + c)$ that approximately satisfies the table.

   a.

x	−0.7854	0.7854	2.3562	3.927	5.4978
y	0	3	0	−3	0

   b.

x	1	2.5708	4.1416	5.7124	7.2832
y	−0.5	0	0.5	0	−0.5

3. Sketch one cycle of the graph of the function defined by $f(x) = 2\sin\left(x + \frac{\pi}{2}\right)$.

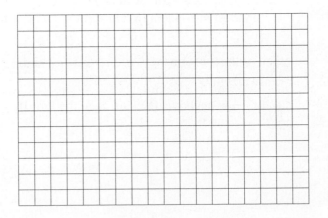

4. Determine an equation for the function defined by the following graph.

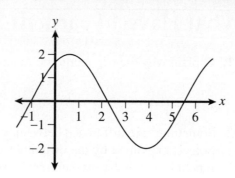

5. Match the given equation to one of the graphs that follow. (Assume that Xscl = 1 and Yscl = 1.)

a. $y = 2\cos(x - 1)$ **b.** $y = 2\sin(x - 1)$

c. $y = 2\cos(x + 2)$ **d.** $y = 2\sin(x + 2)$

i. 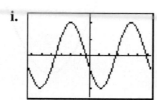   **ii.**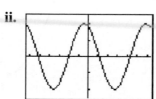

iii. 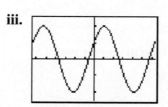   **iv.**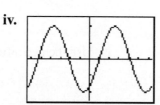

CLUSTER 2 **What Have I Learned?**

1. Explain why the trigonometric functions could be called circular functions.

2. Sometimes the difference between the trigonometric functions and the circular functions is explained by the difference in inputs. The input values for the trigonometric functions are angle measurements, and the input values for the circular functions are real numbers. How does your knowledge of radian measure relate to this?

3. **a.** Estimate the amplitude of the function defined by the following graph.

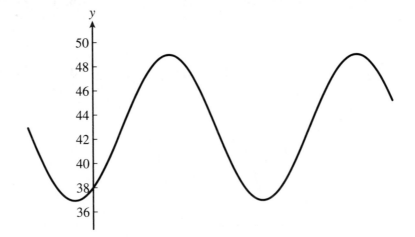

   b. Use the definition of amplitude to answer part a.

4. The period of $y = \sin x$ and $y = \cos x$ is 2π. Explain why this makes sense when sine and cosine are viewed as circular functions.

5. Given a function defined by $y = a\sin(bx + c)$ and $b > 0$, and given the fact that b and c possess opposite signs, determine whether the graph of the function is displaced to the right or to the left. Explain.

CLUSTER 2   # How Can I Practice?

1. Determine the coordinates of the point on the unit circle corresponding to the following central angles. If necessary, round your results to the nearest hundredth.

    a. 36°   **b.** 210°

    c. −90°   **d.** 317°

    e. −144°   **f.** 450°

2. For each of the points on the unit circle determined in Exercise 1, determine the distance traveled along the circle to the point from (1, 0).

    a.   **b.**

    c.   **d.**

    e.   **f.**

3. Convert the following degree measures to radian measures in terms of π.

    a. 18°   **b.** 150°

    c. 390°   **d.** −72°

4. Convert the following radian measures to degree measures.

    a. $\frac{5\pi}{6}$ **b.** 1.7π

    c. -3π **d.** 0.9π

Obtain the following information about each of the functions defined by the equations in Exercises 5–9:

    a. Use your graphing calculator to sketch a graph.

    b. From the defining equation, determine the amplitude. Then use your graph to verify that your amplitude is correct.

    c. From the defining equation, determine the period. Then use your graph to verify that your period is correct.

    d. From the defining equation, determine the displacement. Then use your graph to verify that your displacement is correct.

5. $y = 4\cos 3x$ 6. $y = -2\sin(x - 1)$

7. $s = 3.2\sin(-2x)$ 8. $f(x) = -\cos\left(\frac{x}{2} + 1\right)$

9. $g(x) = 4 - 3\cos(4x - 1)$

10. You rent a cottage on the ocean for a week one summer and notice that the tide comes in twice a day with approximate regularity. Remembering that the trigonometric functions model repetitive behavior, you place a meter stick in the water to measure water height every hour from 6:00 A.M. to midnight. At low tide the height of the water is zero centimeters, and at high tide the height is 80 centimeters.

a. Explain why a sine or a cosine function models this relationship between height of water in centimeters and time in hours.

b. What is the amplitude of this function?

c. Approximate the period of this function. Explain.

d. Determine a reasonable defining equation for this function. Explain.

The bracketed numbers following each concept indicate the activity in which the concept is discussed.

CONCEPT/SKILL	DESCRIPTION	EXAMPLE
The sine function of the acute angle *A* of a right triangle [6.1]	$\sin A = \dfrac{\text{length of the side opposite } A}{\text{length of the hypotenuse}}$	Example 2, Activity 6.1, page 649
The cosine function of the acute angle *A* of a right triangle [6.1]	$\cos A = \dfrac{\text{length of the side adjacent to } A}{\text{length of the hypotenuse}}$	Example 2, Activity 6.1, page 649
The tangent function of the acute angle *A* of a right triangle [6.1]	$\tan A = \dfrac{\text{length of the side opposite } A}{\text{length of the side adjacent to } A}$	Example 2, Activity 6.1, page 649
Complementary angles [6.2]	Complementary angles are two acute angles, the sum of whose measures is 90°.	Angles of 30° and 60° are complementary angles.
Cofunctions related to complementary angles [6.2]	Cofunctions of complementary angles are equal.	$\sin 35° = \cos 55°$
Inverse trigonometric functions [6.3]	The inverse trigonometric functions are defined by $\sin^{-1} x = \theta$ if and only if $\sin \theta = x$ $\cos^{-1} x = \theta$ if and only if $\cos \theta = x$ $\tan^{-1} x = \theta$ if and only if $\tan \theta = x$	Examples 1 and 2, Activity 6.3, pages 662, 663
The domain of the inverse trigonometric functions [6.3]	The domain (inputs) of the inverse trigonometric functions is the set of ratios of the lengths of the sides of a right triangle.	The domain of the inverse sine function is all real numbers from −1 to 1, including both −1 and 1.
The range of the inverse trigonometric functions [6.3]	The range (outputs) of the inverse trigonometric functions is the set of angles.	The range of the inverse tangent function is all angles from −90° to 90°.
Solving right triangles [6.4]	When trigonometric problems involve solving right triangles, employ the following trigonometric problem-solving strategy: 1. Draw a diagram of the situation using right triangles. 2. Identify all known sides and angles. 3. Identify sides and/or angles you want to know. 4. Identify functions that relate the known and unknown. 5. Write and solve the appropriate trigonometric equation(s).	Example 1, Activity 6.4, page 668

Central angle [6.7]	A central angle is an angle whose vertex is the center of a circle.	Problem 2b, Activity 6.7, page 686
The domain of the sine and cosine functions [6.7]	The domain of the sine and cosine functions is all angles, both positive and negative.	The domains are all real numbers.
The range of the sine and cosine functions [6.7]	The range of the sine and cosine functions is all values of N such that $-1 \le N \le 1$.	The range is all real numbers from -1 to 1 inclusive.
The graph of $y = \sin x$ [6.7]	The graph of $y = \sin x$ is a periodic wave. One graph is shown in the Example.	
The graph of $y = \cos x$ [6.7]	The graph of $y = \cos x$ is a periodic wave. One cycle is shown in the Example.	
The period of the sine and cosine functions [6.7]	The period is the number of units required to complete one cycle of the graph of a function.	The period is 2π for $y = \sin x$ and $y = \cos x$.
Radian measure [6.8]	Radian measure is used when the input of a repeating function is better defined by real numbers than angles measured in degrees.	$360°$ is equivalent to 2π radians.
Converting from radian measure to degree measure [6.8]	To convert degree measure to radian measure, multiply the degree measure by $\dfrac{\pi \text{ radians}}{180°}$.	$30° = 30 \cdot \dfrac{\pi}{180} = \dfrac{\pi}{6}$ radians
Converting from degree measure to radian measure [6.8]	To convert radian measure to degree measure, multiply the radian measure by $\dfrac{180°}{\pi \text{ radians}}$.	$\dfrac{2\pi}{3} = \dfrac{2\pi}{3} \cdot \dfrac{180°}{\pi} = \dfrac{360°}{3}$ $= 120°$
The amplitude of trigonometric functions [6.9]	The amplitude of trigonometric functions defined by $y = a\sin(bx)$ or $y = a\cos(bx)$ is defined by $$\frac{1}{2}(M - m),$$ where M represents the maximum function value and m represents the minimum function value.	Example 1, Activity 6.9, page 707

The period of sine and cosine [6.10]	For the trigonometric functions defined by $y = a \sin(bx)$ or $y = a \cos(bx)$, $b > 0$, the period is $\frac{2\pi}{b}$.	The period of $y = \sin(2x)$ is $\frac{2\pi}{2} = \pi$ units.
The frequency of sine and cosine [6.10]	The frequency of trigonometric functions defined by $y = a \sin(bx)$ or $y = a \cos(bx)$, $b > 0$ is the number of cycles completed by the graphs over intervals of length 2π. The frequency of trigonometric functions defined by $y = a \sin(bx)$ or $y = a \cos(bx)$, $b > 0$ is b.	The frequency of $y = \sin(2x)$ is 2; two complete cycles occur in 2π units.
The displacement, or phase shift, of the graph of sine or cosine [6.11]	The displacement, or phase shift, of the graph of $y = a \sin(bx + c)$ $b > 0$ or $y = a \cos(bx + c)$ $b > 0$ is the smallest movement (left or right) necessary for the graph of $y = a \sin bx$ to match the graph of $y = a \sin(bx + c)$ exactly.	Examples 1 and 2, Activity 6.11 page 716
The displacement, or phase shift, of the graph of sine or cosine [6.11]	For the functions $y = a \sin(bx + c)$ and $y = a \cos(bx + c)$, $b > 0$, the phase shift is given by $-\frac{c}{b}$.	The displacement of the function $y = 2\sin\left(3x + \frac{\pi}{2}\right)$ is given by $-\frac{\frac{\pi}{2}}{3} = -\frac{\pi}{6}$. The shift is $\frac{\pi}{6}$ units to the left.

Gateway Review

1. You walk seven miles in a straight line 63° north of east.

   a. Determine how far north you have traveled.

   b. Determine how far east you have traveled.

2. Solve the following triangles.

   a.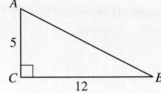

   b.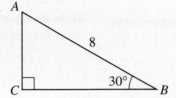

   c.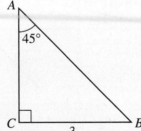

   d.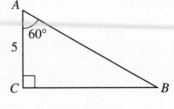

3. **a.** Given $\sin \theta = \frac{6}{10}$, determine $\cos \theta$ and $\tan \theta$ without using your calculator.

   b. Given $\cos \theta = \frac{\sqrt{3}}{2}$, determine $\sin \theta$, $\tan \theta$, and θ without using your calculator.

c. Given $\tan\theta = \frac{8}{5}$, determine $\sin\theta$ and $\cos\theta$ without using your calculator.

4. You are taking your nephew to see the Empire State Building. When you are 100 feet away from the building, you and your nephew look up to see the top. You are 6 feet tall, your nephew is 3 feet tall, and the Empire State Building is 1414 feet high. You notice that even though he is only half your height, your nephew does not have to tilt his head any more than you do. Is your observation correct? Is the angle always independent of people's heights? Explain.

5. In the diagram below, determine the lengths of a, b, c, and h.

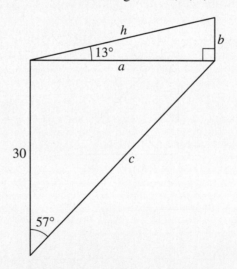

6. In the following diagram, determine x and h. (*Hint:* See Exercises 2c and d.)

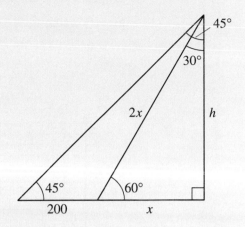

7. Using the following triangles, complete the table without using your calculator.

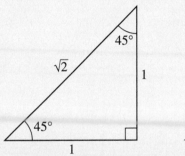

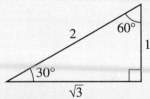

ANGLE θ	SIN θ	COS θ	TAN θ
120°	$\frac{\sqrt{3}}{2}$		
135°			
150°			
180°			
210°			
225°			
240°			
270°			
300°			
315°			
330°			
360°			

8. Determine the amplitude and period of the given functions, and then sketch their graphs. Use your graphing calculator to verify your results.

a. $y = 2 \sin x$ **b.** $y = -2 \sin x$

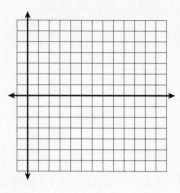

c. $y = \cos 2x$ **d.** $y = \cos 2\pi x$

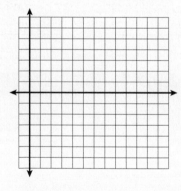
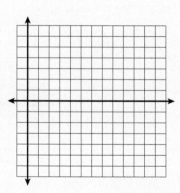

e. $y = \sin \frac{x}{2}$ **f.** $y = \sin \frac{\pi x}{2}$

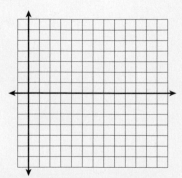
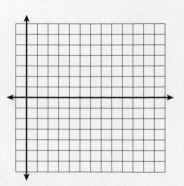

g. $y = \cos x$

h. $y = \frac{2}{3}\cos(2x)$

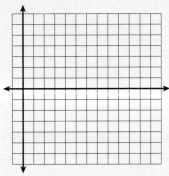

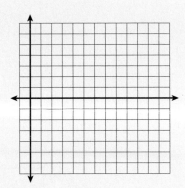

i. $y = \sin(2\pi x - 3\pi)$

j. $y = 3\sin(2\pi x - 3\pi)$

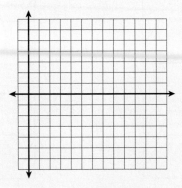

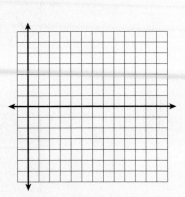

9. Match the given equation to one of the accompanying graphs. (Assume that Xscl = 1 and Yscl = 1.)

a. $y = -3\sin x$

b. $y = 2\sin\left(\frac{\pi x}{2} + \frac{\pi}{2}\right)$

c. $y = 2\sin\left(\frac{\pi x}{2} - \frac{\pi}{2}\right)$

d. $y = -3\cos x$

e. $y = 2\cos\left(\frac{\pi x}{2} + \frac{\pi}{2}\right)$

f. $y = -\cos\left(\pi x - \frac{\pi}{2}\right)$

i.

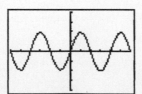

ii.

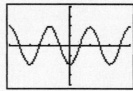

iii.

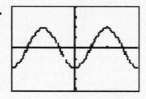

iv.

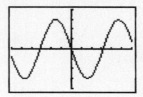

v.

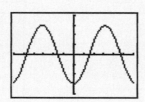

vi.

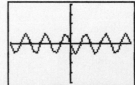

vii.

viii.

Concept Review

Properties of Exponents

The basic properties of exponents are summarized as follows:

> If a is a real number greater than zero and n and m are rational numbers then,
>
> **1.** $a^n a^m = a^{n+m}$ **2.** $\dfrac{a^n}{a^m} = a^{n-m}$ **3.** $\left(a^n\right)^m = a^{nm}$
>
> **4.** $\left(ab\right)^n = a^n b^n$ **5.** $a^0 = 1$ **6.** $a^{-n} = \dfrac{1}{a^n}$

Property 1: $a^n a^m = a^{n+m}$ If you are multiplying two powers of the same base, add the exponents.

Example 1: $x^4 \cdot x^7 = x^{4+7} = x^{11}$

Note that the exponents were added and the base did not change.

Property 2: $\dfrac{a^n}{a^m} = a^{n-m}$ If you are dividing two powers of the same base, subtract the exponents.

Example 2: $\dfrac{6^6}{6^4} = 6^{6-4} = 6^2 = 36$

Note that the exponents were subtracted and the base did not change.

Property 3: $\left(a^n\right)^m = a^{nm}$ If a power is raised to a power, multiply the exponents.

Example 3: $\left(y^3\right)^4 = y^{12}$

The exponents were multiplied. The base does not change.

Property 4: $\left(ab\right)^n = a^n b^n$ If a product is raised to a power, each factor is raised to that power.

Example 4: $\left(2x^2 y^3\right) = 2^3 \cdot \left(x^2\right)^3 \cdot \left(y^3\right)^3 = 8x^6 y^9$

Note that since the base contained three factors, each of those was raised to the third power. The common mistake in an expansion such as this is not to raise the coefficient to the power.

Property 5: $a^0 = 1, a \neq 0$. Often presented as a definition, property 5 states that any nonzero base raised to the zero power is one. This property or definition is a result of property 2 of exponents as follows:

Consider $\dfrac{x^5}{x^5}$. Using property 2, $x^{5-5} = x^0$. However, you know that any fraction in which the numerator and the denominator are equal is equivalent to 1. Therefore, $x^0 = 1$.

Example 5: $\left(\dfrac{2x^3}{3yz^5}\right)^0 = 1$

Given a nonzero base, if the exponent is zero, the value is one.

Property 6: $a^{-n} = \dfrac{1}{a^n}$ Sometimes presented as a definition, property 6 states that any base raised to a negative power is equivalent to the reciprocal of the base raised to the positive power. Note that the negative exponent does not have any effect on the sign of the base. This property could also be viewed as a result of the second property of exponents as follows:

Consider $\dfrac{x^3}{x^5}$. Using property 2, $x^{3-5} = x^{-2}$. If you view this expression algebraically, you have three factors of x in the numerator and five in the denominator. If you divide out the three common factors, you are left with $\dfrac{1}{x^2}$. Therefore, if property 2 is true, then $x^{-2} = \dfrac{1}{x^2}$.

Example 6: Write each of the following without negative exponents.

      a. 3^{-2} **b.** $\dfrac{2}{x^{-3}}$

Solution: **a.** $3^{-2} = \dfrac{1}{3^2} = \dfrac{1}{9}$ **b.** $\dfrac{2}{x^{-3}} = \dfrac{2}{\dfrac{1}{x^3}} = 2 \div \dfrac{1}{x^3} = 2 \cdot x^3 = 2x^3$

A **factor** can be moved from a numerator to a denominator or from a denominator to a numerator by changing the *sign of the exponent.*

Example 7: Simplify and express your results with positive exponents only.

$$\left(\dfrac{x^3y^{-4}}{2x^{-3}y^{-2}z}\right) \cdot \left(\dfrac{4x^3y^2z}{x^5y^{-3}z^3}\right)$$

Solution: Simplify each factor by writing them with positive exponents only.

$$\left(\dfrac{x^6}{2y^2z}\right) \cdot \left(\dfrac{4y^5}{x^2z^2}\right)$$

Now multiply and simplify.

$$\dfrac{4x^6y^5}{2x^2y^2z^3} = \dfrac{2x^4y^3}{z^3}$$

Exercises

Simplify and express your results with positive exponents only.

1. 5^{-3}

2. $\dfrac{1}{x^{-5}}$

3. $\dfrac{3x}{y^{-2}}$

4. $\dfrac{10x^2y^5}{2x^{-3}}$

5. $\dfrac{5^{-1}z}{x^{-1}z^{-2}}$

6. $5x^0$

7. $(a + b)^0$

8. $-3(x^0 - 4y^0)$

9. $x^6 \cdot x^{-3}$

10. $\dfrac{4^{-2}}{4^{-3}}$

11. $(4x^2y^3) \cdot (3x^{-3}y^{-2})$

12. $\dfrac{24x^{-2}y^3}{6x^3y^{-1}}$

13. $\dfrac{(14x^{-2}y^{-3}) \cdot (5x^3y^{-2})}{6x^2y^{-3}z^{-3}}$

14. $\left(\dfrac{2x^{-2}y^{-3}}{z^2}\right) \cdot \left(\dfrac{x^5y^3}{z^{-3}}\right)$

15. $\dfrac{(6x^4y^{-3}z^{-2})(3x^{-3}y^4)}{15x^{-3}y^{-3}z^2}$

Solving 3 × 3 Linear Systems Algebraically

Linear equations such as $3x + 2y - z = 4$ involve three variables, x, y, and z. A solution of such an equation is an ordered triple (x, y, z) such that if the values of x, y, and z in the ordered triple are substituted into the equation, the result is a true statement.

A system of three linear equations in three variables (a 3 × 3 system) such as

$$-x + y + z = -3$$
$$3x + 9y + 5z = 5$$
$$x + 3y + 2z = 4$$

has as its solution all ordered triples (x, y, z) that will make all three equations true.

Solving a 3 × 3 Linear System Algebraically

1. Eliminate one variable using any two of the given three equations to obtain an equation in two variables (or less).

2. Eliminate the same variable using the third equation not used in step 1 and either of the other two equations to obtain a second equation in two variables (or less).

3. Solve the system consisting of the two equations found in steps 1 and 2.

4. Substitute the values obtained in step 3 into any equation involving all three variables to determine the value of the third variable.

5. Check your solution in all three equations.

Example 1: Determine all solutions of

$$-x + y + z = -3$$
$$3x + 9y + 5z = 5$$
$$x + 3y + 2z = 4.$$

Solution: Since the coefficients of x are -1, 3, and 1, they have an LCD of 3. Thus, you may choose to eliminate x using the addition method.

Step 1 Multiply both sides of the first equation by 3 and add it to the second equation.

$$3(-x + y + z = -3) \qquad -3x + 3y + 3z = -9$$
$$3x + 9y + 5z = 5 \quad \text{or} \quad 3x + 9y + 5z = 5$$

The sum is $12y + 8z = -4$.

Step 2 Multiply the third equation by -3 and add to the second

$$3x + 9y + 5z = 5 \qquad 3x + 9y + 5z = 5$$
$$-3(x + 3y + 2z = 4) \quad \text{or} \quad -3x - 9y - 6z = -12$$

The sum is $-z = -7$.

Step 3 The 2 × 2 system resulting from steps 1 and 2 is

$$12y + 8z = -4$$
$$-z = -7$$

The second equation is equivalent to $z = 7$. Substituting this value into the first equation of the new system yields

$$12y + 8(7) = -4$$
$$12y + 56 = -4$$
$$12y = -60$$
$$y = -5$$

Step 4   Using the values of y and z in the third equation of the original system.

$$x + 3(-5) + 2(7) = 4$$
$$x - 15 + 14 = 4$$
$$x - 1 = 4$$
$$x = 5$$

The potential solution is $(5, -5, 7)$ and should be checked in all three of the original equations.

Not every 3×3 linear system has unique solutions; some have multiple solutions. These are called **dependent systems.** Some systems have no solution and are called **inconsistent systems.**

Example 2: Solve the following system.

$$x + 2y + 3z = 5$$
$$-x + y - z = -6$$
$$2x + y + 4z = 4$$

Solution: Since the coefficients of x are 1, -1, and 2, the LCD is 2. Again, you may eliminate x using the addition method.

Step 1   Sum the first two equations to eliminate x.

$$x + 2y + 3z = 5$$
$$-x + y - z = -6$$

The sum is $3y + 2z = -1$.

Step 2   Multiply the second equation by 2 and add it to the third.

$$-2x + 2y - 2z = -12$$
$$2x + y + 4z = 4$$

The sum is $3y + 2z = -8$. The new system is

$$3y + 2z = -1$$
$$3y + 2z = -8$$

Step 3   To solve the new system, multiply the first equation by -1 and add it to the second.

$$-3y - 2z = 1$$
$$3y + 2z = -8$$

The sum is $0 = -7$.

Since $0 = -7$ is a false statement, the conclusion is there is no solution. The original system was an inconsistent system.

Had the sum of the equations in step 3 resulted in a true statement such as $0 = 0$, the conclusion would have been that there were an infinite number of solutions. That is, the system would have been *dependent*.

Exercises

Solve the following systems algebraically. If the system is dependent or inconsistent, state this as the answer.

1. $x + y - z = 9$
   $x + y + z = 5$
   $x - y + 2z = 1$

2. $-2x + y + 4z = 3$
   $x + y - 3z = 2$
   $x - y + 2z = 1$

3. $x + 2y + 3z = 5$
   $-x + y - z = -6$
   $2x + y + 4z = 4$

4. $3x - 2y + 3z = 11$
   $2x + 3y - 2z = -5$
   $x + 4y - z = -5$

5. $x - 4y + z = -5$
   $3x - 12y + 3z = -15$
   $-2x + 8y - 2z = 10$

6. $2x + 3y + 4z = 3$

    $6x - 6y + 8z = 3$

    $4x + 3y - 4z = 2$

7. $x + 2y = 10$

    $-x + 3z = -23$

    $4y - z = 9$

Inequalities Involving Absolute Value

The key to solving absolute value inequalities algebraically is to rewrite them using the following properties.

Absolute Value Properties

For any real number x and $a > 0$,

$$|x| < a \text{ is equivalent to } -a < x < a.$$

For any real number x and $a > 0$,

$$|x| > a \text{ is equivalent to } x > a \text{ or } x < -a.$$

Solving Absolute Value Inequalities

1. Rewrite the inequality with the absolute value isolated.
2. Rewrite the inequality as a compound inequality or pair of inequalities.
3. Solve the resulting inequality(s).

Example 1: Solve $|2x - 3| + 3 \leq 8$.

Solution: Subtract 3 from both sides.

$$|2x - 3| + 3 - 3 \leq 8 - 3 \text{ or } |2x - 3| \leq 5$$

Using the properties,

$$-5 \leq 2x - 3 \leq 5$$

Add 3 to each part.

$$-5 + 3 \le 2x - 3 + 3 \le 5 + 3$$
$$-2 \le 2x \le 8$$

Divide each part by 2.

$$\frac{-2}{2} \le \frac{2x}{x} \le \frac{8}{2} \quad \text{or} \quad -1 \le x \le 4$$

Example 2: Solve $|4x + 3| - 4 > 7$.

Solution: Add 4 to both sides to isolate the absolute value.

$$|4x + 3| - 4 + 4 > 7 + 4$$
$$|4x + 3| > 11$$

Using the absolute value properties.

$$4x + 3 > 11 \quad \text{or} \quad 4x + 3 < -11$$

Solving these inequalities.

$$4x + 3 - 3 > 11 - 3 \qquad 4x + 3 - 3 < -11 - 3$$
$$4x > 8 \qquad\qquad x < -14$$
$$x > 2 \qquad\qquad x < -\frac{7}{2}$$

Exercises

Solve the following inequalities.

1. $|3x - 5| < 5$ **2.** $|x - 3| - 2 \le 3$

3. $|4x - 1| > 3$ **4.** $|2x - 1| - 4 \ge 7$

Solving Equations by Factoring

Many quadratic and higher-order polynomial equations can be solved by using factoring and the zero product rule.

The process is as follows.

Solving an Equation by Factoring

1. Use the addition principle to remove all terms from one side of the equation. This results in the equation being set equal to zero.
2. Combine like terms and then factor.
3. Use the zero product rule to set each factor containing a variable equal to zero and then solve the equations.
4. Check your solutions in the original equation.

Example 1: Solve the equation $x(x + 5) = 0$.

Solution: This equation already satisfies the first two steps in our process, so we simply start at step 3.

$$x = 0 \quad x + 5 = 0$$
$$x = -5$$

Thus, we have two solutions $x = 0$ and $x = -5$. The check is left to the reader.

Example 2: Solve the equation $6x^2 = 16x$.

Solution: Setting the equation equal to zero, $6x^2 = 16x = 0$.
Since there are no like terms, factor the binomial.

$$2x(3x - 8) = 0$$

Using the zero-product principle,

$$2x = 0 \quad \text{or} \quad 3x - 8 = 0$$
$$x = 0 \quad \text{or} \quad 3x = 8$$
$$x = \tfrac{8}{3}$$

The two potential solutions are $x = 0$ and $x = 8/3$. The check is left to the reader.

Example 3: Solve $3x^2 - 2 = -x$.

Solution: Setting the equation equal to zero, $3x^2 + x - 2 = 0$.
Since there are no like terms, factor the trinomial.

$$(3x - 2)(x + 1) = 0.$$

Using the zero-product principle,

$$3x - 2 = 0 \quad \text{or} \quad x + 1 = 0$$
$$3x = 2 \quad \text{or} \quad x = -1$$
$$x = \tfrac{2}{3}$$

The two potential solutions are $x = \tfrac{2}{3}$ and $x = -1$. The check is left to the reader.

Example 4: Solve $3x^3 - 8x^2 = 3x$.

Solution: Setting the equation equal to zero, $3x^3 - 8x^2 - 3x = 0$.
Since there are no like terms, factor the trinomial.

$$x(3x^2 - 8x - 3) = 0$$
$$x(3x + 1)(x - 3) = 0$$

Using the zero-product principle,

$$x = 0 \quad \text{or} \quad 3x + 1 = 0 \quad \text{or} \quad x - 3 = 0$$
$$3x = -1 \quad \text{or} \quad x = 3$$
$$x = -\tfrac{1}{3}$$

The three potential solutions are $x = 0$, $x = -1/3$, and $x = 3$. The check is left to the reader.

Exercises

Solve each of the following equations.

1. $x(x + 7) = 0$

2. $3(x - 5)(2x + 1) = 0$

3. $12x = x^2$

4. $x^2 + 5x = 0$

5. $x^2 - 2x - 63 = 0$

6. $3x^2 - 9x - 30 = 0$

7. $-7x + 6x^2 = 10$

8. $3y^2 = 2 - y$

9. $-28x^2 + 15x - 2 = 0$

10. $4x^2 - 25 = 0$

11. $(x + 4)^2 - 16 = 0$ **12.** $(x + 1)^2 - 3x = 7$

13. $2(x + 2)(x - 2) = (x - 2)(x + 3) - 2$

14. $18x^3 = 15x^2 + 12x$

Solving Quadratic Equations by Completing the Square

The square root property can be used to solve equations of the form $x^2 = a$.

> **Square Root Property**
>
> If $x^2 = a$, where a is a real number then $x = \pm \sqrt{a}$.

Example 1: Solve the equation $(x + 3)^2 = 9$.

Solution: This equation fits the form of the hypothesis of the square root property, where $B = x + 3$. Therefore,

$$x + 3 = \pm 3.$$

You now have two equations to solve $x + 3 = 3$ and $x + 3 = -3$. The solutions are $x = 0, -6$. Both of these values make the original statement true. Hence, both are solutions.

This example illustrates the goal of an algebraic technique of solving quadratic equation known as *completing the square*. The strategy is to rewrite the quadratic equation $ax^2 + bx + c = 0, a \neq 0$ in the form $(x + h)^2 = k$ and solve as in Example 1. This requires an algebraic process know as completing the square.

Consider the binomial $x^2 + 6x$. What term must be added to the binomial to produce a binomial that is a perfect square. The answer is one half the coefficient of the linear term squared. In this case, one half of 6 is 3. $3^2 = 9$, and

$$x^2 + 6x + 9 = (x + 3)^2.$$

This process can be helpful in solving quadratic equations as follows.

Solving a Quadratic Equation by Completing the Square

1. Use the multiplication principle to make the coefficient of x^2 equal to 1.
2. Rewrite the equation with the constant term isolated on one side.
3. Use the addition principle to add one half the coefficient of the linear term to both sides of the equation.
4. Replace the trinomial with its factored form.
5. Apply the square root property.
6. Solve the resulting linear equations.
7. Check your solutions in the original equation.

Example 2: Solve $x^2 - 4x - 5 = 0$ by the completing the square method.

Since the coefficient of x^2 is one, step 1 is not necessary.

Step 2　　　Adding 5 from both sides to isolate the constant term yields

$$x^2 - 4x = 5.$$

Step 3　　　The value need to complete the square is $\left(\frac{1}{2} \cdot (-4)\right)^2 = (-2)^2 = 4$. Using the addition principle to add this to both sides produces

$$x^2 - 4x + 4 = 5 + 4$$
$$x^2 - 4x + 4 = 9.$$

Step 4　　　Replacing the trinomial with its perfect square form,

$$(x - 2)^2 = 9.$$

Steps 5 and 6　Applying the square root principle and solving,

$$x - 2 = \pm 3$$
$$x = 2 \pm 3$$
$$x = -1, 5.$$

The checking of the solutions is left to the reader.

Example 3: Solve $6x + 6 = -x^2$ by the completing the square method.

Step 1　　　Multiply each term by -1 to make the coefficient of x^2 equal to one.

$$-6x - 6 = x^2$$

Step 2　　　Using the addition principle to isolate the constant term.

$$-6 = x^2 + 6x$$

Step 3   The value necessary to complete the square is $\left(\frac{1}{2} \cdot 6\right)^2 = 9$. Completing the square yields

$$-6 + 9 = x^2 + 6x + 9.$$

Step 4   Factoring and simplifying.

$$3 = (x + 3)^2$$

Steps 5 and 6   Taking the square root of both sides and solving.

$$\pm\sqrt{3} = x + 3$$
$$x = -3 \pm \sqrt{3}$$

The check is left to the reader. Note the solutions in this case are real, but not rational.

Exercises

Solve the following quadratic equations using the completing the square method.

1. $x^2 - 6x + 8 = 0$ **2.** $x^2 - 9x + 14 = 0$

3. $-4x = -x^2 + 12$ **4.** $2x^2 + 2x - 24 = 0$

5. $3x^2 + 2x = 1$ **6.** $-\frac{1}{2}x^2 - x + \frac{3}{2} = 0$

7. $10x^2 + 6x = 5$ **8.** $15x^2 - 10x - 3 = 0$

Derivation of the Quadratic Formula

The quadratic formula results from applying the completing the square method to the general quadratic equation $ax^2 + bx + c = 0$, where $a > 0$.

Step 1 Make the coefficient of x^2 equal 1, by multiplying both sides of the equation by $\frac{1}{a}$.

$$x^2 + \frac{b}{a}x + \frac{c}{a} = 0$$

Step 2 Use the addition principle to isolate the constant term on one side of the equals sign.

$$x^2 + \frac{b}{a}x = -\frac{c}{a}$$

Step 3 Complete the square of the binomial. The coefficient of the linear term is $\frac{b}{a}$. The term needed to complete the square is $\left(\frac{1}{2} \cdot \frac{b}{a}\right)^2 = \frac{b^2}{4a^2}$. Using the addition principle to add this term to both sides yields

$$x^2 + \frac{b}{a}x + \frac{b^2}{4a^2} = \frac{b^2}{4a^2} - \frac{c}{a}.$$

Rewriting the right-hand side as a single fraction,

$$x^2 = \frac{b}{a}x + \frac{b^2}{4a^2} = \frac{b^2 - 4ac}{4a^2}.$$

Step 4 Express the left-hand side in factored form.

$$\left(x + \frac{b}{2a}\right)^2 = \frac{b^2 - 4ac}{4a^2}$$

Step 5 Apply the square root property to the resulting equation.

$$x + \frac{b}{2a} = \pm\frac{\sqrt{b^2 - 4ac}}{2a}$$

Step 6 Solve for x.

$$x = -\frac{b}{2a} \pm \frac{\sqrt{b^2 - 4ac}}{2a} \text{ or } x = \frac{-b \pm \sqrt{b^2 - 4ac}}{2a}$$

This equation can be used to solve any quadratic equation in standard form $ax^2 + bx + c = 0 \; a \neq 0$, and is called the **quadratic formula.**

Rational Expressions

You may need to practice skills relating to rational functions to enhance your understanding of these functions

Simplifying Rational Expressions
  1. Factor the numerator and the denominator.
  2. Divide the numerator and the denominator by the common factors.

Example 1: Simplify $\dfrac{x^2 - 10x + 24}{x^2 - 5x + 4}$.

Solution:

Step 1   Factor the numerator and the denominator.

$$\frac{x^2 - 10x + 24}{x^2 - 5x + 4} = \frac{(x-4)(x-6)}{(x-1)(x-4)}$$

Step 2   Divide the numerator and denominator by the common factor.

$$\frac{\dfrac{(x-4)(x-6)}{(x-4)}}{\dfrac{(x-1)(x-4)}{(x-4)}} = \frac{x-6}{x-1}$$

Multiplying or Dividing Rational Expressions
  1. Factor the numerator and denominator of each fraction completely.
  2. Divide out the common factors (cancel).
  3. Multiply remaining factors.
  4. In division, proceed as above after inverting the divisor (the fraction after the division sign).

Example 2: Divide and simplify $\dfrac{x^2 + 3x - 10}{2x} \div \dfrac{x^2 - 5x + 6}{x^2 - 3x}$.

Solution:

Step 1   Rewrite as multiplication.
$$\frac{x^2 + 3x - 10}{2x} \div \frac{x^2 - 5x + 6}{x^2 - 3x} = \frac{x^2 + 3x - 10}{2x} \cdot \frac{x^2 - 3x}{x^2 - 5x + 6}$$

Step 2   Factor each fraction completely.
$$\frac{x^2 + 3x - 10}{2x} \cdot \frac{x^2 - 3x}{x^2 - 5x + 6} = \frac{(x+5)(x-2)}{2 \cdot x} \cdot \frac{x \cdot (x-3)}{(x-2)(x-3)}$$

Step 3   Cancel common factors.
$$\frac{(x+5)(x-2)}{2 \cdot x} \cdot \frac{x \cdot (x-3)}{(x-2)(x-3)} = \frac{(x+5)}{2} \cdot \frac{1}{(x-2)}$$

Step 4   Multiply the remaining fractions.
$$\frac{(x+5)}{2} \cdot \frac{1}{(x-2)} = \frac{x+5}{2x-4}$$

Adding or Subtracting Rational Expressions

1. Find the LCD (least common denominator.).
2. Build each fraction to have the LCD.
3. Add or subtract numerators.
4. Place the numerator over the LCD; and simplify if necessary.

Example 3: Add and simplify $\dfrac{x}{x+1} + \dfrac{3}{(x+1)^2}$.

Solution:

Step 1   Since the denominators are already factored, it is clear that the LCD is $(x+1)^2$.

Step 2   Build each fraction to have the LCD.

$$\frac{x}{x+1} + \frac{3}{(x+1)^2} = \frac{x(x+1)}{(x+1)(x+1)} + \frac{3}{(x+1)^2} = \frac{x^2+x}{(x+1)^2} = \frac{3}{(x+1)^2}$$

Step 3   Add or subtract the numerators.

$$\frac{x^2+x}{(x+1)^2} + \frac{3}{(x+1)^2} = \frac{x^2+x+3}{(x+1)^2}$$

Step 4   Since the numerator can't be factored we are done.

Solving Rational Equations

1. Find the LCD of all fractions in the equation.
2. Multiply both sides of the equation by $\frac{\text{LCD}}{1}$ (clear all denominators).
3. Solve the resulting equation.
4. Check for extraneous roots.

Example 4: Solve $3 - \dfrac{4}{x} = \dfrac{5}{2}$.

Solution:

Step 1   The LCD is $2x$.

Step 2   Multiply both sides of the equation by $\frac{2x}{1}$.

$$\tfrac{2x}{1}\left(3 - \tfrac{4}{x}\right) = \tfrac{2x}{1}\left(\tfrac{5}{2}\right)$$

This is equivalent to $6x - \frac{8x}{x} = \frac{10x}{2}$ or $6x - 8 = 5x$.

Step 3   Solving the resulting equation

$$6x - 6x - 8 = 5x - 6x \quad \text{or} \quad -8 = -x \quad \text{so} \quad x = 8.$$

Step 4   The check is left to the reader.

Exercises

Simplify the following.

1. $\dfrac{3x^2 - 6x}{x^2 + x - 6}$

2. $\dfrac{2x^3 + 4x^2 - 4x}{2x + 4}$

3. $\dfrac{x^2 + 2x - 15}{3 - x}$

Perform the indicated operations and simplify.

4. $\dfrac{4x^2y}{5xz} \cdot \dfrac{15x^6}{8xy^2}$

5. $\dfrac{x^2 + 2x - 15}{3x + 15} \cdot \dfrac{x - 3}{3}$

6. $\dfrac{3}{x^2} + \dfrac{5}{6x}$

7. $\dfrac{2}{x - 5} - \dfrac{3}{x + 3}$

8. $\dfrac{3}{x - 3} + \dfrac{x - 2}{x^2 - 9}$

9. $\dfrac{5}{x^2 - x - 2} - \dfrac{2}{x^2 + 4x + 3}$

10. $\dfrac{x - 3}{x^2 - 3x + 2} - \dfrac{x + 1}{x^2 - 4}$

Solve the following equations.

11. $\dfrac{x}{3} + \dfrac{2x}{7} = 10$

12. $\dfrac{-2}{x} + \dfrac{8}{3} = \dfrac{2}{x}$

13. $\dfrac{x - 2}{x - 4} = \dfrac{x}{x - 1}$

14. $\dfrac{1}{x - 4} + x = \dfrac{-3}{x - 4}$

15. Solve $\dfrac{1}{R_1} + \dfrac{1}{R_2} = \dfrac{1}{R}$ for R.

Complex Fractions

Complex fractions are fractions with a fractional expression in the numerator, the denominator, or both. Examples include

$$\frac{\frac{2}{5} + \frac{1}{3}}{7}, \quad \frac{x + 3}{\frac{x}{x+1} - 2}, \quad \text{and} \quad \frac{x + \frac{1}{x} - 3}{x^3 - x - \frac{2}{x^2}}.$$

There are two methods commonly used to simplify complex fractions. The first is to express the numerator and denominator as single fractions then divide.

Simplifying a Complex Fraction by Simplifying the Numerator and Denominator

1. Express the numerator as a single fraction.
2. Express the denominator as a single fraction.
3. Divide the numerator by the denominator.
4. Simplify, if possible.

Example 1: Simplify $\dfrac{1 - \frac{7}{16}}{3 - \frac{2}{5}}$.

Solution:

Step 1   Simplify the numerator. $1 - \dfrac{7}{16} = \dfrac{16}{16} - \dfrac{7}{16} = \dfrac{9}{16}.$

Step 2   Simplify the denominator. $3 - \dfrac{2}{5} = \dfrac{3}{1} - \dfrac{2}{5} = \dfrac{15}{5} - \dfrac{2}{5} = \dfrac{13}{5}.$

Step 3   Divide the numerator by the denominator. $\dfrac{9}{16} \div \dfrac{13}{5} = \dfrac{9}{16} \cdot \dfrac{5}{13} = \dfrac{45}{208}.$

Step 4   Since the fraction cannot be simplified, the simplified result is $\dfrac{45}{208}.$

Example 2: Simplify $\dfrac{\frac{1}{x} + \frac{2}{x^2}}{2 + \frac{1}{x^2}}$.

Solution:

Step 1   Simplify the numerator. $\dfrac{1}{x} + \dfrac{2}{x^2} = \dfrac{x}{x^2} + \dfrac{2}{x^2} = \dfrac{x + 2}{x^2}$

Step 2   Simplify the denominator. $2 + \dfrac{1}{x^2} = \dfrac{2}{1} + \dfrac{1}{x^2} = \dfrac{2x^2}{x^2} + \dfrac{1}{x^2} = \dfrac{2x^2 + 1}{x^2}$

Step 3   Divide the numerator by the denominator.

$$\frac{x + 2}{x^2} \div \frac{2x^2 + 1}{x^2} = \frac{x + 2}{x^2} \cdot \frac{x^2}{2x^2 + 1} = \frac{x + 2}{2x^2 + 1}$$

Step 4   The result in step 3 is simplified.

The second method of simplifying a complex fraction is to multiply the numerator and denominator by the LCD of the entire fraction.

Simplifying a Complex Fraction by Multiplying by the LCD

1. Determine the LCD of the numerator fractions and denominator fractions.

2. Multiply the numerator and denominator by 1 in the form $\frac{LCD}{LCD}$.

3. Simplify, if possible.

Example 3: Simplify $\dfrac{1 - \frac{7}{16}}{3 - \frac{2}{5}}$.

Solution:

Step 1   The only denominators are 16 and 5. Since there are no common factors the LCD is 80.

Step 2 $\dfrac{80\left(1 - \frac{7}{16}\right)}{80\left(3 - \frac{2}{5}\right)} = \dfrac{80 - \frac{80 \cdot 7}{16}}{240 - \frac{80 \cdot 2}{5}} = \dfrac{80 - 5 \cdot 7}{240 - 16 \cdot 2} = \dfrac{80 - 35}{240 - 32} = \dfrac{45}{208}.$

Step 3   Since the fraction is simplified, $\dfrac{45}{208}$ is the desired result.

Example 4: Simplify $\dfrac{\frac{3}{n - 5} - 2}{1 - \frac{4}{n - 5}}$.

Step 1   The LCD is $n - 5$.

Step 2

$$\frac{(n - 5) \cdot \left(\frac{3}{n - 5} - 2\right)}{(n - 5) \cdot \left(1 - \frac{4}{n - 5}\right)} = \frac{\frac{3(n - 5)}{n - 5} - 2(n - 5)}{(n - 5) - \frac{4(n - 5)}{n - 5}} = \frac{3 - 2n + 10}{n - 5 - 4} = \frac{-2n + 13}{n - 9}$$

Step 3   Since the numerator and denominator have no common factors the simplified result is $\dfrac{-2n + 13}{n - 9}$.

Exercises

Simplify the following complex fractions.

1. $\dfrac{\frac{1}{2} - \frac{1}{4}}{\frac{5}{8} + \frac{3}{4}}$

2. $\dfrac{\frac{5}{6y}}{\frac{10}{3xy}}$

3. $\dfrac{\frac{8x^2y}{3z^3}}{\frac{4xy}{9z^5}}$

4. $\dfrac{3 - \frac{1}{x}}{1 - \frac{1}{x}}$

5. $\dfrac{\frac{x^2}{y} - y}{\frac{y^2}{x} - x}$

6. $\dfrac{4 + \frac{6}{n+1}}{7 - \frac{4}{n+1}}$

7. $\dfrac{\frac{1}{y-2} + \frac{3}{x}}{\frac{5}{x} - \frac{4}{xy-2x}}$

8. $\dfrac{\frac{x}{x+1} - 1}{\frac{x+1}{x-1}}$

9. $\dfrac{1 + \frac{x}{x+1}}{\frac{2x+1}{x-1}}$

10. $\dfrac{\frac{x+1}{x-1} + \frac{x-1}{x+1}}{\frac{x+1}{x-1} - \frac{x-1}{x+1}}$

Radicals and Fractional Exponents

You may need to practice skills relating to radical functions to enhance your understanding of these functions.

Translating Radical Expressions to Expressions Using Rational Exponents

1. The power of the expression becomes the numerator of the exponent.
2. The index (root) becomes the denominator of the exponent.

For example, $\sqrt[3]{x^2} = x^{\frac{2}{3}}$. In reverse, $a^{\frac{5}{4}} = \sqrt[4]{a^5}$.

When a $\sqrt{}$ is written without an index, the index is assumed to be 2.

Fractional exponents also obey the laws of exponents as outlined on page A-1

1. $a^n a^m = a^{n+m}$ **2.** $\dfrac{a^n}{a^m} = a^{n-m}$ **3.** $(a^n)^m = a^{nm}$

4. $(ab)^n = a^n b^n$ **5.** $a^0 = 1$ **6.** $a^{-n} = \dfrac{1}{a^n}$

In Examples 1–6, the following steps are used:

1. Write each expression using rational exponents.
2. Apply the appropriate property of exponents.
3. Write the expression using radical notation.

Example 1: $\sqrt[3]{x^2} \cdot \sqrt[4]{x} = x^{\frac{2}{3}} \cdot x^{\frac{1}{4}} = x^{\left(\frac{2}{3}+\frac{1}{4}\right)} = x^{\left(\frac{8}{12}+\frac{3}{12}\right)} = x^{\frac{11}{12}} = \sqrt[12]{x^{11}}$

Example 2: $\dfrac{\sqrt[5]{x^4}}{\sqrt{x^7}} = \dfrac{x^{\frac{4}{5}}}{x^{\frac{7}{10}}} = x^{\left(\frac{4}{5}-\frac{7}{10}\right)} = x^{\frac{1}{10}} = \sqrt[10]{x}$

Example 3: $\left(\sqrt[3]{\sqrt{x^5}}\right) = \left(x^{\frac{5}{2}}\right)^{\frac{1}{3}} = x^{\frac{5}{2} \cdot \frac{1}{3}} = x^{\frac{5}{6}} = \sqrt[6]{x^5}$

Example 4: $\sqrt[3]{ab^2} = (ab^2)^{\frac{1}{3}} = a^{\frac{1}{3}} \cdot b^{\frac{2}{3}} = \sqrt[3]{a} \cdot \sqrt[3]{b^2}$

Example 5: $\left(\sqrt[4]{x}\right)^0 = \left(x^{\frac{1}{4}}\right)^0 = x^0 = 1$

Example 6: $x^{\frac{-2}{3}} = \dfrac{1}{x^{\frac{2}{3}}} = \dfrac{1}{\sqrt[3]{x^2}}$

Exercises

Rewrite the following using exponents.

1. a. $\sqrt[5]{x^4}$ **b.** $\sqrt[6]{x^3}$

c. $\sqrt[3]{(x+y)^2}$ **d.** $\sqrt[3]{(a-b)^3}$

Rewrite these expressions using a radical.

2. a. $9x^{\frac{3}{2}}$ **b.** $(9x)^{\frac{3}{2}}$

    c. $\left(4x^2 - 9y^2\right)^{\frac{1}{2}}$ **d.** $\left(x - y\right)^{\frac{4}{5}}$

Simplify and express your results in radical form, if appropriate.

  3. a. $\sqrt{x^{12}}$ **b.** $\sqrt[5]{6.87^5}$

    c. $\sqrt{\sqrt{a^2b}}$ **d.** $\left(\sqrt[3]{a^4bc^3}\right)^{30}$

    e. $x^{\frac{1}{4}} \cdot x^{\frac{3}{8}}$ **f.** $\dfrac{x^{\frac{1}{2}}}{x^{\frac{1}{3}}}$

    g. $\left(x^{\frac{-2}{5}}\right)^{\frac{1}{4}}$ **h.** $\left(2x^{\frac{1}{3}}\right)^{0}$

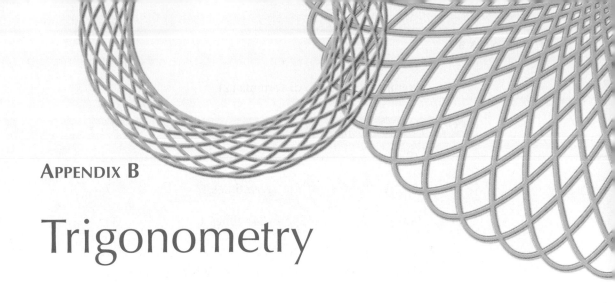

Trigonometry

In *degrees*, a protractor measures angles from 0 to 180. In *radians*, the angles range in value from 0 to π:

Protractor in degrees   **Protractor in radians**

The fundamental idea given above is that the measure of a straight angle can be taken to be either 180 degrees or π radians.

$$180° \equiv \pi \text{ radians} \qquad \textbf{(1)}$$

All other angles are done proportionately. The following table gives some examples. The last line of the table is useful for converting *any* angle from degrees to radians:

ANGLE (in degrees)	REASONING	CALCULATIONS	ANGLE (in radians)
90	90 is *one half* of 180	$\frac{1}{2} \cdot \pi$	$\frac{\pi}{2}$
60	60 is *one third* of 180	$\frac{1}{3} \cdot \pi$	$\frac{\pi}{3}$
45	45 is *one fourth* of 180	$\frac{1}{4} \cdot \pi$	$\frac{\pi}{4}$
30	30 is *one sixth* of 180	$\frac{1}{6} \cdot \pi$	$\frac{\pi}{6}$
120	120 is *two thirds* of 180	$\frac{2}{3} \cdot \pi$	$\frac{2\pi}{3}$
1	1 is *one one hundred eightieth* of 180	$\frac{1}{180} \cdot \pi$	$\frac{\pi}{180}$

$$1° \equiv \pi/180 \text{ radians} \qquad \textbf{(2)}$$

Examples of the use of formula (2).

ANGLE (in degrees)	REASONING	CALCULATIONS	ANGLE (in radians)
12	12 is *twelve* times 1	$12 \cdot \frac{\pi}{180}$	$\frac{\pi}{15}$
7	7 is *seven* times 1	$7 \cdot \frac{\pi}{180}$	$\frac{7\pi}{180}$
345	345 is *345* times 1	$345 \cdot \frac{\pi}{180}$	$\frac{85\pi}{34}$

The angle to wrap around a *full circle* is *twice* a straight angle of 180°, so it is 360° or 2π radians. You can also have angles that wrap around a circle more than once! (Think of a fishing reel or spool of wire, with the string or wire wrapped around many times.)

Equivalence 1 also enables us to convert angles from radians to degrees. Study these examples:

ANGLE (in degrees)	REASONING	CALCULATIONS	ANGLE (in radians)
$\frac{2\pi}{3}$	*Two-thirds* of π	$\frac{2}{3} \cdot 180$	120
7π	*Seven* times π	$7 \cdot 180$	1260
1	From (1), $\pi \equiv 180°$ Divide both sides of this equivalenceby π.	$345 \cdot \frac{\pi}{180}$	$\frac{85\pi}{34}$

The last line of the preceding table gives us an equivalence useful in converting from radians to degrees:

$$1 \text{ radian} \equiv \frac{180°}{\pi} \qquad \textbf{(3)}$$

Since $\pi \approx 3.14$, equivalence (3) shows that 1 radian $\approx 57.3°$. This is worth seeing on a protractor:

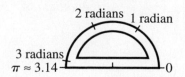

To convert 5 radians: $5 \cdot \dfrac{180}{\pi} = \dfrac{900}{\pi}$ degrees.

To convert 0.6 radians: $0.6 \cdot \dfrac{180}{\pi} = \dfrac{108}{\pi}$ degrees.

The answers just given can be written approximately: 5 radians $\approx 286.5°$ and 0.6 radians $\approx 34.4°$.

Exercises

In Exercises 1–9, convert the given angle from degrees to radians or vice versa.

1. $30°$
2. $135°$
3. $\dfrac{2\pi}{5}$ radians

4. $150°$
5. $\dfrac{5\pi}{3}$ radians
6. 1.5 radians

7. $27°$
8. $\dfrac{2}{3}$ radian
9. $450°$

10. How many times would you have to wrap a length of string around a circle to mark off an angle of 4π radians? 12π radians? 15π radians? 7 radians? 2000 radians?

11. Recall that the circumference of a circle, C, is given by the formula $C = 2\pi r$, where r is the radius of the circle. A **unit circle** is one whose radius is 1. Explain why the circumference of a unit circle equals the radian measure of the angle needed to wrap once around the circle.

12. a. Explain why a central angle of 1 radian in a unit circle subtends an arc whose length is 1 unit. (*Hint:* see Exercise 11.)

 b. Explain why a central angle of t radians in a unit circle subtends an arc whose length is t units.

 c. Explain why a central angle of t radians in a circle of radius r subtends an arc whose length is tr units.

13. Label the following radian measures on the circle:
$$\frac{\pi}{4}, \frac{\pi}{2}, \frac{5\pi}{4}, \frac{3\pi}{2}, 2\pi.$$

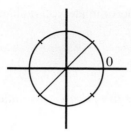

14. Label the following radian measures on the circle:
$$\frac{\pi}{3}, \frac{\pi}{2}, \frac{2\pi}{3}, \frac{4\pi}{3}, \pi.$$

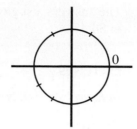

15. Label the following radian measures on the circle:
$$\frac{\pi}{6}, \frac{\pi}{2}, \frac{7\pi}{6}, \frac{4\pi}{3}, \frac{11\pi}{6}.$$

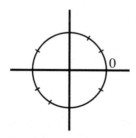

16. Locate approximately the following radian measures on the circle:
$$1, \frac{\pi}{6}, 2, 0.6, 5, 3, \frac{3\pi}{4}, \frac{3\pi}{2}, 1.4.$$

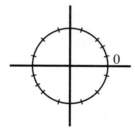

Trigonometric Functions in Right Triangles

For an angle θ in a right triangle (as pictured), the basic trigonometric functions (sine, cosine, tangent) are defined by

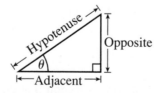

$$\sin\theta = \frac{\text{opposite}}{\text{hypotenuse}} \qquad \cos\theta = \frac{\text{adjacent}}{\text{hypotenuse}} \qquad \tan\theta = \frac{\text{opposite}}{\text{adjacent}}$$

The acronym **SOH CAH TOA** summarizes this; for example, **SOH** tells you that **s**ine equals **o**pposite over the **h**ypotenuse. Using the accompanying triangle,

$$\sin\theta = \frac{\text{opp}}{\text{hyp}} = \frac{6}{10} = \frac{3}{5}$$

$$\cos\theta = \frac{\text{adj}}{\text{hyp}} = \frac{8}{10} = \frac{4}{5}$$

$$\tan\theta = \frac{\text{opp}}{\text{adj}} = \frac{6}{8} = \frac{3}{4}$$

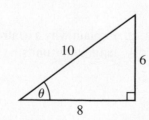

Frequently, you must use the **Pythagorean theorem** for right triangles. As you recall, the theorem says that in a right triangle, $c^2 = a^2 + b^2$. For example, to determine the values of trigonometric functions in the given triangle, you first use the Pythagorean theorem to determine the missing side:

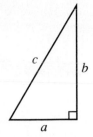

$$2^2 = 1^2 + x^2 \quad \text{or} \quad 4 = 1 + x^2 \quad \text{or} \quad x^2 = 3$$
$$x = \sqrt{3}$$

Then, as before, $\sin \theta = \dfrac{\text{opp}}{\text{hyp}} = \dfrac{\sqrt{3}}{2}$ $\quad \cos \theta = \dfrac{\text{adj}}{\text{hyp}} = \dfrac{1}{2}$

$$\tan \theta = \dfrac{\text{opp}}{\text{adj}} = \dfrac{\sqrt{3}}{1} = \sqrt{3}$$

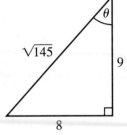

Exercises

In Exercises 1–6, find sin θ, cos θ, and tan θ.

1.

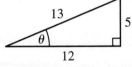

2.

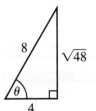

3.

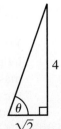

4.

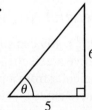

5.

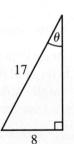

6.

7. Using the Pythagorean theorem and SOH CAH TOA, show that $\sin^2 \theta + \cos^2 \theta = 1$ for any angle θ in a right triangle.

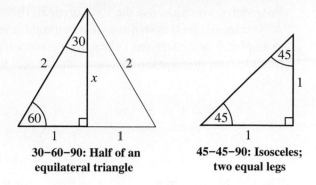

**30–60–90: Half of an
equilateral triangle**   **45–45–90: Isosceles;
two equal legs**

The third side in each triangle can be found using the Pythagorean theorem. The proportions are:

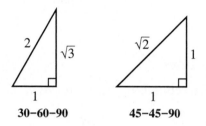

30–60–90   **45–45–90**

From the triangles, you obtain these important values for trigonometric functions:

ANGLE (in degrees)	REASONING	CALCULATIONS	ANGLE (in radians)
30	$\dfrac{1}{2}$	$\dfrac{\sqrt{3}}{2}$	$\dfrac{\sqrt{3}}{3}$
45	$\dfrac{\sqrt{2}}{2}$	$\dfrac{\sqrt{2}}{2}$	1
60	$\dfrac{\sqrt{3}}{2}$	$\dfrac{1}{2}$	$\sqrt{3}$

Trigonometric Functions for More General Angles

When an angle θ is larger than 90° (in radians, θ, $\frac{\pi}{2}$), you can still evaluate sine, cosine, and tangent. You work in an xy plane, make the positive x axis the initial side of the angle, and make a *reference triangle* by dropping a perpendicular from a point on the terminal side of the angle to the x axis. (See the figure.) For positive angles, you rotate *counterclockwise* to find the terminal side.

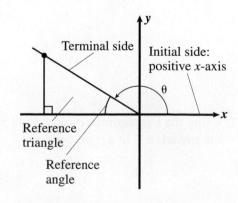

You then use SOH CAH TOA on the *reference triangle*. **Warning:** The adjacent and opposite sides may be *negative* in this situation, depending on the quadrant in which the terminal side of the angle lies.

For example, in the figure below, note the negative sign for the adjacent side. (It lies on the *negative x* axis.) You have

$$\sin \theta = \frac{\text{opp}}{\text{hyp}} = \frac{4}{5}$$

$$\cos \theta = \frac{\text{adj}}{\text{hyp}} = -\frac{3}{5}$$

$$\tan \theta = \frac{\text{opp}}{\text{adj}} = \frac{4}{(-3)} = -\frac{4}{3}$$

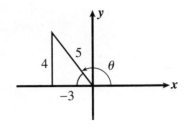

You also need to use the Pythagorean theorem at times. In the figure below, you have $6^2 = (-2) + y^2$ or $y^2 = 32$. Since y must be *negative* (do you see why?), $y = -\sqrt{32} = -4\sqrt{2}$. You now find

$$\sin \theta = \frac{\text{opp}}{\text{hyp}} = \frac{-4\sqrt{2}}{6} = \frac{-2\sqrt{2}}{3}$$

$$\cos \theta = \frac{\text{adj}}{\text{hyp}} = \frac{-2}{6} = \frac{-1}{3}$$

$$\tan \theta = \frac{\text{opp}}{\text{adj}} = \frac{-4\sqrt{2}}{(-2)} = 2\sqrt{2}$$

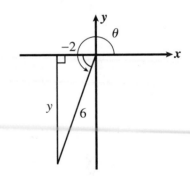

As noted previously, reference triangles with angles of 30°, 60°, or 45° show up frequently because of the symmetry involved. To find the values of the three trigonometric functions for $\theta = 5\pi/3$ radians, we convert to degrees and sketch the angle and the reference triangle:

$$\frac{5\pi}{3} = \frac{5}{3} \cdot 180° = 300°$$

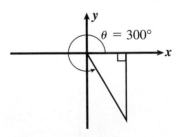

As noted previously, reference triangles with angles 30°, of 30°, 60°, or 45° show up frequently because of the symmetry involved. To find the values of the three trigonometric functions for $\theta = 5\pi/3$ radians, we convert to degrees and sketch the angle and the reference triangle:

$$\frac{5\pi}{3} = \frac{5}{3} \cdot 180° = 300°$$

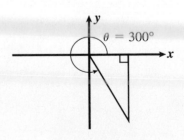

To help remember the ± signs, the phrase *All Students Take Calculus* is useful. The four words go in the four quadrants of the *xy* plane:

The *All* means *all* trigonometric functions are positive in the quadrant I; **S** for **S**tudents means *Sine* is positive in quadrant II; **T** for **T**ake means *Tangent* is positive in quadrant III; **C** for **C**alculus means *Cosine* is positive in quadrant IV.

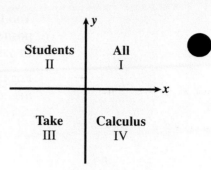

Note that for *negative* angles we locate the terminal side by rotating *clockwise* from the positive *x* axis. The figure at the right shows that

$$\sin \frac{-3\pi}{4} = \frac{-1}{\sqrt{2}} = \frac{-\sqrt{2}}{2};$$

and by similar calculations,

$$\cos \frac{-3\pi}{4} = \frac{-\sqrt{2}}{2} \quad \text{and} \quad \tan \frac{-3\pi}{4} = 1.$$

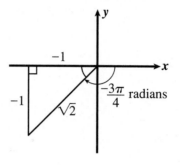

Exercises

In Exercises 1–4, find the remaining side of the reference triangle and evaluate sin θ, cos θ, and tan θ.

1.

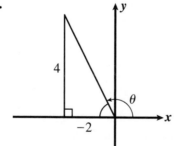

2.

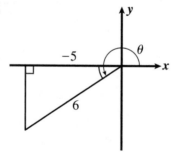

3.

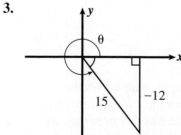

4.

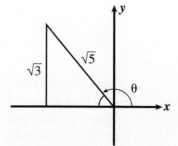

In Exercises 5–10, sketch the angle and reference triangle and then work out the values of the three trigonometric functions for that angle.

5. $\theta = \frac{\pi}{2}$ radians

6. $\theta = 225°$

7. $\theta = -150°$

8. $\theta = 330°$

9. θ is fourth quadrant angle whose sine is $-\frac{6}{8}$.

10. θ is fourth quadrant tangent angle whose tangent is $-\frac{3}{5}$.

11. Show that for θ in *any* quadrant, the relationship $\sin^2\theta + \cos^2\theta = 1$ holds.

Dealing with Special Angles: 0°, ±90°, ±180°

For multiples of 90° (equivalently, multiples of $\frac{\pi}{2}$ radians), the reference triangle degenerates to a straight line. Either the adjacent or opposite side degenerates to 0:

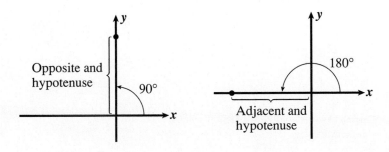

For simplicity, notice that for ordinary angles, when we mark a point on the terminal side of an angle θ, the *x value* gives the value of the *adjacent* side, and the *y* value gives the *opposite*. Let us also use r for the length of the hypotenuse:

Therefore, we could have defined

$$\sin\theta = \frac{\text{opp}}{\text{hyp}} = \frac{y}{\text{hyp}} = \frac{y}{r} \qquad \cos\theta = \frac{\text{adj}}{\text{hyp}} = \frac{x}{\text{hyp}} = \frac{x}{r} \qquad \tan\theta = \frac{\text{opp}}{\text{adj}} = \frac{y}{x} \qquad (4)$$

We use $\dfrac{y}{r}$, $\dfrac{x}{r}$, and $\dfrac{y}{x}$ for the special angles to find the values of the trigonometric functions. When we have a multiple of 90°, we mark a point on the terminal side and label it with its x and y numbers. Use 0 and ± 1 for simplicity, then use formula 4. Here are the figures for -90 degrees and 180 degrees.

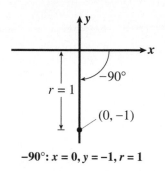

−90°: $x = 0, y = -1, r = 1$

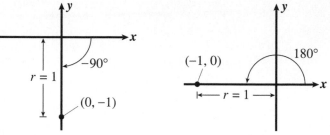

180°: $x = -1, y = 0, r = 1$

Now the trigonometric values are:

$$\sin 180° = \frac{\text{opp}}{\text{hyp}} = \frac{y}{r} = \frac{0}{1} = 0 \qquad \sin(-90°) = \frac{\text{opp}}{\text{hyp}} = \frac{y}{r} = \frac{-1}{1} = -1$$

$$\cos 180° = \frac{\text{adj}}{\text{hyp}} = \frac{x}{r} = -\frac{1}{1} = -1 \qquad \cos(-90°) = \frac{\text{adj}}{\text{hyp}} = \frac{x}{r} = \frac{0}{1} = 0$$

$$\tan 180° = \frac{\text{opp}}{\text{adj}} = \frac{y}{x} = \frac{0}{-1} = 0 \qquad \tan(-90°) = \frac{\text{opp}}{\text{adj}} = \frac{y}{x} = \frac{-1}{0} = undefined$$

If you use only 0 and ± 1 for x and y values of these special angles, r always equals 1. Also, whenever the x value is 0, the tangent is *undefined*, since the formula then involves division by 0.

Exercises

In Exercises 1–6, for each given value of θ draw the angle, label a point on the terminal side, and use formulas 4 to get the values of sin θ, cos θ, and tan θ.

1. θ = 90°

2. θ = −π radians

3. θ = 720°

4. θ = −630°

5. θ = $\dfrac{7\pi}{2}$ radians

6. θ = 23π radians

Getting Started with the TI-83 Plus

ON-OFF

To turn on the TI-83 Plus, press the ⊆ON⊇ key. To turn off the TI-83 Plus, press ⊆2nd⊇ and then ⊆ON⊇.

In general, to access any of the white commands, press the black or gray white key. To access the gold commands, press ⊆2nd⊇ and then the black or gray white key below the desired command. Similarly, to access any of the green commands or symbols, press ⊆ALPHA⊇ followed by the appropriate black or gray white key.

Contrast

To adjust the contrast on your screen, press and release the ⊆2nd⊇ key and hold ⊆▲⊇ to darken and ⊆▼⊇ to lighten.

Mode

The ⊆MODE⊇ key controls many calculator settings. The activated settings are highlighted. For most of your work in this course, the settings in the left-hand column should be highlighted.

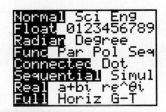

To change a setting, move the cursor to the desired setting and press ⊆ENTER⊇.

The Home Screen

The home screen is used for calculations.

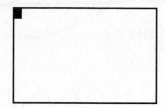

You may return to the home screen at any time by using the QUIT command. This command is accessed by pressing (2nd) (MODE). All calculations in the home screen are subject to the order of operations.

Enter all expressions as you would write them. Always observe the order of operations. Once you have typed the expression, press (ENTER) to obtain the simplified result. Before you press (ENTER), you may edit your expression by using the arrow keys, the delete command (DEL), and the insert command (2nd) (DEL).

Three keys of special note are the reciprocal key (X⁻¹), the caret key (^), and the negative key (⊖).

The reciprocal command (X⁻¹) will invert the number in the home screen.

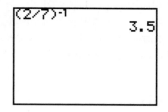

The caret key (^) is used to raise numbers to powers

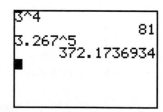

The negative key is different from the minus key. To enter a negative number use the gray white key (⊖), not the blue (−) key.

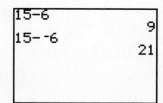

A table of keys and their functions follows.

KEY	FUNCTION DESCRIPTION
ON	Turns calculator on or off.
CLEAR	Clears the current line on the text screen.
ENTER	Executes a command.
(−)	Calculates the additive inverse.
MODE	Displays current operating settings.
DEL	Deletes the character at the cursor.
^	Symbol used for exponentiation.
ANS	Storage location of the last calculation.
ENTRY	Retrieves the previously executed expression.

ANS and ENTRY

The last two commands in the table can be real time savers. The result of your last calculation is always stored in a memory location known as ANS. It is accessed by pressing (2nd) (−) or it can be automatically accessed by pressing any operation button.

Suppose you want to evaluate $12.5\sqrt{1 + 0.5 \cdot (0.55)^2}$. It could be evaluated in one expression and checked with a series of calculations using ANS.

```
12.5√(1+0.5*.55^
2)
          13.41203983
■
```

```
1+.5*.55^2
              1.15125
√(Ans)
          1.072963187
Ans*12.5
          13.41203983
■
```

The ENTRY command recalls the last expression. Even if you have pressed (ENTER), you can edit the previous expression. The (2nd) (ENTER) sequence will recall the previous expression for editing.

Suppose you want to evaluate the compound interest expression $P\left(1 + \frac{r}{n}\right)^{nt}$, where P is the principal, r is the interest rate, n is the number of compounding periods annually, and t is the number of years, when $P = \$1000$, $r = 6.5\%$, $n = 1$, and $t = 2, 5,$ and 15 years.

Using the ENTRY command, this expression would be entered once and edited twice.

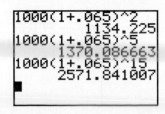

```
1000(1+.065)^2
          1134.225
1000(1+.065)^5
          1370.086663
1000(1+.065)^15
          2571.841007
■
```

Note that there are many last expressions stored in the ENTRY memory location. You can repeat the ENTRY command as many times as you want to retrieve a previously entered expression.

Functions and Graphing with the TI-83 Plus

"Y =" Menu

Functions of the form $y = f(x)$ can be entered into the TI-83 Plus using the "Y ="menu. To access the "Y =" menu press the ⟨Y=⟩ key. Type the expression $f(x)$ after Y_1 using the ⟨X,T,θ,n⟩ key for the variable x and press ⟨ENTER⟩.

For example, enter the function $f(x) = 3x^5 - 4x + 1$.

```
Plot1  Plot2  Plot3
\Y1■3X^5-4X+1
\Y2=
\Y3=
\Y4=
\Y5=
\Y6=
\Y7=
```

Note the = sign after Y_1 is highlighted. This shows Y_1 is selected to be graphed. The highlighting may be turned on or off by using the arrow keys to move the cursor to the = sign and then pressing ⟨ENTER⟩.

```
Plot1  Plot2  Plot3
\Y1=3X^5-4X+1
\Y2=
\Y3=
\Y4=
\Y5=
\Y6=
\Y7=
```

Once the function is entered in the Y = menu, function values may be evaluated in the home screen.

For example, given $f(x) = 3x^5 - 4x + 1$, evaluate $f(4)$. In the home screen, press ⟨VARS⟩.

```
VARS  Y-VARS
1■Window…
2:Zoom…
3:GDB…
4:Picture…
5:Statistics…
6:Table…
7:String…
```

Move the cursor to Y-VARS and press ⟨ENTER⟩.

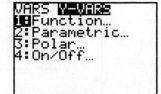 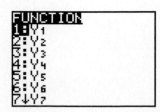

Press (ENTER) again to select Y₁. Y₁ now appears in the home screen.

To evaluate $f(4)$, press ([) (4) (]) after Y₁ and press (ENTER).

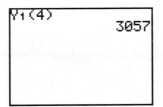

Tables of Values

If you are interested in viewing several function values for the same function, you may want to construct a table.

Before constructing the table, look at the settings in the Table Setup menu. To do this, press (2nd) (WINDOW).

TblStart tells you where the table of values will start. ΔTbl tells you the increment from one input value to the next. Set the TblStart at -2 and the ΔTbl at 0.5, then press (2nd) (GRAPH) to access the following table. (Make certain that the $=$ sign following Y₁ is highlighted.)

X	Y₁	
-2	-87	
-1.5	-15.78	
-1	2	
-.5	2.9063	
0	1	
.5	-.9063	
1	0	

X= -2

Use the (▲) and (▼) keys to view other values in the table.

If the input values of interest are not evenly spaced, you may want to choose the ask mode for the independent variable from the Table Setup menu.

The resulting table is blank, but you can fill it by choosing any values you like for x and pressing (ENTER) after each.

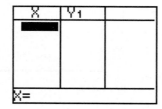 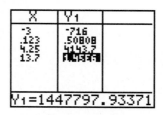

Note that the number of digits shown in the output is limited by the table width, but if you want more digits, move the cursor to the desired output and more digits appear at the bottom of the screen.

Graphing a Function

Once a function is entered in the "Y =" menu and activated it can be displayed and analyzed. For this discussion we will use the function $f(x) = -x^2 + 10x + 12$. Enter this as Y_1.

The Viewing Window

The viewing window is the portion of the rectangular coordinate system that is displayed when you graph a function.

Xmin defines the left edge of the window.

Xmax defines the right edge of the window.

Xscl defines the distance between horizontal tick marks.

Ymin defines the bottom edge of the window.

Ymax defines the top edge of the window.

Yscl defines the distance between vertical tick marks.

In the standard viewing window, Xmin $= -10$, Xmax $= 10$, Xscl $= 1$, Ymin $= -10$, Ymax $= 10$, and Yscl $= 1$.

To select the standard viewing window, press (ZOOM) (6).

If you press the (GRAPH) key now, you will view the following:

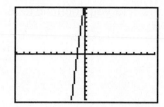

Is this an accurate and or complete picture of your function, or is the window giving you a misleading impression? You may want to use your table function to view the output values from −10 to 10.

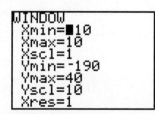 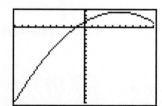

The table indicates that the minimum output value is −188 and the maximum output value is 37. Press (WINDOW) and reset the settings to approximately the following;

Xmin = −10, Xmax = 10, Xscl = 1, Ymin = −190, Ymax = 40, Yscl = 10

WINDOW
 Xmin=■10
 Xmax=10
 Xscl=1
 Ymin=⁻190
 Ymax=40
 Yscl=10
 Xres=1

The new graph gives us a much more complete picture of the behavior of the function on the interval (–10, 10).

Specific points on the curve can be viewed by activating the trace feature. While in the graph window, press (TRACE). A flashing cursor will appear on the curve at approximately the midpoint of the screen.

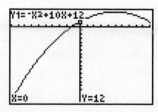

The left arrow key, ◄, will move the cursor toward smaller input values. The right arrow key, ►, will move the cursor toward larger input values. If the cursor reaches the edge of the window and you continue to move the cursor, the window will adjust automatically.

Zoom Menu

The Zoom menu offers several options for changing the window very quickly.

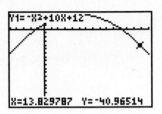 

The features of each of the commands are summarized in the following table.

ZOOM COMMAND	DESCRIPTION
1: ZBox	Draws a box to define the viewing window.
2: Zoom In	Magnifies the graph near the cursor.
3: Zoom Out	Increases the viewing window around the cursor.
4: ZDecimal	Sets a window so that Xscl and Yscl are 0.1.
5: ZSquare	Sets equal size pixels on the x and y axes.
6: ZStandard	Sets the window to standard settings.
7: ZTrig	Sets built-in trig window variables.
8: ZInteger	Sets integer values on the x and y axes.
9: ZoomStat	Sets window based on the current values in the stat lists.
0: ZoomFit	Replots graph to include the max and min output values for the current Xmin and Xmax.

Solving Equations Graphically Using the TI-83 Plus

The Intersection Method

This method is based on the fact that solutions to the equation $f(x) = g(x)$ are input values of x that produce the same outputs for the functions f and g. Graphically these are the x-coordinates of the intersection points of $y = f(x)$ and $y = g(x)$.

The following procedure illustrates how to use the intersection method to solve $x^3 + 3 = 3x$ graphically.

Step 1   Enter the left-hand side of the equation as Y_1 in the "Y =" editor and the right-hand side as Y_2.

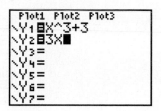

Step 2   Examine the graphs to determine the number of intersection points.

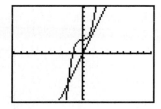

You may need a couple of windows to be certain of the number of intersection points.

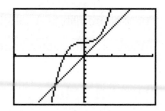

Step 3   Access the Calculate menu by pushing ⌜2nd⌝ ⌜TRACE⌝, then choose option 5: intersect.

 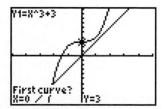

The cursor will appear on the first curve in the center of the window.

Step 4   Move the cursor close to the desired intersection point and press ⌜ENTER⌝.

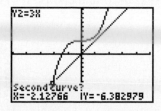

The cursor will now jump vertically to the other curve.

Step 5   Repeat step 4 for the second curve.

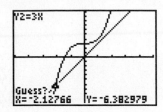

Step 6   To use the cursor's current location as your guess, press (ENTER) in response to the question on the screen that asks Guess? If you want to move to a better guess value, do so before you press (ENTER).

The coordinates of the intersection point appear below the word Intersection

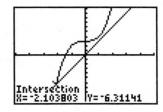

The *x*-coordinate is a solution to the equation.

If there are other intersection points, repeat the process as necessary.

Using the TI-83 Plus to Determine the Linear Regression Equation for a Set of Data

Example:

INPUT	OUTPUT
2	2
3	5
4	3
5	7
6	9

Enter the data into the calculator as follows:

1. Press (STAT) and choose EDIT.

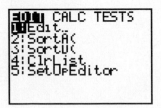

2. The TI-83 Plus has six built-in lists, L1, L2, ... , L6. If there is data in L1, clear the list as follows:

   a. Use the arrows to place the cursor on L1 at the top of the list. Press (CLEAR) followed by (ENTER), followed by the down arrow.

   b. Follow the same procedure to clear L2 if necessary.

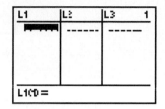

   c. Enter the input into L1 and the output into L2.

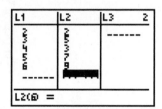

To see a scatterplot of the data proceed as follows.

1. STAT PLOT is the 2nd function of the (Y=) key. You must press (2nd) before pressing (Y=) to access the STAT PLOT menu.

2. Make sure plot 2 and plot 3 are off and choose plot 1. Highlight On, the scatter plot symbol and the square mark as indicated below. L1 is the Xlist and L2 is the Ylist. L1 is the 2nd function of the 1 button and L2 is the 2nd function of the 2 button, and so on.

3. Press (Y=) and clear or deselect any functions currently stored.

4. To display the scatterplot, have the calculator determine an appropriate window by pressing (ZOOM) and then (9) (ZoomStat).

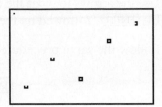

Calculate the linear regression equation as follows.

1. Press (STAT) and right arrow to highlight CALC.

2. Choose 4: LinReg(ax + b). LinReg(ax + b) will be pasted to the home screen. To tell the calculator where the data is, press (2nd) and (1) (for L1), then (,), then (2nd) and (2) (for L2) so the display looks like this:

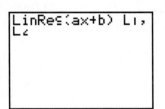

3. Press (,) and then press (VARS).

4. Right arrow to highlight Y-VARS.

5. Choose 1, FUNCTION.

6. Choose 1 for Y_1 (or 2 for Y_2, etc.).

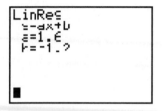

7. Press (ENTER).

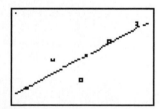

The linear regression equation for this data is $y = 1.6x - 1.2$.

8. To display the regression line, press (GRAPH).

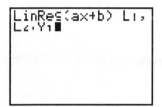

9. Press the (Y=) key to see the equation.

Selected Answers

Chapter 1

Activity 1.1 Exercises: 1. a. Weight is input, and height is output; **c.** Height is input, and weight is output; **3. a.** Yes, each numerical grade will correspond to one letter grade; **b.** No, the letter grades may correspond to several different numerical grades; **5. a.** Yes, in this table each amount of snowfall is paired with one elevation; **b.** Yes, in this table each quantity of snow is paired with one elevation; **7. a.** Yes, each input value has one output value; **b.** No, the one input 5 is paired with four different outputs; **9. a.** The input is x, the output is $g(x)$ or y, The function name is g, y equals g of x; **c.** The input is 6, The output is 3.527, The name is f, f of 6 equals 3.527; **e.** The input is price, The output is sales tax, The name is T, Sales tax is a function of price; **10. a.** 1600; 2400; **c.** $f(6) = 2400$

Activity 1.2 Exercises: 1. x is the input, $h(x)$ is the output, h is the function, $h(x) = 0.08x$; **3.** $f(2) = -1$, $f(-3.2) = -11.4$, $f(\pi) \approx 1.2832$, $f(a) = 2a - 5$
5. $f(2) = 4$, $f(-3.2) = 4$, $f(\pi) = 4$, $f(a) = 4$;

7.

x	h(x)
10	0.1
20	0.05
30	0.033
40	0.025

9. a. The distance traveled is three times the number of hours I have hiked; **b.** The input is hours. The output is distance; **d.** $h(t)$ is the dependent variable since distance is the output; **f.** $h(7) = 3(7) = 21$. $(7, 21)$;
h. The practical domain depends on the individual and is probably real numbers from 0 to about 8. Using this domain, the range is real numbers from 0 to 24; **10. a.** The domain is $-2 \le x \le 3$, and the range is $-1 \le y \le 2$. **c.** The domain is all real numbers. The range is $y \ge 0$.
11. a. domain $\{-2, 0, 5, 8\}$, range $\{4, 3, 8, 11\}$;
12. a. $0 \le m \le 157{,}200$ for your car; for all cars $0 \le m \le 250{,}000$; **13. a.** $0 \le C \le 100$; **b.** $32 \le F \le 212$

Activity 1.3 Exercises: 1. a. 3.7, 6.4, 9.1, 11.8;
b. $f(2500) = 7.75$ inches of snow; **c.** $f(-2000) = -4.4$. It has no meaning in this context. -2000 would mean 2000 feet below sea level, but -4.4 inches of snow is not

possible; **d.** Xmin $= 0$, Xmax $= 5000$, Ymin $= 0$, Ymax $= 15$; **e.** Yes, any vertical line will intersect the graph no more than once; **f.** increasing; **g.** It is the same, 7.75.
3. a. 7238; 11,494; 17,157; 24,429; **c.** The practical domain is $0 \le r \le 25$. The practical range is $0 \le h \le 65{,}450$;
e. $f(r) = \dfrac{4}{3}\pi r^3$; **g.**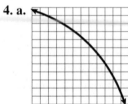

i. Using the vertical line test, I can see that no vertical line crosses the graph more than once. The graph represents a function.
4. a. 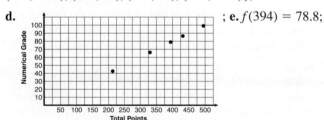 **b.** The graph is a horizontal line;
5. a. This is a function;
b. This is not a function;
6. Xmin $= -9$, Xmax $= 9$, Ymin $= -254{,}000$, Ymax $= 54{,}000$

How Can I Practice? 1. a. Yes, because none of the total points values is repeated; **b.** Yes. For each numerical grade in the table, there is one value of total points; **c.** $f = \{(432, 86.4), (394, 78.8), (495, 99), (330, 66), (213, 42.6)\}$;
d.

; **e.** $f(394) = 78.8$;

f. 78.8; **g.** $f(213) = 42.6$ **h.** 42.6; **i.** $n = 330$; **2.** This is a function; **3.** This could be a function, depending on how activity level is measured; **4.** This is a function; **5.** This is not a function; **6.** This is a function; **7.** This is not a function; **8.** This is a function; **9.** This is a function; **10.** This is not a function; **11. a.** $c = 107h$; **b.** $f(h) = 107h$; **c.** 214, 428, 749, 856, 1177; **d.** \$321, $(3, 321)$; **e.** $h = 5$ hours; **f.** $f(h)$ or c is the output variable. This is the variable that depends on the number of credits taken; **g.** h is the independent variable. It is the input variable; **h.** For each value of input, there is one value of output; **i.** Assuming there are no half-credit courses, the practical domain is all whole numbers from 0 to 11, depending on the college; **j.** The horizontal axis represents the input,

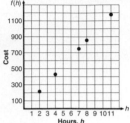

k. $f(h)$ is a function because the graph passes the vertical line test; **m.** \$963; **12. a.** $p(3) = 13$; **b.** $p(-4) = -1$; **c.** $p\left(\frac{1}{2}\right) = 8$; **d.** $p(0) = 7$; **13. a.** $t(2) = -3$; **b.** $t(-3) = 18 + 9 - 5 = 22$; **14. a.** 48, 59, 67, 73, 75; **b.** $f(85) = 0.27(85) + 48.3 = 71$. The life expectancy for male born in 1985 is 71 years;

c.   ; **d.** The graph is increasing. It is rising to the right; **e.** They are the same;

15. a. domain $\{3, 4, 5, 6\}$, range $\{5, 8, 10\}$; **b.** domain $\{0, 50, 100, 150, 200\}$, range $\{19.95, 23.45, 26.95, 30.45, 33.95\}$; **c.** domain $-3 \leq x \leq 4$, range $-1 \leq y \leq 3$; **d.** domain $-3 \leq x \leq 3$, range $0 \leq y \leq 4$; **e.** domain is all real numbers, range is all real numbers; **16. a.** The net profit increases during the first two quarters of 1993. The net profit then decreases for about 2.5 quarters, and then it increases through the final quarter of 1994; **b.** The annual income rises rather steadily for three years; in the fourth year, it rises sharply. Then it suffers a sharp decline during the next year. During the last year, the income recovers to about the point it was originally.

Activity 1.5   Exercises: 1. a. $\frac{26.1 - 22.8}{1990 - 1950} = \frac{3.3}{40} = 0.0825$ years/year; **b.** The median age of a man at the time of his first marriage is increasing at an average rate of .0825 years/year; **3.** $\frac{26.8 - 25.1}{90} = \frac{1.7}{90} \approx 0.019$ years/year; **5. a.** It means that the median age of a man at the time of his first marriage is decreasing; **b.** 1910–1920, 1920–1930, or 1940–1950; **c.** The graph would go down to the right; **8. a.** $\frac{1519 - 1476}{1998 - 1997} = 43$ new hotels/year; **b.** $\frac{1402 - 1519}{1999 - 1998} = -117$ new hotels/year; **c.** The rate of new hotel construction increased from 1997 to 1998 then decreased from 1998 to 1999; **d.** The graph is going down to the right. The rate of new hotel construction is decreasing;

9. a. $\frac{760 - 668}{1970 - 1960} = \frac{92}{10} = 9.2$ gal/year;

b. $\frac{520 - 668}{1990 - 1960} = \frac{-148}{30} \approx -4.93$ gal/year;

c. $\frac{538 - 530}{1997 - 1995} = \frac{8}{2} = 4$ gal/year;

d. $\frac{552 - 668}{1999 - 1960} = \frac{-116}{39} \approx -2.97$ gal/year;

e. It means that from 1960 to 1999, the average fuel consumption per year of a passenger car in the U.S. decreased by nearly 3 gal/year.

Activity 1.6   Exercises: 1. a. yes, linear; $m = 10$; **b.** no, not linear; **c.** Yes, linear $m = \frac{-9}{4}$; **3. a.** Yes, the rate of change is a constant -3; **b.** No, between weeks 1 and 2 the slope is -5. Between weeks 2 and 3, the slope is -4; **c.** Yes, the slope is 0 for all pairs of points; **5. a.** $m = \frac{5 - (-7)}{0 - 2} = \frac{12}{-2} = -6$; **b.** $(0, 5)$; **c.** $f(x) = -6x + 5$; **d.** $\left(\frac{5}{6}, 0\right)$; **7. a.** Yes, the slope is a constant; **b.** $m = \frac{3000 - 3500}{5 - 0} = \frac{-500}{5} = -100$ feet/second; **c.** The jet is losing altitude; **d.** $(0, 3500)$; **e.** $h = -100t + 3500$; **f.** $(35, 0)$; The jet lands in 35 seconds. **8. a.** $(-1/2, 0)$; **b.** $(6, 0)$

Activity 1.7   Exercises: 1. a. $(0, 35)$; The vertical intercept occurs where the input $x = 0$; **b.** $m = \frac{40 - 35}{100 - 0} = \frac{5}{100} = 0.05$; The mileage charges are \$0.05 per mile; **c.** $c = 0.05x + 35$; **2. a.** $m = \frac{145 - 75}{4 - 2} = \frac{70}{2} = 35$ mph; **b.** $d = 35t + b$ $75 = 35(2) + b$ $75 = 70 + b$ $b = 5$ $d = 35t + 5$; **3. a. i.** $m = 1$, **ii.** $(0, -2)$, **iii.** $y = x - 2$; **c. i.** $m = -2$, **ii.** $(0, 6)$, **iii.** $y = -2x + 6$; **5. a.** $y = \frac{1}{2}x - 1$; **b.** $y = -\frac{4}{3}x + 1$; **c.** $m = \frac{6 - (-3)}{2 - (-4)} = \frac{6 + 3}{2 + 4} = \frac{9}{6} = \frac{3}{2}$, $y = \frac{3}{2}x + b$, $6 = \frac{3}{2}(2) + b$, $6 = 3 + b$, $b = 3$, $y = \frac{3}{2}x + 3$; **d.** $-5 = 3(2) + b$, $b = -11$, $y = 3x - 11$; **7. a.** $m = \frac{31.42 - 0}{5 - 0} = 6.28$; **b.** $(0, 0)$; **c.** $C = 6.28r$; **d.** $C = 2\pi r$; **e.** Yes, π is approximately 3.14, so 2π is approximately 6.28.

Activity 1.8   Exercises: 1. a. $y = 2x - 3, m = 2, (0, -3)$; **b.** $y = -x - 2, m = -1, (0, -2)$; **c.** $y = \frac{2}{3}x - \frac{7}{3}, m = \frac{2}{3}, \left(0, -\frac{7}{3}\right)$; **d.** $y = \frac{1}{2}x + 2, m = \frac{1}{2}, (0, 2)$; **e.** $y = 4, m = 0, (0, 4)$; **3. a.**

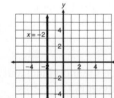

b. This is not a function. It does not pass the vertical line test; **c.** $x = -2$; **d.** The slope is undefined; **e.** vertical: none; horizontal: $(-2, 0)$; **5. a.** $f(x) = 2000$; **b.** 2000, 2000, 2000;

c.  ;

d. The slope is zero. This means the fee does not change; **e.** The graph is a horizontal line through $(0, 2000)$; **6. a.** $250w$; **b.** $200d$; **c.** $250w + 200d = 10,000$; **d.** $d = \frac{10,000 - 250w}{200} = 50 - \frac{5}{4}w$; **e.** $(40, 0)$; The maximum number of washers I can purchase is 40.

Activity 1.9   Exercises:
1. a.

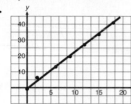

; **b.** (Answers will vary.) Yes, the points are very close to a line; **c.** $f(x) = 2.299x - 0.761$; **d.** 22.229; **e.** 56.714;

f. $f(10)$ is more accurate. 10 is within the given data. 25 is not. $f(10)$ uses interpolation. $f(25)$ uses extrapolation;
2. b. $f(x) = 1.150x + 8.579$; **c.** The slope of the line is 1.150. This means that the average debt per person is increasing at an average rate of $1150 per year; **d.** The regression line predicts an average debt of $13,179 in 1990. This is 179 above the actual, 13,000, an error of slightly less than 1.4%; **e.** $26,979; **f.** extrapolation

How Can I Practice?   1. a. $g(x) = 2x - 3$;
b. $h(x) = -2x - 3$; **c.** $x = 2$; **d.** none; **e.** $f(x) = -2x + 3$;
f. $y = -2x$; **g.** $y = 2$; **h.** none; **i.** none; **2. a.** 28, 40, 52, 60, 68;
b. Yes; **c.** 4; **d.** $c = f(m) = 4m + 20$;
e.

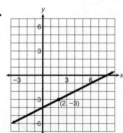

; **f.** This represents the monthly charge. **g.** $(0, 20)$; It indicates that the initial rental cost is $20; **h.** $(-5, 0)$. It has no practical meaning in this case;

i. $65 = 4m + 20$ or $45 = 4m$ or $m = 11.25$. I can keep the graphing calculator for 11 months; **3. a.** 1.5; **b.** 1.5; **c.** $s(t)$ is a linear function because the rate of change is constant;
4. $m = \frac{12 - 8}{-5 - 3} = -\frac{1}{2}$; **5.** $m = -4$; **6.** $m = \frac{2}{5}$;
7. $y = -7x + 4$; **8.** $y = 2x + 10$;
9. $y = 5$; **10.** $x = -3$; **11.** $y = -\frac{1}{2}x - 2$; **12.** $y = \frac{1}{3}x - 3$;
13.

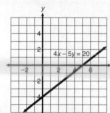

; **14.**

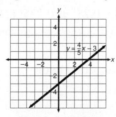

;

15.

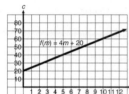

; **16.**

;

17. a.

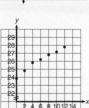

; **b.** $y = 0.322x + 24.156$;
c. $y = 27.054$;
d. $y = 30.596$;

18. a.

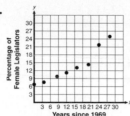

; **b.** $f(x) = 0.657x + 4.533$;
c. In 1986, $x = 17$;
$f(17) = 15.7\%$. In 2010, $x = 41$; $f(41) = 31.5\%$;
d. 1986 because it is interpolated from the data. The 2010 percentage was a result of extrapolation, which is usually less accurate.

Activity 1.10   Exercises:
1. a. Numerically   Graphically

x	y₁	y₂
0	-3	7
1	2	12
2	7	17
3	12	22
4	17	27

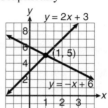

Algebraically (substitution method)
$y = 2x + 3$ $y = -x + 6$

$$2x + 3 = -x + 6 \qquad y = -1 + 6$$
$$\underline{+ x \qquad\qquad + x} \qquad\qquad y = 5$$
$$3x + 3 = 6 \qquad\quad \text{The answer is } (1, 5)$$
$$\underline{- 3 - 3}$$
$$3x = 3$$
$$\frac{3}{3} = \frac{3}{3}$$
$$x = 1$$

c. Numerically   Graphically

x	y₁	y₁
0	-3	7
1	2	12
2	7	17
3	12	22
4	17	27

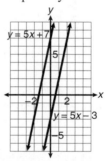

Algebraically (substitution method)
$y = 5x - 3$ $y = 5x + 7$

$$5x - 3 = 5x + 7 \quad \text{There is no solution;}$$
$$\underline{- 5x \qquad - 5x}$$
$$-3 = 7 \text{ false}$$

2. a. $s = 17.2 + 1.5n$; **b.** $s = 9.6 + 2.3n$; **c.** in the year 2011;
3. a. $c = 3560 + 15n$; **b.** $c = 2850 + 28n$; **c.** $n = 54.6$ or 55 months; **d.** dealer 1's system;
5. a.

t, NUMBER OF YEARS SINCE 1975	LIFE EXPECTANCY FOR WOMEN	LIFE EXPECTANCY FOR MEN
0	76.74	69.11
25	79.89	73.34
50	83.04	77.56
100	89.34	86.01
200	101.94	102.91
175	98.79	98.69
180	99.42	99.53
177	99.04	99.02

; **b.** 177 years after 1975, year 2152, the life expectancy for both men and women will be 99 years; **c.** $(177.44, 99.10)$;

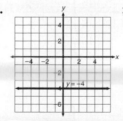

d. $0.126t + 76.74 = 0.169t + 69.11$
$$7.63 = 0.043t$$
$$177.44 = t$$
$$E = 0.126(177.44) + 76.74 = 99.10;$$
7. a. $2 = x, y = 4$; **b.** $16 = y, x = -39$; **c.** $y = \frac{1}{2}, x = \frac{3}{4}$;
d. $1 = -3$ no solution

Activity 1.11   Exercises: 1. a. $x = -5$; **b.** $x = -2$;
c. $\frac{4}{7} = x$; **3. a.** $y = 5, (1, 5)$; **b.** $y = -3, (2, -3)$;
5. a. $8x + 5y = 106$
$$x + 6y = 24$$
b. $x = -6y + 24$
$$8(-6y + 24) + 5y = 106$$
$$-48y + 192 + 5y = 106$$
$$-43y = -86$$
$$y = 2$$
$$x = -6(2) + 24 = 12$$
$(12, 2)$ Each centerpiece costs \$12, and each glass costs \$2.
c.

$8x + 5y = 106$	$8x + 5y = 106$
$-8(x + 6y) = -8(24)$	$-8x - 48y = -192$

$$\underline{-8x - 48y = -192}$$
$$-43y = -86$$
$$y = 2$$
$$x + 6(2) = 24$$
$$x + 12 = 24$$
$$x = 12$$

d.

Activity 1.12   Exercises: 1. $(0, -3, 5)$; **3.** $(-5, 3, 1)$;
5. a. dependent; **b.** inconsistent

Activity 1.13   Exercises: 1. $l + w + d \le 61$;
3. $C(A) < C(B)$; **5.** $24,650 < i \le 59,750$;

7. $x > -2$

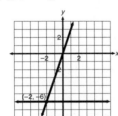

9. $x < 4$

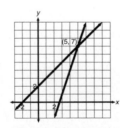

11. $x \ge 8$

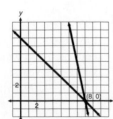

13. $x \ge -0.4$

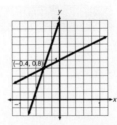

15. $1 < x < 2$

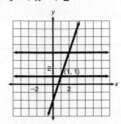

17. a. $-14.25t + 598.69 < 200$; **b.** $t > 27.98$; 28 years
after 1985, the year 2013; **18. a.** $C = 40.95 + 0.19n$;
b. $C = 19.95 + 0.49n$;
c. $19.95 + 0.49n < 40.95 + 0.19n$; **d.** $n < 70$;
19. a. $150 + 60n$; **b.** $150 + 60n \le 1200$; **c.** $n \le 17.5$
The maximum number of boxes that can be placed in the
elevator is 17; **20.** $57.5 \le w \le 70$;
21. a. $-79.8 \le F \le 134$ **b.** $-79.8 \le 1.8C + 32 \le 134$

Activity 1.14   Exercises:
1. e. $1.20 + .90(11) = \$11.10$;

3. a. $f(x) = \begin{cases} 2.5x & x \le 15,000 \\ 37,500 + 3(x - 15,000) & 15,000 \le x \le 21,000 \\ 55,500 + 4(x - 21,000) & x > 21,000 \end{cases}$

b.
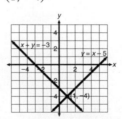
; **d.** 23,375 books;
5. a. $f(x)$: 8, 7, 6, 5, 4, 3, 2, 1, 0, 1, 2;
$g(x)$: 2, 1, 0, 1, 2, 3, 4, 5, 6, 7, 8.

Activity 1.15   Exercises: 1. a. 0.25 cm;
b. $|x - 8| \le 0.25$; **c.** [number line 7.75 to 8.25] $7.75 \le x \le 8.25$;
4. a. [number line -8 to 5] $x = -8$ or $x = 2$;
c. [number line 165 to 190] $170 \le x \le 180$;
e. [number line -30 to 0] $-25 < x < -15$;
9. a. $|x - 0.25| \le 0.025$; **b.** $0.225 \le x \le 0.275$; accept-
able thickness of the sheet of steel [number line 0.20 to 0.30]

How Can I Practice?
1. a. $(1, -4)$

b. $(3, 2)$

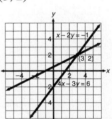

c. $(-4, -5)$

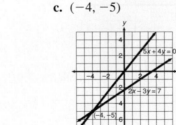

d. $-2 = 6$ inconsistent;
no solution

2. a. $(1, -4)$; **b.** $(3, 2)$; **c.** $(-4, -5)$; **d.** inconsistent; **3.** $(1, 5, -2)$

4.

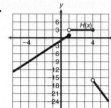

5. a. $x = 11$ or $x = -3$

b. $2/3 \le x \le 10/3$

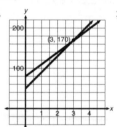

c. $x > 9$ or $x < -4$

6. a. $x \ge 1.8$ **b.** $x > -6$; **c.** $1 \le x < 5$; **7. a.** $t + d = 80$, $0.50t + 0.75d = 52$; **b.** $d = 48, t = 32$; **c.**

8. a. $y = 80 + 30x$, $y = 50 + 40x$;

b.

COLUMN 2	COLUMN 3
140	130
200	210
260	290
320	370

; c. ;

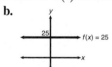

d. 3 hours; the total cost is $170;
e. $x = 3, y = 80 + 90 = 170$; **f.** I will use Towne Truck; its graph is below World Transport for $x = 6$; **9. a.** $y = 2x$; **b.** $x = 7, y = 14, z = 6$; **c.** It checks; **10. a.** $-5 < x \le 6$; **b.** $x < -5$ or $x \ge 3$; **c.** $-3 \le x < 4$; **11. a.** $x \ge 1367$; **b.** $1034 \le x \le 1500$; **12. a.** $|x - 8| \le 1.2$; **b.** $6.8 \le x \le 9.2$; **c.** $|x - 8| > 1.2$; **d.** $x > 9.2$ or $x < 6.8$;

Gateway Review   1. a. Yes, it is a function; **b.** No, it is not a function. There are two different outputs paired with 2; **c.**   Yes, it is a function;

2. 20, 36, 44, 60, 76; **a.** Yes, for each input there is one output; **b.** The input is x, the number of hours worked;

c. The dependent variable is $f(x)$, the total cost; **d.** Negative values would not be realistic domain values. A negative number of hours worked does not make sense; **e.** The rate of change is $8 per hour; **f.** The rate of change is $8 per hour; **g.** The rate of change between any two points is $8 per hour; **h.** The relationship is linear; **i.** $f(x) = 8x + 20$; **j.** The slope is the hourly rate I charge, $8 per hour; **k.** $(0, 20)$ is the vertical intercept. This represents the fertilizer cost; **l.** $f(4) = 8(4) + 20 = 52$; **m.** $8x + 20 = 92$ or $8x = 72$ or $x = 9$; I need to work 9 hours for $92 pay; **3. a.** $f(-2) = 14, g(-2) = 10$; **b.** $-6 + (-5) = -11$; **c.** $24 - 16 = 8$; **d.** $36(-2) = -72$; **4. a.** This represents a linear function; **b.** This represents a linear function; **c.** This does not represent a linear function; **d.** This represents a linear function; **5. a.** $m = \dfrac{9 + 3}{-4 - 5} = \dfrac{12}{-9} = \dfrac{-4}{3}$; **b.** $m = \frac{3}{7}$;

c. $m = \frac{1}{2}$

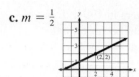

; 6. a. $y = 4$; **b.** $y = 2x + 5$; **c.** $-14 = -3(6) + b, b = 4$, $y = -3x + 4$; **d.** $-2 = 2(7) + b, b = -16$, $y = 2x - 16$; **e.** $x = 2$; **f.** $0 = -5(4) + b, b = 20, y = -5x + 20$; **g.** $16 = 4(2) + b, b = 8, y = 4x + 8$; **h.** $y = \frac{-1}{2}x + 5$;

7.

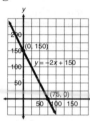

$y = \frac{-2}{5}x + 2$;

8. a. $f(x) = 300,000 - 10,000x$; **b.** $m = -10,000$. The building depreciates $10,000 per year; **c.** $(0, 300,000)$; The original value is $300,000; **d.** $(30, 0)$; It takes 30 years for the building to fully depreciate; **9. a.** $(0, -3)$; **b.** $(0, -3)$; **c.** $(0, -3)$; **d.** The graphs all intersect at the point $(0, -3)$; **e.** The results are the same; **10. a.** $m = -2$; $(0, 1)$; **b.** $m = -2$; $(0, -1)$; **c.** $m = -2$; $(0, -3)$; **d.** The graphs are parallel lines; **11. a.** $m = -3$; $(0, 2)$; **b.** $m = -3$; $(0, 2)$ $m = -3$; $(0, 2)$**c.** $m = -3$; $(0, 2)$; **d.** The graphs are all the same.; **e.** the slopes; **f.** the slopes and the y-intercepts; **g.** The results are the same; **12. a.** $(0, 150), (75, 0)$

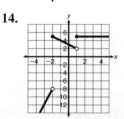

; c. The domain and range are all real numbers; **d.** $w(t) = -2t + 150$; **e.** They are the same; **f.** The vertical intercept is $(0, 150)$. It indicates the person's initial weight of 150 pounds. The horizontal intercept $(75, 0)$ indicates that after 75 weeks of weight loss, the person weighs nothing;

g. The practical domain is $0 \le t \le 15$. The practical range is $120 \le w(t) \le 150$; **13. a.** $f(x) = 25$;
b.   horizontal line through $(0, 25)$;

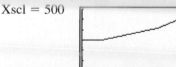

c. The slope is 0;

14.

15. a. $f(x) = \begin{cases} 1500 & \text{if } x \le 10,000 \\ 1500 + 0.29(x - 10,000) & \text{if } 10,000 < x < 40,000; \\ 2100 + 0.04(x - 40,000) & \text{if } x > 40,000 \end{cases}$

b. Xmin $= 0$, Xmax $= 50,000$, Ymin $= 0$, Ymax $= 3000$, Xscl $= 500$

c. $f(25,000) = 1500 + 0.02(15,000) = 1800$;
d. $x = 66,250$; **16. a.** $y = 1040x + 7900$ or $t(n) = 1040n + 7900$; **b.** 1040; The number of finishers increased at a rate of 1040 per year; **c.** $(0, 7900)$ The model

indicates that there were 7900 finishers in 1994; **d.** pretty well; **e.** 14,140; **f.** I used extrapolation because I am predicting outside the original data; **g.** No, 2024 is farther from the data than 2000. The farther removed we are from the data, the more likely our prediction is incorrect; **17. a.** $(3, -1)$; **b.** $(-1, 6)$; **c.** 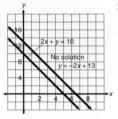 ; **d.** $4 = 4$; This is a dependent system. Any pair of numbers that satisfies one equation satisfies both equations;

18. $x = 4.50$, $y = 0.75$; **19. a.** $(0, 1, 2)$; **b.** $(-3, 1, 0)$; **c.** $(0.5, 0.25, -0.5)$; **d.** $(12, 7, 9)$; **20. a.** It is a good deal; **b.** Answers will vary. I would not take advantage of this. I don't give away many pictures;

21. a. 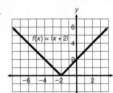 ; **b.** increasing $x > -2$, decreasing $x < -2$; **c.** The domain is all real numbers; **d.** The range is $y \geq 0$; **e.** g is the reflection through the x-axis; **f.** f shifts the graph of $y = |x|$ two units to the left. h shifts the graph of $y = |x|$ two units up; **22. a.** $x = 28$ or $x = 18$; **b.** $x = -5$ or $x = -19$; **c.** $x = \frac{11}{2}$ or $x = -\frac{1}{2}$; **d.** $x = \frac{1}{5}$ or $x = 1$;

23. a. $2.3 \leq x \leq 2.7$ **b.** $x > -3$ or $x < -7$

c. $3 < x < 6$ **d.** $\dfrac{-20}{3} \leq x \leq \dfrac{-16}{3}$

24. a. $|x - 453| \leq 8$; **b.** $445 \leq x \leq 461$

Chapter 2

Activity 2.1 Exercises:

1. a.

COLUMN 2	COLUMN 3	COLUMN 4
100	250	350
100	500	600
100	750	850
100	1000	1100
100	1250	1350

b. $C(x) = 12.50x + 100$;

c.

COLUMN 2	COLUMN 3	COLUMN 4
3	2250	292.50
6	4500	585
9	6750	877.50
12	9000	1170
15	11,250	1462.50

d. $R(x) = 0.13(750)(0.15x) = 14.625x$; **e.** $P(x) = 2.125x - 100$; **f.** $0 \leq x \leq$ the number of people the banquet room will accommodate. **g.** Set $P(x) = 0$ and solve for x, $x = 47.06$; 48 people must attend; **h.** Set the profit equal to 500 and solve for x, $500 = 2.125x - 100$; $x = 282.35$; 283 must attend; **3.** $(f + g)(x)$: 4, -6, 1, 6, 2, 0; $(f - g)(x)$: 2, -4, -1, 8, -4, -8; **5. a.** $5x - 2$; **b.** $x^2 + 3x - 8$; **c.** $-x + 30$; **d.** $-3x^2 + 8x - 3$; **e.** $15x - 15$; **f.** 11; **g.** $5x^2 + 5x - 9$; **h.** $-6x^2 + 15x - 4$; **i.** $-5x + 25$; **j.** $-33x - 10$; **7. a.** $4x^2 - x$; **b.** $4x^2 - 5x - 9$; **c.** $3 + 4x$; **d.** $4x^2 - 5x + 3$; **9. a.** 1, -9, -11, -5, 9, 31; **b.** $(f - g)(x) = x^2 + 5x - 5$; **c.** The answers check.

Activity 2.2 Exercises:

1. a. $= -x + 1$ **b.** $= -x - 5$

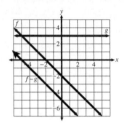

c. $= x + 5$

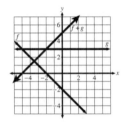

2. a. $= 2x - 5$ **b.** $= -2x + 1$

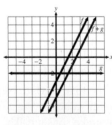

3.

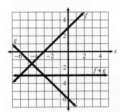

Activity 2.3 Exercises: 1. a. $(3x)(2x) = 6x^2$; b. $2x - 3$; c. $(2x - 3)(3x) = 6x^2 - 9x$; 3. a. $50 + x$; b. $7.50 - 0.05x$;

c. $C(x) = (50 + x)(7.50 - 0.05x)$; **d.** The domain is $0 \le x \le 50$;

e. 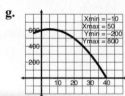; **f.** $C(x) = (50 + x)(7.50 - 0.05x) = 375 - 2.50x + 7.50x - 0.05x^2 = -0.05x^2 + 5x + 375$;

g. The graphs are the same;

5.

x	x^2	$3x$	-5
x	x^3	$3x^2$	$-5x$
3	$3x^2$	$9x$	-15

$(x + 3)(x^2 + 3x - 5) = x^3 + 3x^2 - 5x + 3x^2 + 9x - 15 = x^3 + 6x^2 + 4x - 15$; **7. a.** $6x^2 + 19x + 10$;
b. $6x^2 - 19x + 10$; **c.** $4x^2 + 5x - 6$; **d.** $4x^2 - 5x - 6$;
9. a. $9x^2 - 12x + 4$; **b.** $25x^2 - 4$; **c.** $x^4 - 25$; **d.** The outer product and inner product are opposites. Their sum is 0;
10. $f(x) \cdot g(x) = 2x^2 - x - 3$;

b.

x	$f(x)$	$g(x)$	$f(x) \cdot g(x)$
0	1	-3	-3
1	2	-1	-2
2	3	1	3
3	4	3	12
4	5	5	25

c. Answers may vary depending on the choices of x.

Activity 2.4   Exercises: 1. 5.66075×10^{11};
3. a. 3×10^{21}; **b.** 4.5×10^{16};
c. $9{,}000{,}000{,}000{,}000{,}000{,}000{,}000{,}000{,}000$;
d. $\frac{9 \times 10^{27}}{4.5 \times 10^{16}} = 2 \times 10^{11}$; **5. a.** $\frac{6.35 \times 10^8}{2.27 \times 10^9}$; **b.** 2.797×10^{-1};
7. 1; **9.** 10; **11.** $\frac{4}{x^4}$; **15.** $\frac{2y^4}{5x^4}$; **17.** $3x^{-3} = \frac{3}{x^3}$; **19.** $\frac{-1}{2a^2b^3}$

How Can I Practice?   1. a. $x = 30$; **b.** $N = f(t) = 30 + t$;
c. $C = g(t) = 20 - 0.5t$; **d.** $N = f(t)$: 30, 32, 34, 36, 38, 40;
$C = g(t)$: 20, 19, 18, 17, 16, 15;
e. $R(t)$: 600, 608, 612, 612, 608, 600;
f. $R(t) = f(t) \cdot g(t) = (30 + t)(20 - 0.5t) = -0.5t^2 + 5t + 600$;

g. 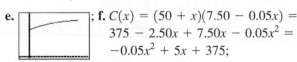; **h.** \$612.50 is the maximum revenue if 35 couples attend;
i. 35 tickets must be sold to obtain the maximum revenue;

2. a. $3x - 1$; **b.** $-x + 5$; **c.** $2x^2 + x - 6$; **d.** 2; **e.** 0;
f. $3x + 6$; **3. a.** $-3x + 5$; **b.** $x^4 - x^3 - 5x^2 + 9x - 4$; **c.** 0;
d. $-x^2 - 7x + 14$; **4. a.** $5x - 2$; **b.** $2x^2 - 2x - 8$;
c. $-2x + 12$; **d.** $2x^2 - 13x - 8$; **e.** $5x^2 - x + 2$; **5. a.** x^4;
b. x^9; **c.** $6x^8$; **d.** x^5y^6z; **e.** $10x^6y^5z^8$; **f.** $-30a^5b^3$;
6. a. $x^2 - 7x + 10$; **b.** $4x^2 + 25x - 21$; **c.** $4x^2 - 9$;
d. $x^3 + x^2 - 11x + 10$; **e.** $2x^3 + x^2 + 3x + 2$;

f. $-2x^2 - 5x - 21$; **g.** $11x^2 - 2x$;
h. $-x^5 - x^3 + 3x^2 + 2x - 1$; **i.** $9x^2 + 30x + 25$;
j. $4x^2 - 28x + 49$; **k.** $x^3 + 12x^2 + 48x + 64$;
l. $25x^2 - 49$; **7. a.** 2650, 3025, 3400, 3775, 4150;
b. $f(t) = 75x + 2650$; **c.** 1500, 2125, 2750, 3375, 4000;
d. $f(t) = 125x + 1500$; **e.** 50, 50, 50, 50, 50; **f.** $h(t) = 50$;
g. $k(t) = 200x + 4200$;

h.

$f(t)$	$g(t)$	$h(t)$	$k(t)$
2875	1875	50	4800
3550	3000	50	6600
4000	3750	50	7800
4525	4625	50	9200

i. It will equal \$10,000 in 2019; **8. a.** 17, -3, -7, 5, 33, 77;
b. $(f - g)(x) = 2x^2 + 2x - 7$; **c.** The answers check;
9. a. $\frac{2}{x^3}$; **b.** $9x^2$; **c.** $\frac{1}{3^4}$; **d.** 1; **e.** $2x^7$; **f.** $-6x^4y^8$; **g.** 4; **h.** $\frac{2}{3x^4}$; **i.** $\frac{5y^2}{x^4}$;
j. $\frac{-15}{x^5}$; **k.** $-2x$; **l.** $\frac{a^4}{b^4c^5}$; **10. a.** 10,080,000; **b.** I own approximately 0.2 of a square mile; **c.** 3.4439×10^{14} cubic miles; **11. a.** 2.75×10^8; 8.61×10^7;
b. $\frac{2.75 \times 10^8}{8.61 \times 10^7} = \frac{2.75 \times 10}{8.61} = \frac{27.5}{8.61} = 3.20$ or approximately 1 in 3;
12. $\frac{7.198 \times 10^6}{3.636 \times 10^6} = \frac{7.2 \times 10^6}{3.6 \times 10^6} = 2 \times 10^0 = 2$; The zero property of exponents was used; **13. a.** $A = 35 + 12x + x^2$;
b. $(7 + 4)(5 + 4) = 99$ or $35 + 12(4) + 4^2 = 99$;
$99 - 35 = 64$ sq ft; **c.** $A = \pi r^2$ $A = \pi(r + 2)^2$ or
$A = \pi r^2 + 4\pi r + 4\pi$

Activity 2.5   Exercises: 1. a. $g(2) = 200$. The radius of the slick is 200 ft 2 hours after the spill;
b. $f(g(2)) = f(200) = \pi(200)^2 = 125{,}664$; The area of the oil slick is 125,664 sq ft 2 hours after the spill;
c. $f(g(10)) = f(1000) = \pi(1000)^2 = 1{,}000{,}000\pi$. This is the area of the slick after 10 hours;
d. $f(g(t)) = f(100t) = \pi(100t)^2 = 10{,}000\pi t^2$;
e. $f(g(10)) = 10{,}000\pi(10)^2 = 1{,}000{,}000\pi$. The results are the same; **3.** $= -18x^2 + 18x - 3$;
b. $= -6t^2 + 6t + 2$; **5. a.** $L(x) = .99x$; **b.** $D(x) = .90x$;
c. $S(x) = L(D(x)) = L(0.90x) = 0.99(0.90x) = 0.891x$;
d. $S(500) = 0.891(500) \approx 446$; **e.** $D(x)$ needs to be improved because the airbags fail 10% of the time, whereas the seat belts fail only 1% of the time

Activity 2.6   Exercises: 1. a. $f(x) = 5280x$;
b. $f(5) = 26{,}400$; **c.** $g(w) = 12w$; **d.** $g(26{,}400) = 316{,}800$;
e. $g(f(x)) = g(5280x) = 12(5280x) = 63{,}360x$;
f. $g(f(5)) = 65{,}560(5) = 316{,}800$; 5 miles is equivalent to 316,800 inches; **3. a.** $f(x) = x - 1500$; **b.** $g(x) = 0.9x$;
c. $g(f(20{,}000)) = g(20{,}000 - 1500) = 0.9(18{,}500) = \$16{,}650$; The price of a \$20,000 car with a \$1500 rebate and a 10% discount is \$16,650; **5. a.** 8; **b.** 4;
c. 15; **d.** 2

Activity 2.7   Exercises: 1. a. $f(-1) = 6$; **b.** $g(12) = 9$;
c. $f(x - 3) = 2x - 6 + 8 = 2x + 2$;
d. $g(2x + 8) = 2x + 5$; **3.** $f(g(x)) = x^{18}$
$g(f(x)) = g(x^6) = -x^{18}$; **5. a.** 10; **b.** 12; **c.** 4; **d.** 256;

e. 1.903654; **f.** 4; **g.** $\dfrac{1}{\sqrt{25}} = \dfrac{1}{5}$; **h.** $\dfrac{1}{(\sqrt[3]{27})^2} = \dfrac{1}{3^2} = \dfrac{1}{9}$;

7. a. $x^{\frac{1}{2}}$; **b.** $x^{\frac{3}{4}}$; **c.** $(x + y)^{\frac{1}{3}}$; **d.** $a^{\frac{2}{5}}b^{\frac{3}{5}}$; **9. a.** -4; **b.** 0; **c.** 3;

10. a. $t = f(L) = 2\pi\left(\frac{L}{32}\right)^{\frac{1}{2}}$; **b.** $t = 6.28\sqrt{\frac{4}{32}} \approx 2.22$ seconds

Activity 2.8   Exercises: 1. a. 7; **b.** 4; **c.** x; **d.** x;
2. a. $h^{-1} = \{(3, 2), (4, 3), (5, 4), (6, 5)\}$;
b. $h(3) = 4$ $h^{-1}(h(3)) = h^{-1}(4) = 3$;
c. $h^{-1}(5) = 4$ $h(h^{-1}(5)) = h(4) = 5$;

3. a.

x	$r^{-1}(x)$
2	0
3	1
4	2
2	3

b. No, because the input value 2 is paired with two different
output values, 0 and 3; **c.** No, the interchange of the input
and output values does not result in a function;
5. a. $m = \dfrac{158.88 - 79.44}{100 - 50} = 1.5888$; **b.** You have 1.5888
Canadian dollars for one U.S. dollar; **c.** $f(x) = 1.5888x$;
d. $f(3000) = 1.5888(3000) = \$4766.40$; If I have \$3000
U.S., then I can exchange it for \$4766.40 Canadian;
e. $m = \dfrac{100 - 50}{158.88 - 79.44} \approx 0.6294$; **f.** You have 0.6294 U.S.
dollars for one Canadian dollar; **g.** $g(x) = 0.6294x$;
h. $g(6000) = \$3776.40$; If I have \$6000 Canadian, then I
can exchange it for \$3776.40 U.S;
i. $f(g(x)) = f(0.6294x) = 1.5888(0.6294x) = x$;
$g(f(x)) = g(1.5888x) = 0.6294(1.5888x) = x$

Activity 2.9   Exercises: 1. a. $P = f(S) = 0.05S + 250$;
b. $f(6000) = 0.05(6000) + 250 = 550$. The weekly salary
for \$6000 worth of sales is \$550; **c.** $S = g(P) = \dfrac{P - 250}{0.05}$;
d. $g(400) = \dfrac{400 - 250}{0.05} = 3000$. A weekly salary of \$400
means I sold \$3000 worth of merchandise;
e. $g(f(8000)) = 8000$; **3. a.** $y = 3x - 4$ so, $x = 3y - 4$ or
$y = f^{-1}(x) = \dfrac{x + 4}{3}$; **b.** $z = \dfrac{w - 4}{2}$ or $w = g^{-1}(x) = 2z + 4$;
c. $t = \dfrac{5}{s}$ or $s = \dfrac{5}{t}$;

5.

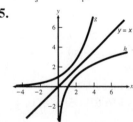

g and h are inverses. The graphs
of g and h are symmetric with
respect to the line $y = x$.

7. a. $g^{-1}(x) = \frac{3x - 6}{4}$ **b.**

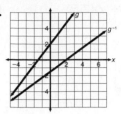

c. Yes, because the
graphs are reflec-
tions in the line
$y = x$;

d. $g^{-1}\left(\dfrac{6 + 4x}{3}\right) = \dfrac{3\left(\frac{6 + 4x}{3}\right) - 6}{4} = \dfrac{4x}{4} = x$.

Yes, because $g^{-1}(g(x)) = x$;

9. a.

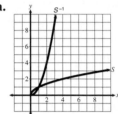

; b. ;

c. The area of the square is the
input. The length of the side of
the square is the output; **d.** The
length of the side of the square is
the input. The area of the square
is the output; **e.** Given the length of the side, we can determine the area of the granite top.

How Can I Practice?   1. a. $f(-1) = -3$; **b.** $g(5) = 7$;
c. $f(x + 2) = (x + 2)^2 - 4 = x^2 + 4x$;
d. $g(x^2 - 4) = x^2 - 4 + 2 = x^2 - 2$;
e. $f(x^2 - 4) = (x^2 - 4)^2 = x^4 - 8x^2 + 12$;
f. $x = y + 2$ or $y = g^{-1}(x) = x - 2$;
2. a. $f(-2) = -6$; **b.** $g(-1) = 2$; **c.** $g(-2) = -2$;
d. $f(4 + x - x^2) = (4 + x - x^2) - 4 = x - x^2$;
e. $= -x^2 + 9x - 16$; **f.** $x = y - 4$ or $y = f^{-1}(x) = x + 4$;
3. $-1, 3, 2, 1, 0$; **4. a.** x^8; **b.** x^3y^3; **c.** $32x^{20}y^5$; **d.** $432x^{10}y^4$;
e. 25; **f.** -25; **g.** $2x^2$; **h.** $-8x^3$; **i.** $-8x^{22}$; **j.** 125; **k.** 16; **l.** -3;
m. $\frac{27}{64}$; **n.** 1.59; **o.** x^2y^4; **p.** $x^{\frac{7}{6}}$; **q.** $2x^{\frac{1}{3}}y^{\frac{1}{3}}$;
5. a. $g(3x^2) = -2(3x^2)^3 = -54x^6$;
b. $f(-2x^3) = 3(-2x^3)^2 = 12x^6$;
c. $g(48) = -2(48)^3 = -221,184$;
6. a. $s(4x - 1) = (4x - 1)^2 + 4(4x - 1) - 1 =$
$16x^2 - 8x + 1 + 16x - 4 - 1 = 16x^2 + 8x - 4$;
b. $t(x^2 + 4x - 1) = 4(x^2 + 4x - 1) - 1 =$
$4x^2 + 16x - 5$; **c.** $x = 4y - 1$ or $y = T^{-1}(x) = \frac{x + 1}{4}$;
7. a. $p(\sqrt{x + 2}) = \dfrac{1}{\sqrt{x + 2}}$; **b.** $c\left(\frac{1}{x}\right) = \sqrt{\frac{1}{x} + 2}$;
c. $x = \frac{1}{y}$ or $y = p^{-1}(x) = \frac{1}{x}$;
8. $\{(6, 4), (-9, 7), (1, -2), (0, 0)\}$;
9. $f(g(x)) = f\left(\frac{x + 3}{2}\right) = 2\left(\frac{x + 3}{2}\right) - 3 = x$
$g(f(x)) = g(2x - 3) = \dfrac{(2x - 3) + 3}{2} = x$; Because
$f(g(x)) = g(f(x)) = x, f$, and g are inverse functions;
10. a. $x = 4y + 3$ or $y = f^{-1}(x) = \frac{x - 3}{4}$;

b.

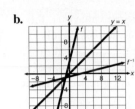

; **c.** The intercepts of f are $(0, 3)$ and $\left(-\frac{3}{4}, 0\right)$. The intercepts of f^{-1} are $\left(0, -\frac{3}{4}\right)$ and $(3, 0)$; **d.** The slope of the graph of f is 4. The slope of the graph of the inverse is $\frac{1}{4}$;

e.

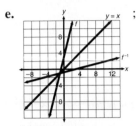

;

11. a. 41, 42.8, 44.6, 47;

b.

41	42.8	44.6	47
0	3	6	10

;

c. $f^{-1}(x) = \frac{x - 41}{0.6}$; **d.** The population will be 46 million in 1998; **e.** The graphs are symmetrical about the line $y = x$; **f.** The horizontal intercept of the function is the vertical intercept of its inverse. The vertical intercept of the function is the horizontal intercept of its inverse

HORIZONTAL INTERCEPT	VERTICAL INTERCEPT
$(-68.3, 0)$	$(0, 41)$
$(41, 0)$	$(0, -68.3)$

;

g. Function: $m = 0.6$, Inverse function: $m = \frac{1}{0.6}$ or 1.67, The slopes of the two functions are reciprocals; **h.** The population will be 50 million in 2005; **12. a.** 495.6 euros; **b.** $345.28; **c.** $g(f(x)) = g(.00826x) = .9865(.00826x) = .008148x$; **d.** $g(f(60,000)) = 0.008148(60,000) = 488.88; **13.** $x = \frac{4}{3}$; **14. a.** $b = f(x) = x^2$, $V = g(b) = 10b$; **b.** $V = g(f(x)) = g(x^2) = 10x^2$

Gateway Review 1. $2x^2 - 2x - 1$; **b.** $-x^2 + 5x - 4$; **c.** $4x^2 - 13x + 3$; **d.** $x^3 - 7x^2 + 13x - 15$; **e.** $-11x + 11$; **f.** $2x^4 - 5x^3 + 4x^2 + 7x - 4$; **2. a.** $6x^8$; **b.** $16x^6y^2$; **c.** $-2x^5y^3$; **d.** $-10x^5y^5z^4$; **e.** $9x^6y^2$; **f.** $-125x^3y^3$; **g.** $2x^3$; **h.** 2; **i.** 1; **j.** $\frac{3y^3}{2z^3}$; **k.** $-5x^{-8} = \frac{-5}{x^8}$; **l.** $-4x^2$; **m.** $(-5)^3(x^{-3})^3 = -125x^{-9} = \frac{-125}{x^9}$; **n.** $x^{\frac{4}{5} + \frac{1}{2}} = x^{\frac{8}{10} + \frac{5}{10}} = x^{\frac{13}{10}}$; **o.** $x^{\frac{2}{3} \cdot 3} = x^2$; **3. a.** -20; **b.** $4x + 1$; **c.** $f(3) - g(3) = 16 - (-3) = 19$; **d.** $(6x - 2)(-2x + 3) = -12x^2 + 22x - 6$; **e.** $= -12x + 16$; **f.** $g(10) = -2(10) + 3 = -17$; **g.** $x = 6y - 2$ or $y = f^{-1}(x) = \frac{x + 2}{6}$; **4. a.** $x^2 - 4x + 5$; **b.** $= 3x^3 - 5x^2 + 11x - 6$; **c.** $= 9x^2 - 15x + 9$; **d.** $g(5) = 3(5) - 2 = 13$; **5. a.** 7; **b.** 4; **c.** 81; **d.** 3.214; **e.** 9; **f.** 32; **g.** $\frac{1}{16}$; **6. a.** $f(x) = 2(0.01)x^2 = 0.02x^2$;

b. $g(x) = 4(0.004)(x)(3x) = 0.048x^2$; **c.** $(f + g)(x) = 0.02x^2 + 0.048x^2 = 0.068x^2$; **d.**

$f(x)$	g	$(f + g)(x)$
0.08	0.19	0.27
0.32	0.77	1.09
0.72	1.73	2.45
1.28	3.07	4.35
2.00	4.80	6.80

;

e. $f(5) = 0.50$, $g(5) = 1.2$, $(f + g)(5) = 1.70$ For a box whose base is 5 in. by 5 in., the cost of the top and bottom of the box is $0.50, the cost of the 4 sides is $1.20, and the total cost of the box is $1.70; **7. a.** $f(x) = 12x + 300$; **b.** $g(x) = 25.95x$; **c.** $h(x) = 25.95x - (12x + 300) = 13.95x - 300$; **d.** 22 hats must be sold, because 21 hats is not quite enough. The solution was obtained graphically; **e.** $f(50) = 900$; the cost of producing 50 hats is $900. $g(50) = 1297.50$; the revenue from 50 hats is $1297.50. $h(50) = 397.50$; the profit from selling 50 hats is $397.50; **f.** The profit is the difference between the revenue and cost functions; **8. a.** $f(x) = 60(110) + x(110 - 2x)$; **b.** $f(x) = 6600 + 110x - 2x^2$; **c.** integers $0 \le x \le 30$; **d.** $f(15) = 7800$. At regular price the cost is $8250, so the savings are $450; **9. a.** -10; **b.** 41; **c.** 2; **d.** 5; **10. a.** integers $0 \le x \le 30$; **b.** $f(22) = 2864.40$; **c.** $f(3.75t) = 150(3.75t) - 0.9(3.75t)^2 = 562.50t - 12.65625t^2$; **d.** The input variable is t; **e.** $f(g(4)) = 2047.50; **f.** $3500 = 562.50t - 12.65625t^2$ t is about 7.5 hours (determined graphically); **11. a.** $\frac{2y - 3}{5} = x$ or $y = f^{-1}(x) = \frac{5x + 3}{2}$; **b.** The slope of f is $\frac{2}{5}$. The slope of f^{-1} is $\frac{5}{2}$. The slopes are reciprocals; **12.** $f(g(x)) = f\left(\frac{1 - x}{2}\right) = -2\left(\frac{1 - x}{2}\right) + 1 = x$ $g(f(x)) = g(-2x + 1) = \frac{1 - (-2x + 1)}{2} = \frac{2x}{2} = x$ Since $f(g(x)) = g(f(x)) = x$, f, and g are inverses; **b.**

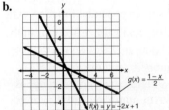

; **c.** f and g are symmetric with respect to the line $y = x$;

13. a. Yes, the ratio (change in cost)/(change in number of tickets) is constant; **b.** $f(x) = 5.5x$; **c.** The cost of one ticket is $5.50. This represents the slope; **d.**

TOTAL COST	NUMBER OF TICKETS
$11.00	2
$27.50	5
$38.50	7
$66.00	12

;

e. $g(x) = \frac{2}{11}x$; **f.** The slope is 2/11. The slopes are reciprocals; **g.** $f\left(\frac{2}{11}x\right) = 5.5\left(\frac{2}{11}\right)x = x$, $g(5.5x) = \frac{2}{11}(5.5x) = x$ The functions are inverses because $f(g(x)) = g(f(x)) = x$.

Chapter 3

Activity 3.1   Exercises: 1. a. top table: 0.008, 0.04, 0.2, 1, 5, 25, 125; bottom table: 125, 25, 5, 1, 0.2, 0.04, 0.008;

b.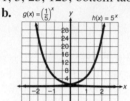

c.

BASE, b	GROWTH OR DECAY FACTOR	x-INTERCEPT	y-INTERCEPT	HORIZONTAL ASYMPTOTE	INCREASING OR DECREASING
5	growth	none	(0, 1)	$y = 0$	increasing
$\frac{1}{5}$	decay	none	(0, 1)	$y = 0$	decreasing

3. a.

The graphs of f, g, and h are all increasing exponential functions with a growth factor of 2. The domain of each function is the set of all real numbers. The range of f is the set of all positive real numbers. The range of g is $y > 1$. The range of h is $y > -3$. The x-axis ($y = 0$) is the horizontal asymptote of the graph of f. The horizontal asymptote for g is $y = 1$. The horizontal asymptote for h is $y = -3$. The graph of g is the graph of f shifted 1 unit up. The graph of h is the graph of f shifted 3 units down; **b.** The graph of f is a decreasing exponential function with a decay factor of $\frac{3}{4}$. The graph of g is an increasing exponential function with a growth factor of $\frac{4}{3}$. The graphs of f and g are reflections in the y-axis; **c.** The graph of f is an increasing exponential function with a growth factor of 10. The graph of g is the graph of f reflected in the x-axis; **d.** The graph of f is an increasing exponential function with a growzth factor of 3. The graph of g is a decreasing exponential function with a decay factor of $\frac{1}{3}$. The graphs of f and g are reflections in the y-axis; **5. a.** 24.948, 26.944, 29.099; **b.** 7.5, 5.625, 4.21875; **7. a.** The domain is all real numbers, and the range is $y > 2$. **b.** The domain is all real numbers, and the range is $y > -4$. **c.** The domain is all real numbers, and the range is $y < 0$. **9. a.** 2, 4, 8, 16, 32; **b.** 256; **c.** The data is exponential with a growth factor of 2; **d.** (Answers may vary.) The practical domain is the set of nonnegative integers from 0 to 10. The practical range is the set of whole-number powers of 2 up to 2^{10}.

Activity 3.2   Exercises: 1. a. $P = 148.0(0.9973)^t$; **b.** Yes; substituting 5 for t yields $P = 148.0(0.9973)^5 = 146.01$; **c.** $P = 148.0(0.9973)^{12} = 143.3$ million; **2. a.** iii; **b.** i; **c.** ii;

4. a.  ; **b.** For the same value of x, the graph of g is 3 times the distance from the x-axis that the graph of f is.; **c.** The graph of h is the graph of g reflected in the x-axis;

5. a. $f(-2) = 3 \cdot 4^{-2} = 3 \cdot \frac{1}{4^2} = \frac{3}{16}$;
b. $f\left(\frac{1}{2}\right) = 3 \cdot 4^{\frac{1}{2}} = 3 \cdot 2 = 6$;
c. $f(2) = 3 \cdot 4^2 = 3 \cdot 16 = 48$;
d. $f(1.3) = 3 \cdot 4^{1.3} \approx 18.1886$; **7. a.** 2.5 ppm; **b.** 2.5, 1.75, 1.225, 0.8575, 0.6003, 0.4202;

c. 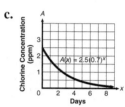 ; **d.** $A(3) = 2.5(0.70)^3 = 0.8575$ ppm; **e.** Chlorine should be added in 1.4 days.

Activity 3.3   Exercises:

1.

GROWTH FACTOR	GROWTH RATE	DECAY FACTOR	DECAY RATE
1.02	2%	0.77	23%
1.029	2.9%	.32	68%
2.23	123%	0.953	4.7%
1.34	34%	.803	19.7%
1.0002	.02%	0.9948	.52%

2. a. Bozeman: $P(t) = 27,509(1.0196)^t$, Butte: $P(t) = 32,370(0.9971)^t$;
b. Bozeman: $P(5) = 27,509(1.0196)^5 = 30,313$ Butte: $P(5) = 32,370(0.9971)^5 = 31,903$; **c.** The population of Bozeman will be 55,018 when $t \approx 35.7$;
d. The populations will be equal when $t \approx 7.3$.;
4. a. 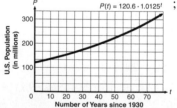 ;

b. The growth factor is $b = 1.0125$; the growth rate is $0.0125 = 1.25\%$; **c.** $P(70) = 120.6(1.0125)^{70} = 287.7$ million. The prediction is a little higher;
5. a. $V(t) = 20,000(0.85)^t$; **b.** The decay rate is $15\% = 0.15$;
c. The decay factor is 0.85;
d. $V(5) = 20,000(0.85)^5 = 8874.11$;
e. ; **f.** The value will be $10,000 when $t \approx 4.3$ years.

Activity 3.5   Exercises: 1. a. $A = 25,000\left(1 + \frac{0.045}{4}\right)^{4t}$;
c. approximately 21.4 years; **e.** approximately 21.2 years;
3. a. $A = 1900e^{0.06 \cdot 2} = \$2142.24$; **b.** approximately 11.5
years; **5. a.** $y = 20e^{-0.0244 \cdot 20} = 12.277$; **b.** The half-life is
approximately 28.4 years; **c.** The decay factor is
$b = e^{-0.0244} = 0.9759$. The decay rate is 0.0241 or 2.41%;
5. a. $y = 20e^{-0.0244 \cdot 20} = 12.277$; **b.** approximately 28.4
years; **c.** The decay factor is $b = e^{-0.0244} = 0.9759$. The
decay rate is 0.0241 or 2.41%; **7. a.** In 1996, $x = 9$, so
$A = 36.2 \cdot e^{0.14 \cdot 9} = 127.62$ billion dollars;
b.

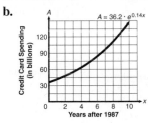

c. The vertical intercept is
(0, 36.2). There was
\$36,200,000,000 in holiday
credit-card spending in 1987;
d. 1993 ($x = 5$);
e. approximately 4.95 years

Activity 3.6   Exercises: 1. a. The data is not linear. The
output seems to be increasing exponentially;
b.

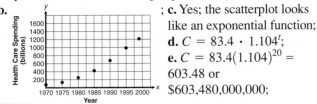

c. Yes; the scatterplot looks
like an exponential function;
d. $C = 83.4 \cdot 1.104^t$;
e. $C = 83.4(1.104)^{20} =$
603.48 or
\$603,480,000,000;

f. The growth factor, b, is 1.104; **g.** 0.104 or 10.4%; **h.** 1996
($t = 26$); **i.** approximately 7 years; **3. a.** the set of all real
numbers; **b.** the set of all positive real numbers;
c. $y = a \cdot b^x$ is positive for all values of x; **d.** $y = a \cdot b^x$ is
never negative; **e.** (0, a)

How Can I Practice?   1. a. $C = 17,000(1.04)^t$;
b. The growth rate is 0.04; the growth factor is 1.04;
c. $C = 17,000(1.04)^3 = 19,122.69$; **d.** 14.5 years;
2. a. graph i is function b; **b.** graph ii is function c;
c. graph iii is function a; **3.** Graph i is function g because it
is decreasing and the only growth factor between 0 and 1 is
0.47 in function g, Graph ii is function h because it is
positive and increasing with the growth factor of 1.47,
Graph iii is function f because it is the only one that yields
negative output; **4. a.** 13.01, 33.18, 84.61; **b.** 1.26, 0.76,
0.46; **c.** 216, 7776, 279,836; **5. a.** $y = 2(2.55)^x$;
b. $y = 3.5(0.6)^x$; **c.** $y = \frac{1}{6}(36)^x$;
6. a. $f(0) = 1.3$; decreasing; **b.** $f(0) = 0.6$; increasing;
c. $f(0) = 3$; decreasing; **7. a.** Yes; **b.** The constant ratio is
2.5 or $\frac{5}{2}$; **c.** $y = 2(2.5)^x$; **8.** Plan 1: $S = 22,000 + 1000x$,
Plan 2: $S = 22,000(1.04)^x$; **b.** Plan 1: \$22,000, \$23,000,
\$25,000, \$27,000, \$32,000, \$37,000, Plan 2: \$22,000,
\$22,800, \$24,747, \$26,766, \$32,565, \$39,621;
c. It depends. If I plan to be with the company less than ten
years, I would take plan 1, because it takes plan 2 about
nine years to catch up. If I expect to be with the company

for a long time, say twenty years, I would choose plan 2,
because by then I would be better off by more than \$6000
per year; **9. a.** $N = f(t) = 2e^{0.075t}$;
b. $N = f(8) = 2e^{0.075(8)} = 3.6442$; **c.**

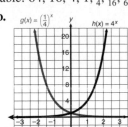

d. 14.6482 or 15 weeks; **10. a.** top
table: $\frac{1}{64}, \frac{1}{16}, \frac{1}{4}$, 1, 4, 16, 64; bottom
table: 64, 16, 4, 1, $\frac{1}{4}, \frac{1}{16}, \frac{1}{64}$;

b.

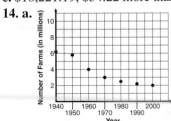

c. Row 1: 4, growth, none, (0, 1),
x-axis, increasing; Row 2: $\frac{1}{4}$,
decay, none, (0, 1), x-axis,
decreasing; **11. a.** \$415;
b. 0.0118 or 1.18% per
month; **c.** 1.0118;
d. $f(x) = 415(1.0118)^x$;
e. (0, 415); **f.** This represents
the initial balance on the card;
g. \$466.65; **h.** With no payments, I exceed my credit limit
during the sixteenth month; **12. a.** The ratios are all 1.02;
b. 1.02; **c.** $w(t) = 12.50(1.02)^x$; **d.** 2%; **e.** \$14.94; **f.** about
35 years; **13. a.** $A = 10,000(1.01)^{12t}$; **b.** \$18,166.97;
c. approximately 6 years; **d.** $A = 10,000e^{0.12t}$;
e. \$18,221.19, \$54.22 more than in part b;
14. a.

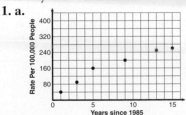

b. An exponential decay
model would better
model the data. The data
is decreasing, but not at a
constant rate;
c. $y = 6.24(0.9795)^x$;
d. $6.24(0.9795)^{70} = 1.46$
million farms; **e.** 0.9795;
f. $0.9795 = 1 - r$; $r = 1$ $r = 1 - 0.9795 = 0.0205$ or
2.05%; **g.** 34 years

Activity 3.7   Exercises: 1. a. 5; **b.** 3; **c.** −1; **d.** −6; **e.** 0;
f. 2; **g.** $\frac{1}{2}$; **h.** $\frac{1}{2}$; **i.** 0; **j.** 0; **k.** 5; **l.** −2; **m.** 0; **3. a.** $\log_3 9 = 2$;
b. $\log_{121} 11 = \frac{1}{2}$; **c.** $\log_4 27 = t$; **d.** $\log_b 19 = 3$;
5. a. $x = 0.5119$; **b.** $x = 2.771$; **c.** $x = -5.347$

Activity 3.8   Exercises: 1. a. $x > 0$; **b.** the set of all real
numbers; **c** $x > 1$; **d.** $0 < x < 1$; **e.** $x = 1$; **f.** $x = 10$;
3. $y = \log_2 x$; **4. a.** $x > 0$; **b.** the set of all real numbers;
c. $x > 1$; **d.** $0 < x < 1$; **e.** $x = 1$; **f.** $x = e$;
5. $n = \dfrac{\log(1000) - \log(34,000)}{\log(1 - 0.40)} \approx 6.9$ years

Activity 3.9   Exercises:
1. a.

b. It could very well be
logarithmic. It increases
more slowly as the
input increases;

c. $R = 23.530 + 84.378 \ln(t)$; **d.** Yes, it is a very good fit;
e. 2004 is 19 years after 1985, so evaluate R when $t = 19$.
$R = 23.530 + 84.378 \ln(19) = 272$ per 100,000
population; **3. a.** $f(0) = 27$ in. This is the pressure
in the eye of the storm; **b.** 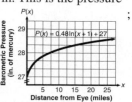 ;
c. As you move away
from the hurricane's
eye, the pressure
increases quickly at first
and then more slowly.

Activity 3.10   Exercises: 1. a. $\log_b 3 + \log_b 7$;
b. $\log_3 3 + \log_3 13 = 1 + \log_3 13$; **c.** $\log_7 13 - \log_7 17$;
d. $\log_3 x + \log_3 y - \log_3 3 = \log_3 x + \log_3 y - 1$;
3. a. ; **b.** The graphs are the same. This is not
surprising because the log of a product
is the sum of the logs;
5. a. $7.4 \log(15) = 8.7$ or 9 cars; **b.** 9, 13, 21; **c.** The sum
of the sales from the smaller ads exceeds the sales from the
larger ad by 1; **d.** Pretty close. 15 times 50 equals 750, so I
would have expected the sum of the sales from the smaller
ads to equal the sales from the largest. The error is due to
rounding; **e.** Forget about the giant ad. It is a waste of
money; **7. a.** $\log_2 245$; **b.** $\log \sqrt[4]{\dfrac{x^3}{z^5}}$; **c.** $\ln \dfrac{2^2 5^2 z^4}{5^3} = \ln \dfrac{4z^4}{5}$;
d. $\log_5 \dfrac{x^2 + 3x + 2}{x^2 + 6x + 9}$

Activity 3.11   Exercises: 1. a. 111,700 arrests in 2000;
b. There will be 50,000 arrests 9 years after 1990, or 1999;
3. a. Yes; 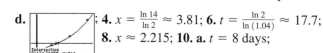 ; **b.** $A(t) = 352.65(1.006)^t$;
c. $t = \dfrac{\ln\left(\frac{705.3}{352.65}\right)}{\ln(1.006)} \approx 116$; 116 years
after 1990 would be the year 2106;
d. 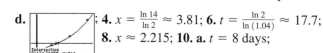 ; **4.** $x = \dfrac{\ln 14}{\ln 2} \approx 3.81$; **6.** $t = \dfrac{\ln 2}{\ln(1.04)} \approx 17.7$;
8. $x \approx 2.215$; **10. a.** $t = 8$ days;
b. $\frac{1}{5}P_0 = P_0 e^{-0.086t}$ or $t = \dfrac{\ln(0.2)}{-0.086} = 19$ days

Activity 3.12   Exercises: 1. $x = 2^5 = 32$;
3. $x = 5^{\frac{5}{3}} - 2 = 12.62$; **5.** $x = 303.17$;
7. $m = 8.8 + 5.1 \log(200) = 20.5$;
9. $[H^+] = 10^{-2.4} = 0.00398$ moles/liter

How Can I Practice?   1. a. $\log_4 16 = 2$;
b. $\log_{10}(0.0001) = -4$; **c.** $\log_3\left(\frac{1}{81}\right) = -4$; **2. a.** $2^5 = 32$;
b. $5^0 = 1$; **c.** $10^{-3} = .001$; **d.** $e^1 = e$; **3. a.** $x = 4^{-3} = \frac{1}{64}$;
b. $b = 2$; **c.** $y = 3$; **4. a.** -1; -0.5; 0; 1; 2; 3;

b. 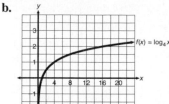 ; **d.** (1, 0); **e.** $x > 0$;
f. all real numbers;
g. The y-axis ($x = 0$) is a
vertical asymptote;
h. $f(32) = 2.5$;
i. $x = 90.5$;

5. a. $\log_b x + 2 \log_b y - \log_b z$;
b. $\frac{3}{2} \log_3 x + \frac{1}{2} \log_3 y - \log_3 z$; **c.** $\log_5 x + \frac{1}{2} \log_5 (x^2 + 4)$;
d. $\frac{1}{3} \log_4 x + \frac{2}{3} \log_4 y - \frac{2}{3} \log_4 z$; **6. a.** $\log \dfrac{x\sqrt[3]{y}}{\sqrt{z}}$;
b. $\log_3 (x + 3)^3 z^2$; **c.** $\log_3 \sqrt[3]{\dfrac{x}{y^2 z^4}}$; **7. a.** $\dfrac{\log 17}{\log 5} = 1.76$;
b. $\frac{1}{3} \cdot \dfrac{\log 41}{\log 13} = 0.4826$; **8. a.** $x \approx 0.0067$; **b.** $x = 646.08$;
9. a. $x = \dfrac{\log 17}{\log 3} = 2.5789$; **b.** $x = \dfrac{\ln(14)}{1.7} \approx 1.55$;
c. $x = \dfrac{\log 2}{\log 4 - \log 3} = 2.409$;
10. a. 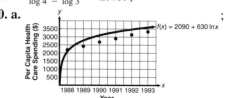 ;
c. $E = f(18) = 2090 + 630 \ln(18) = 3910.93$;
d. $x = 9.38$, which you round up to 10. The year is 1997;
e. $2090 + 630 \ln(x) = 3500$; **f.** $x = e^{2.238} = 9.38$; The
year is 1997. It is the same.

Gateway Review   1. a. $\frac{1}{8}$, $\frac{1}{2}$; 1; 8; 16; 64; 512;
b. 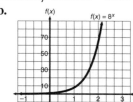 ; **c.** The function is increasing,
because $b = 8 > 1$;
d. all real numbers; **e.** $y > 0$;
f. There is no x-intercept. The
y-intercept is (0, 1); **g.** There is
one horizontal asymptote, the
x-axis, $y = 0$;
h. The domain and range are the same. The graphs are
reflections in the y-axis. f is increasing; g is decreasing;
i. h is moved upward 5 units; **j.** $x = 8^y$; $y = \log_8 x$;

2.

BASE, b	GROWTH OR DECAY FACTOR	x-INTERCEPT	y-INTERCEPT	HORIZONTAL ASYMPTOTE	INCREASING OR DECREASING
6	growth	none	(0, 1)	$y = 0$	increasing
$\frac{1}{3}$	decay	none	(0, 1)	$y = 0$	decreasing
2.34	growth	none	(0, 5)	$y = 0$	increasing
0.78	decay	none	(0, 3)	$y = 0$	decreasing
2	growth	(2, 0)	(0, −3)	$y = −4$	increasing

3.

all reals	all reals	all reals	$x > 0$	$x > 3$
$y > 0$	$y > 2$	$y > -5$	all reals	all reals

4. a. The table is approximately exponential. The growth factor is about 1.55; **b.** $y = 10 \cdot 1.55^x$; **5. a.** 15,000, 15,225, 15,453, 15,685, 15,920, 16,159; **b.** $y = 15,000(1.015)^x$; **c.** $y = \$16,897$; This is reasonable if you assume that 15,000 is a reasonable starting salary and that the 1.5% salary increase per year remains constant;

d. $x = \log_{1.015} 2 = \dfrac{\ln(2)}{\ln(1.015)} = 46.6$ years;

6. a. $A = 5000e^{0.065(8)} = \$8410.14$; **b.** $t \approx 13.5$ years;

7. a.

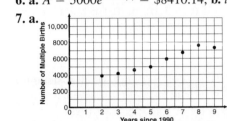

b. It would be better modeled by an exponential model; **c.** $f(x) = 3061(1.112)^x$; **d.** 1.112; **e.** 11.2%; **f.** $f(22) = 31,635$; **g.** approximately 6.5 years; **8. a.** 125; **b.** 27; **c.** $\frac{1}{32}$; **d.** 25; **e.** -2; **f.** 4; **g.** -3; **9. a.** $\log_6 36 = 2$; **b.** $\log_{10} 0.000001 = -6$; **c.** $\log_2 \frac{1}{32} = -5$; **10. a.** $;3^4 = 81$ **b.** $7^0 = 1$; **c.** $10^{-4} = 0.0001$; **d.** $e^1 = e$; **11. a.** $x = \frac{1}{125}$; **b.** $b = 4$; **c.** $y = 6$; **12. a.** $-3, -2, -1, 0, 1, 2$;

b.

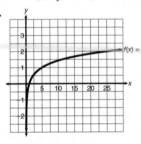 **; c.**

d. (1, 0); **e.** $x > 0$; **f.** all real numbers; **g.** It has a vertical asymptote at $x = 0$. The function gets closer and closer to the y-axis but does not cross it;

h. $f(23) = 1.948$; **i.** $x = 52.416$;

13. a. $\dfrac{\log 21}{\log 7} = 1.56$; **b.** $\dfrac{\log\left(\frac{8}{9}\right)}{\log 15} = -0.0435$;

14. a. $3 \log_2 x + \log_2 y - \left(\frac{1}{2}\right) \log_2 z$;

b. $\left(\frac{1}{3}\right)(4 \log x + 3 \log y - \log z)$; **15. a.** $\log \dfrac{x\sqrt[3]{y}}{z^3}$;

b. $\log \sqrt[3]{\dfrac{x}{y^2 z}}$; **16. a.** $3 + x = \dfrac{\log 7}{\log 3}$; $x = -1.23$;

b. $4x + 9 = 2^4$; $x = 1.75$; **c.** $x \approx 341.5$;

17. a.

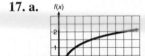 **; c.** 2.87744;

d. $2.319 = \dfrac{\log x}{2 \log 2}$; $x = 24.9$;

18. a. New York 18.98 million; Florida 15.98 million;

b.

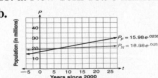

c.

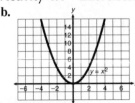 Florida's population will equal that of New York sometime in the year 2009;

d. $t = 19$ years Florida's population will exceed 25 million in the year 2019.

Chapter 4

Activity 4.1   Exercises: 1. a. 9, 4, 1, 0, 1, 4, 9;

b.

 ; c. Yes; **d.** 1; **e.** The domain is all real numbers. The range is all real numbers greater than or equal to 0; **3. a.** $a = -2, b = 0, c = 0$; **b.** $a = \frac{2}{5}, b = 0, c = 3$; **c.** $a = -1, b = 5, c = 0$;

d. $a = 5, b = 2, c = -1$; **5. a.** f opens upward; g opens downward; both pass through (0, 0); **b.** Both f and h open upward; **c.** h is g shifted up two units; both open upward; **d.** Both f and g open upward. The low point of f is 3 units below the x-axis; the low point of g is 3 units above the x-axis; **e.** f opens upward with a vertical intercept at (0, 1); h opens downward with a vertical intercept at (0, -1); both are symmetric with respect to the y-axis; **7. a.** downward; **b.** (0, -4); **9. a.** upward; **b.** (0, 3); **11. a.** downward; **b.** (0, -7); **13. a.** The graph of $y = \frac{3}{5}x^2$ is wider than the graph of $y = x^2$; **b.** The graph of $y = x^2$ would have a greater output value.

Activity 4.2   Exercises: 1. a. upward; **b.** $x = 0$; **c.** (0, -3); **d.** (0, -3); **3. a.** upward; **b.** $x = -2$; **c.** ($-2, -7$); **d.** (0, -3); **5. a.** upward; **b.** $x = -1.5$; **c.** ($-1.5, 1.75$); **d.** (0, 4); **7. a.** upward; **b.** $x = 0.25$; **c.** (0.25, -3.125); **d.** (0, -3); **9. a.** (1, 0), (6, 0); **b.** D: all real numbers; R: $g(x) \leq 6.25$; **c.** $x < 3.5$; **d.** $x > 3.5$; **11. a.** (3.46, 0), ($-3.46, 0$); **b.** D: all real numbers; R: $y \geq -12$; **c.** $x > 0$; **d.** $x < 0$; **13. a.** ($-1, 0$), (3, 0); **b.** D: all real numbers; R: $g(x) \leq 4$; **c.** $x < 1$; **d.** $x > 1$; **15. a.** (0.2, 0), (1, 0); **b.** D: all real numbers; R: $y \leq 0.8$; **c.** $x < 0.6$; **d.** $x > 0.6$; **17. a.** 149 ft; **b.** 6.05 sec; **c.** It indicates the position of the arrow when it is shot; **d.** The practical domain is 0 sec $\leq x \leq 6.05$ sec. The practical range is 0 ft $\leq h(x) \leq 149$ ft; **e.** ($-0.05, 0$), (6.05, 0) The first has no meaning. The second indicates the time in seconds it takes for the arrow to hit the ground;

19. a. (30, 200); **b.** $x = -\dfrac{b}{2a} = \dfrac{120}{4} = 30; C(30) = 200$ The vertex is (30, 200); **c.** They are the same; **d.** minimum point; **e.** The cost of production is minimized when 30 statues are produced; **f.** (0, 2000); It costs $2000, even if no statues are produced.

Activity 4.3   Exercises: 1. $x = 6$ or $x = -2$; **3.** $x = 9$ or $x = -5$; **5.** $x = -11$ or $x = -1$; **7.** $x = \pm 5$; **9.** $x = -3$ or $x = 1$; **11.** $x = 7$ or $x = -4$; **13. a.** $-2 < x < 6$; **b.** $x < -2$ or $x > 6$; **15. a.** $d(55) = 181.5$ ft; **b** $0.04v^2 + 1.1v = 200$; $v \approx 58$ mph

Activity 4.4 **Exercises: 1.** $6x^5(2 - 3x^3)$;
3. $2x(x^2 - 7x + 13)$; **5.** $(x + 3)(x - 2)$; **7.** $(2x - 3)(x + 5)$;
9. $x^2(8x + 1)(x - 6)$; **11.** $x = \frac{1}{3}$ or $x = -4$;
13. $x = 9$ or $x = -2$; **15. a.** 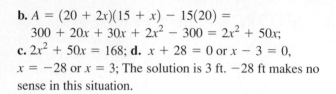 ;

b. $A = (20 + 2x)(15 + x) - 15(20) =$
$300 + 20x + 30x + 2x^2 - 300 = 2x^2 + 50x$;
c. $2x^2 + 50x = 168$; **d.** $x + 28 = 0$ or $x - 3 = 0$,
$x = -28$ or $x = 3$; The solution is 3 ft. -28 ft makes no
sense in this situation.

Activity 4.5 **Exercises:**
1. a. 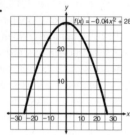 ; **b.** (0, 28) represents the vertex or turning point of the
arch; **c.** $x \approx \pm26.5$; The
intercepts are (26.5, 0) and
(−26.5, 0); **d.** The intercepts
are the same; **e.** The river is
approximately 2(26.5) or 53
ft wide; **f.** No; the highest
point of the arch is 28 ft
above the water;
g. $-0.04x^2 + 28 = 20$; **h.** $x = \pm14.14$ ft Place the pole
14.14 ft to the right or left of the center; **3.** $x = -\frac{1}{2}$;
5. $x = \frac{6 \pm \sqrt{12}}{4} = 2.37$ or 0.63; **7.** $x = \frac{-3 \pm \sqrt{33}}{4} = 0.69$, or
-2.19; **9.** (0, 0) and (−2, 0); **11.** There are no x-intercepts;
13. a. $d = 2.5$ million particles per ft^3; **b.** The minimum
occurs at the vertex. $r = \frac{-b}{2a} = \frac{16}{4} = 4$ or 400 revolutions per
minute $d(4) = 2(4)^2 - 16(4) + 34 = 2$ million particles
per ft^3; **c.** $r = 11$ 1100 rpm is the speed of the engine.

Activity 4.6 **Exercises: 1. a.** ;

b $h(t) = -15.9752t^2 + 52.8875t + 2.5536$;
c. Yes; the curve touches nearly every data
point. **d.** all real numbers from 0 to 3.36
seconds; **e.** real numbers from 0 to 46.33 ft;
f. The ball reaches 35 ft on the way up after 0.81 seconds. It
reaches 35 ft again on the way down, approximately 2.50
seconds after it was struck; **g.** There are only two solutions,
so I got them all; **3. a.** $y = 0.086x^2 - 0.842x + 32.487$;
b. approx. 650 ft; **c.** $0 = 0.086x^2 - 0.842x - 247.513$
Using the quadratic formula, a speed of 58.8 mph requires a
stopping distance of 280 ft.

How Can I Practice?
1.

VALUE OF a	VALUE OF b	VALUE OF c
5	0	0
$\frac{1}{3}$	3	−1
−2	1	0

;

2. a. downward, **b.** $x = 0$; **c.** (0, 4); **d.** (0, 4); **3. a.** upward. **b.**
$x = 0$; **c.** (0, 0); **d.** (0, 0); **4. a.** downward, **b.** $x = 1$; **c.** (1, 10);
d. (0, 7); **5. a.** upward, **b.** $x = \frac{1}{2}$; **c.** $\left(\frac{1}{2}, -1\right)$; **d.** (0, 0);
6. a. upward; **b.** $x = -3$; **c.** (−3, 0); **d.** (0, 9); **7. a.** upward;
b. $x = \frac{1}{2}$; **c.** $\left(\frac{1}{2}, \frac{3}{4}\right)$; **d.** (0, 1); **8. a.** (−2, 0), (2, 0); **b.** D: all real
numbers; R: $y \leq 4$, **c.** $x < 0$; **d.** $x > 0$; **9. a.** (2, 0), (3, 0);
b. D: all real numbers; R: $(y \geq -0.25)$; **c.** $x > 2.5$;
d. $x < 2.5$; **10. a.** (0.91, 0), (−2.91, 0); **b.** D: all real numbers; R: $y \leq 11$; **c.** $x < -1$; **d.** $x > -1$; **11. a.** none;
b. D: all real numbers; R: $y \geq 1.427$; **c.** $x > 1.61$;
d. $x < 1.61$; **12.** (0.75, 26.125); **13. a.** $9a^2(a^3 - 3)$;
b. $6x^2(4x - 1)$; **c.** $4x(x - 5)(x + 1)$; **d.** cannot be factored;
e. $(x - 8)(x + 3)$; **f.** $(y + 5)^2$; **14. a.** 5, 6.05, 7.2, 8.45, 9.8,
11.25, $x \approx 1.2$; **b.** 6.75, 6.16, 5.59, 5.04, 4.51, −1, $x \approx 0.8$;
c. 0, −2, 2, 12, 28, 50, $x = 2$; **15. a.** $x = 1.2, -1.2$;
b. $x = 0.8, 6.2$; **c.** $x = -\frac{1}{3}, 2$; **16. a.** $-8 < x < 2$;
b. $x < -8$ or $x > 2$;   **17. a.** $x = 0, 2$; **b.** $x = 9, -2$;
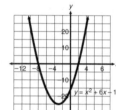
c. $x = 3, 1$; **d.** $x = 4, 4$;
e. $x = 6, -4$; **f.** $y = 5, -3$;
g. $a = 3, -2$; **h.** $x = \frac{1}{4}, -2$;
18. a. 105 feet; **b.** Using the
calculator to solve
$-16t^2 + 80t + 5 = 0$,
$t \approx 5.06$ seconds;
c. $-16t^2 + 80t + 5 = 101$; **d.** $t = 2, 3$; The ball reaches a
height of 101 ft after 2 seconds on the way up and 1 second
later on the way down;

19. a. $0 \leq x \leq 100$;
b. $h(50) = 0.01(50)^2 - 50 + 35 = 10$ ft;
20. a. ;

b. $y = -2.096x^2 - 2.25x + 9.038$;
c. 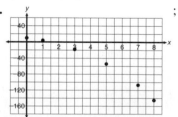 **d.** Predicted values are very close to the
actual values; **e.** −33.5, −181;
f. $x = 4.330, -5.403$

Activity 4.7 **Exercises: 1.** $5i$; **3.** $6i$; **5.** $4i\sqrt{3}$; **7.** $\frac{3}{4}i$;

9. $-5 + 10i$; **11.** $3 - 2i$; **13.** $10 + 5i$; **15.** $x = \frac{1}{3} \pm \frac{\sqrt{80}}{6}i$;

17. $x = 1, -3.5$; **19.** 2 real solutions; **21.** 1 real solution;

23. 2 complex solutions

Activity 4.8 **Exercises: 1. a.** Length: 225, 200, 175, 150,
125, 100, 75, 50, 25; Area: 5625, 10,000, 13,125, 15,000,
15,625, 15,000, 13,125, 10,000, 5625; **b.** 125 ft by 125 ft;
c. $l = 250 - w$; **d.** $A = f(w) = w(250 - w)$; **e.** $w = 125$;
$f(125) = 15,625$; **f.** $w(250 - w) = 0$, $w = 0$ or $w = 250$
There can be no rectangle constructed with these dimensions;
g. domain: $0 < w < 250$; range: $0 < A < 15,625$;
3. 4 mph north, 6 mph east.

How Can I Practice? **1.** $7i$; **2.** $3i\sqrt{5}$; **3.** $11i$; **4.** $i\sqrt{15}$;
5. $4i\sqrt{7}$; **6.** $5i\sqrt{5}$; **7.** $\frac{4}{5}i$; **8.** $\sqrt{\frac{4}{7}}i = \frac{2}{\sqrt{7}}i$; **9.** $4 + 3i$;
10. $-5 + 6i$; **11.** $6 - 4i$; **12.** $-12 - 24i$; **13.** $5 + 10i$;
14. a. $a = 3, b = -1, c = -7$; $b^2 - 4ac = 85$; two real
solutions; $x = 1.70, -1.37$; **b.** $a = 1, b = -4, c = 10$;
$b^2 - 4ac = -24$; two complex solutions; $x = 2 \pm i\sqrt{6}$;
c. $a = 2, b = -5, c = -3$; $b^2 - 4ac = 49$; two real
solutions; $x = 3, -0.5$; **d.** $a = 9, b = -6, c = 1$;
$b^2 - 4ac = 0$; one real repeated solution; $x = \frac{1}{3}$; **15. i.** The
discriminant is 0. The graph only touches the x-axis indicating that there is one, repeated solution; **ii.** The discriminant
is negative. The graph does not intersect the x-axis, indicating that there is no real solution; **iii.** The discriminant is positive. The graph intersects the x-axis twice, indicating that
there are two real solutions; **16. a.** $(20 + x)(30 + 1.5x)$;
b. $A(x) = 1.5x^2 + 60x + 600$; **c.** $1.5x^2 + 60x$;
d. $1.5(2)^2 + 60(2) = 126$ ft^2;
e. $1.5x^2 + 60x = 264$; **f.** $x = -\frac{132}{3}$ or $x = 4$ Reject the
negative; the answer is $x = 4$ feet; **g.** No, the negative value
does not make sense.

Activity 4.10 **Exercises: 1. a.** 2, 32, 64, $y = kx$, $8 = k1$,
so $k = 8$ or $y = 8x$; **b.** $\frac{1}{8}$, 27, 216, $y = x^3$, $1 = k1^3$, so $k = 1$
or $y = x^3$; **3.** $12 = k2^2$, so $k = 3$. So $y = 3x^2$. When $x = 8$,
$y = 3(8)^2 = 192$; **5.** $d = kt^2$, $20 = k(2)^2$, so $k = 5$. Now
$d = 5t^2$, so in 2.5 seconds the skydiver travels
$d = 5(2.5)^2$ or 31.25 meters

7. 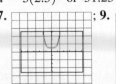 ; **9.** ;

11. $f(x)$ is increasing for $x > 0$; **13.** $y = x^2$ is rising more
slowly than $y = x^3$ for $x > 1$; **15.** $y = -2x^3$ is decreasing
and goes through $(0, 0)$, whereas $y = 2x^3 + 1$ is increasing
and does not pass through the origin. Both have a similar S-
like shape.

Activity 4.11 **Exercises: 1.** $(0, 0), (-1, 0), (-2, 0)$;
3. $(2, 0), (-2, 0), (3, 0), (-3, 0)$;
5.

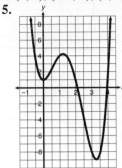

There are two minimum points
$(0, 1)$ and $(3.28, -8.91)$ and one
maximum point $(1.22, 4.23)$;
7. No; as x increases without
bound, y increases without bound;
9. a. increase; **b.** decreasing; **c.** 1

Activity 4.12 **Exercises:**

1. a.

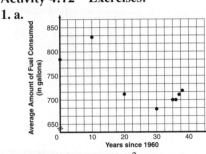

The data does not
appear to be linear
because as the
input increases, the
output increases
and decreases. No
line would be close
to all of the points;

b. quadratic: $g = 0.045t^2 - 4.619t + 810.368$;
cubic: $g = 0.023t^3 - 1.267t^2 + 13.309t + 787.351$;
quartic: $g = -0.00096t^4 + 0.098t^3 - 3.079t + 26.505t + 784.064$;

3. a.

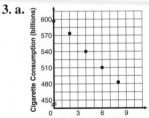

; **b.** $y = -14.15x + 597.4$;

c. $y = -0.125x^2 - 13.15x + 596.4$; **d.** There appears to
be no difference between the two models. Both fit closely
to the data; **e.** The linear model predicts 314.4, and the
quadratic predicts 283.4; **f.** We are predicting quite far
outside our practical domain, so I am not very confident in
either model's prediction.

How Can I Practice? **1.** $y = kx^2$, $45 = k3^2$, so $k = 5$
$y = 5(6)^2 = 180$; **2. a.** double; **b.** $k = 1080$; k represents
the speed at which the sound of thunder travels; **3.** $v = kt$,
$60 = k3$, $k = 20$ $v = 20(4) = 80$ ft/sec;

4. a.

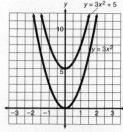

b.

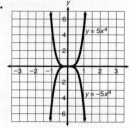

These are the same shape and size; however, $y = 3x^2 + 5$ is shifted up five units;

These are the same shape but are reflections of each other in the x-axis;

c.

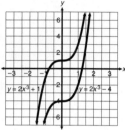

d.

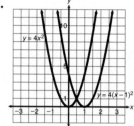

These are the same shape and size, but $y = 2x^3 - 4$ is shifted vertically five units below $y = 2x^3 + 1$;

These are the same shape and size, but $y = 4(x - 1)^2$ is shifted horizontally one unit to the right of $y = 4x^2$;

5. a. i. $(0, 0)$, ii. $(0, 0), (-4, 0), (2, 0)$, iii. $(-2.4, 16.9), (1, -5)$;
b. i. $(0, 3)$, ii. $(-1, 0), (1.57, 0)$, iii. $(0.79, 4.2)$;

6. a.

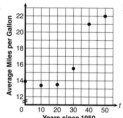 ;

b. $y = 0.00520x^2 - 0.0733x + 13.618$; **c.** 27.9 mpg;
d. 2014; **7. a.**

$W(t)$ becomes negative after 10 min; **b.** $W(0) = 10$ gal;
c. 60.8 gallons, found by using the CALC menu on the graphing calculator; **d.** 7.19 min

Gateway Review   1. a. up; **b.** $x = 0$; **c.** $(0, 2)$; **d.** $(0, 2)$;
2. a. down; **b.** $x = 0$; **c.** $(0, 0)$; **d.** $(0, 0)$; **3. a.** down;
b. $x = 0$; **c.** $(0, 4)$; **d.** $(0, 4)$; **4. a.** up; **b.** $x = \frac{1}{4}$; **c.** $\left(\frac{1}{4}, -\frac{1}{8}\right)$;
d. $(0, 0)$; **5. a.** up; **b.** $x = -\frac{5}{2}$; **c.** $(-2.5, -0.25)$; **d.** $(0, 6)$;
6. a. up; **b.** $x = \frac{3}{2}$; **c.** $(1.5, 1.75)$; **d.** $(0, 4)$; **7. a.** up; **b.** $x = 1$;
c. $(1, 0)$; **d.** $(0, 1)$; **8. a.** down; **b.** $x = 2.5$; **c.** $(2.5, 0.25)$;
d. $(0, -6)$; **9. a.** $(-3, 0), (-1, 0)$; **b.** D: all real numbers;
R: $g(x) \geq -1$; **c.** $x > -2$; **d.** $x < -2$; **10. a.** $(-3, 0), (1, 0)$;
b. D: all real numbers, R: $f(x) \geq -4$; **c.** $x > -1$;
d. $x < -1$; **11. a.** $(0.382, 0), (2.62, 0)$; **b.** D: all real numbers, R: $y \geq -1.25$; **c.** $x > 1.5$; **d.** $x < 1.5$;

12. a. $(-3.22, 0), (-0.775, 0)$; **b.** D: all real numbers,
R: $h(x) \geq -3$; **c.** $x > -2$; **d.** $x < -2$; **13. a.** $(2, 0), (-2, 0)$;
b. D: all real numbers, R: $y(x) \leq 8$; **c.** $x < 0$; **d.** $x > 0$;
14. a. $\left(\frac{1}{3}, 0\right), (1, 0)$; **b.** D: all real numbers, R: $f(x) \leq \frac{1}{3}$;
c. $x < \frac{2}{3}$; **d.** $x > \frac{2}{3}$; **15. a.** none; **b.** D: all real numbers,
R: $g(x) \geq 5$; **c.** $x > 0$; **d.** $x < 0$; **16.** $x = -2$; **17.** $x = 2, 3$;
18. $x = -0.51, 6.51$; **19.** $x = -5, 2$; **20.** $x = \pm 1.1$;
21. $x = -0.21, -4.79$; **22. a.** $9a^2(a^3 - 3)$; **b.** $6x^2(4x - 1)$;
c. $4x(x - 5)(x + 1)$; **d.** cannot be factored;
e. $(x - 8)(x + 3)$; **f.** $(t + 5)^2$; **23.** $x = \pm 3$;
24. $x = \pm 6$; **25.** $x = 3, 4$; **26.** $x = -3, 9$; **27.** $x = 0, -1$;
28. $a = 1, b = 5, c = 3$; $x = -0.7, -4.3$;
29. $a = 2, b = -1, c = 3$; $x = 0.25 \pm 1.2i$;
30. $a = 1, b = 0, c = -81$; $x = \pm 9$;
31. $a = 3, b = 5, c = -12$; $x = -3, \frac{4}{3}$;
32. $a = 2, b = -3, c = -5$; $x = -1, 2.5$;
33. From the grapher: $(0.42, 0), (3.58, 0)$
$$x = \frac{-(-8) \pm \sqrt{(-8)^2 - 4(2)(3)}}{2(2)} = \frac{8 \pm \sqrt{40}}{4} = 3.58,$$

0.42; **34. a.** $7i$; **b.** $4i\sqrt{3}$; **c.** $3i$; **d.** $i\sqrt{23}$; **e.** $\frac{\sqrt{5}}{3}i$; **f.** $\frac{\sqrt{17}}{4}i$;

35. a. $-5 + 17i$; **b.** $5 - 16i$; **c.** $32 + 12i$; **d.** $27 + 6i$;
36. two real solutions; **37.** two real solutions; **38.** two real solutions; **39.** two complex solutions;

40. $x = \dfrac{-2 \pm \sqrt{2^2 - 4(3)(2)}}{2(3)} = \dfrac{-2 \pm \sqrt{-20}}{6} =$
$\dfrac{-2 \pm 2i\sqrt{5}}{2(3)} = \dfrac{-1 \pm i\sqrt{5}}{3}$

41. a. $-2 < x < 3$; **b.** $x < -2$ or $x > 3$; **42. a.** $y = 20$;
b. $y = 32$; **c.** $y = 40$; **43. a.** $(2, 0)$; **b.** D: all real numbers,
R: all real numbers; **c.** increasing for all real numbers;
44. a. $(-1, 0)$; **b.** D: all real numbers, R: all real numbers;
c. decreasing for all real numbers; **45. a.** $(-1.68, 0), (1.68, 0)$;
b. D: all real numbers, R: $y \geq -8$; **c.** inc: $x > 0$ dec:
$x < 0$; **46. a.** $(0, 0), (-1.26, 0)$; **b.** D: all real numbers,
R: $y \geq -1.19$; **c.** inc: $x > -0.8$ dec: $x < -0.8$;
47. a. none; **b.** D: all real numbers, R: $y \geq 5$; **c.** inc: $x > 0$
dec: $x < 0$; **48. a.**

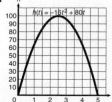

the practical domain is $0 \leq x \leq 5$;

b. $(2.5, 100)$; The ball reaches its highest level, 100 ft, 2.5 sec after being struck; **c.** $(0, 0)$ The ball is on the ground when the club makes contact with it; **d.** $(0, 0), (5, 0)$; The ball is on the ground when the club makes contact, $t = 0$, and returns to the ground 5 sec later; **e.** I am assuming that the elevations are the same; **49 a.** $(-5, 6)$; **b.** $(0, 0), (-5, 6), (-10, 0)$; **c.** $y = -0.24x^2 - 2.4x$; **50. a.** vertex: $(2.5, h(2.5))$ or $(2.5, 105)$ The maximum height iis 105 feet; **b.** Set $h(t) = 0$;

$t = 5.06$ sec; **51. a.** $s(44) = 122.5$ ft away; **b.** $v = -51.78$ or 19.78; Reject the negative. 19.78 ft/sec = 13.5 mph

Chapter 5

Activity 5.1   Exercises: 1. a. The average speed

$$= \frac{20 \text{ km}}{1 \text{ hr } 15 \text{ min}} = \frac{20 \text{ km}}{1.25 \text{ hr}} \approx 16 \text{ km/hr};$$

b. 20, 16, 13.33, 11.43, 10, 8.89, 8; **c.** $s = f(t) = \frac{20}{t}$;
d. i. the set of all nonzero real numbers; **ii.** (Answers will vary.) $1 \le t \le 5$; Because 20 km/hr (when $t = 1$) is fast for a distance runner and 4 km/hr (when $t = 5$) is slow for a distance runner, most times will fall between these values;

iii.   ;

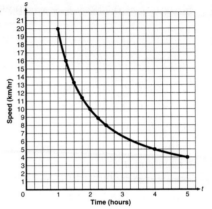

e. The average speed decreases, approaching 0;
f. The average speed increases without bound;
3. a. $D = f(N) = \frac{1400 - 200}{N} = \frac{1200}{N}$; **b.** 1200, 600, 400, 200, 100, 50; **c.**   ;

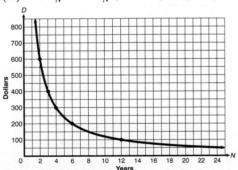

d. Decrease. As N gets larger, D gets smaller.

Activity 5.2   Exercises: 1. a. (Answers will vary.) If a person is 6 ft = 72 in. tall and weighs 200 lb,

$$B = \frac{705(200)}{72^2} = 27.2.; \textbf{b. } B = \frac{119,850}{h^2}; \textbf{c. } 0 < h < 84$$

This works unless the person is over 7 ft tall; **d.** 33.3, 29.3, 25.9, 23.1, 20.7, 18.7;

e.

 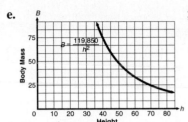

; **f.** Body-mass index B gets smaller and it actually approaches 0. Yes, this makes sense in this case because if you get taller without gaining weight, you should be getting skinnier;

g. $69.2 < h < 79.4$;

3. a.   ; **b.**   ; **c.**   ;

d.

i. graph b; **ii.** graph c; **iii.** graph a; **iv.** graph d;

5. a. all nonzero real numbers;

b.   **c.**

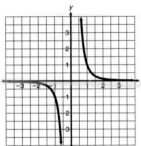 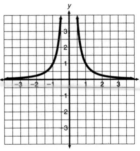

7. a. $y = \frac{2}{x}$; Table: 4, 2, $\frac{1}{3}$ **b.** $y = \frac{8}{x^3}$; Table: 64, $\frac{1}{27}$; **9.** $I = \frac{120}{R}$; $I = \frac{120}{15} = 8$ amps;

11. a.

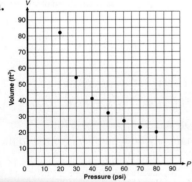

b. No. Using $P = 20$, $V = 82$; $k = 20^2(82) = 32,800$. If $V = \frac{32,800}{P^2}$, then $P = 30$ would yield $V = \frac{32,800}{30^2} = 36.44$ (not very close); **c.** Yes. Using $P = 20$, $V = 82$; $k = 20(82) = 1640$. $V = \frac{1640}{30} = 54.67$ $V = \frac{1640}{40} = 41$

Activity 5.3 Exercises:

1. a. $V = \dfrac{25{,}000 + 55{,}000}{P - 1.5 - 0.40 - 0.60} = \dfrac{80{,}000}{P - 2.5}$;

b. —, 160,000, 32,000, 10,667, 3555.56;
c. V decreases; **d.** $V(2) = -160{,}000$. A price of $2 per cubic meter is not practical; **e.** $P > 2.5$;
f.

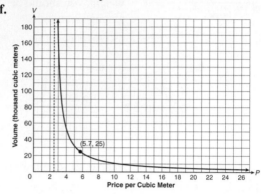

3. a. domain: all real numbers except $x = 7$; vertical asymptote: $x = 7$;

b. domain: all real numbers except $x = 25$; vertical asymptote: $x = 25$;

c. domain: all real numbers except $x = 5$; vertical asymptote: $x = 5$;

d. domain: all real numbers except $x = 14$; vertical asymptote: $x = 14$;

e. domain: all real numbers except $x = -2.5$; vertical asymptote: $x = -2.5$;

5. a. $x = 5$; **b.** $f(x)$ gets large, approaching infinity. $g(x)$ gets small, approaching negative infinity; **c.** $f(x)$ gets small, approaching negative infinity. $g(x)$ gets large, approaching infinity.

Activity 5.4 Exercises:

1. a. $500 + 600 + 500 + 400 = \$2000$; **b.** $2000 + 50n$;
c. $m = \dfrac{50n + 2000}{n}$; **d.** $m = \dfrac{50(100) + 2000}{100} = \$70$;
e. The practical domain is whole numbers from 1 to the size of your class, say 250; **f.** 90, 70, 63.33, 60, 58;
g. Yes, $m = 50$ is the horizontal asymptote. It makes sense because as the number of attendees increases, the fixed costs attributed to each person get smaller and smaller;
h.

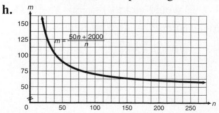

2. a. i. $x \neq -2$,
ii. $x = -2$,
iii.

iv. $y = 4$;

b. i. $x \neq -1$,
ii. $x = -1$,
iii.

iv. $y = -1$;

c. i. all real numbers except 4,
ii. $x = 4$,
iii.

iv. $y = 0$;

d. i. $x \neq \frac{1}{2}$,
ii. $x = \frac{1}{2}$,
iii.

iv. $y = 15$;

3. a. $x = 1$ **b.** $x = \frac{7}{9} \approx 0.778$ **c.** $x = -0.859$

5. a. $15d^2 = 1500$; $d^2 = 100$; $d = \pm 10$ ft, but only 10 makes sense; **b.** $8000 = \dfrac{1500}{d^2}$; $8000d^2 = 1500$; $d^2 = 0.1875$; $d = 0.433$ ft; **7. a.** $W = \dfrac{3(3 - A)}{A}$;
b. $W = \dfrac{3(3 - 2)}{2} = 1.5$ min; **c.** $0 < A \le 3$ A can't be negative, nor can W be negative; **d.** 0, 0.6, 1.5, 3, 6, 15, 42, 87; **e.** The horizontal intercept is $(3, 0)$. This means that if the average service time is 3 minutes and the average arrival time to the restaurant before they get in line is 3 minutes, the average waiting time is 0 minutes; **f.** $A = 0$;
g.

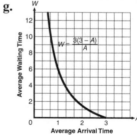

8. a. $R(30) = 7.85$ prey per week; **b.** $61.3 = n$, $n \approx 62$ prey/sq mi; **c.** $17.3 = n$, Putting the two together, $18 \le n \le 62.$;

d.

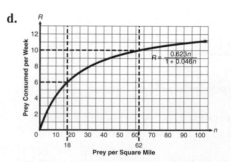

Activity 5.5 Exercises: 1. a. $5T = 90$, $T = 18$ min;
b. $t_2 T + t_1 T = t_1 t_2$, $T = \dfrac{t_1 t_2}{t_1 + t_2}$; **c.** $T = \dfrac{20(15)}{20 + 15} = \dfrac{300}{35} = 8.57$ min; **d.** $3t_2 = 40$, $t_2 = 13.3$ min; **e.** $14T = 150$, $T = 10.7$ min; **3.** $t = \dfrac{72 \pm \sqrt{72^2 - 4(70)}}{2}$, $t \approx 71$ or 1; The only value that makes sense in this situation is $t = 71$ minutes;
5. $21 = 2t$, $t = 10.5$ hrs

Activity 5.6   Exercises:

1. a. $h = \dfrac{330}{1 - \frac{40}{770}} = 348.08$ Hz   The pitch I hear is higher

than the actual pitch; **b.** $h = a \div \dfrac{770 - s}{770}, h = \dfrac{770a}{770 - s}$;

c. $h = \dfrac{770(330)}{770 - 40} \approx 348.08$ Hz; The results are the same.;

d. $h = \dfrac{770(330)}{770 - 60} \approx 357.89$ Hz; **3. a.** $= \dfrac{x - 4}{2x} \cdot \dfrac{1}{x - 4} = \dfrac{1}{2x}$;

b. $= \dfrac{2 - x}{2x} \cdot \dfrac{4x^2}{(2 - x)(2 + x)} = \dfrac{2x}{2 + x}$; **c.** $= \dfrac{3x^2 - 6x}{x^2 - 4x + 3}$;

d. $= \dfrac{x}{2x + 4}$;

How Can I Practice?   1. The graphs of f and g are reflections of each other about the x-axis (and the y-axis).; **2.** The graph of g is the same as f, but shifted 5 units to the right; **3.** They are similar, but the graph of g is closer to the x-axis, and the graph of f is closer to the y-axis; **4. a.** $T =$ time in hours, $s =$ speed in mph, $T = \dfrac{145}{s}$; **b.** $0 < s < 80$; **c.** all real numbers except 0; **5. a.** domain: all real numbers except $x = -5$, vertical asymptote: $x = -5$; **b.** domain: all real numbers except $x = \frac{13}{2}$, vertical asymptote: $x = \frac{13}{2}$; **c.** domain: all real numbers except $x = \frac{8}{5}$, vertical asymptote: $x = \frac{8}{5}$; **d.** domain: all real numbers except $x = 0.5614$, vertical asymptote: $x = 0.5614$; **6.** $4000^2 \cdot 100 = k = 1.6 \cdot 10^9$, $w = \dfrac{1.6 \cdot 10^9}{d^2} = \dfrac{1.6 \cdot 10^9}{(4500)^2} \approx 79.01$ lb; **7. a.** all positive integers, with some realistic upper limit, depending on the specific situation;

b.

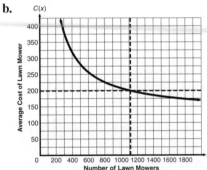

$199x = 132x + 75{,}250, x = 1123$ mowers;

8. a. $t = \dfrac{-(-14) \pm \sqrt{14^2 - 4(0.15)(0.125)}}{2(0.15)} = 93.32$ min or

0.0089 min. Only 93.32 is practical,

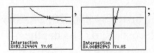

b. The drug will be at its highest concentration 0.913 min after injection,

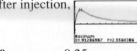

9. a. $x = -0.25$,   **b.** $x = \dfrac{50}{17} \approx 2.94$,

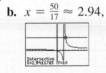

c. $x = \dfrac{-116}{40} \approx -2.9$,   **d.** $x = \dfrac{5.8}{2.4 - (0.3)5.8} = \dfrac{290}{33}$

≈ 8.788

10. a. (Answers will vary.) 200 lb man ≈ 90.9 kg,

$W = \dfrac{90.9}{\left(1 + \frac{15}{6400}\right)^2} = 90.475$ kg; **b.** $W = \dfrac{70}{\left(1 + \frac{h}{6400}\right)^2}$;

c. 70, 69.78, 67.86, 52.36, 45.94, 40.64, 10.66, 4.11;
d. The weight decreases; **e.** 1.3317 kg; **f.** (Answers will vary some at the upper end.) The domain is $0 \le h \le 40{,}000$;

g.

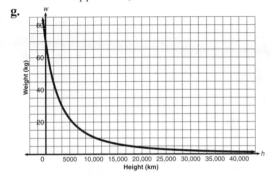

h. $h = 2650.9668$;

11. a. $R_3 = 12$ ohms;

b. $R = \dfrac{R_1R_2R_3}{R_1R_2 + R_2R_3 + R_1R_3}$;

c. $R = \dfrac{4(6)(12)}{4(6) + 4(12) + 6(12)} = \dfrac{288}{24 + 48 + 72} = \dfrac{288}{144} = 2$ ohms;

12. a. $x = 1.3$,   **b.** $x = -1$,

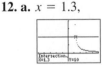

c. $x = 5$,   **d.** $x = \dfrac{1}{4}$

13. a. $s = \dfrac{2(15.3)}{\frac{15.3}{45} + \frac{15.3}{40}} = \dfrac{30.6}{.34 + .3825} = 42.4$ mph;

b. $s = \dfrac{2dr_1r_2}{d(r_1 + r_2)} = \dfrac{2r_1r_2}{r_1 + r_2}$; **c.** $s = \dfrac{2(45)(40)}{(40 + 45)} = 42.4$ mph

The results are the same; **14. a.** $= \dfrac{4x + 2}{x - 3}$;

b. $= \dfrac{(x + 5)(x - 5)}{(x + 5)} = x - 5$; **c.** $= \dfrac{1}{x + 2} \cdot \dfrac{x + 2}{x + 3} = \dfrac{1}{x + 3}$

Activity 5.7 **Exercises: 1. a.** 5.48; **b.** 2.45; **c.** 169; **d.** 27; **3. a.** (0, 12) In 1993 there was $12 billion in new student loans; **b.** Table: 12, 18.8, 21.6, 23.8, 25.6, 27.2, 28.7, 30; The equation matches very well, with the possible exception of 1994, when it is off by $0.8 billion;

c. 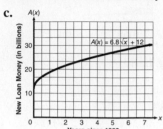 ; **d.** In 2005, $x = 12$, so $A(12) = 6.8\sqrt{12} + 12$ = $35.6 billion in new student loan money;

5. 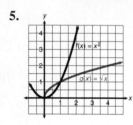 ; **b.** No, the graph of f is below the graph of g for $0 < x < 1$; **c.** Yes; the graph of f is above the graph of g for $x > 1$;

7. a. i. The domain is all real numbers such that $4x + 8 \geq 0$ or $x \geq -2$; **ii.** The x-intercept is $(-2, 0)$, and the y-intercept is $(0, -\sqrt{8})$; **iii.** 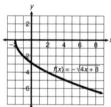 ;

b. i. The domain is all real numbers such that $5 - x \geq 0$ or $x \leq 5$; **ii.** The x-intercept is $(5, 0)$. The y-intercept is $(0, \sqrt{5})$; **iii.** 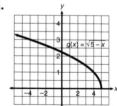 ;

c. i. The domain is all real numbers such that $x^2 + 9 \geq 0$ or all real numbers; **ii.** There is no x-intercept. The y-intercept is $(0, 3)$; **iii.** 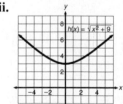 ;

9. $d = \sqrt{12^2 + 24^2 + 17^2} = \sqrt{1009} \approx 31.8$ in. It will not fit; **10. a.** $s = \sqrt{30(0.85)l} = \sqrt{25.5l}$; **b.** $s = \sqrt{25.5(90)} = 47.9$ mph; **c.** 0 feet $\leq l \leq$ 300 feet is possible; **d.** 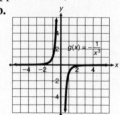 ;

e. The length of the skid marks is approximately 192 feet.

Activity 5.8 **Exercises: 1. a.** $x = 4$; **b.** $\sqrt{x + 1} = -4$; This can't happen. There is no solution. Equation b has no solution. The left side of equation b will always be greater than 1, so no solution is possible; **c.** $x = -2$; but $x = 1$ does not check;

3. a. $x = -4$ checks, **b.** $x = 2$ checks,

c. $x = 1$; $x = -4$ does not check ;

5. $L = 32\left(\frac{1.95}{2\pi}\right)^2$, $L = 3.08$ feet; **7. a.** $V = \sqrt{\frac{1000(10)}{3}} = 57.7$ mph; **b.** $P = \frac{14700}{1000} = 14.7$lb/ft^2; **9. a.** $A = \sqrt{\frac{70 \cdot 200}{3131}} = 2.11$; **b.** $w = \frac{3131A^2}{h}$

Activity 5.9 **Exercises: 1. a.** 4; **b.** 2; **c.** -3; **d.** 5; **e.** $\frac{1}{6}$; **f.** not real; **g.** 10; **h.** not real; **3.** The difference is $\sqrt[3]{1450} - \sqrt[3]{1280} = 0.46$ in.; **5. a.** all real numbers; **b.** $x \geq 3$; **c.** all real numbers; **d.** $x \leq 2$; **7. a.** $x = 64$; **b.** $x = 81$; **9. a.** $r = \sqrt[3]{\frac{3 \cdot 40}{4\pi}} = 2.12$ cm; **b.** $V = \frac{4\pi(3.5)^3}{3} = 179.6$ ft^3; **c.** $V = \frac{4\pi r^3}{3}$

How Can I Practice? **1. a.** $x = 98$; **b.** $x = 41$; **c.** $x = 12$; **d.** $x = \pm\sqrt{61}$; **e.** no solution; **f.** $x = 10$; **g.** $x = 10.5$; **h.** $x \approx 0.95$; **2. a.** $x \leq 6$; **b.** all real numbers; **c.** $x \geq 2$ or $x \leq -2$; **3.** Length is approximately 7.71 in.; **4.** $r = \sqrt[3]{\frac{3V}{4\pi}}$, $V = 620$, $r = 5.29$ cm; **5.** $v = 100$, $100 = \sqrt{64d}$; $d = 156.25$ feet; **6.** $x = 6.5$ in.; **7.** The graphs are reflections about the line $x = 2$.

Gateway Review **1. a.** $d = \frac{1200}{w}$; **b.** 40, 34.286, 30, 24, 20; **c.** As width increases, the depth decreases; **d.** The depth is 12 ft, not enough room for any theater set; **e.** No. Division by 0 is undefined; **f.** (Answers may vary.) $30 \leq w \leq 60$; **g.** a rational function; **h.** all real numbers except 0; **i.** $w = 0$; **j.** $d = 0$; as w increases, d approaches 0; **2. a.** 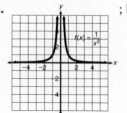 ; **b.** ;

c. The graphs have the same horizontal and vertical asymptotes. $f(x) = \frac{1}{x^2}$ is symmetrical with respect to the y-axis. $f(x) = -\frac{1}{x^3}$ is symmetrical with respect to $(0, 0)$ in quadrants II and IV. f is always positive. g is both positive and negative;

3. a. $y = \frac{k}{x}$; $12 = \frac{k}{10}$; $k = 120$; $y = \frac{120}{30} = 4$;

b. $l = \frac{k}{d^2}$; $32 = \frac{k}{16}$; $k = 512$; $l = \frac{512}{100} = 5.12$ dB;

c. $h = \frac{V}{r^2}$; $8 = \frac{V}{4}$; $32 = v$; $h = \frac{32}{25} = 1.28$ in.;

4. a. H: $y = 0$, V: $x = 0$, No y-intercept, No x-intercept;

b. H: $y = 0$, V: $x = 3$, $\left(0, -\frac{4}{3}\right)$, No x-intercept; **c.** H: $y = 2$, V: $x = -2$, $(0, 0)$, $(0, 0)$;

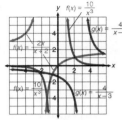

5. a. $f(n) = 45n + 600$; **b.** $f(100) = \$5100$;

c. $A(n) = \frac{45n + 600}{n}$; **d.** $A(100) = \frac{45(100) + 600}{100}$ $A(100) = \$51$;

e. 57, 51, 49, 48, 47.40; **f.** $50 = \frac{45n + 600}{n}$

$50n = 45n + 600$; $5n = 600$; $n = 120$ people, ;

g. $0 < n < $ size of restaurant; **h.** The vertical asymptote is $n = 0$. Zero people cannot attend the event. There would not be an event;

i. The horizontal asymptote is $A(n) = 45$. As the number of people attending increases, the average cost approaches \$45;

6. a. $\frac{4}{x - 2} = 6$; $4 = 6x - 12$; $16 = 6x$; $x = \frac{8}{3}$;

b. The solution is the x-coordinate of the x-intercept;

7. a. $x = -\frac{7}{5} = -1.4$,   **b.** $x = 1$,

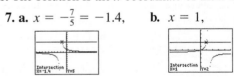

8. $\frac{1}{20} + \frac{1}{15} = \frac{1}{x}$ $60x\left(\frac{1}{20} + \frac{1}{15}\right) = 60x\left(\frac{1}{x}\right)$ $3x + 4x = 60$ $7x = 60$ $x = 8.57$ min; **9.** $x = 67.5$, $2x = 135$ min or $2\frac{1}{4}$ hrs; **10. a.** $x = 10.2$; **b.** $x = 33$; **11. a.** $S = \frac{C}{1 - r}$ $S(1 - r) = C$ $S - Sr = C$ $S - C = Sr$ $r = \frac{S - C}{S}$;

b. $bc - 4ab = -3ac$ $b(c - 4a) = -3ac$ $b = \frac{-3ac}{c - 4a}$;

12. a. $\frac{b + 2a}{2b + a}$; **b.** $\frac{x + 2}{x - 2}$;

13. a. $f = \frac{1}{\frac{1}{4} + \frac{1}{3}} = \frac{1}{\frac{3}{12} + \frac{4}{12}} = 1 \div \frac{7}{12} = \frac{12}{7} = 1.71$ m;

b. $f = \frac{1}{\frac{1}{p} + \frac{1}{q}} = \frac{1}{\frac{q}{pq} + \frac{p}{pq}} = \frac{1}{\frac{p + q}{pq}} = \frac{pq}{p + q}$;

c. $f = \frac{4(3)}{4 + 3} = \frac{12}{7} = 1.71$ m; The values are the same;

14. a. $x \geq -4$;

b.

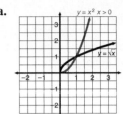

; **c.** The output is increasing; **d.** $y \geq 0$; **e.** The x-intercept is $(-4, 0)$. The y-intercept is $(0, 2)$; **f.** g has the same shape but is shifted 8 units to the right; **g.** The graphs are reflected through the x-axis;

15. a.

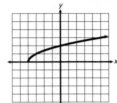

; **b.** $y = \sqrt{x}$; $x = \sqrt{y}$; $y = x^2$; $f^{-1}(x) = x^2$; $x \geq 0$;

d. The graphs are reflections in $y = x$;

e. $(f^{-1}(f(x)) = f^{-1}(\sqrt{x}) = (\sqrt{x})^2 = x$;

16. a. i. $x \geq 0$, $y \geq 4$,   **b. i.** $x \geq -4$, $y \geq 0$,

     ii. $(0, 4)$ only,   **ii.** $(0, 2)$ and $(-4, 0)$,

     iii.   **iii.**

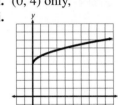

17. a. $x = 6$; **b.** $x = 6$; **c.** $x = \frac{23}{5} = 4.6$; **d.** $x = -1$ -1 does not check. There is no solution; **e.** $x = 27$;

18. a. all real numbers; **b.** $x \geq 6$; **c.** $x \geq -1$;

19. $36 = 1.5h$, $h = 24$ ft; **20.** $d = 153.76$ ft

Chapter 6

Activity 6.1   Exercises: 1. a. 0.6000; **b.** 0.8000; **c.** 0.8000; **d.** 0.6000; **e.** 0.7500; **f.** 1.3333; **3. a.** Given $\tan B = \frac{7}{4}$, the side opposite angle B is 7; the side adjacent to angle B is 4. Using the Pythagorean theorem, I determine that the hypotenuse is $\sqrt{65}$.; **b.** $\sin B = \frac{7}{\sqrt{65}} = 0.8682$;

c. $\cos B = \frac{4}{\sqrt{65}} = 0.4961$; **5. a.** the sine function;

b. $\sin B = \frac{y}{c}$; **7. a.** $x = 9.1$; **b.** $x = 85.6$; **c.** $x = 61.4$;

8. a.

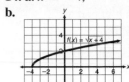

;

b. $3 \cdot 7 = 21$ in. The increase in height from one end of the ramp to the top of the stairs is $\frac{21}{12} = 1.75$ ft.;

c. $x = 20.1$ ft The ramp needs to be at least 20.1 feet long. Therefore, the donated ramp will not be long enough to meet the code. Alternative approach: $15 \sin 5 = 1.3$ ft. The 3 steps must measure at most 1.3 ft high for the 15-foot ramp to satisfy the code. Each solution suggests ways to think about modifications to either the ramp or the steps (or both) that could be used to meet the code.

Activity 6.2 **Exercises: 1. a.** $s + w = 0.68$ miles;
b. These calculations confirm the result in part a;
3. a.

90 − x	sin x	cos (90 − x)
83	0.1219	0.1219
73	0.2924	0.2924
66	0.4067	0.4067
57	0.5446	0.5446
42	0.7431	0.7431
23	0.9205	0.9205
13	0.9744	0.9744

b. The table in part a illustrates the property that cofunctions of complementary angles are equal.

Activity 6.3 **Exercises: 1. a.** $\theta = 30°$; **b.** $\theta = 64.62°$;
c. $\theta = 67.04°$; **d.** $\theta = 63.82°$; **e.** $\theta = 66.80°$; **f.** $\theta = 64.62°$;
g. $\theta = 22.28°$; **h.** $\theta = 20.76°$;
2.

V	θ	E	N
32	65°	13.5	29.0
23.3	59.0°	12	20
4.1	54°	2.4	3.3
26	43.8°	18.8	18
4.5	45°	3.2	3.2

5. a.

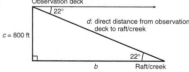

; **b.** Because grade is rise over run, $\tan\theta = 0.1$.
$\theta = \tan^{-1}(0.1) = 5.7°$
The ramp makes an angle of 5.7° with the horizontal;
c. $y = 15\sin(5.7) \approx 1.5$ ft. The elevation changes 1.5 ft from one end of the ramp to the other.

Activity 6.4 **Exercises: 1. a.** The side adjacent to the 57° angle is 4.2 ft. The hypotenuse is 7.8 ft. The other acute angle is 33°; **b.** The hypotenuse is 19.0. The angle adjacent to side 18 is 18.4°. The other acute angle is 71.6°; **c.** The other leg is 7.9 in. The angle adjacent to side 9 in. is 41.4°. The other angle is 48.6° **3. a.**

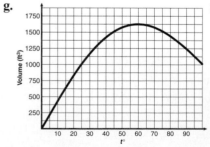

b. The direct distance, d, from the observation deck to the raft is approximately $\frac{800}{\sin(22°)} = 2135$ ft. If you could walk straight down the cliff and straight across at the base of the cliff to the creek, the distance would be approximately $b + c = 2780$ ft., where $b = \frac{800}{\tan 22}$.

Activity 6.5 **Exercises: 1. a.** The slope is $\frac{5}{100}$ or $\frac{1}{20}$ or 0.05; **b.** $A = \tan^{-1}(0.05) = 2.86°$ The highway makes an angle of 2.86° with the horizontal. This angle is called the angle of elevation; **c.** I would be 264 ft above sea level after 1 mile; 1 mile is equivalent to 5280 ft. If x represents the number of feet above sea level after walking 1 mile, then $x = 5280 \cdot \sin(2.86°) = $ approximately 264 ft;

3. Using the following diagram:

The two equations are (a) $\tan 25° = \frac{5}{x+y}$ and (b) $\tan 30° = \frac{5}{y}$. Solving equation (b), $y = 8.7$ miles. Then equation (a) becomes $\tan 25° = \frac{5}{x+8.7}$. Solve this equation for x. $(x + 8.7)\tan 25° = 5$, $x\tan 25° = 5 - 8.7\tan 25°$,
$x = \dfrac{(5 - 8.7\tan 25°)}{\tan 25°}$, $x \approx 2.0$ miles. The runway is approximately 2 miles long.

Project Activity 6.6 **Exercises: 1. a.** Let A represent the area of the trapezoidal cross section. The height of the cross section is h, and the two bases are 5 and $5 + 2x$ respectively. The
area is then determined by the formula $A = \frac{1}{2}h(10 + 2x)$, which, after simplifying, is $A = h(5 + x)$ or $A = 5h + hx$;
b. $h = 5\sin t$, $x = 5\cos t$, $A = 5(\sin t)(5 + 5\cos t)$ or $A = 25\sin t(1 + \cos t)$ or, $A = 25\sin t + 25\sin t(\cos t)$;
c.

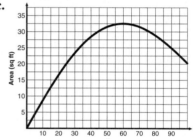

; **d.** The graph in part c indicates that the area of the trapezoidal cross section (output) is the greatest when the angle t is 60°;
e. The area is approximately 32.5 sq ft as read from the graph in part c; **f.** Let V represent the volume. Then, $V = 50 \cdot A$, where A is the cross-section area. In terms of t, the volume is $V = 1250\sin t(1 + \cos t)$;
g.

; **h.** The graph indicates the greatest value for the volume between 0° and 90° is approximately 1625 ft³ when the angle t is 60°; **i.** The angle is the same, namely 60° in this scenario.

How Can I Practice? **1. a.** $\frac{8}{15}$; **b.** $\frac{15}{8}$; **c.** $\frac{15}{17}$; **d.** $\frac{8}{17}$; **e.** $\frac{8}{17}$; **f.** $\frac{15}{17}$;
2. a. 0.731; **b.** 0.574; **c.** 0.601; **d.** 5.671; **3.** $\cos A = \frac{12}{13}$
and $\tan A = \frac{5}{12}$; **4.** $\sin B = \frac{7}{\sqrt{65}}$ and $\cos B = \frac{4}{\sqrt{65}}$;
5. a. $\theta = 48.6°$; **b.** $\theta = 23.5°$; **c.** $\theta = 74.1°$; **d.** $\theta = 16.6°$;
e. $\theta = 44.2°$; **f.** $\theta = 13.7°$; **6.** $BC = 4.8$ cm, $AC = 3.6$ cm, $\angle B = 37°$, $\angle C = 90°$; **7.** $A = \arctan\left(\frac{10}{15}\right) = 33.7°$
Therefore, I should buy the 35° trusses; **8. a.** $\theta = 20.6$;
b. $D = \sqrt{16^2 + 6^2} = 17.1$ ft.;

9. a.

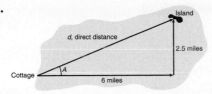

The direct distance, d, from the cottage to the island is $d = \sqrt{2.5^2 + 6^2} = 6.5$ miles; **b.** $A = \arctan\left(\frac{2.5}{6}\right) = 22.6°$; I should direct my boat 22.6° north of east to get from the cottage to the island in the shortest distance.

Activity 6.7   Exercises: 1. a. (0.31, 0.95); **b.** (0.64, −0.77); **c.** (0, −1); **d.** (−0.36, 0.93); **e.** (−0.85, −0.53); **f.** (0.26, 0.97); **g.** (0.34, −0.94); **2. a.** $\frac{72}{360} \cdot 2\pi = 1.26$; **b.** $\frac{310}{360} \cdot 2\pi = 5.41$; **c.** $\frac{270}{360} \cdot 2\pi = 4.71$; **d.** $\frac{111}{360} \cdot 2\pi = 1.94$; **e.** $\frac{212}{360} \cdot 2\pi = 3.70$; **f.** $\frac{435}{360} \cdot 2\pi = 7.59$; **g.** $\frac{-70}{360} \cdot 2\pi = -1.22$; Distance is 1.22; **4. a.** The graph looks like the cosine function reflected in the x-axis; **b.** The motion is the x coordinate starting at (−1, 0) and moving counterclockwise;

5. a.

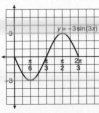

b. Yes, the number of hours of daylight is cyclical; **c.** The graph has the same shape; **d.** South. The number of hours of daylight is greater from October to February, winter in the Northern Hemisphere.

Activity 6.8   Exercises: 1. a. $45 \cdot \frac{\pi}{180} = \frac{\pi}{4} = 0.785$; **b.** $140 \cdot \frac{\pi}{180} = \frac{7\pi}{9} = 2.443$; **c.** $330 \cdot \frac{\pi}{180} = \frac{11\pi}{6} = 5.760$; **d.** $-36 \cdot \frac{\pi}{180} = \frac{-\pi}{5} = -0.628$

3.

0°	30°	45°	60°	90°	135°	180°	210°	270°	360°
0	π/6	π/4	π/3	π/2	3π/4	π	7π/6	3π/2	2π

5. a. $y = 2\cos x$ is $y = \cos x$ stretched vertically by a factor of 2. $y = \cos(2x)$ is $y = \cos x$ compressed horizontally by a factor of 2; **b** $y = \cos\left(\frac{1}{3}x\right)$ is $y = \cos x$ stretched horizontally by a factor of 3. $y = \cos(3x)$ is $y = \cos x$ compressed horizontally by a factor of 3;

7.   **9.**

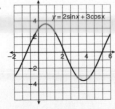

max: 3,   max: 3.606,
min: −3,   min: −3.606,
period: $\frac{2\pi}{3}$;   period: 2π

10. a. The bill is highest for December and January. The amount of the bill is approximately $300; **b.** The bill is lowest for June and July. The amount of the bill is approximately $125; **c.** The largest value is $325; **d.** The period is 6 billing periods or 12 months; **e.** The graph will be stretched vertically by a factor of 1.05. This will not affect the period of the function; **f.** The amount of the bills for the summer months would increase. The graph would flatten out as the monthly charges become more equal.

Activity 6.9   Exercises:

1.

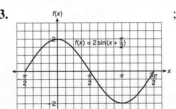

; 3. a. The maximum value is 100; **b.** The minimum value is −100; **5.** $x = 150\cos\theta$; $y = 150\sin\theta$

Activity 6.10   Exercises: 1. a. amp: 3; period: $\frac{2\pi}{1.5} = \frac{4\pi}{3}$; **b.** amp: 0.5; period: $\frac{2\pi}{2} = \pi$; **c.** amp: 2.3; period: $\frac{2\pi}{0.4} = 5\pi$; **d.** amp: 36; period: $\frac{2\pi}{2\pi} = 1$; **3. a.** $y = -15\sin(2x)$; **b.** $y = 1.3\cos(0.7x)$; **5. a.** graph is iii; **b.** graph is iv; **c.** graph is i; **d.** graph is ii.

Activity 6.11   Exercises: 1. a. amplitude: 0.7, period: π, displacement: $\frac{-\frac{\pi}{2}}{2} = -\frac{\pi}{4}$; **b.** amplitude: π, period: 2π, displacement: $\frac{-(-1)}{1} = 1$; **c.** amplitude: 2.5, period: 5π, displacement: $\frac{-\frac{\pi}{3}}{0.4} = -\frac{5\pi}{6}$; **d.** amplitude: 15, period: 1, displacement: $\frac{-(-0.3)}{2\pi} = \frac{3}{20\pi} = 0.0477$;

3.

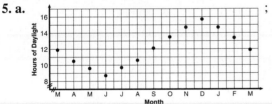

; 5. a. graph is iii; **b.** graph is iv; **c.** graph is i; **d.** graph is ii.

How Can I Practice?   1. a. (0.81, 0.59); **b.** (−0.87, −0.5); **c.** (0, −1); **d.** (0.73, −0.68); **e.** (−0.81, −0.59); **f.** (0, 1); **2. a.** 0.63 units; **b.** 3.67 units; **c.** 1.57 units clockwise; **d.** 5.53 units; **e.** 2.51 units clockwise; **f.** 7.85 units; **3. a.** $18 \cdot \frac{\pi}{180} = \frac{\pi}{10}$; **b.** $18 \cdot \frac{\pi}{180} = \frac{\pi}{10}$; **c.** $390 \cdot \frac{\pi}{180} = \frac{13\pi}{6}$; **d.** $-72 \cdot \frac{\pi}{180} = \frac{-2\pi}{5}$; **4. a.** $\frac{5\pi}{6} \cdot \frac{180}{\pi} = 150°$; **b.** $1.7\pi \cdot \frac{180}{\pi} = 306°$; **c.** $-3\pi \cdot \frac{180}{\pi} = -540°$; **d.** $0.9\pi \cdot \frac{180}{\pi} = 162°$;

5. amplitude: 4, period: $\frac{2\pi}{3}$, displacement: 0,

6. amplitude: 2; period: $\frac{2\pi}{1} = 2\pi$, displacement: 1,

; ;

7. amplitude: 3.2,
period: $\frac{2\pi}{2} = \pi$,
displacement: 0,

8. amplitude: 1,
period: $\frac{2\pi}{\frac{1}{2}} = 4\pi$,

 ;

displacement: -2,

 ;

9. amplitude: 3,
period: $\frac{2\pi}{4} = \frac{\pi}{2}$,
displacement: $\frac{-(-1)}{4} = \frac{1}{4}$,

 ;

10. a. because of the repetitive nature of the height of the water as a function of time; **b.** amplitude = $\frac{\text{range}}{2} = \frac{80 - 0}{2} = 40$;

c. The period is approximately 12 hours, because high tide occurs twice a day; **d.** Let x represent the number of hours since midnight. $y = a \sin(bx + c) + d$, $a = 40$, the amplitude, period $= \frac{2\pi}{b} = 12$, $b = \frac{\pi}{6}$, displacement $= 3 = \frac{-c}{b} = \frac{-c}{\frac{\pi}{6}}$, $-c = 3 \cdot \left(\frac{\pi}{6}\right)$, $c = \frac{-\pi}{2}$, Vertical shift $= d = 40$, $y = 40 \sin\left(\frac{\pi}{6}x - \frac{\pi}{2}\right) + 40$, Other equations are possible.

Gateway Review   1. a. $N = 7\sin(63°) = 6.24$ miles;
b. $E = 7\cos(63°) = 3.18$ miles; **2. a.** side $c = 13$, angle $A = 67.4°$, angle $B = 22.6°$; **b.** side $a = 6.93$, side $b = 4$, angle $A = 60°$; **c.** side $b = 3$, side $c = 4.24$, angle $B = 45°$; **d.** side $a = 8.66$, side $c = 10$, angle $B = 30°$;
3. a. $\cos\theta = \frac{8}{10}$, $\tan\theta = \frac{6}{8}$; **b.** $\sin\theta = \frac{1}{2}$, $\tan\theta = \frac{1}{\sqrt{3}}$, $\theta = 30°$;
c. $c^2 = 5^2 + 8^2 = 89$, $c = \sqrt{89}$, $\sin\theta = \frac{8}{\sqrt{89}}$, $\cos\theta = \frac{5}{\sqrt{89}}$;
4. No; there is a difference, but it is so small that it is difficult to see. For me: $\theta = \tan^{-1}\left(\frac{1408}{100}\right) = 85.9375°$.
For my nephew: $\theta = \tan^{-1}\left(\frac{1411}{100}\right) = 85.9461°$;
5.

$$\tan 57° = \frac{a}{30} \qquad \tan 13° = \frac{b}{46.2}$$
$$30\tan 57° = a \qquad b = 46.2\tan 13°$$
$$a = 46.2 \qquad b = 10.7$$
$$\cos 57° = \frac{30}{c} \qquad \cos 13° = \frac{46.2}{h}$$
$$c = \frac{30}{\cos 57°} \qquad h = \frac{46.2}{\cos 13°}$$
$$c = 55.1 \qquad h = 47.4;$$

6. So, $(2x)^2 = x^2 + (200 + x)^2$ $x = \dfrac{400 \pm \sqrt{480{,}000}}{4}$

The negative does not make sense, thus $x = \dfrac{400 \pm 400\sqrt{3}}{4}$

$x = 100 + 100\sqrt{3}$; $h = 300 + 100\sqrt{3}$;

7.

SIN θ	COS θ	TAN θ
$\frac{\sqrt{3}}{2}$	$-\frac{1}{2}$	$-\sqrt{3}$
$\frac{1}{\sqrt{2}}$	$-\frac{1}{\sqrt{2}}$	-1
$\frac{1}{2}$	$-\frac{\sqrt{3}}{2}$	$-\frac{1}{\sqrt{3}}$
0	-1	0
$-\frac{1}{2}$	$-\frac{\sqrt{3}}{2}$	$\frac{1}{\sqrt{3}}$
$-\frac{1}{\sqrt{2}}$	$-\frac{1}{\sqrt{2}}$	1
$-\frac{\sqrt{3}}{2}$	$-\frac{1}{2}$	$\sqrt{3}$
-1	0	undef.
$-\frac{\sqrt{3}}{2}$	$\frac{1}{2}$	$-\sqrt{3}$
$-\frac{1}{\sqrt{2}}$	$\frac{1}{\sqrt{2}}$	-1
$-\frac{1}{2}$	$\frac{\sqrt{3}}{2}$	$-\frac{1}{\sqrt{3}}$
0	1	0

;

8. a. amplitude: 2, period: 2π,

b. amplitude: 2, period: 2π;

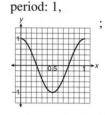 ; ;

c. amplitude: 1, period: π,

d. amplitude: 1, period: 1,

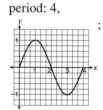 ; ;

e. amplitude: 1, period: 4π,

f. amplitude: 1, period: 4,

; ;

g. amplitude: 1, period: 2π,

h. amplitude: $\frac{2}{3}$, period: π,

; ;

i. amplitude: 1, period: 1,

j. amplitude: 3, period: 1,

 ; ;

9. a. graph is vi; **b.** graph is ii; **c.** graph is iv; **d.** graph is v; **e.** graph is vii; **f.** graph is viii.

Glossary

A **addition of functions** *see* sum function.

argument Another name for the input, or independent variable, of a function.

average rate of change, or simply, the **rate of change** The ratio $\dfrac{\Delta t}{\Delta w}$, where Δt represents the change in output and Δw represents the change in input.

axis of symmetry A vertical line that separates the graph of a parabola into two mirror images.

C **change of base formula** $\log_b x = \dfrac{\log_a x}{\log_a b}$, where $b > 0, b \neq 1$, is the formula used to change logarithms of one base to logarithms of another base.

coefficient The numerical multiplier of a variable.

common logarithms Base 10 logarithms.

complex numbers Numbers of the form $a + bi$ such that a and b are real numbers and $i = \sqrt{-1}$.

composition function, $f \circ g$ The function that is created when the output of the function, g, becomes the input for a second function, f. The rule is given symbolically by $(f \circ g)(x) = f(g(x))$.

consistent system of linear equations A system with exactly one solution.

constant function A function in which there is no change in the output. The graph of a constant function is a horizontal line.

constant of proportionality A constant, k, that gives the rate of variation in the direct proportional relationship $y = kx^n$.

constant term A term that does not change in value.

continuous compounding of an investment Occurs when the period is so short it is essentially an instant in time. The formula for continuous compounding is $A = Pe^{rt}$.

cubic A third-degree polynomial function having the general equation $y = ax^3 + bx^2 + cx + d$, where $a, b, c,$ and d are real numbers and $a \neq 0$.

D **decay factor of an exponential function** The number, b, in the equation $y = a \cdot b^x$, where $0 < b < 1$ and a is the amount when $x = 0$.

decreasing function A function in which the output decreases in value as the input increases. The graph goes down to the right.

degree of a polynomial function The exponent of the term with the largest exponent.

dependent variable of a function The output variable.

DERIVE The name of a computer-based algebra system.

difference function, $f - g$ The function that is created from two functions, f and g, by the rule $(f - g)(x) = f(x) - g(x)$.

direct variation between two variables A relationship in which as the independent variable (input) increases in value, the dependent variable (output) increases. Also the independent variable decreases as the dependent variable decreases.

discriminant The expression $b^2 - 4ac$ under the radical of the quadratic formula. The value of the discriminant determines the type of solutions of the equation $ax^2 + bx + c = 0$.

domain of a function The set of all possible input values of a function.

E **exponential function** A function of the form $y = a \cdot b^x$, where the independent variable, x, is the exponent.

extraneous solution A potential solution that is not really a solution to the original equation or problem.

F **function** A relationship between the input and the output such that for each input value there is exactly one output value.

G **general form of a linear equation** $Ax + By = C$, where A, B, and C are real numbers.

growth factor of the exponential function $y = a \cdot b^x$ The number b, where $b > 1$ and a is the amount when $x = 0$.

H **horizontal intercepts** All points of the graph of the function whose y-coordinate is 0. (*see* x-intercept)

I **identity function** The function in which the output value is always identical to the input value.

imaginary unit $i = \sqrt{-1}$.

inconsistent system of linear equations A linear system with no solution. Graphically, two parallel lines represent the system.

increasing function A function in which the output increases in value as the input increases. The graph goes up to the right.

input variable The independent variable.

inverse functions Two functions f and g related such that $f(g(x)) = x = g(f(x))$. Graphically, these functions are mirror images in the line $y = x$.

inverse variation between two variables A relationship in which as the independent variable (input) increases in value, the dependent variable (output) decreases. Also, the independent variable decreases as the dependent variable increases.

irrational number Any real number that cannot be written as a rational number.

L **linear function** Any function in which the rate of change, or slope, is constant.

linear term The term of a polynomial function of the form bx, where b is a real number.

logarithm function A function of the form $\log_b x = y$, where the base $b > 0$, $b \neq 1$.

M   **magnitude** The relative size of a number or quantity, expressed as a distance or absolute value (and is therefore not negative).

mathematical model A function that best fits the actual data and can be used to predict output values for input values not in the table.

N   **natural logarithm** A logarithm to the base e. The logarithm is written as $\log_e x = \ln x$.

O   **ordered pair** A pair of values, separated by a comma and enclosed in a set of parentheses. The input value is written to the left of the output value.

output variable The dependent variable.

P   **parabola** The graph of a quadratic function (second-degree polynomial function). The graph is U-shaped, opening either upward or downward.

piecewise function A function in which the function rule for determining the output is given separately, or in pieces, for different values of the input.

polynomial function Any function defined by a finite number of terms of the form ax^n, where a is a real number and n is a nonnegative integer.

practical domain The set of all input values that make sense in a problem situation.

practical range The set of all output values that make sense in a problem situation.

product function, $f \cdot g$ The function that is created from two functions f and g, by the rule $(f \cdot g)(x) = f(x) \cdot g(x)$

profit function A function that is common in the business world and is defined by Profit = Revenue − Cost.

Q   **quadratic formula** The formula $x = \dfrac{-b \pm \sqrt{b^2 - 4ac}}{2a}$ that represents the solutions to the quadratic equation $ax^2 + bx + c = 0$.

quadratic function A second-degree polynomial function defined by an equation of the form $f(x) = ax^2 + bx + c$, where a, b, and c are real numbers and $a \neq 0$.

quartic function A fourth-degree polynomial function defined by an equation of the form $f(x) = ax^4 + bx^3 + cx^2 + dx + e$, where a, b, c, d, and e are real numbers and $a \neq 0$.

R   **radical function** Any function involving a radical (square root, cube root, and so on)

radicand The expression under the radical.

range The collection of all values of the dependent variable.

ratio The growth or decay factor, b, of the exponential function $y = a \cdot b^x$, where $b > 0$ and $b \neq 1$.

rational equation An equation composed of fractions where the numerators and denominators are polynomials, with the variable appearing somewhere in a denominator.

rational function Any function that can be defined as the ratio of two polynomial functions.

real numbers All numbers that are either rational or irrational.

S   **slope of a line** The constant rate of change of output to input.

slope-intercept form of the equation of a line The equation $y = mx + b$, where m represents the slope of the line and $(0, b)$ is the vertical intercept.

solution of a system of equations in two variables The ordered pair of numbers (x, y) that make both equations true.

sum function, $f + g$ The function that is created from two functions f and g by the rule $(f + g)(x) = f(x) + g(x)$.

system of linear equations in two variables A pair of equations that can be written in the form $y = ax + b$ or $y = cx + d$, and where a, b, c, and d are real numbers.

V **variation** How the dependent variable changes when the independent variable changes. (*see* direct variation or inverse variation.)

vertex The turning point of the graph of a parabola having coordinates $\left(\dfrac{-b}{2a}, f\left(\dfrac{-b}{2a}\right)\right)$, where a and b are determined from the equation $f(x) = ax^2 + bx + c$. The vertex is the highest or lowest point of a parabola.

vertical asymptote A vertical line that a curve approaches, getting progressively closer, without ever touching that vertical line.

vertical intercept The point of the graph of the function whose x-coordinate is 0. (*see* y-intercept)

X **x-intercepts** All points of the graph of the function whose y-coordinate is 0. (*see* horizontal intercept.)

Xmax The largest value of input visible in the window of a graphing calculator.

Xmin The smallest value of input visible in the window of a graphing calculator.

Y **y-intercept** All points of the graph of the function whose x-coordinate is 0. (*see* vertical intercept.)

Ymax The largest value of output visible in the window of a graphing calculator.

Ymin The smallest value of output visible in the window of a graphing calculator.

Z **zero-product rule** The algebraic rule that says if a and b are real numbers such that $a \cdot b = 0$, then either a or b, or both, must be equal to zero.

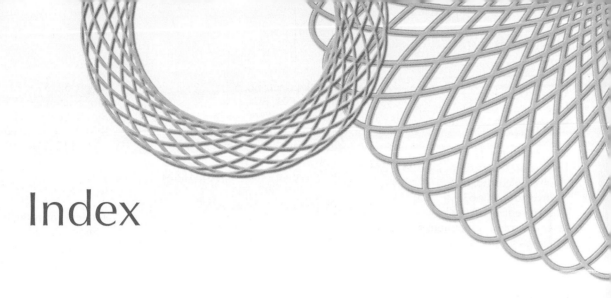

Index